*Edited by*
*Alexander Schuhmacher,*
*Markus Hinder and*
*Oliver Gassmann*

**Value Creation in the Pharmaceutical Industry**

## *Related Titles*

Mollah, A., Long, M., Baseman, H. (eds.)

**Risk Management Applications in Pharmaceutical and Biopharmaceutical Manufacturing**

2013
ISBN: 978-0-470-55234-6, also available in electronic formats

Chaguturu, R. (ed.)

**Collaborative Innovation in Drug Discovery**
**Strategies for Public and Private Partnerships**

2014
ISBN: 978-0-470-91737-4, also available in electronic formats

Rajagopal, R.

**Sustainable Value Creation in the Fine and Speciality Chemicals Industry**

2014
ISBN: 978-1-118-53967-5, also available in electronic formats

Meyer, H., Schmidhalter, D.R. (eds.)

**Industrial Scale Suspension Culture of Living Cells**

2014
ISBN: 978-3-527-33547-3, also available in electronic formats

Behme, S.

**Manufacturing of Pharmaceutical Proteins**
**From Technology to Economy**
**2nd Edition**

2015
ISBN: 978-3-527-33766-8, also available in electronic formats

*Edited by*
*Alexander Schuhmacher,*
*Markus Hinder and*
*Oliver Gassmann*

# Value Creation in the Pharmaceutical Industry

## The Critical Path to Innovation

WILEY-VCH
Verlag GmbH & Co. KGaA

**Editors**

***Prof. Dr. Alexander Schuhmacher***
Reutlingen University
School of Applied Chemistry
Alteburgstr. 150
72762 Reutlingen
Germany

***Prof. Dr. Markus Hinder***
Novartis Pharma AG
Novartis Institutes for BioMedical Research
Postfach, Forum 1
4002 Basel
Switzerland

***Prof. Dr. Oliver Gassmann***
University of St. Gallen
Institute of Technology Management
Dufourstr. 40a
9000 St. Gallen
Switzerland

**Cover**
Image © Kadmy / fotolia.



**Library of Congress Card No.:** applied for

**British Library Cataloguing-in-Publication Data**
A catalogue record for this book is available from the British Library.

**Bibliographic information published by the Deutsche Nationalbibliothek**
The Deutsche Nationalbibliothek lists this publication in the Deutsche Nationalbibliografie; detailed bibliographic data are available on the Internet at <http://dnb.d-nb.de>.



**Print ISBN:** 978-3-527-33913-6
**ePDF ISBN:** 978-3-527-69343-6
**ePub ISBN:** 978-3-527-69341-2
**Mobi ISBN:** 978-3-527-69342-9
**oBook ISBN:** 978-3-527-69340-5

**Cover Design** Bluesea Design, McLeese Lake, Canada
**Typesetting** SPi Global, Chennai, India
**Printing and Binding** Markono Print Media Pte Ltd, Singapore

Printed on acid-free paper

# Contents

# List of Contributors

*Martin A. Bader*
BGW AG
Management Advisory Group
Varnbüelstrasse 13
9000 St. Gallen
Switzerland

*and*

Technische Hochschule
Ingolstadt
THI Business School
Esplanade 10
85049 Ingolstadt
Germany

*Ulrich A. K. Betz*
Merck KGaA
Frankfurter Str. 250
64293 Darmstadt
Germany

*Michael Buckley*
Independent Principal
Consultant
Livermore
CA

*Rob Caldwell*
Abbvie Inc. R466
AP13A-3, 1 North waukegan
Road
North Chicago
IL 6000064
USA

*John Darbyshire*
Mooshagweg 10
CH 4123 Allschwil
Switzerland

*Oliver Gassmann*
University of St. Gallen
Institute of Technology
Management
Dufourstrasse 40a
9000 St. Gallen
Switzerland

*Paul Germann*
AbbVie Deutschland GmbH &
Co KG
Knollstraße 50
67061
Ludwigshafen am Rhein
Germany

***Joachim M. Greuel***
Bioscience Valuation BSV GmbH
Am Zigeunerbergl 3
82491 Grainau
Germany

***Antal K. Hajos***
Procelsis Holding UG
Am Farnberg 3
79289 Horben
Germany

***Galina Hesse***
Sanofi-Aventis Deutschland
GmbH
Industriepark Hoechst
65926 Frankfurt am Main
Germany

***Markus Hinder***
Novartis Pharma AG
Novartis Institutes for
Biomedical
Research
Postfach, Forum 1
4002 Basel
Switzerland

***Aubyn Howard***
Château Chavagnac
07610 Lemps
Ardèche
France

*and*

Psychosynthesis Coaching
Limited
London
United Kingdom

***Paul Kamudoni***
Institute of Medicines
Development
Duffryn House
CF 23 6NP
Cardiff
UK

***Werner Kramer***
Biomedical & Scientific
Consulting
55130
Mainz- Germany

***Carol A. Krech***
AG Management Advisory
Group
Varnbüelstrasse 13
9000 St. Gallen
Switzerland
Technische Hochschule
Ingolstadt
THI Business School
Esplanade 10
85049 Ingolstadt
Germany

***Gezim Lahu***
Takeda Pharmaceuticals
International
Thurgauerstrasse 130
CH 8152 Glattpark-Opfikon
(Zürich)
Switzerland

***Qiang Liu***
Takeda California, Inc.
10410 Science Center Drive
San Diego
CA 92121
USA

***Stephan Luther***
IHPH – Institute for Hygiene and Public Health
WHO Collaborating Centre for Health Promoting Water Management and Risk Communication
Medical Geography & Public Health Workgroup
University of Bonn
Sigmund-Freud-Straße 25
D-53105 Bonn
Germany

***Nigel McCracken***
Debiopharm Group
Forum "après-demain"
Chemin Messidor 5-7
Case postale 5911
1002 Lausanne
Switzerland

***Pierre A. Morgon***
AJ Biologics
Level 4, Menara Atlan
161B Jalan Ampang
50450 Kuala Lumpur
Malaysia

*and*

Mérieux Développement
17 rue Bourgelat
69002 Lyon
France

*and*

Theradiag SA
14 rue Ambroise Croizat
77183 Croissy Beaubourg
France

*and*

Eurocine Vaccines AB
Fogdevreten 2
17165 Solna
Sweden

*and*

MRGN Advisors
Rue du Mont-Blanc 4
Case postale 2067
1211 Genève 1
Switzerland

***Hannah Nawi***
AJ Biologics
Level 4, Menara Atlan
161B Jalan Ampang
50450 Kuala Lumpur
Malaysia

***Sanjay Patel***
Takeda California, Inc.
10410 Science Center Drive
San Diego
CA 92121
USA

***Gerrit Reepmeyer***
Ockham Razor Ventures
33717 Woodward Avenue #143
Birmingham
MI 48009
USA

***Sam Salek***
University of Hartfordshire
School of Life and Medical
Sciences
Department of Pharmacy
Pharmacology and Postgraduate
Medicine
College Lane
Hartfield
Herts
AL10 9AB
UK

*and*

Institute of Medicines
Development
Duffryn House
CF 23 6NP
Cardiff
UK

***Mathias Schmidt***
Takeda California, Inc.
10410 Science Center Drive
San Diego
CA 92121
USA

***Peter Schmitz***
IHPH – Institute for Hygiene
and Public Health,
WHO Collaborating Centre for
Health Promoting Water
Management and Risk
Communication,
Medical Geography & Public
Health Workgroup,
University of Bonn
Sigmund-Freud-Straβe 25
D-53105 Bonn, Germany

***Alexander Schuhmacher***
Reutlingen University
School of Applied Chemistry
Alteburgstrasse 150
72762 Reutlingen
Germany

***Sviataslau Sivagrakau***
Goethe University Frankfurt
Theodor-W.-Adorno-Platz 1
60323 Frankfurt
Germany

***Petter Veiby***
Takeda Pharmaceuticals
International
40 Landsdowne Street
Cambridge
MA 02139

***Axel Wiest***
Merck Serono
Frankfurter Straße 250
64293 Darmstadt
Germany

# Foreword

The main driver for sustainable profitable growth in the pharmaceutical industry is innovation.

Research and development (R&D) is the primary source for product innovation; it is the lifeblood of our industry. Product innovation is the result of long-term investments, which at the same time takes place in a challenging and dynamic environment, as pharmaceutical companies are highly pressurized by long development times, shorter commercialization of the intellectual property (IP) rights, and cost pressure by the public healthcare sector. R&D requires commitment, flexibility and perseverance: We need to invest and believe, learn from setbacks, scrutinize, and execute.

With the objective to provide new and differentiated therapies to patients and society, our biopharmaceutical company, AbbVie, is devoted to having a remarkable impact on patients' lives, especially in areas with a high unmet medical need AbbVie's approach to innovation builds on track record of developing breakthrough science. For example, our work in immunology has benefited over 850,000 patients with rheumatoid arthritis, psoriasis, Crohn's disease and other chronic autoimmune conditions. And, at the beginning of 2015, we received the approval for our oral, interferon-free treatment option for patients with chronic hepatitis C, which provides a very high probability of cure. Decades of research, extensive investment, and collaboration across functions and countries have made this kind of innovation possible.

With this book, Alexander Schuhmacher, Markus Hinder, and Oliver Gassmann provide a unique overview of the success-critical components necessary to deliver pharmaceutical innovation. The authors have compiled an overview of today's state-of-the-art pharmaceutical R&D processes and the challenges the industry is facing. All authors are thought leaders in the industry and academia and offer a wide range of experience. They give the reader comprehensive insights into research-, development-, and business-related subjects of the pharmaceutical industry. In addition, each author provides his or her personal view of how the industry might evolve in the future and how current issues and challenges can be addressed to increase overall productivity.

Starting with two epidemiological and pharmacoeconomic analyses, this book provides first-class strategic and operational insights into drug discovery, translational and clinical development up to more managerial aspects such as portfolio or IP management. As the industry is increasing its integration and as some companies need to adopt a more effective approach to their R&D organizations, the chapters on open innovation and new innovation models give a brilliant summary of some key drivers in pharmaceutical R&D today and tomorrow.

This book is a exclusive compilation of challenges and state-of-the-art solutions within the pharmaceutical R&D process. In its wider significance, the book deals with the critical path towards value creation in the pharmaceutical industry. Thus, the book is targeted at R&D managers, business managers, researchers, drug developers, marketing leaders, and sales managers – in a nutshell to all innovators in the pharmaceutical sector. In addition, it should be a valuable platform for academics, educational organizations, and university students who are interested in today's world of pharmaceutical innovation.

Ludwigshafen, September 2015

*Dr Friedrich Richter*
*Vice President*
*Global Drug Product Development*

# 1
# Introduction to the Book

*Alexander Schuhmacher, Oliver Gassmann, and Markus Hinder*

"Value Creation in the Pharmaceutical Industry: The Critical Path to Innovation" is intended to review the current state of the art and to provide cutting-edge knowledge in the pharmaceutical research and development (R&D) process. All authors are well-known experts in their field of activity and provide first-hand scientific, regulatory, management, or business information. They share their personal vision on how their field of expertise will or need to develop to finally keep pace with the changes that will happen in the pharmaceutical industry.

With this book, we examine the situation of pharmaceutical innovation from three different perspectives:

- Technically from the sequence of R&D
- Operationally when we answer the question of what can be done to increase R&D efficiency
- Strategically by examining environmental factors and trends that may influence pharmaceutical R&D in the future

Due to its unique structure and content, we expect that this book will be a way to update knowledge and spark new ideas for R&D managers, industry specialists, academics, and other stakeholders interested in pharmaceutical R&D.

As depicted in Figure 1.1, this book addresses the critical path of value creation in the pharmaceutical industry from the view points of research, development, and business. The articles on epidemiology, antibodies, and drug discovery may be assigned best to the section on "research." At the interface of "research" to the part of "development," we provide the articles on preclinical safety, translational medicine (TM), and drug costs. The section on "development" is represented by the articles on pharmaceutical, clinical, and translational development. A more holistic view on pharmaceutical R&D with an interface to the topic "business" is provided by the texts on portfolio management, financing of R&D, open innovation, licensing, outsourcing, innovation models, leadership in R&D, and management of intellectual assets. The business part is represented by the articles on marketing, vaccines, and pharmacoeconomics.

We will start this introduction by describing in general how does the highly regulated and standardized R&D process in the pharmaceutical industry look like.

*Value Creation in the Pharmaceutical Industry: The Critical Path to Innovation*, First Edition.
Edited by Alexander Schuhmacher, Markus Hinder, and Oliver Gassmann.

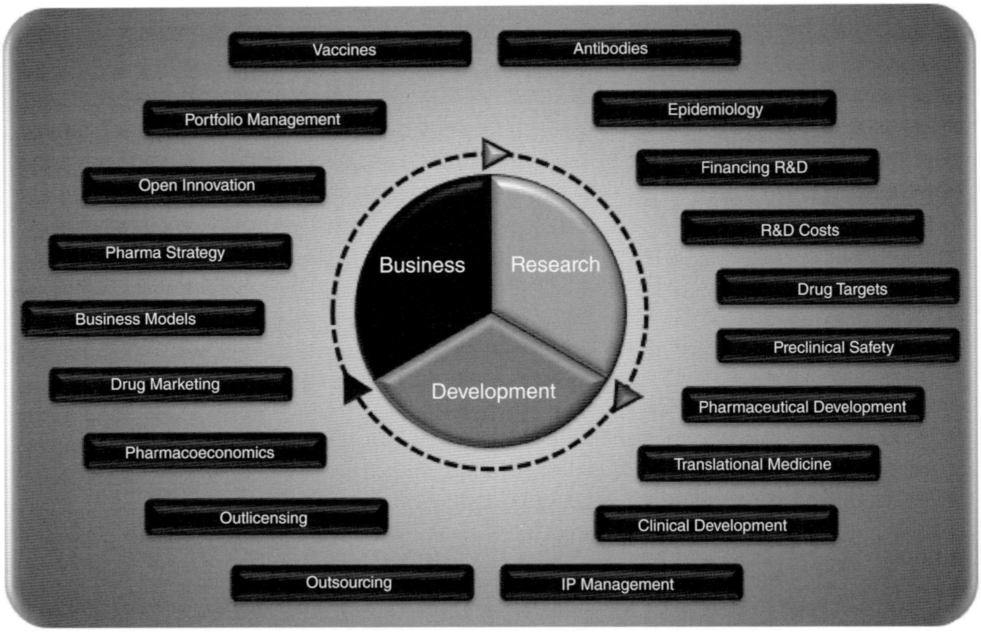


**Figure 1.1** The pharmaceutical innovation hemisphere.

The first step of pharmaceutical R&D is the identification and validation of a suitable drug target that, if up- or downregulated, activated, or inhibited, may play a role in a disease. Thus, an in-depth understanding of the disease and its molecular mechanism is key to search for new drug targets. In a next step, researchers search for lead compounds that potentially influence the drug target in the aforementioned way. If a lead compound is discovered, researchers optimize the potential of the compound to become a drug candidate. In preclinical development, it is analyzed as to whether this candidate can be used in the human situation, and it undergoes a series of preclinical testing primarily intended to understand how the compound works and as to whether it is safe in animal models. Next, safe drug candidate can be used for test series in the human situation. First and in view of the US market, an Investigational New Drug Application (IND) needs to be filed at the Food and Drug Administration (FDA). In the following years, several clinical trials are conducted to analyze the efficacy and safety profile of the drug candidate. Principally, the clinical trial process is conducted in three phases. In phase I, the drug candidate is tested in a small group of healthy volunteers to analyze its pharmacokinetic. Phase II trials are conducted to analyze the safety and the efficacy of the drug candidate in a selected group of patients that have the disease under investigation. In the phase III trials, the drug candidate is tested in large groups of patients to provide statically well-founded data on the efficacy and safety of the drug and the overall risk–benefit ratio. Finally, a new drug application (NDA) is filed to get market approval for the new drug. The FDA reviews all data and assesses the benefit versus the risk of the drug candidate and decides as to whether an approval can be granted.

Today, this R&D process lasts on average for about one to two decades and is related with a very low probability of success (PoS) from discovering a new drug candidate to its launch to the first market. The complexity of drug R&D combined with the increasing permeation by technology, the costs related with failed drugs, and the capitalization of costs over the long timelines are the main drivers of the enormous high costs that need to be invested per new molecular entity (NME). Today, the average costs per NME are probably above USD 2 billion (Figure 1.2).

With a total of 22 chapters, this book reviews the whole value chain of pharmaceutical R&D from drug discovery to marketing of a new drug. In detail, this book starts with three chapters that set the stage for the pharmaceutical industry, namely, epidemiology, healthcare needs, and a definition of value in the pharmaceutical business and the shrinking R&D efficiency.

First, **Stephan Luther** provides an overview and introduction on "*Global Epidemiological Developments.*" He reviews the basic models which describe the burden of disease in different geographical areas and under different socioeconomic and climate conditions. Based on these factors, he reviews the healthcare needs

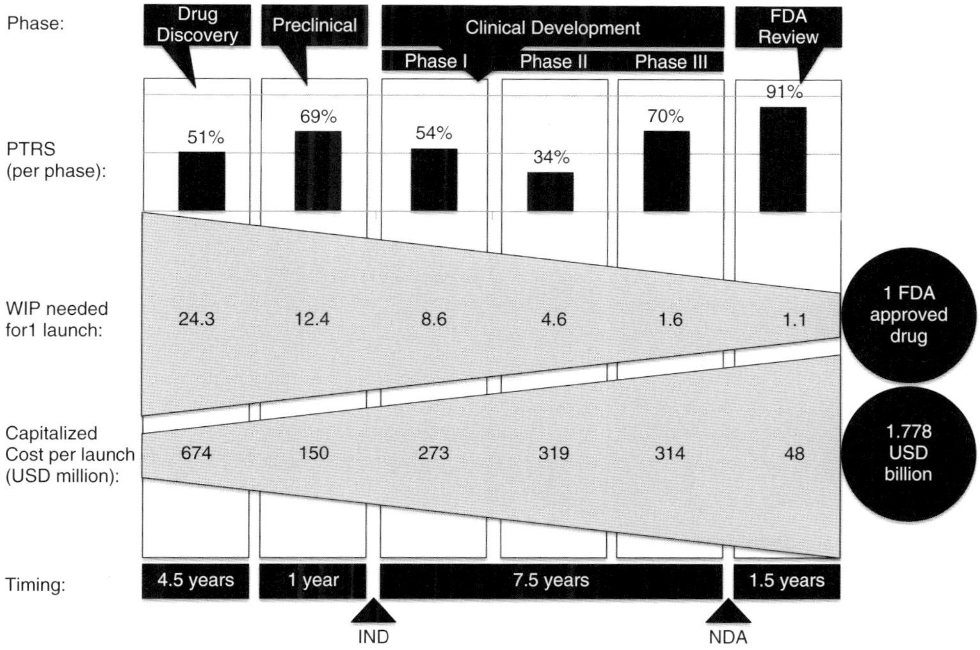


**Figure 1.2** The traditional R&D phase model; IND (Investigational New Drug), NDA (New Drug Application), FDA (Food and Drug Administration), PTRS (probability of technical and regulatory success), WIP (work in progress), USD (U.S. Dollar), data derived from Paul S. et al. (2010)

for specific areas and provides an outlook on likely future developments worldwide. The chapter explains why on a global scale, there is a shift from communicable disease to noncommunicable disease and why in addition to mortality the disability-adjusted life years (DALYs) will become a prominent estimate for the overall global burden of disease.

The chapter by **Sam Salek and Paul Kamudoni** on "*The Value of Pharmaceutical Innovation: Concepts and Assessment*" introduces the reader to the different concepts of value and its assessment. Importantly the authors describe value from different perspectives of the multiple stakeholders in the healthcare sector and how the concept of value has evolved over time. They review on how value is assessed today and describe the consequences for pharmaceutical R&D. Based on recent developments, they provide an outlook how the pharmaceutical industry and regulatory agencies decision makers in the healthcare sector can work together in a more unified and transparent way to improve outcomes for the patients and the healthcare systems.

In the following chapter ("*A Review of the Pharmaceutical R&D Efficiency: Costs, Timelines, and Probabilities*"), we review the efficiency of the current R&D, namely, the R&D costs, the cycle times, the PoS of pharmaceutical R&D, and the number of NMEs that have been launched in past years. **Alexander Schuhmacher, Oliver Gassmann, and Markus Hinder** describe the traditional R&D phase model, highlight the reasons of the low success rates, and answer the question of why pharmaceutical R&D takes so long. We also detail the drivers of the enormous R&D costs and summarize our research on the question of how much does an NME cost today. Finally, further impact factors on R&D efficiency are discussed.

It is realistic to say that today more than USD 2 billion is required to bring one NME to the market, in a process that takes one to two decades. Accordingly, and as provided by the chapter of **Sviataslau Sivagrakau** ("*Financing Pharmaceutical Innovation*"), drug development is concentrated almost exclusively in advanced economies. The United States is the global leader with market share of 60% in scientifically novel new drugs. Globally, two-thirds of the investments in biomedical R&D come from the industry, whereas one-third are publicly funded. Since the 1980s, multinational pharmaceutical companies lost their dominance in providing NME. Whereas in the 1980s the big players originated three quarters of all NMEs, they have lost today's majority market share to smaller companies. In light of this, the financing landscape has become more fragmented and includes venture capital, university funds, public and charity grants, alliances, private–public partnerships, corporate and state venture capital, acquisitions by larger firms, and initial public offerings for companies with late-stage compounds. The last 5 years exhibited very favorable market conditions for exchange-listed drug developers: high valuations and strong industry-level performance. On the other hand, funding at early stages, particularly translational phase, remains scarce.

The next seven book chapters focus more specifically on the R&D process and the related potential of value creation in the phases of drug discovery and preclinical and clinical development.

In *"Challenges and Options for Drug Discovery,"* **Werner Kramer** analyzes the different approaches taken in research–discovery and compares historical promises and delivery in these disciplines. He identifies the key obstacles, which need to be overcome to provide sustained success in the discovery space. He proposes and describes a new model, which unites scientific scrutiny, decisions based on understanding of human and molecular physiology. This includes the weighing of target-related safety and efficacy and the stringent application of decision trees in the assessment of projects. The chapter is especially valuable because the author does not forget to build the bridge to neighboring discipline TM.

The transition of a new molecule from animals to humans is a key event in the development of a new medicine. On the one hand, this is important to ensure adequate clinical safety for study participants. On the other hand, TM up to clinical proof of concept can offer precious information on a molecule's mode of action, pharmacokinetics (PKs), and pharmacodynamics (PDs) and its therapeutic potential. **Gezim Lahu and John Darbyshire** review in *"Translational Medicine: Enabling the Proof of Concepts"* the overall process and show the benefit of established and emerging tools and skills to enable informed and better decision making. By embedding TM into the bigger context between drug discovery and development, they provide a perspective of how TM can become a value driver in both directions.

In their chapter *"Preclinical Safety and Risk Assessment,"* **Paul Germann and Rob Caldwell** review the state of the art in preclinical safety assessment. The authors give an overview of today's preclinical test strategies to support drug candidate testing in the early phases of drug development. They also provide an overview on the general components that are required for regulatory acceptable preclinical data package. Furthermore, the authors give information on the interaction of therapeutic use, route of application, treatment duration, and therapeutic indication and its influence on the safety assessment and the determination of the therapeutic window.

In the next book chapter on pharmaceutical development, it is described that more and more biological molecules (peptides, proteins, monoclonal antibodies (mAbs)) in not only oncological indications but also general medicine indications (e.g., diabetes, rheumatoid arthritis, or psoriasis) are reaching the market today and more will follow in the future. Thus, **Galina Hesse**'s chapter (*"Developing Commercial Solutions for Therapeutic Proteins"*) describes the challenges associated with this switch from low-molecular-weight orally available molecules to parenterals. The chapter describes all key steps of pharmaceutical development from formulation over devices and quality design to ensure a successful and target product profile-driven formulation development.

Clinical development's remit is under full transformation, from a discipline whose primary deliverable was drug approval to an area of expertise where new and neglected aspects of the right use of medications come together. **Markus Hinder and Alexander Schuhmacher** review in *"The Evolution of Clinical Development: From Technical Success to Clinical Value Creation"* how the road to technical success has continued and will continue to evolve and which new aspects need to be integrated during the process of clinical development to finally provide a drug which benefits the individual and society. The chapter highlights and provides concrete examples how this process will become more patient-centric, successful, and efficient by integrating knowledge early on the different customer's needs in the field.

Finally and in a more holistic way, **Nigel McCracken** describes the role of "*Translational Development*" within today's pharmaceutical business to help translate basic research into clinical utility. He also outlines the multidisciplinary tools and the highly collaborative approaches that are used to deliver a specific development solution designed to maximize the risk–benefit ratio of a new drug. Therefore, he answers the following questions: what is translational development, and where does it fit into the drug development process? What are the main types of activities where translational development provides a value for the development of a new drug?

In a next step to examine the critical path of value creation in the pharmaceutical industry, the book focuses more on business-related aspects of pharmaceutical R&D.

In a book chapter with the title *"40 Years of Innovation in Biopharmaceuticals: Will the Next 40 Years Be as Revolutionary?,"* **Mathias Schmidt** and colleagues present that the invention of recombinant DNA technology and the ability to generate mAbs have revolutionized the pharmaceutical industry and the way serious diseases are treated. The chapter will review milestones and innovations along the success path of mAbs and other biologics and will critically challenge successes and setbacks. They also introduce the next wave of innovation for mAbs that is focusing on miniaturized antibodies, novel binding scaffolds, bispecific antibodies, oral availability, antibody–drug conjugates, permeation of the blood–brain barrier, and targeting of biologics to the cytoplasm.

**Pierre Morgon**'s first chapter in this book ("Vaccines: *Where Inertia, Innovation, and Revolution Create Value, Simultaneously and Quietly*") provides an overview on the space and role of vaccines within the healthcare sector, the emergence of novel immunization approaches, the drivers of immunization, and its fast growth as a product segment. In addition, we will analyze why innovation is needed along the whole value chain of the vaccine business and who will be a player that might drive this business in the future.

In his second book chapter ("*The Patient-Centric Pharma Company: Evolution, Reboot, or Revolution?*"), **Pierre Morgon** provides new trends that are affecting the players in the healthcare sector and that are driving the increasing focus on real-life patient data in the course of the process of clinical development. And we will illustrate why patient satisfaction should be the ultimate performance indicator of healthcare procurement.

Subsequently, the authors of the Chapters 15–21 address the more management-related topics that impact the value-creation potential of pharmaceutical R&D. Chapters 15–19 review the impact of open innovation, outsourcing, out-licensing, new business models, and leadership styles on the efficiency and productivity of pharmaceutical R&D and illustrate the newest trends in these fields. Chapters 20 and 21 describe portfolio management and the management of intellectual property (IP) rights as key success factors.

With the book chapter on open innovation ("*The Pharmaceutical Industry Is Opening Its R&D Boundaries*"), we provide a sound basis to understand the complex of open innovation. It begins with a comparison of closed versus open innovation and an insight into the open innovation process. **Alexander Schuhmacher and Ulrich Betz** offer an overview of the more traditional elements of open innovation in the pharmaceutical industry, such as target scouting, research collaborations, drug licensing, outsourcing, and joint ventures. In addition, they provide examples of new open innovation initiatives in the pharmaceutical business, for example, new frontier sciences, drug discovery alliances, private–public partnerships, innovation incubators, virtual R&D, crowdsourcing, open-source innovation, innovation camps, and fluctuating open teams. Finally, the role of open innovation in new R&D business models is examined, and the open and virtual innovation model "knowledge leverager" is explained in detail.

While in-licensing is a key source of new drug development, out-licensing does not play a central role in pharmaceutical companies' R&D strategies yet. **Oliver Gassmann** and colleagues describe in their book chapter "*Out-Licensing in Pharmaceutical Research and Development*" how out-licensing can contribute to an increase in R&D efficiency. Therefore, the authors address the following questions: What is the relevance of R&D collaborations in the pharmaceutical industry? What are the drivers of out-licensing? And how is out-licensing managed in R&D organizations?

Outsourcing has originally been established as an off-the-shelf service in the pharmaceutical industry decades ago to reduce R&D costs and to increase R&D flexibility. **Antal K. Hajos** ("*Trends and Innovations in Pharmaceutical R&D Outsourcing*") describes the fundamental changes in the clinical research organization (CRO) industry that have happened in the past years and the development of outsourcing as a strategic partnership option. In addition, he illustrates how the CRO industry has started to become more differentiated into global players, specialist, and niche providers.

Next, the book chapter on "*New Innovation Models in Pharmaceutical R&D*" illustrates the consequences and measures that have been taken in the past years as a result of the historically low success rates in R&D. We outline the development in pipeline sizes of multinational pharmaceutical companies as we illustrate the R&D investments and the measures that have been taken in the past years to reduce the R&D costs. Finally, **Alexander Schuhmacher, Markus Hinder, and Oliver Gassmann** discuss and review some R&D models that were developed to increase R&D efficiency.

In his book chapter on "*The Influence of Leadership Paradigm and Styles on Pharmaceutical Innovation,*" **Aubyn Howard** sets the topic of innovation within the context of leadership. He shows how both collective leadership paradigms and individual leadership styles influence the process of innovation in the pharmaceutical industry. Furthermore, he shows how the challenges that the industry is facing are contextualized within a wider process of transformation and evolution within organizations and society today. Finally, we are providing information of how leadership paradigms and styles can impact the capacity of pharmaceutical companies to enable innovation.

With Chapters 20 and 21, the authors complete our view on pharmaceutical R&D by providing their insights on both portfolio and IP management in today's pharmaceutical world.

As it is becoming more and more expensive and risky to develop an NCE (New Chemical Entity), pharmaceutical companies need to make sure that their portfolio of drug candidates is well balanced financially and risk-wise. According to the authors **Joachim Greuel and Axel Wiest**, a primary goal of portfolio management is to ensure that an entire R&D portfolio is successful while allowing individual projects to fail. With their book chapter "*The Role of Modern Portfolio Management in the Pharmaceutical Industry*," they describe that since H. M. Markowitz was awarded the Nobel Memorial Prize in Economic Sciences in 1990 for his pioneering work in modern portfolio theory, portfolio management has been a main pillar in asset allocation and financial investment. However, the benefit of portfolio management for the pharmaceutical industry is still controversial. Although most larger pharmaceutical companies report to have implemented portfolio management processes, some question whether a portfolio management system leads to a higher R&D productivity. The authors put a hypothesis forward suggesting that pharmaceutical portfolio management is not only important to allocate resources and optimize project management from a strategic perspective. It may be seen as a crucial enabling element in the entire pharmaceutical innovation process.

Finally, the purpose of the book chapter "*Patent Management Through the innovation Lifecycle*" is to provide an overview of the role of patents in today's pharmaceutical business. The authors **Martin Bader and Oliver Gassmann** provide best practice examples of the pharmaceutical industry and outline how patent management is done in the environment of low R&D efficiency. Therefore, the authors answer the questions of what are the challenges that companies face when managing patents and how can patents be managed throughout the product life cycle.

## Reference

Paul, S.M. *et al.* (2010) *Nat. Rev. Drug Discovery*, **9**, 203–214.

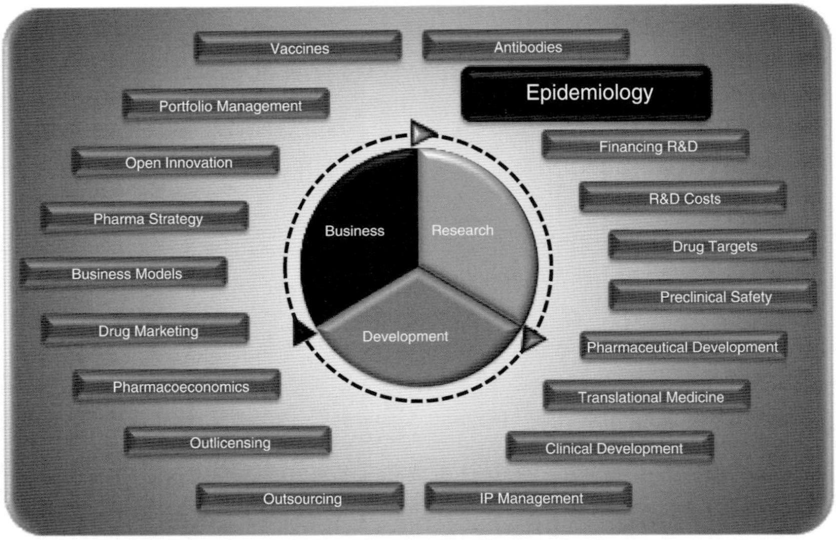

Vaccines
Antibodies
Epidemiology
Portfolio Management
Financing R&D
Open Innovation
R&D Costs
Pharma Strategy
Business
Research
Drug Targets
Business Models
Preclinical Safety
Drug Marketing
Development
Pharmaceutical Development
Pharmacoeconomics
Translational Medicine
Outlicensing
Clinical Development
Outsourcing
IP Management

# 2
# Global Epidemiological Developments

*Stephan Luther and Peter Schmitz*

## 2.1
## Introduction

The purpose of this chapter is to provide an overview of the changing patterns of diseases in the world. The reader will learn how the global burden of disease (GBD) can be quantified, how it is distributed in the different regions of the world, and what changes can be expected in the future. Moreover the underlying causes for the observed patterns and the special characteristics of selected diseases will be reviewed.

The structure of disease burden in populations is the result of diverse factors, including cultural, social, economic, and environmental aspects. Health geography analyzes medical questions by application of geographical methods and combines the mentioned aspects and various determinants under a spatial perspective to develop a holistic understanding of health and diseases. This understanding is essential to anticipate future developments and needs on a global scale.

The following questions will be explored in this chapter:

What are the underlying mechanisms (causes and drivers) of disease patterns and dynamics of epidemiology in populations?
How can the burden and structure of diseases in a population be quantified?
What are neglected diseases, and where can one find them?
What is the status of infectious diseases in the world?
Why is antimicrobial resistance a global problem?
Does climate change have an impact on health and diseases?
What are the challenges derived from epidemiological transition?

The second section will provide the theoretical background to understand the disease patterns in different populations. On this basis the GBD and the appropriate tools for its quantification will be described. A detailed analysis of the diverse disease groups will be followed by a review of the emerging challenge of antimicrobial resistance (AMR). The last section will explain some of the significant drivers of permanent changes in global disease structure and spread.

*Value Creation in the Pharmaceutical Industry: The Critical Path to Innovation,* First Edition.
Edited by Alexander Schuhmacher, Markus Hinder, and Oliver Gassmann.

## 2.2
## Model of Epidemiological Transition

In 1971 Abdel Omran published his theory of epidemiological transition (Omran, 1971). The theory uses fertility, mortality, and morbidity as the basic indicators to describe the epidemiological state of a population and the change in patterns of disease in a generalized manner. The demographic, economic, and sociologic consequences of the interaction of health and disease are the core elements of the model so that it is possible to describe the state of epidemiological transition for a population or usually a country.

The epidemiological transition describes the change in fertility, mortality, and morbidity in populations over time as a result of reduced incidences of infectious diseases and an increase of noncommunicable diseases (NCDs) like cardiovascular diseases or cancer as the main causes of morbidity and mortality.

Originally, Omran (1971) described different modes of transition with a temporary perspective. He described a classical/western model, as it was seen in England, Wales, or Sweden (1971), where a high mortality (>30/1000 persons) and fertility (>40/1000 persons) declined to low mortality (<10/1000 persons) and low fertility (<20/1000 persons), resulting in a growing total population. As a variation of the classical/western model, Omran described two further models of epidemiological transition. First, Japan is the archetype of an accelerated transition model, where the decline of mortality from more than 30/1000 persons to less than 10/1000 ran off in a much shorter time. Second, the contemporary/delayed model describes the still ongoing transition of most low-income countries as then Ceylon (today Sri Lanka) and Chile, where mortality declines were observed not before the end of World War II and were propelled by imported medical products, technology, and knowledge.

Since 1971 the model of epidemiological transition has been modified several times by Omran himself (Omran, 1983, 1998) and by other authors (Gaziano, 2010; Harper and Armelagos, 2010; Olshansky and Ault, 1986; Rogers and Hackenberg, 1987) to face criticism (Caselli, Meslé, and Vallin, 2002; Mackenbach, 1994; Smallman-Raynor and Phillips, 1999; Carolina and Gustavo, 2003). Five stages of transition can be described today by means of mortality rates and disease structure (Harper and Armelagos, 2010; Omran, 1971).

1) *Paleolithic baseline*: Pre-neolithic cultures consisted of small hunter-gatherer populations in a preagricultural world, with low mortality and fertility rates. They consumed a varied diet with only few nutritional deficiencies. Infections such as tapeworm, body lice, and pinworm; typhoid; and staph infections were the most prevalent diseases.
2) *Phase of pestilence and famine*: This phase was an agricultural world after the Neolithic Revolution, with a diet relying mainly on crops. Epidemics, famines and malnutrition, wars, and maternity complications were the main causes of mortality. Malaria, smallpox, measles, and tuberculosis (TB) were the common infectious diseases of this age, resulting in high and fluctuating mortality

and high fertility (30–50 births per 1000 persons; 30 and more deaths per 1000 persons; life expectancy 20–40 years).

3) *Phase of receding pandemics*: As mortality declined (life expectancy 30–50 years), population growth became exponential. This phase was observed in early modern times till the twentieth century in Western Europe or the United States.
4) *Phase of degenerative and man-made diseases*: Fertility became the main factor for population growth, as mortality declined continuously (life expectancy > 50 years). The main characteristics of this phase are a large population size, increased life expectancy, and a varied diet with common overnutrition. Progress in public health concepts, medicine, and health care became increasingly important as hygiene knowledge was applied and vaccines and antimicrobials reduced morbidity and mortality. Disease patterns are now dominated by degenerative diseases such as cardiovascular disease, stroke, diabetes, cancer, allergies, and sexually transmitted infections. This phase is still ongoing in some lower- and middle-income countries (India, Mexico, Thailand).
5) *Phase of AMRs*: (Re-)emergence of infectious diseases is observed, as well as the development of antimicrobiological resistances. Disease spreading becomes globalized due to global traffic and transportation. Obesity and delayed chronic diseases are the characteristics of this fifth phase. These phenomena have been observed from the late twentieth century onward, affecting especially high-income countries but on a more global scale than during former phases/transitions. Alongside the loss of antimicrobial efficacy, novel infections are spreading faster than ever due to the globalized exchange of goods and travel of persons. Chronic diseases are delayed in their onset to higher ages. Mental and behavioral disorders are gaining importance as disabling diseases in this phase of transition.

These five phases reflect an idealistic transition that does not occur everywhere in this manner. In reality these transitions may not occur at all or may occur at a different speed, or certain transition phases might even be skipped completely.

The reliability of the epidemiological transition theory is dependent on the scale of observation. Patterns of transition can be observed at a country level but are possibly different when looking at data from smaller administrative units of the same country (Santosa *et al.*, 2014).

The changing structure of diseases in populations is caused by several determinants that can be grouped into four categories:

1) Ecologic determinants of mortality, for instance, the interdependencies of environment, vector diffusion, and transmission of vector-borne diseases like malaria
2) Environmental determinants, like climatic factors
3) Socioeconomic, demographic, political, and cultural determinants like the level and equality of income, educational level, ethnicity, cultural habits (e.g., burial traditions), and policy regulations

4) Medical and public health determinants, like improved sanitation, immunization, availability of medicines, efficacy of the health-care system, and availability of health insurance (Santosa *et al.*, 2014).

High mortality rates were an important factor for population dynamics, in particular with regard to the slow population growth until 1650. Thereafter, in the middle of the eighteenth century, the decline of mortality gained momentum, accompanied by an increase in life expectancy and therefore an increase in total population. Over a long time (two centuries in the classical model), fertility remained high and led, in combination with declining mortality, to population growth till fertility reached the low level of mortality. The accelerated model of epidemiological transition describes the same path in a much shorter time, as this was observed for Japan. Medical and public health determinants became important in industrialized, western societies not before the dawn of the twentieth century. Developed countries exported these medical technologies and public health programs to developing countries and thus changed their path of epidemiological transition. Socioeconomic determinants led to a change in appreciation of children. In the early stages of the epidemiological transition, children were perceived as a socioeconomic asset rather than a liability. This changed following the classical model of epidemiological transition, primarily due to socioeconomic determinants, leading to reduced fertility in the third phase of epidemiological transition. Thus, fertility as well as mortality is declining today, but with a little time lag. In lower- and middle-income countries, this change of socioeconomic determinants, and thus the decline in fertility rate, is still an ongoing process. At the same time mortality is dramatically reduced due to imported medical technology and public health measures. This leads to a widening of the gap between fertility and mortality and, as a consequence, to tremendous population growth (e.g., 2.65% for 2010–2015 in sub-Saharan Africa; 0.25% for 2010–2015 in Western Europe (United Nations, 2014c)). The decline in infectious diseases in populations is particularly beneficial to newborns, children, and women in the reproductive age, who are most vulnerable to infectious disease complications. Women in their reproductive age have a higher mortality risk than men. From the age of 50 years onward, women have a smaller mortality risk than men. The availability of anti-infectives, improved water, sanitation, and hygiene (WASH) standards, improved access to health care, and the reduction of infections lead to a life expectancy in women in the reproductive phase that is comparable to that of men.

The theory of epidemiological transition has been confirmed or improved by different researchers (Gaziano, 2010; Harper and Armelagos, 2010; Olshansky and Ault, 1986; Rogers and Hackenberg, 1987; Salomon and Murray, 2002). When all-cause mortality declines, a systematic change in the overall cause composition occurs in different age groups and females and males. In addition, income has a pronounced impact on the cause composition (Salomon and Murray, 2002). By grouping mortality causes into three different groups – communicable, maternal, perinatal, and nutritional diseases (Group 1), NCDs (Group 2), and injuries

(Group 3) – it can be shown that with increasing income (per capita), mortality ratios due to Group 1 diseases are decreasing as compared to Group 2 and Group 3 diseases. It is worthwhile mentioning that injuries (Group 3) increase more rapidly with increasing income than NCDs. When considering gender and different age groups, the results are much more heterogeneous (Salomon and Murray, 2002). In the first year of life, the epidemiological transition produces a shift from Group 1 causes to Group 2 causes, regardless of gender. This effect is independent of income and all-cause mortality. In children under the age of 15 years, there is a shift from Group 1 causes to a nearly equal mix of Group 2 and Group 3 causes. For males, injuries gain importance from the age of 5 years and beyond. This results in a more diverse transition for young adults between 15 and 44 years. In this age group declining overall mortality for males is associated with a shift from injuries to NCDs. With rising income, this effect is reversed. The transitional effect on females is comparable. A decline in overall mortality leads to an increase in NCDs, followed by an increase in injuries and a decline in NCDs. Rising income underlines the change from NCDs to injuries. From the age of 50 years onward, mortality rates for different disease groups are not dependent on income or gender anymore, but solely on the all-cause mortality rate (Salomon and Murray, 2002).

In summary, the epidemiological transition describes the relationship and change of the age- and gender-specific disease structure, in connection with mortality, fertility, and different determinants. The theory of epidemiological transition can be used to develop cause-of-death models and aid in the estimation of age- and cause-specific mortality rates in developing countries where health and death registers are missing.

## 2.3 Global Burden of Diseases

In the last 20 years, the global distribution of diseases has changed tremendously. The mean global life expectancy had increased from 65.3 years in 1990 to 71.5 years in 2013 (Anonymous, 2015). The median age of the world population rose from 24 years in 1990 to 28.5 years in 2010 and is predicted to reach 36.1 years in 2050 (United Nations, 2013b). Simultaneously the world population is growing and will probably reach 9.6 billion people by 2050 (United Nations, 2014c). The leading causes of mortality have changed from communicable diseases to NCDs. As a consequence it is important to measure disease burden and disability rather than exclusively mortality. However, there are regions in which these general statements are not true. Sub-Saharan Africa is one of the examples where the generalizations mentioned earlier do not apply.

In order to quantify the burden of disease in addition to just measuring mortality, disability-adjusted life years (DALYs) are used. The strength of DALYs is the combination of lost years due to premature mortality and lost years due to disability as a consequence of illness.

*One DALY equals one lost year of healthy life.*

*(see Box 2.1)*

**Box 2.1: DALYs**

DALYs can be understood as a measure to assess the disease burden of a population at a given time. They can be summarized or specifically calculated for one disease and consist of years of life lost (YLLs) due to premature mortality and years lived with disability (YLDs).

$$\text{DALY} = \text{YLL} + \text{YLD}$$

To calculate the YLL due to premature death, the highest known life expectancy for the age group of the deceased person is used, regardless of country or gender. For the calculation of YLL, the mean duration of an illness is needed. The amount of disability has been assessed for 291 health outcomes in the Global Burden of Disease Study 2010 (Murray *et al.*, 2012).

DALYs were first calculated for the Global Burden of Diseases, Injuries, and Risk Factors 1990 Study (Murray and Lopez, 1996; World Bank, 1993). The World Health Organization (WHO) published updated estimates for global DALYs for the years 1999, 2000, 2001, 2002, and 2004. The latest comprehensive publication of newly calculated DALYs stems from the GBD 2010 report (Murray *et al.*, 2012). This update comprises country-specific values and aggregate numbers for 21 world regions and 7 super regions. The global DALYs remained stable from 1990 to 2010. However, they decreased by 23% due to population increase when calculated as DALYs per 1000 persons. Thus globally DALYs added up to 361 DALYs/1000 persons in 2010 (Murray *et al.*, 2012).

### 2.3.1
### Trends in the Distribution of Disease Burden

Since 1990 the global DALYs for children under 5 years declined from 41 to 25% in 2010. This represents the efforts to fulfill Millennium Development Goal (MDG) number 4. The amount of YLL varies between high- and low-income countries. In high-income countries YLL accounts for half of DALYs, whereas in low-income countries YLL accounts for 80% of DALYs. In addition there has been a shift in the composition of DALYs from 1990 to 2010. The DALY portion of communicable, maternal, neonatal, and nutritional disorders declined by more than 10%. During the same time frame, the amount of NCDs increased by nearly the same amount (Murray *et al.*, 2012). This reflects the epidemiological transition as described in Section 2.2.

In 2010 the worldwide leading causes for DALYs were, in descending order, ischemic heart disease (4th in 1990, +29%), lower respiratory tract infections (LRIs) (1st in 1990, −44%), stroke (5th in 1990, +19%), diarrheal diseases (2nd in

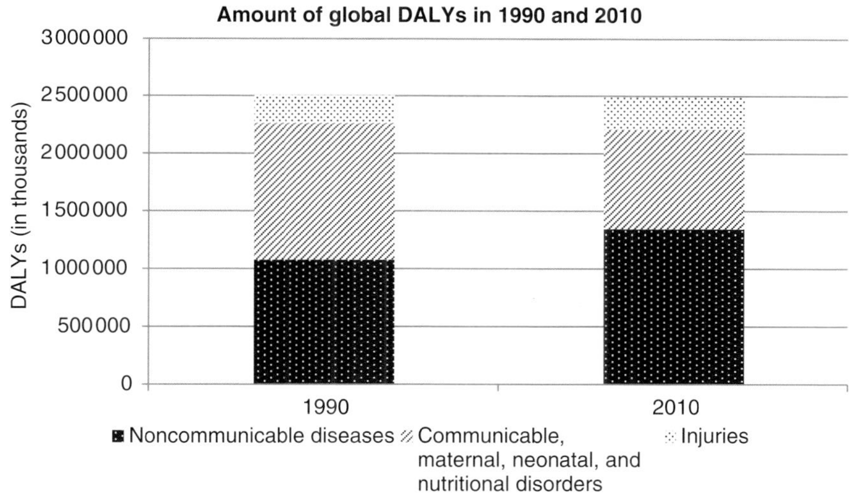


**Figure 2.1** Composition of global DALYs in 1990 and 2010 (Murray *et al.*, 2012).

1990, −51%), and HIV/AIDS (33rd in 1990, +351%). Equally important the DALYs due to major depressive disorder increased by 37% (15th to 11th rank) (Murray *et al.*, 2012). For YLD mental and substance use disorders were the leading causes in 2010 with an emphasis on depressive disorders, accounting for 40.5% of DALYs in this disease group (Whiteford *et al.*, 2013). The highest increase of total global DALYs in the past 20 years has been registered for HIV/AIDS (353%), diabetes (69%), and the group of "other" musculoskeletal diseases (50%), excluding rheumatoid arthritis, osteoarthritis, low back and neck pain, and gout.

Although there is a change in the global DALY composition from maternal, neonatal, and nutritional disorders to NCDs and injuries (Figure 2.1), a different pattern can be observed on a regional scale. As shown in Figure 2.2, in developed or high-income countries, NCDs account for more than 80% of all DALYs, whereas in sub-Saharan Africa and South Asia, communicable, maternal, neonatal, and nutritional disorders account for more than 40% of all DALYs. In high-income countries (mainly Western Europe, Canada, the United States, Australia, New Zealand, Japan, South Korea, Southern Latin America), where demographic and epidemiological transitions are much more advanced, cardiovascular diseases, cancer, musculoskeletal disorders, and mental disorders account for nearly 60% of all DALYs. Mental disorders like depression can have a very disabling effect on an individual and therefore have been one of the top three causes of disability in every region of the world. Depression can cause fatigue and limited ability to work and can even lead to suicide at school age. Furthermore, neglected tropical diseases (NTDs), excluding malaria, are responsible for about 1% of global DALYs. They show a huge regional variability (by a factor of 961) and are highest in central sub-Saharan Africa. Most of NTDs do not cause premature death but long-lasting disability. Thus, quantification of disability from these diseases proves to be important (see Section 2.4).

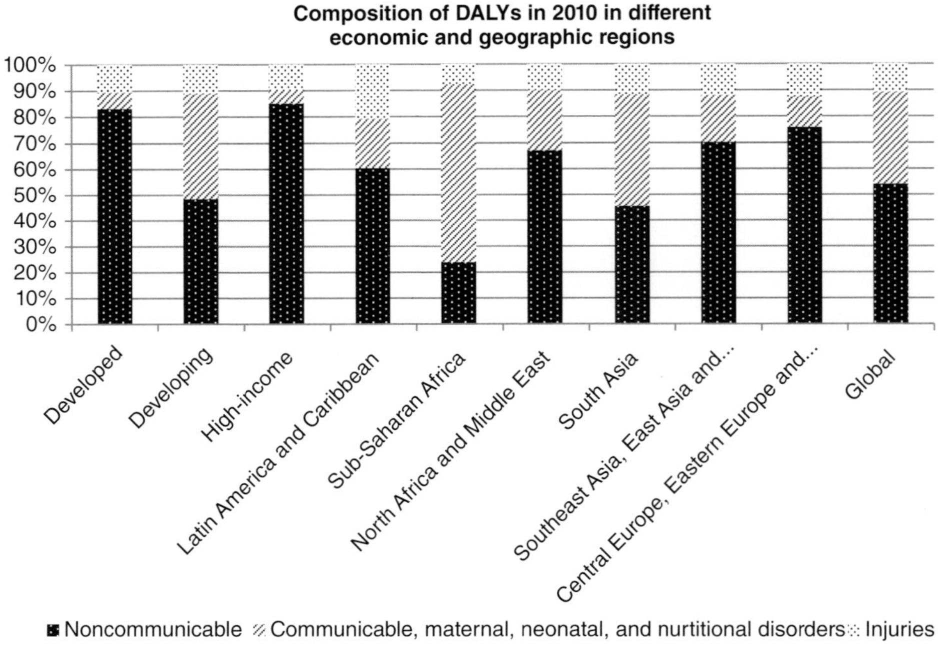


**Figure 2.2** 2010 Composition of DALYs for different economic and geographic regions (Institute for Health Metrics and Evaluation (IHME), 2013c).

Oceania, South Asia, and sub-Saharan Africa are impacted by a huge share of total DALYs (18, 16, and 20%) due to diarrhea, LRIs, and other common infectious diseases. All these regions have to deal with HIV/AIDS and TB. The southern part of sub-Saharan Africa is especially burdened by these two diseases. About 40% of total DALYs are due to HIV/TB in this region.

As described in Omran's theory (1971) of epidemiological transition, which was further developed by Salomon and Murray (2002), the stage of a population in the path of epidemiological transition also describes the age-related disease structure in it. Phase 2 of the transition is characterized by a high amount of infectious diseases in children and young adults. This phenomenon can be observed especially in the sub-Saharan African countries. The shift to a decreased mortality rate and higher life expectancy is accompanied by a shift in disease structure to NCDs, as we can see in the high-income countries (Table 2.1).

Considering the projections for global population development till 2050, this has tremendous implications for global health. In addition to an observed transition in DALYs and mortality rates, a transition in global risk factors can be observed. Globally DALYs from hypertension, the leading risk factor now, have increased by 30% since 1990. Childhood underweight was the leading risk factor for DALYs, causing high DALY amounts by communicable diseases in children in 1990. Being still the leading risk factor in central sub-Saharan Africa, on a global perspective, it decreased to rank 8. In contrast, a high body

**Table 2.1** 2010 DALY rate, change of rate since 1990, and three leading causes in geographic and socioeconomic regions (Institute for Health Metrics and Evaluation (IHME), 2013b; Institute for Health Metrics and Evaluation (IHME), 2013a).

| Geographic/ socioeconomic region | 2010 DALYs (per 100 000) | Change since 1990–2010 in percentage | Top 3 leading causes |
|---|---|---|---|
| Developed | 29 756 | −4.07 | Ischemic heart disease, stroke, low back pain |
| Developing | 37 467 | −27.63 | Lower respiratory tract infections, diarrheal diseases, ischemic heart disease |
| High-income | 26 127 | −7.20 | Ischemic heart disease, low back pain, stroke |
| Latin America and Caribbean | 28 508 | −17.42 | Forces of nature, interpersonal violence, ischemic heart disease |
| Sub-Saharan Africa | 66 893 | −31.08 | Malaria, HIV/AIDS, lower respiratory infections |
| North Africa and Middle East | 28 606 | −32.00 | Ischemic heart disease, stroke, major depressive disorder |
| South Asia | 42 155 | −36.48 | Lower respiratory tract infections, preterm birth complications, diarrheal diseases |
| Southeast Asia, East Asia, and Oceania | 26 027 | −25.11 | Stroke, ischemic heart disease, COPD |
| Central Europe, Eastern Europe, and Central Asia | 39 430 | 0.36 | Ischemic heart disease, stroke, low back pain |
| Global | 36 027 | −23.52 | Ischemic heart disease, lower respiratory tract infections, stroke |

mass index (BMI), a risk factor for cardiovascular and circulatory diseases and diabetes, was the sixth leading risk factor contributing to global overall DALYs in 2010.

The GBD approach is a comprehensive approach to measure the burden of ill health in populations. However it needs to be acknowledged that for less than one-third of all deaths worldwide, the cause of death is known (Quinn and Samet, 2010). Thus, the validity of calculated, disease-specific DALYs can be questioned. In addition there are even much more implications of disease and premature death on societies, social networks, and social coherence which have not yet been quantified. For example, the loss of knowledge, family support, and active and supporting individuals that form a society is not computable. The financial burden on the health-care system and on the individual (considering that in many countries health care is financed out of the pocket of each individual) due to chronic diseases and YLD can be calculated to some extent. However the GBD approach gives a consistent and global perspective on disease distribution and composition that helps to assess and compare the differences in different countries.

## 2.4 Infectious Diseases

Classification of infectious diseases can be a challenge depending on what the perspective and the goals of this classification are (Table 2.2). Usually there are four common classifications for infectious diseases, which are important to know for further understanding.

As shown in Section 2.3, infectious diseases are responsible for about 25% (without maternal and neonatal disorders and nutritional deficiencies) of all DALYs. Globally HIV/AIDS, malaria, TB, LRIs, diarrhea, and meningitis are of importance (Table 2.3).

Correspondent to the epidemiological transition, these infectious diseases have their greatest impact, concerning DALYs and death, on young age groups with declining impact from children under the age of 5 years to adults up to the age of 50 years. Some of these diseases are addressed by the MDGs, which have to be achieved at the end of 2015, and will be addressed by the Sustainable Development Goals (SDGs) for the time of 2015–2030 (United Nations, 2014b). As already mentioned in Section 2.3, HIV/AIDS infections have notably increased in the past two decades and are still one of the major global public health issues. Worldwide approximately 35 million people lived with HIV in 2013 and 2.1 million became infected in the same year (World Health Organization, 2014c). Sub-Saharan Africa is the region with the highest burden due to HIV/AIDS. In 2013, about 24.7 million people were living there with HIV, and in 2010, 116 people out of 100 000 died from it. The second most affected region in 2010 was Central/Eastern Europe and Central Asia, burdening about 20 deaths per 100 000 due to HIV (Institute for Health Metrics and Evaluation (IHME), 2013b). At present, HIV infections cannot be healed, but treatment with multiple antiretroviral (ARV) drugs can stabilize the disease for long periods of time, leading to a considerably higher life expectancy in patients. According to the WHO, less than half of the patients receive ARV

**Table 2.2** Classification of infectious diseases (Nelson, Masters, and Graham, 2001).

| Classification | Elements |
|---|---|
| Clinical classification: according to most common or important clinical manifestations or affected organ system | Diarrheal diseases, respiratory diseases, central nervous system infection, cardiovascular infection, sepsis |
| Microbiological classification | Bacterial, viral, fungal, parasitic, prion |
| Epidemiological – means of transmission/reservoir of organism | Contact, food- or waterborne, airborne, vector-borne, perinatal |
| Epidemiological – reservoir of organism | Human, animal (zoonoses), soil, water |

therapy (World Health Organization, 2014c). Especially in low-income countries, less than a quarter of children and only a third of adults living with HIV are getting the ARV therapy they need, contradicting the MDG 6b to make HIV/AIDS treatment universally accessible. As for every infectious disease, besides extensive prevention measures, early, valid, and cost-effective testing is an essential prerequisite to contain further spread by infected people. By up to 6 weeks after infection, patients show the highest amount of infectivity without antibodies being detectable yet, and furthermore, for post-exposure prophylaxis, ARV drugs have to be used within 72 h of exposure (World Health Organization, 2014c). Although ARV therapy decreases the probability of opportunistic infections in patients with HIV by reducing the viral load under the level of detection, these patients remain carriers of HIV that can still be transmitted. But there is increasing evidence that early ARV therapy reduces the risk of transmission (Cohen *et al.*, 2012) between 38% in a population perspective (Tanser *et al.*, 2013) and up to 96% concerning sexual transmission (Cohen *et al.*, 2011).

Similar to HIV/AIDS, malaria is one of the main infectious diseases responsible for global deaths and DALYs (Table 2.3), especially in the sub-Saharan African region. It is also mentioned in the MDG 6c. In sub-Saharan Africa, malaria was responsible for 123 deaths in 100 000 people compared to South Asia, the second ranked region with 3.7 deaths in 100 000 people. Children are mainly affected by

**Table 2.3** Impact of major infectious diseases, as share of global DALYs and global deaths (Institute for Health Metrics and Evaluation (IHME), 2013b; Anonymous, 2015).

| | Percent of total DALYs 2010 (%) | Total deaths 2013 (mio.) | Description |
|---|---|---|---|
| HIV/AIDS | 3.29 | 1.34 | Retrovirus, transmitted by contact with body fluids, causing immunodeficiency and danger of opportunistic diseases like pneumonia, fungal infections, tuberculosis, or cryptosporidiosis |
| Malaria | 3.33 | 854 600 | Parasite, vector-borne, severe illness, fever, anemia, neurological complications, death |
| Tuberculosis (TB) | 1.99 | 1.29 | Bacteria, airborne, lungs affected, cough, fever |
| Lower respiratory tract infections | 4.65 | 2.65 | Mostly bacterial and viral infections, airborne, affecting trachea, lungs (pneumonia), and bronchi (bronchitis). Causing fever, cough, and difficulty to breathe |
| Diarrhea | 3.61 | 1.26 | Bacteria, viruses, parasites, food- and waterborne, person to person contact, ≥3 loose or liquid stools per day, dehydration, electrolyte imbalances |

this vector-borne disease. About 10% of global deaths at the age under 5 years are due to malaria (Institute for Health Metrics and Evaluation (IHME), 2013b). Even if increased prevention measures reduced the malaria diffusion, risks are coming also from global transportation, as non-immune travelers from malaria-free areas are especially vulnerable. The *Anopheles* mosquitoes are the vectors, carrying the malaria parasites *Plasmodium falciparum, Plasmodium vivax, Plasmodium malariae,* and *Plasmodium ovale. Plasmodium knowlesi* occurring in Southeast Asia usually causes malaria in monkeys. Only in recent years there were reports of infected humans. Chemoprophylaxis with different drugs is available but not effective against *Plasmodium knowlesi.* In addition, long-term prophylaxis without or few side effects is not available. To stop diffusion of malaria, vector control is the main and most effective tool. As treatment WHO recommends an artemisinin-based combination therapy (ACT) which acts via its toxic effect on *Plasmodium* gametes. The use of ACT is seen by the WHO as an important contribution to stop further spread of the *Plasmodium* parasite (World Health Organization, 2015f).

In sub-Saharan Africa, former states of the Soviet Union, and Southeast Asia, TB is still a threat, although TB is addressed in the MDG 6c. According to the WHO, 9 million people were infected with TB in 2013. Especially in the context of HIV/AIDS infections, TB is frequent and a huge threat. One-fourth of HIV-related deaths are due to TB. Multidrug resistance (not responding at least to isoniazid and rifampicin) is increasingly becoming a problem in TB therapy (multidrug-resistant TB (MDR-TB)). Countries which are especially affected comprise India, China, and the Russian Federation. Second-line drugs are not available everywhere and must be taken up to 2 years for treatment, thus making it more costly and increasing the risk of side effects. But in some cases even second-line drugs are useless because of bacterial resistance. In these cases extensively drug-resistant TB (XDR-TB) is present. The WHO estimates that about 9.0% of MDR-TB cases are XDR-TB. A MDR-TB-sensitive quick test for a diagnosis within 2 h will be coming on the market and will help to detect even less infectious forms of TB that cannot be detected with classical sputum smear microscopy (World Health Organization, 2015h).

LRIs are a heterogeneous group of different infections affecting the lower respiratory tract (LRT). LRIs with the highest global death toll comprise bacterial infections due to *Haemophilus influenzae* type B, causing pneumonia; pneumococcal infections, which are the major cause of pneumonia; and viral infections with influenza (affecting upper and LRT) or respiratory syncytial virus. Children below the age of 5 years are severely affected by LRIs and pneumonia in particular. Thus, pneumonia was the major mortality cause for children under the age of 5 years in 2013 (770 000) (Anonymous, 2015). But for the most part of LRI-caused deaths, no etiologic agent has been identified (1.7 million deaths in 2013). Especially affected regions include Central Africa, parts of the Arabian Peninsula, Afghanistan, and Papua New Guinea (Institute for Health Metrics and Evaluation (IHME), 2013c).

Populations at high risk for LRIs are children under the age of 5 years, hospitalized persons or patients with immunodeficiency/immunosuppression, and people consuming tobacco. Pneumonia is usually treated with the antibiotic amoxicillin.

Diarrheal diseases made up for 3.6% of global DALYs, with an emphasis on sub-Saharan Africa (6.8% of DALYs in this region, 64 deaths/100 000) and South Asia (5.3% of DALYs in this region, 44 deaths/100 000) (Institute for Health Metrics and Evaluation (IHME), 2013b). Important for diarrheal diseases are mortality figures (diarrhea being the second leading cause of death in children under the age of 5 years). Besides causing life-threatening dehydration in children, recurrent diarrheal diseases result in a high amount of YLDs. The outcome depends on health-care availability and risk factors like under- and malnutrition and the lack of safe drinking water. hygiene, and sanitation. Dividing DALYs in YLLs and YLDs caused by diarrheal diseases, a different geographical pattern can be observed. The YLLs/100 000 show a similar pattern as the mortality rate. Switching to YLDs/100 000, the highest rate can be found in sub-Saharan Africa (195/100 000), followed by North Africa and the Middle East (146/100 000), Latin America and Caribbean (137/100 000), and South Asia (129/100 000) (Institute for Health Metrics and Evaluation (IHME), 2013b). Diarrheal diseases are usually caused by viruses, bacteria, and parasites. For more than half of the deaths caused by diarrhea, no etiology has been identified. For diarrheal infections, for which a specific pathogen is known, the most important (concerning death cause) pathogens are the rota- and adenoviruses; the bacteria *Shigella* (shigellosis), enterotoxigenic *Escherichia coli*, *Clostridium difficile*, and *Vibrio cholerae* (cholera); and the protozoan parasite *Cryptosporidium* (cryptosporidiosis) (Anonymous, 2015). Depending on a quick and valid diagnosis of the responsible pathogen, several medical treatments are available. Viral diarrhea is self-limiting and usually symptomatically treated with oral rehydration solutions (ORSs) or intravenous fluid in severe cases. Rotavirus infections have been prevented with vaccines since 1999. Nitazoxanide is administered usually against *Cryptosporidium* infections but it is also effective in reducing the duration of rotavirus infections (Rossignol *et al.*, 2006). In addition to symptomatic therapy, bacterial diarrhea can be treated with antibiotics.

### 2.4.1
### (Re-)emerging Infectious Diseases

There is no internationally accepted definition of (re-)emerging infectious diseases. According to the CDC edited scientific journal *Emerging Infectious Diseases*, these diseases are:

> ( … ) whose incidence in humans has increased in the past 2 decades or threatens to increase in the near future, have been defined as "emerging." These diseases, which respect no national boundaries, include
>
> • New infections resulting from changes or evolution of existing organisms

- Known infections spreading to new geographic areas or populations
- Previously unrecognized infections appearing in areas undergoing ecologic transformation
- Old infections reemerging as a result of antimicrobial resistance in known agents or breakdowns in public health measures.

*(Centers for Disease Control and Prevention, n.d..)*

The WHO is more specific in their description concerning these diseases and mentions HIV/AIDS, Ebola virus disease, hepatitis C, Lyme disease, hantavirus pulmonary syndrome, and severe acute respiratory syndrome (SARS) as some of 30 diseases which have emerged in the past two decades, emphasizing that they are incurable and impose high suffering and death. Besides the earlier mentioned diseases, diseases which were deemed to be controlled in the past reemerge due to drug resistance. These diseases comprise malaria, TB, and bacterial pneumonia (World Health Organization, 2015c).

When diseases reach new geographical regions or populations, they are also called *emerging diseases*, even if they were already abundant in other regions. Thus, some researches speak of emerging diseases in case they reach the consciousness of the wealthy and developed countries (Farmer, 1996). This resembles the definition of neglected diseases. They have already been known, but ignored by global science and politics because they were affecting only a small geographical region or a small population. It is one of the inherent traits of infectious diseases that they change and adapt over time, especially in a globalized world where diseases spread much faster, population density is high, intensive contact between humans and animals through livestock farming is frequent, and the acquisition and destruction of tropical forests and other untouched landscapes increase. One of the additional triggers for the (re-)emergence of infectious diseases is the loss of biodiversity (Keesing *et al.*, 2010). Thus, a static list of emerging diseases is not feasible, because anthropologic and environmental dynamics will continuously lead to new strains, serotypes, or new infectious microbes at all. Military conflicts can also provoke (re-)emergence of nearly eradicated diseases, like polio, which regained significance through the conflict in the Middle East. Another phenomenon was observed in 2010 in Congo where ineffectiveness of common polio vaccinations was detected (Drexler *et al.*, 2014).

Changes that can occur are:

- Change in distribution
- Change in resistance to known treatment
- Change of virulence or transmissibility
- Change in clinical expression.

(Petersen, Chen, and Schlagenhauf, 2011)

Diseases that have not changed and have not been recognized till now are in many cases zoonoses. RNA viruses in particular are causative agents of these diseases. Changes in treatment resistance will be discussed in Section 2.6.

A number of examples of newly emerging diseases have been reported in the last few years. Avian influenza, hantavirus pulmonary syndrome, and SARS are typical examples of recently emerging infectious diseases. A novel coronavirus emerged in southern China in 2002 as a zoonotic disease in consequence of close animal–human contacts by professionals processing food or restaurant chefs and was finally traced back to bats (Chan *et al.*, 2015). It caused over 8000 infections and over 800 deaths, after it was imported to Hong Kong and was then spread by air travel to 30 countries all over the world. Coronaviruses, able to affect humans, have been detected in several animals. Through fast detection of cases during the outbreak and effective public health measures, more human to human infections of SARS have been prevented since July 2003 (Hilgenfeld and Peiris, 2013). In 2012 a new unknown human coronavirus emerged in the Middle East causing the Middle East respiratory syndrome (MERS). Camels are suspected to be intermediate transmitters of the virus and bats as the primary hosts. Till February 2015, 971 cases of MERS had been identified all over the world and resulted in 356 deaths (World Health Organization, 2015g).

The 2014 emergence of Ebola virus disease in western Africa (Sierra Leone, Liberia, Guinea; about 26 000 cases and about 10 700 deaths till April 2015) (World Health Organization, 2015a) is an example of infectious disease emergence in a new geographical region. Until 2014 Ebola virus disease was mainly seen in Central and East Africa (Uganda, Democratic Republic of Congo) with case counts up to 425 and 224 deaths (World Health Organization, 2015b). Ebola virus disease is a viral, zoonotic disease, with fruit bats as its suspected initial reservoir (Vogel, 2014). Ebola is transmitted by direct person to person contact and contact with contaminated objects through skin lesions or mucous membranes or by consumption of meat from infected animals like chimpanzees and gorillas (bushmeat) (Centers for Disease Control and Prevention, 2015). Because of the similarity of symptoms with other hemorrhagic fever diseases, only an expensive laboratory analysis can confirm an Ebola infection. The Ebola outbreak in West Africa got more global relevance due to global air travel. Infected health professionals have been treated in the United States, Spain, Germany, and the United Kingdom. The first human to human Ebola infection on US and European grounds has occurred: two US nurses were infected with Ebola during the treatment of a health professional who got infected in Liberia. The first Ebola infection in Europe (Spain) occurred in the same context. The unusual tremendous outbreak in West Africa was also a consequence of an unprepared health-care system with scarce resources to manage an outbreak of that dimension, in addition to the lack of experience to handle Ebola, which has never been diagnosed before in that region. Furthermore traditional beliefs of the population neglected the existence of the Ebola virus as a cause for the Ebola virus disease, making the implementation of public health measures much more difficult (World Health Organization, 2014d).

The emergence of MDR-TB and methicillin-resistant *Staphylococcus aureus* (MRSA) is covered in Section 2.6.

### 2.4.2 Neglected Tropical Diseases

Similar to (re-)emerging infectious diseases, definitions of NTDs are not consistent.

The WHO states:

> Neglected tropical diseases are a medically diverse group of infections caused by a variety of pathogens such as viruses, bacteria, protozoa, and helminths ( … ).
>
> *(World Health Organization Regional Office for the Western Pacific (WHOWPRO) (2015a))*

> ( … ) (They) are a diverse group of diseases with distinct characteristics that thrive mainly among the poorest populations. ( … ).
>
> *(World Health Organization Regional Office for the Western Pacific (WHOWPRO) (2015a))*

The CDC defines this disease group as follows:

> NTDs are a group of infectious diseases that are the source of tremendous suffering because of their disfiguring, debilitating, and sometimes deadly impact. They are called neglected because they have been largely wiped out in the more developed parts of the world and persist only in the poorest, most marginalized communities and conflict areas.
>
> *(Centers for Disease Control and Prevention, 2011)*

The *PLOS Neglected Tropical Diseases* journal defines NTDs as follows:

> The NTDs are defined as a group of poverty-promoting chronic infectious diseases, which primarily occur in rural areas and poor urban areas of low-income and middle-income countries. They are poverty-promoting because of their impact on child health and development, pregnancy, and worker productivity, as well as their stigmatizing features.
>
> *(PLOS Neglected Tropical Diseases, n.d..)*

Institutions like the WHO, CDC, and IHME and the *PLOS Neglected Tropical Diseases* journal are naming several different diseases of this disease group. Therefore in this chapter we will address the intersection of such diseases that all these institutions name in common (Table 2.4).

NTDs are of importance not only because of the overall number of deaths, DALYs, and geographical heterogeneous distribution but also due to their high frequency in the lowest socioeconomic groups. These diseases occur mainly in low-income and in few middle-income countries of the tropics and can result in catastrophic health expenditures; they are also responsible for household poverty and financial burden of health-care systems. Most of these diseases cause low

**Table 2.4** Selection of some NTDs (not comprehensive) (Hotez *et al.*, 2014; Institute for Health Metrics and Evaluation (IHME), 2013b).

| | Percent of global DALYs 2010 (%) | Total deaths | Expected number of cases in 2010 | Description |
|---|---|---|---|---|
| *Protozoa* | | | | |
| Human African trypanosomiasis (sleeping sickness) | 0.02 | 9 110 | 37 000 | Vector-borne, transmitted by tsetse fly, fever, weakness, pain in the joints. Later, affecting the central nervous system, causing neurological changes, psychiatric disorders, coma, seizures, and finally death |
| Leishmaniases | 0.1 | 51 564 | 76 000 | Caused by *Leishmania* parasite, vector-borne, transmitted by female phlebotomine sandflies, commonly in the form of visceral (fever, weight loss, enlargement of the spleen and liver, anemia), cutaneous (skin lesions, ulcers, scars), and mucocutaneous leishmaniasis (destruction of mucous membranes of the nose, mouth, and throat) |
| *Bacteria* | | | | |
| Trachoma | 0.01 | 0 | 4.4 million | Caused by *Chlamydia trachomatis*, transmitted through personal contact and flies, affecting the eye, can become chronic, eyelid can turn inward, then leading to irreversible blindness |
| *Helminth* | | | | |
| Cysticercosis/ taeniasis | 0.02 | 1 160 | 1.4 million | Taeniasis caused by tapeworm *Taenia solium*. Infection of larval stages causes cysticercosis, can affect the central nervous system, causing epilepsy |
| Food-borne trematodiases | 0.1 | 0 | 16 million | Infection caused by trematodes, food-borne, worms colonizing different organs, mainly the liver, inadequate availability of medicines to treat clonorchiasis and opisthorchiasis |

**Table 2.4** *Continued.*

| | Percent of global DALYs 2010 (%) | Total deaths | Expected number of cases in 2010 | Description |
|---|---|---|---|---|
| Onchocerciasis (river blindness) | 0.02 | 0 | 30.4 million | Caused by the filarial worm *Onchocerca volvulus*, vector-borne, transmitted by bites of black flies which breed in rivers, eye and skin disease, can lead to irreversible blindness |
| Schistosomiasis | 0.1 | 11 650 | 252 million | Caused by larval forms (cercariae) of schistosomes, waterborne, affecting the urinary tract, genitals, and intestines, abdominal pain, diarrhea, and blood in urine and stool |
| Intestinal nematode infections (hookworm, ascariasis, trichuriasis) | 0.2 | 2 676 | 1.723 billion | Infection caused by helminths, transmitted by eggs in human feces, contaminating soil in areas with poor sanitation. Causing diarrhea, abdominal pain, general malaise, weakness. Anemia by hookworms |
| *Virus* | | | | |
| Dengue | 0.03 | 14 661 | 179 000 (393 million (Bhatt *et al.*, 2013)) | Mosquito-borne (same species as with chikungunya), flu-like illness, 30-fold increase of incidence in the last 50 years, vaccine is highly anticipated (probably 2015) |
| Rabies | 0.1 | 26 404 | 1 100 | Zoonotic disease, transmitted in saliva of rabid animals, encephalitis, myelitis, paralysis, anxiety, confusion, agitation |

mortality, but high and long-lasting disability. To prevent long-lasting suffering and permanent damage due to NTDs, early treatment, and therefore early diagnosis, is required. Because clinical examination and laboratory resources are traditionally often not available in the rural areas where these diseases are common, there is a huge need for cost-effective, simple, and quick diagnostic tools. According to the WHO, a key role in defeating NTDs is the capacity for vector management to control the transmission of pathogens, especially in NTD-endemic countries (World Health Organization, 2015d). The WHO is emphasizing the need for research and development of new diagnostic tools, medicines, vaccines, and methods for vector control (World Health Organization, 2015d), particularly because of the old nature of available medicines. Some of them are about 50 years old and only 10 of 1556 newly developed medicines from 1975 to 2004 have been for the treatment of some few NTDs (Chirac and Torreele, 2006). To eliminate this deficit, the pharmaceutical industry is already supporting other researchers by opening parts of their compound libraries. The Drugs for Neglected Diseases initiative is facilitating these processes (World Health Organization, 2015g). In addition, the WHO has adopted the resolution WHA66.12 of the World Health Assembly 2013 (World Health Organization, 2013b) concerning NTDs. In addition to prevention programs, control programs for the elimination of NTDs are under way. For instance, the elimination of schistosomiasis in China or elimination of onchocerciasis by the 47 member states of the African WHO region is in progress. As with other diseases, NTDs will be addressed in the SDG 3.3 to keep them in focus of global politics and

> by 2030 end the epidemics of ( … ) neglected tropical diseases ( … ).
>
> *(United Nations, 2014a)*

Considering the rise of NCDs, NTDs get significance in addition, because many NTDs are the underlying cause of NCDs. Thus, some kinds of cancer, cardiovascular disease, and liver disease result from former NTD infections (Hotez and Daar, 2008).

## 2.5 Noncommunicable Diseases

Due to advances in the fight against communicable diseases and the aging of the global population (Section 2.3), NCDs have gained much more importance for global public health measures (Table 2.5).

The naming of this disease group is not an etiologic one. The term *noncommunicable diseases* is used to differentiate it from communicable or infectious diseases and usually excludes mental and behavioral disorders. In our understanding we will include mental and behavioral disorders and physical diseases like cancer, cardiovascular disease, and so on in the term. These diseases have a long-lasting course; often they are incurable and become chronic. But they are not transmittable.

Especially in this group of diseases, the DALY concept is suitable to reflect the different outcomes in diseases like cardiovascular diseases (e.g., ischemic heart disease, stroke) which contribute to a high amount of deaths and mental and behavioral disorders (like unipolar depressive disorders) with no direct causal deaths, but a high amount of long-lasting disability. This can be captured adequately by means of assessing DALYs. Mental and behavioral disorders like

**Table 2.5** Major NCDs based on DALYs, deaths, and change of DALYs per 100 000 since 1990 (Institute for Health Metrics and Evaluation (IHME), 2013b; Murray *et al.*, 2012).

| Disease | Percent of total DALYs 2010 (%) | Percent of total deaths 2010 (%) | Change of DALYs per 100 000 in percentage from 1990 to 2010 (%) |
|---|---|---|---|
| *Supergroup* | | | |
| Cardiovascular and circulatory diseases | 12.0 | 30.0 | −5.7 |
| Cancer | 7.6 | 15.0 | −2.1 |
| Mental and behavioral disorders | 7.4 | 0.4 | — |
| *Subgroup* | | | |
| Ischemic heart disease | 5.2 | 13.3 | −0.6 |
| Stroke | 4.1 | 11.1 | −8.5 |
| Low back and neck pain | 4.6 | 0 | 9.4 |
| Unipolar depressive disorders | 3.0 | 0 | 5.8 |
| Chronic obstructive pulmonary disease (COPD) | 3.1 | 5.5 | −24.6 |
| Cirrhosis of the liver | 1.3 | 2.0 | −1.9 |
| Other musculoskeletal disorders | 1.2 | 0.2 | 15.4 |
| Alzheimer's disease and other dementias | 0.5 | 0.9 | 53.3 |
| Diabetes mellitus | 1.9 | 2.4 | 30.0 |
| Trachea, bronchus, and lung cancers | 1.3 | 2.9 | 4.5 |
| Prostate cancer | 0.1 | 0.5 | 23.9 |
| Parkinson's disease | 0.1 | 0.2 | 34.9 |

depression or anxiety were among the top 10 causes of disability in nearly every world region.

As for communicable diseases, overall global DALYs for NCDs have risen since 1990, but decreased a little, concerning DALYs per 100 000. Neoplasms have decreased per 100 000, as well as cardiovascular and circulatory diseases, chronic respiratory diseases, liver cirrhosis, and digestive diseases. On the other hand, in four disease groups one can observe a rise of DALYs per 100,000 since 1990: 'Neurological disorders', 'Mental and behavioural disorders', 'musculoskeletal disorders' and 'diabetes, urogenital, blood, and endocrine diseases'. In this context, diabetes, mental and behavioral disorders, and neurological disorders are the main drivers (Murray *et al.*, 2012).

The mean age and life expectancy of the global population are rising and increasingly more and more people reach advanced ages. Thus, the proportion of older people who are rather affected by NCDs is increasing. Mental and behavioral disorders occur mainly in the life span of teenagers up to 50 years of age. In the middle ages up to 65 years, mental and behavioral disorders, cancer, and musculoskeletal disorders are the main NCD causes for DALYs. From 65 years and beyond, cardiovascular and circulatory diseases, as well as neurological disorders, gain significance (Murray *et al.*, 2012). From a geographical perspective, NCDs prevail in high-income countries, but are becoming more and more important in low- and middle-income countries. Lung cancer mortality, for instance, due to a reduced prevalence of smoking has decreased in high-income countries (Hunter and Reddy, 2013). Globally, the WHO assumes an increase from 38 million NCD-caused deaths in 2012 to 52 million NCD-caused deaths by 2030. The majority of premature deaths by NCDs occur in low- and middle-income countries (82%) with a higher proportion before the age of 70 years (48%) (World Health Organization, 2014b). Thus, low- and middle-income countries are facing a double burden of NCDs and infectious diseases (Bygbjerg, 2012). Based on the knowledge of the interconnections between NCDs and infectious diseases, there are several possible options to practice bilateral prevention: (i) Fighting malnutrition and tobacco and alcohol use would have a positive effect on TB incidence as well as diabetes, chronic obstructive pulmonary disease (COPD), and some cancers. (ii) Vaccination against human papilloma virus and hepatitis B could prevent common cancers. Especially in low-income countries, maternal and child health is at risk due to the interaction between malnutrition and infectious diseases, increasing the risk for diabetes, metabolic diseases, and cardiovascular diseases later in life (Bygbjerg, 2012).

Several modifiable risk factors for NCDs have been identified in the last decades. Among the most important ones that promote cardiovascular diseases, diabetes, obesity, and neck and back pain are insufficient physical exercise, dietary risks, a diet lacking fruits and vegetables, high energy intake, high BMI, harmful alcohol use, and tobacco use. As most NCDs are multifactorial, these factors can increase the risk of multiple diseases simultaneously. Since 1990 DALYs due to risk factors like high blood pressure, unhealthy diet, high BMI, and high fasting plasma glucose have risen. All of them are clearly related to several NCDs. These findings apply on a global scale. In contrast, sub-Saharan Africa is facing risk factors like childhood underweight, suboptimal breastfeeding, household air pollution, and iron deficiency, leading to a different structure of disease burden with less impact by NCDs (Institute for Health Metrics and Evaluation, 2013).

Besides the aforementioned lifestyle measures, a better understanding of the microbiome of the gut seems to be a viable path for an enhanced treatment of several NCDs like diabetes, cancer, metabolic syndrome, atherosclerosis, and inflammatory bowel disease (Bartlett, 2015).

Mental and behavioral disorders, in contrast, have much less obvious risk factors. Their etiology is not well understood until today. Consensus exists on an etiologic trinity of biologic (genetic), social, and psychological factors. Often mental disorders come as a comorbidity of physical illness and can lead to stigma and loss of social integrity and poverty. Data from the last decade hint toward the fact that living in big cities might constitute a risk factor as well (Dekker *et al.*, 2008; Lederbogen *et al.*, 2011; Peen *et al.*, 2010).

Often, psychiatric disorders or physical NCDs have their origins in young ages of an individual. Therefore, prevention should and could already start during childhood. On the other hand the often long-lasting nature of NCDs results in a long-time treatment. As a result, drug therapy for cardiovascular diseases in low- and middle-income countries can lead to high out-of-pocket health expenditure (Hunter and Reddy, 2013). Thus, the WHO demands in their global targets a reduction in the burden of NCDs till 2025 and an increased availability and affordability of basic medicines (e.g., aspirin, statins, angiotensin-converting enzyme inhibitors, thiazide diuretics, long-acting calcium-channel blockers, beta-blockers, metformin, insulin, bronchodilators, and steroid inhalants) (World Health Organization, 2014b) to treat major NCDs (Target 9).

The health-care systems today are rather prepared for episodic treatment of acute illnesses than for long-lasting NCD management. Considering an aging population, frequent and easy accessibility of health-care services will be however necessary.

## 2.6 Antimicrobial Resistance

The antibacterial age started with the discovery of penicillin in 1928 and got a push in the 1950s and 1960s with the production of penicillin on an industrial scale, in addition to the discovery of far more antibacterial drugs (Lewis, 2012). After the detection of penicillin, the survival rate of patients infected with pneumococcal pneumonia (*Streptococcus pneumoniae*) increased tremendously (from 10 to 90%). But then the decline of discoveries propelled the risk of antibacterial resistances (ABRs); thus from 2008 to 2012, only two new systemic antibacterial agents have been approved by the US Food and Drug Administration (Boucher *et al.*, 2013). In the 1960s the emergence of the first MRSA strains has been observed and remained mainly a problem of hospital-acquired infections. This has changed in the last decade and community-acquired MRSA infections were increasingly reported. ABR is still mainly found in bacteria from common health-care- associated and community-acquired infections. Common infections like pneumonia; cystitis; infections in neonatal and intensive care, cancer treatment, and organ transplant; and infections of surgical sites after operations will become more dangerous and expensive when oral antibacterial treatment is not sufficient anymore or antibacterial drugs are losing their efficacy at all (Laxminarayan *et al.*, 2013). There are not enough data available to assess the global burden due to ABR. On an individual level, ABR causes prolonged suffering, worse clinical outcome, more consumed health resources, and probably death. According to the US CDC, about 23 000 patients die in the United States due to AMR infections every year (Centers for Disease Control and Prevention, 2013). As a consequence of the growing number of resistant bacteria, the WHO declared nine combinations of pathogen resistance to a specific drug as global public health threats. Six of them are shown in Table 2.6. The data have been collected from national institutions of the WHO member states, resulting in a reported range, not a mean value. Unfortunately not

**Table 2.6** Selection of pathogen antibiotic resistances in WHO regions (World Health Organization, 2014a).

| Pathogen resistance | Typical disease | AFR (%) | AMR (%) | EMR (%) | EUR (%) | SEAR (%) | WPR (%) |
|---|---|---|---|---|---|---|---|
| *Escherichia coli* – third-generation cephalosporins | Urinary tract infections, bloodstream infections | 2–70 | 0–48 | 22–63 | 3–82 | 16–68 | 0–77 |
| *Klebsiella pneumoniae* – carbapenems | Pneumonia, bloodstream infections, urinary tract infections | 0–4 | 0–11 | 0–54 | 0–68 | 0–8 | 0–8 |
| Methicillin-resistant *Staphylococcus aureus* (MRSA) – beta-lactam antibacterial drugs | Wound infections, bloodstream infections | 12–80 | 21–90 | 10–53 | 0.3–60 | 10–26 | 4–84 |
| Nontyphoidal *Salmonella* – fluoroquinolones | Food-borne diarrhea, bloodstream infections | 0–35 | 0–96 | 2–49 | 2–3 | 0.2–4 | 0–14 |
| *Shigella* species – fluoroquinolones | Diarrhea | 0–3 | 0–8 | 3–10 | 0–47 | 0–82[a] | 3–28 |
| *Neisseria gonorrhoeae* – third-generation cephalosporins | Gonorrhea, sexually transmitted infection, affecting genitals | 0–12 | 0–31 | 0–12 | 0–36 | 0–5 | 0–31 |
| *Streptococcus pneumonia* – penicillin | Pneumonia, meningitis, otitis | 3–16 | 0–48 | 13–34 | 0–61 | 47–48 | 0–47 |

a) Based on publications. AFR: African region; AMR: region of the Americas; EMR: Eastern Mediterranean region; EUR: European region; SEAR: Southeast Asia region; and WPR: Western Pacific region.

all member states reported data on resistances, demonstrating the lack of surveillance data.

A major driver regarding the emergence and spread of ABRs comes from their use in livestock farming. ABR in bacteria of food-producing animals can either be transmitted by direct contact to humans or enter the food chain. As runoff from manured agricultural surfaces, these bacteria can also enter ground and surface waters, where drinking water may be collected. Finally these resistant strains may additionally affect human health (Silbergeld, Graham, and Price, 2008; Woolhouse and Ward, 2013).

Furthermore AMR is observed in other disease-specific surveillance programs, which are not only restricted to ABR. The WHO had reported 84 000 cases of MDR-TB in 2012, with only 21% of the 450 000 by WHO-estimated MDR-TB cases. It is assumed that globally 3.6% of new TB cases each year are multidrug-resistant (MDR) cases (World Health Organization, 2014a). Regionally these cases have their focus in Eastern Europe and Central Asia. Because of a lack in surveillance and reliable, quick diagnostic tools, MDR-TB is most probably underreported. On a regional scale there is an emphasis on new MDR-TB cases in Eastern Europe, the Russian Federation, Kazakhstan, Kyrgyzstan, Azerbaijan, and Uzbekistan, with up to about 35% MDR-TB cases of all TB cases. TB is one of the diseases addressed in the MDGs and will be addressed in the SDGs. Considering that, control of MDR-TB is crucial (World Health Organization, 2014a). Recent data indicate that spectinamides may be an upcoming answer for MDR-TB (Cully, 2014).

Already in the 1970s resistance to the antimalarial drug chloroquine has emerged in Southeast Asia, followed by the emergence of mefloquine resistance in the 1980s in Cambodia and Thailand. Then in the 1990s *P. falciparum* developed resistance against amodiaquine and sulfadoxine–pyrimethamine. AMR against the antimalarial medicine artemisinin is suspected or confirmed in Cambodia, Myanmar, Thailand, and Vietnam. One of the challenges in antimalarial drug resistance surveillance is the detection of resistance, which is not possible with routine laboratory procedures (World Health Organization, 2014a) and is therefore not feasible in rural regions of low- and middle-income countries of the tropics. This in turn means that the regions with the highest exposure are excluded from the statistic. Today antimalarial drugs are usually a combination of artemisinin and one of the other common agents (ACT), ensuring that the parasite is susceptible to at least one of the agents. In case that these combinations lose their efficacy (Taylor and Juliano, 2014), already achieved goals to reduce the spread of malaria will be lost and likely an increased mortality rate will follow.

For HIV, an increase in transmitted drug resistance against antiretroviral treatment (ART) has been observed in the last decade, especially in low- and middle-income countries. There, HIV resistance to ART ranged from 4.8 to 6.8% in 2007. Moreover Australia, Europe, Japan, and the United States have reported cases of transmitted ART resistance. In these countries 10–17% of patients without prior ART are infected with ART-resistant virus strains (World Health Organization, 2014a). The development of resistance against ART is highly correlated with availability and use of ART.

For influenza A virus, antiviral drug resistance is reported for several serotypes on a global scale. The global public health burden due to annual epidemics and periodic pandemics of influenza strains requires a better surveillance of resistance development in influenza virus. At the present, vaccines remain the best tool for influenza prevention. Because of the widespread resistance to adamantanes, neuraminidase inhibitors remain as primary treatment drug after infection (World Health Organization, 2014a).

A further challenge of health-care systems in industrialized countries is the emerging resistance of diverse fungal infections like *Aspergillus* (Barnes *et al.*,

2014) or *Candida* against the common antifungal drugs of triazole class and the newest class of antifungal drugs, the echinocandins (World Health Organization, 2014a).

Thus, perhaps the constant development of newer antimicrobials is for now the main option to keep up with the evolution of microorganisms. This is facing selective pressure due to use, misuse, and overuse of antimicrobial drugs for humans and animals and thus is challenging global public health.

Furthermore, illustrating the global significance of AMR and ABR, the topic was addressed by the US Administration in 2014, releasing its "National Strategy for Combating Antibiotic-Resistant Bacteria" accompanied by appropriate funding (Bartlett, 2015). As part of the country's G7 presidency, Germany is addressing the AMR topic in the G7 group of nations in 2015 (Burger, 2015).

## 2.7 Dynamics

Geographic patterns of diseases, as well as the demographic and socioeconomic distribution of diseases, vary over time. This affects AMRs, as well as all the earlier mentioned disease groups.

There are four primary factors responsible for the leading causes of years of healthy life lost:

- Aging populations
- Increases in NCDs
- Shifts from fatal causes to disabling causes, especially due to mental and behavioral disorders (depression, anxiety, substance abuse) and musculoskeletal conditions like low back pain and neck pain
- Change in risk factors.

The structure of the world population, as well as the mortality cause structure (Figure 2.3) (World Health Organization, 2013a), will change tremendously till 2050. The overall global population is estimated to increase to 9.6 billion, producing pressure on food production (intensified agriculture and livestock farming) and increasing the need for safe drinking water. This will have remarkable impact on child health in poorer populations (McMichael, 2013). At the same time the life expectancy will rise to about 76 years and the mean age will change to 36.1 years. This leads to a larger population with a greater share of aged people, which get older than ever before (United Nations, 2013a).

In addition the population will live much tighter together. In 2014 54% of the world's population lived in urban areas. In 2050 this amount will increase to 66% (United Nations, 2014d), leading to areas of high population density and thus giving many possibilities of interaction for the exchange of microorganisms and psychological stress.

Unhealthy diet is a risk factor for NCDs like diabetes, cardiovascular diseases, and cancer. Global changes in diet, with an increase in daily calories, animal fats

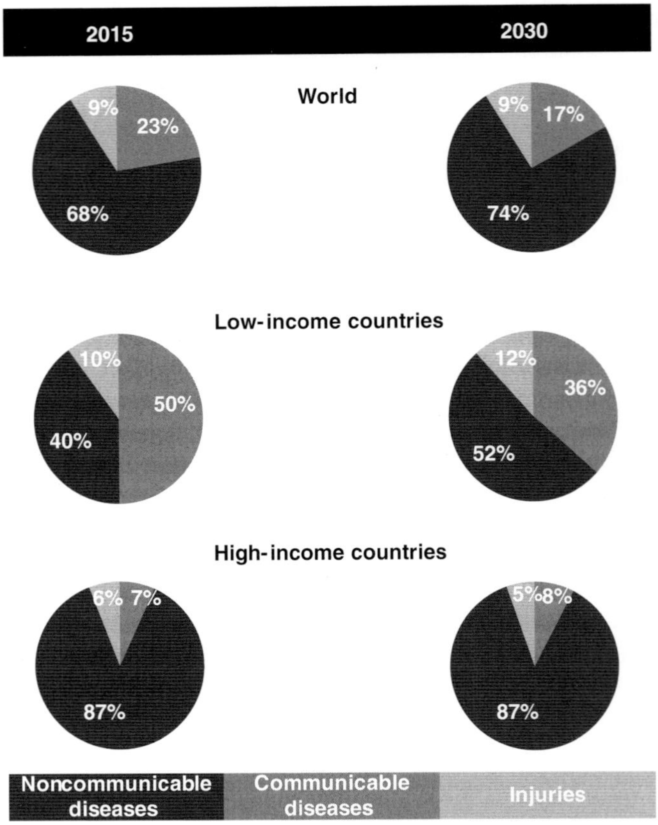


**Figure 2.3** Composition of causes of death in 2015 and 2030 in three different regions.

and meats, and energy-dense and nutrient-poor foods, can increase the frequency of those diseases (Popkin, 2006). The effects of unhealthy diets will be amplified by the lack of physical activity and improved transportation systems and communications technologies. The increase in daily unhealthy calorie intake is correlated with an increase in income per capita. In addition, in low-income countries, the impact of economic growth, market integration, foreign direct investment, and urbanization has a threefold greater impact on the epidemiological transition than aging alone (Stuckler, 2008).

Changes in infectious diseases are the consequence of mutations in microorganisms and subsequently man-made evolutionary pressure and forced selection. Viral mutations of nearly extinct diseases demand for new vaccines as evidenced in the case of a poliomyelitis outbreak in 2010 in the Republic of Congo. There, a lethality of 47% was observed due to the fact that vaccinated persons showed lower antibody titers against a mutated strain of poliovirus (PV1-RC2010) (Drexler *et al.*, 2014).

Infectious diseases have adapted to their environment as a result of a coevolution with mankind. A low population density, as it has been found before the

Neolithic Revolution and is still found in parts of rural tropical regions, like in sub-Saharan Africa, has different requirements on the survival mechanisms of infectious disease agents. Once pathogens have crossed from animals to humans, both are vulnerable to infections by the same pathogen. Given a low human population density, this is an advantage for pathogen spreading because of a larger target population (humans and animals). Vector-borne diseases can travel long distances. Long-lasting immunity in such populations would reduce the pool of available victims, but a long course of disease provides a longer time window for future infections. With migration and settlement of humans into temperate regions and the rise of population density, some of these characteristics were not necessary anymore; thus crowd epidemic diseases could develop. Such diseases are far more acute, have a changed route of transmission, and are confined to humans alone. This partly explains the geographic distribution of diseases into mostly tropical diseases (HIV, Chagas disease, cholera, dengue, trypanosomiasis, malaria, leishmaniasis) and mostly temperate diseases (hepatitis B, influenza A, measles, pertussis, rotavirus A, syphilis, tetanus, and TB) (Wolfe, Dunavan, and Diamond, 2007).

Also climate change has effects on health. It affects health in three dimensions:

1) A direct impact on extreme weather events (precipitation, storms, temperature) affects natural systems like parts of the environment, water, air, and disease vectors and alters human systems like agriculture and occupational health (IPCC, 2014; McMichael, 2013).
2) Climate change and human-promoted environmental degradation will change the environment in which diseases will spread or can emerge.
3) Mosquito-borne diseases will spread in areas where extreme floods become more frequent. Overall warmer climate leads to warmer climates in higher elevations.

In consequence, the geographical habitats of vectors like the *Anopheles* and *Aedes* mosquitoes will extend to higher elevations and regions so far not affected will become vulnerable for the spread of vector-borne diseases like malaria or dengue. Furthermore leishmaniasis is linked to environmental changes such as deforestation, building of dams, irrigation schemes, and urbanization (World Health Organization, 2015e). The increasing loss of biodiversity can result in an increase in infectious disease prevalence. The probability for parasites like the trematode that causes schistosomiasis to end up in unsuitable intermediate hosts is correlated with the amount of biodiversity. A high amount of biodiversity therefore reduces the risk of infection to humans (Keesing *et al.*, 2010).

Extent, speed, and use of regional and global transportation systems make nearly every place on Earth accessible for humans and microorganisms, thus increasing possibilities of exchange between pathogens, vectors, hosts, and vulnerable populations. As seen from the 2012 MERS emergence or the 2014 Ebola virus disease outbreak in West Africa, accelerated global travel can turn a regional danger into a global threat.

Military conflicts like the Syrian civil war have a tremendous impact on public health measures, resulting in the reemergence of infectious diseases, like the 2013 polio outbreak in eastern Syria (Aylward and Alwan, 2014). The migration of refugees out of such regions will raise pressure on the public health sector of destination countries.

Intensive surveillance is one of the key tools to assess the risk of diseases and AMR (Vernet *et al.*, 2014) and to stay informed about changes in disease structure and emergence of new (infectious) diseases. Knowing that animals are a major source of emerging viral diseases, Wolfe, Dunavan, and Diamond (2007) proposed the establishment of a global infectious disease origin initiative, where people with high levels of exposure to wild animals and farm animals are systematically monitored because of their risk to get infected by animal viruses. In addition, an in-depth analysis on a regional scale is essential. As we have seen, the validity of global values of disease burden, prevalence, and mortality is questionable on a regional scale. Thus regional population, gender and age structure, population distribution, migration, high-resolution surveillance, economic characteristics, policy regulations, mobility and transport systems, the health-care system, health-care regulations, availability of health-care technologies and pharmaceuticals, physical geography, and cultural beliefs and practices have to be accounted for to assess the real impact and consequences of diseases and interventions in societies.

## References

Anonymous (2015) Global, regional, and national age–sex specific all-cause and cause-specific mortality for 240 causes of death, 1990–2013: a systematic analysis for the Global Burden of Disease Study 2013. *Lancet*, **385** (9963), 117–171.

Aylward, R.B. and Alwan, A. (2014) Polio in Syria. *Lancet*, **383** (9916), 489–491.

Barnes, R.A. *et al.* (2014) Antifungal resistance: more research needed. *Lancet*, **384** (9952), 1427.

Bartlett, J. (2015) Predictions for the Field of Infectious Diseases for 2015, http://www.medscape.com/viewarticle/838729 (accessed 09 February 2015).

Bhatt, S. *et al.* (2013) The global distribution and burden of dengue. *Nature*, **496** (7446), 504–507.

Boucher, H.W. *et al.* (2013) 10 × '20 progress—development of new drugs active against gram-negative bacilli: an update from the Infectious Diseases Society of America. *Clin. Infect. Dis. Off. Publ. Infect. Dis. Soc. Am.*, **56** (12), 1685–1694.

Burger, L. (2015) Germany in Push to Encourage Antibiotics Development. Medscape, http://www.medscape.com/viewarticle/839943 (accessed 20 April 2015).

Bygbjerg, I.C. (2012) Double burden of non-communicable and infectious diseases in developing countries. *Science*, **337** (6101), 1499–1501.

Carolina, M.S. and Gustavo, L.F. (2003) Epidemiological transition: model or illusion? A look at the problem of health in Mexico. *Soc. Sci. Med.*, **57** (3), 539–550.

Caselli, G., Meslé, F., and Vallin, J. (2002) Epidemiologic transition theory exceptions. *Genus*, **58** (1), 9–51.

Centers for Disease Control and Prevention (2011) Neglected Tropical Diseases, http://www.cdc.gov/globalhealth/ntd/ (accessed 22 April 2015).

Centers for Disease Control and Prevention (2013) Antibiotic Resistance Threats in the United States, http://www.cdc.gov/drugresistance/pdf/ar-threats-2013-508.pdf (accessed 09 February 2015).

Centers for Disease Control and Prevention (2015) Ebola Virus Disease – Transmission, http://www.cdc.gov/vhf/ebola/transmission/ (accessed 22 April 2015).

Centers for Disease Control and Prevention (n.d.) Background and Goals – Emerging Infectious Disease Journal – CDC, http://wwwnc.cdc.gov/eid/page/background-goals (accessed 22 April 2015).

Chan, J.F.W. *et al.* (2015) Middle east respiratory syndrome coronavirus: another zoonotic betacoronavirus causing SARS-like disease. *Clin. Microbiol. Rev.*, **28** (2), 465 – 522.

Chirac, P. and Torreele, E. (2006) Global framework on essential health R&D. *Lancet*, **367** (9522), 1560 – 1561.

Cohen, M.S. *et al.* (2011) Prevention of HIV-1 infection with early antiretroviral therapy. *N. Engl. J. Med.*, **365** (6), 493 – 505.

Cohen, M.S. *et al.* (2012) Antiviral agents and HIV prevention: controversies, conflicts, and consensus. *AIDS (London, England)*, **26** (13), 1585 – 1598.

Cully, M. (2014) Antibacterial drugs: redesigned antibiotic combats drug-resistant tuberculosis. *Nat. Rev. Drug Discovery*, **13** (4), 256 – 257.

Dekker, J. *et al.* (2008) Psychiatric disorders and urbanization in Germany. *BMC Public Health*, **8** (1), 17.

Drexler, J.F. *et al.* (2014) Robustness against serum neutralization of a poliovirus type 1 from a lethal epidemic of poliomyelitis in the Republic of Congo in 2010. *Proc. Natl. Acad. Sci. U.S.A.*, **111** (35), 12889 – 12894.

Farmer, P. (1996) Social inequalities and emerging infectious diseases. *Emerging Infect. Dis.*, **2** (4), 259 – 269.

Gaziano, J. (2010) Fifth phase of the epidemiologic transition: the age of obesity and inactivity. *JAMA*, **303** (3), 275 – 276.

Harper, K. and Armelagos, G. (2010) The changing disease-scape in the third epidemiological transition. *Int. J. Environ. Res. Public Health*, **7** (2), 675 – 697.

Hilgenfeld, R. and Peiris, M. (2013) From SARS to MERS: 10 years of research on highly pathogenic human coronaviruses. *Antiviral Res.*, **100** (1), 286 – 295.

Hotez, P.J. and Daar, A.S. (2008) The CNCDs and the NTDs: blurring the lines dividing noncommunicable and communicable chronic diseases. *PLoS Negl. Trop. Dis.*, **2** (10), e312.

Hotez, P.J. *et al.* (2014) The global burden of disease study 2010: interpretation and implications for the neglected tropical diseases N. de Silva, ed.. *PLoS Negl. Trop. Dis.*, **8** (7), e2865.

Hunter, D.J. and Reddy, K.S. (2013) Noncommunicable diseases. *N. Engl. J. Med.*, **369** (14), 1336 – 1343.

Institute for Health Metrics and Evaluation (2013) *The Global Burden of Disease: Generating Evidence, Guiding Policy*, IHME, Seattle, WA, http://www.healthdata.org/sites/default/files/files/policy_report/2013/GBD_GeneratingEvidence/IHME_GBD_GeneratingEvidence_FullReport.pdf (accessed 18 February 2015).

Institute for Health Metrics and Evaluation (IHME) (2013a) GBD Arrow Diagram, http://vizhub.healthdata.org/irank/arrow.php (accessed 22 April 2015).

Institute for Health Metrics and Evaluation (IHME) (2013b) GBD Cause Patterns, http://vizhub.healthdata.org/gbd-cause-patterns/ (accessed 22 April 2015).

Institute for Health Metrics and Evaluation (IHME) (2013c) GBD Compare, http://vizhub.healthdata.org/gbd-compare/ (accessed 22 April 2015).

IPCC (2014) Climate Change 2014: Impacts, Adaptation, and Vulnerability. Part A: Global and Sectoral Aspects. Contribution of Working Group II to the Fifth Assessment Report of the Intergovernmental Panel on Climate Change, Cambridge, https://ipcc-wg2.gov/AR5/images/uploads/WGIIAR5-PartA_FINAL.pdf (accessed 11 August 2015).

Keesing, F. *et al.* (2010) Impacts of biodiversity on the emergence and transmission of infectious diseases. *Nature*, **468** (7324), 647 – 652.

Laxminarayan, R. *et al.* (2013) Antibiotic resistance-the need for global solutions. *Lancet Infect. Dis.*, **13** (12), 1057 – 1098.

Lederbogen, F. *et al.* (2011) City living and urban upbringing affect neural social stress processing in humans. *Nature*, **474** (7352), 498 – 501.

Lewis, K. (2012) Antibiotics: recover the lost art of drug discovery. *Nature*, **485** (7399), 439 – 440.

Mackenbach, J.P. (1994) The epidemiologic transition theory. *J. Epidemiol. Community Health*, **48** (4), 329–331.

McMichael, A.J. (2013) Globalization, climate change, and human health. *N. Engl. J. Med.*, **368** (14), 1335–1343.

Murray, C.J.L. and Lopez, A.D. (1996) Global Burden of Disease, http://www.hup.harvard.edu/catalog.php?isbn=9780674354487 (accessed 09 April 2013).

Murray, C.J.L. *et al.* (2012) Disability-adjusted life years (DALYs) for 291 diseases and injuries in 21 regions, 1990–2010: a systematic analysis for the Global Burden of Disease Study 2010. *Lancet*, **380** (9859), 2197–2223.

Nelson, N.M.H., Masters, C.F., and Graham, N.M.H. (2001) *Infectious Disease Epidemiology: Theory and Practice*, Aspen Publishers Inc., Gaithersburg, MD.

Olshansky, S.J. and Ault, A.B. (1986) The fourth stage of the epidemiologic transition: the age of delayed degenerative diseases. *Milbank Q.*, **64** (3), 355–391.

Omran, A.R. (1971) The epidemiologic transition: a theory of the epidemiology of population change. *Milbank Mem. Fund Q.*, **49** (4), 509–538.

Omran, A.R. (1983) The epidemiologic transition theory. A preliminary update. *J. Trop. Pediatr.*, **29** (6), 305–316.

Omran, A.R. (1998) The epidemiologic transition theory revisited thirty years later. *World Health Stat. Q.*, **51** (2–4), 99–119.

Peen, J. *et al.* (2010) The current status of urban-rural differences in psychiatric disorders. *Acta Psychiatr. Scand.*, **121** (2), 84–93.

Petersen, E., Chen, L.H., and Schlagenhauf, P. (2011) *Infectious Diseases: A Geographic Guide* 1. Auflage, John Wiley & Sons, Ltd., Chichester, West Sussex.

PLOS Neglected Tropical Diseases (n.d.) Journal Scope, http://www.plosntds.org/static/scope (accessed 22 April 2015).

Popkin, B.M. (2006) Global nutrition dynamics: the world is shifting rapidly toward a diet linked with noncommunicable diseases. *Am. J. Clin. Nutr.*, **84** (2), 289–298.

Quinn, T.C. and Samet, J.M. (2010) Epidemiologic approaches to global health. *Epidemiol. Rev.*, **32** (1), 1–4.

Rogers, R.G. and Hackenberg, R. (1987) Extending epidemiologic transition theory: a new stage. *Soc. Biol.*, **34** (3–4), 234–243.

Rossignol, J.-F. *et al.* (2006) Effect of nitazoxanide for treatment of severe rotavirus diarrhoea: randomised double-blind placebo-controlled trial. *Lancet*, **368** (9530), 124–129.

Salomon, J.A. and Murray, C.J.L. (2002) The epidemiologic transition revisited: compositional models for causes of death by age and sex. *Popul. Dev. Rev.*, **28** (2), 205–228.

Santosa, A. *et al.* (2014) The development and experience of epidemiological transition theory over four decades: a systematic review. *Global Health Action*, **7**, http://www.ncbi.nlm.nih.gov/pmc/articles/PMC4038769/ (accessed 17 February 2015), 56–71.

Silbergeld, E.K., Graham, J., and Price, L.B. (2008) Industrial food animal production, antimicrobial resistance, and human health. *Annu. Rev. Public Health*, **29** (1), 151–169.

Smallman-Raynor, M. and Phillips, D. (1999) Late stages of epidemiological transition: health status in the developed world. *Health Place*, **5** (3), 209–222.

Stuckler, D. (2008) Population causes and consequences of leading chronic diseases: a comparative analysis of prevailing explanations. *Milbank Q.*, **86** (2), 273–326.

Tanser, F. *et al.* (2013) High coverage of ART associated with decline in risk of HIV acquisition in rural KwaZulu-Natal, South Africa. *Science*, **339** (6122), 966–971.

Taylor, S.M. and Juliano, J.J. (2014) Artemisinin combination therapies and malaria parasite drug resistance: the game is afoot. *J. Infect. Dis.*, **210** (3), 335–337.

United Nations (2013a) World Population Prospects: The 2012 Revision Key Findings and Advance Tables, http://esa.un.org/unpd/wpp/Documentation/pdf/WPP2012_%20KEY%20FINDINGS.pdf (accessed 22 April 2015).

United Nations (2013b) World Population Prospects: The 2012 Revision–Median Age, https://data.un.org/Data.aspx?d=PopDiv&f=variableID%3a41%3bcrID%3a900%3btimeID%3a1041%2c1061%2c1101&c=2,4,6,7&s=_crEngNameOrderBy:asc,_timeEngNameOrderBy:desc,_

varEngNameOrderBy:asc&v=1 (accessed 22 April 2015).

United Nations (2014a) Report of the Open Working Group of the General Assembly on Sustainable Development Goals, http://www.un.org/ga/search/view_doc.asp?symbol=A/68/970&Lang=E (accessed 22 April 2015).

United Nations (2014b) The Road to Dignity by 2030: Ending Poverty, Transforming all Lives and Protecting the Planet, Synthesis Report of the Secretary-General on the Post-2015 Sustainable Development Agenda, United Nations General Assembly, http://www.un.org/ga/search/view_doc.asp?symbol=A/69/700&Lang=E (accessed 22 April 2015).

United Nations (2014c) World Population Prospects, the 2012 Revision, http://esa.un.org/wpp/unpp/panel_indicators.htm (accessed 27 April 2015).

United Nations (2014d) World Urbanization Prospects (Highlights), United Nations, New York, http://esa.un.org/unpd/wup/Highlights/WUP2014-Highlights.pdf (accessed 22 April 2015).

Vernet, G. *et al.* (2014) Surveillance for antimicrobial drug resistance in under-resourced countries. *Emerging Infect. Dis.*, **20** (3), 434–441.

Vogel, G. (2014) Are bats spreading Ebola across sub-Saharan Africa? *Science*, **344** (6180), 140.

Whiteford, H.A. *et al.* (2013) Global burden of disease attributable to mental and substance use disorders: findings from the Global Burden of Disease Study 2010. *Lancet*, **382** (9904), 1575–1586.

Wolfe, N.D., Dunavan, C.P., and Diamond, J. (2007) Origins of major human infectious diseases. *Nature*, **447** (7142), 279–283.

Woolhouse, M.E. and Ward, M.J. (2013) Sources of antimicrobial resistance. *Science*, **341** (6153), 1460–1461.

World Bank (1993) *World Development Report 1993: Investing in Health* Published for the World Bank,, Oxford University Press, New York, https://openknowledge.worldbank.org/handle/10986/5976 (accessed 10 August 2015).

World Health Organization (2013a) Global Health Estimates Summary Tables: Projection of Death by Cause, Age and Sex, by World Bank Income Group, http://www.who.int/healthinfo/global_burden_disease/en/ (accessed 10 August 2015).

World Health Organization (2013b) Sixty-Sixth World Health Assembly Agenda Item 16.2 Neglected Tropical Diseases, http://www.who.int/neglected_diseases/mediacentre/WHA_66.12_Eng.pdf (accessed 22 April 2015).

World Health Organization (2014a) Antimicrobial Resistance: Global Report on Surveillance, http://apps.who.int/iris/bitstream/10665/112642/1/9789241564748_eng.pdf?ua=1 (accessed 10 August 2015).

World Health Organization (2014b) Global Status Report on Noncommunicable Diseases 2014, http://apps.who.int/iris/bitstream/10665/148114/1/9789241564854_eng.pdf (accessed 26 February 2015).

World Health Organization (2014c) HIV/AIDS, Fact sheet N°360, WHO, http://www.who.int/mediacentre/factsheets/fs360/en/ (accessed 22 April 2015).

World Health Organization (2014d) Statement on the 1st Meeting of the IHR Emergency Committee on the 2014 Ebola Outbreak in West Africa, http://www.who.int/mediacentre/news/statements/2014/ebola-20140808/en/ (accessed 22 April 2015).

World Health Organization (2015a) Ebola Situation Report – 15 April 2015, http://apps.who.int/ebola/current-situation/ebola-situation-report-15-april-2015 (accessed 22 April 2015).

World Health Organization (2015b) Ebola Virus Disease, Fact sheet N°103, WHO, http://www.who.int/mediacentre/factsheets/fs103/en/ (accessed 22 April 2015).

World Health Organization (2015c) Emerging Diseases, http://www.who.int/trade/glossary/story022/en/ (accessed 12 April 2015).

World Health Organization (2015d) Investing to Overcome the Global Impact of Neglected Tropical Diseases. Third WHO Report on Neglected Tropical Diseases 2015, World Health Organization, Geneva, http://apps.who.int/iris/bitstream/10665/152781/1/9789241564861_eng.pdf?ua=1 (accessed 22 April 2015).

World Health Organization (2015e) Leishmaniasis, Fact sheet N°375, WHO, http://www.who.int/mediacentre/factsheets/fs375/en/ (accessed 22 April 2015).
World Health Organization (2015f) Malaria, Fact sheet N°94, WHO, http://www.who.int/mediacentre/factsheets/fs094/en/ (accessed 22 April 2015).
World Health Organization (2015g) Middle East Respiratory Syndrome Coronavirus (MERS–CoV): Summary of Current Situation, Literature Update and Risk Assessment, http://www.who.int/csr/disease/coronavirus_infections/mers-5-february-2015.pdf?ua=1 (accessed 22 April 2015).
World Health Organization (2015h) Tuberculosis, Fact sheet N°104, WHO, http://www.who.int/mediacentre/factsheets/fs104/en/ (accessed 22 April 2015).
World Health Organization (2015a) Regional Office for the Western Pacific, (WHOW-PRO), http://www.wpro.who.int/topics/neglected_tropical_diseases/en/

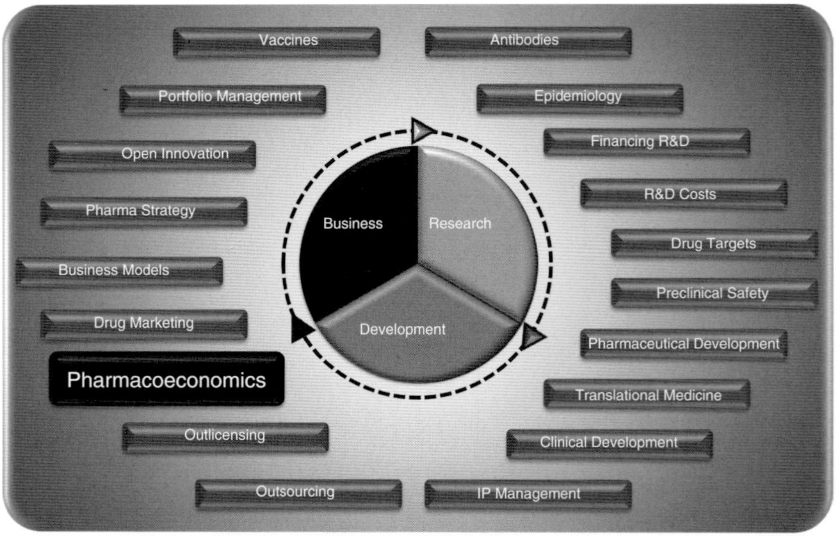
Vaccines
Antibodies
Portfolio Management
Epidemiology
Financing R&D
Open Innovation
R&D Costs
Pharma Strategy
Business
Research
Drug Targets
Business Models
Preclinical Safety
Drug Marketing
Development
Pharmaceutical Development
Pharmacoeconomics
Translational Medicine
Outlicensing
Clinical Development
Outsourcing
IP Management

# 3
# The Value of Pharmaceutical Innovation: Concepts and Assessment

*Sam Salek and Paul Kamudoni*

## 3.1
## Introduction

Much of the extension in life expectancy and the improvement in quality of life (QoL) in the twenty-first century have been attributed to innovations in medicines that have dealt with various life-threatening diseases such as tuberculosis, diabetes, and HIV. The life expectancy of a baby born in the western world today is around 100 years. On the other hand, the process of bringing new medicines from the assay to the clinic not only is complex, costly, and lengthy but also spans multiple stakeholders including the pharmaceutical industry, patients, prescribers, regulators, and payers. Thus, ultimately a new medicine must demonstrate value to all these stakeholders, especially to obtain regulatory approval, formulary listing, or reimbursement, gain acceptability by the prescribers, and ultimately be consumed by patients.

The purpose of this chapter is to discuss the concept of value, and its assessment, for new medicines, from the perspective of the key stakeholders involved in progressing a new medicine from assay to clinic. The need for adopting a broad understanding/perspective of value during the pharmaceutical research and development processes is emphasized.

Specifically, this chapter will explore the following issues:

1) Definitions/conceptualization of concept of value for new medicines
2) The assessment of value for new medicines, from the point of view of drug regulatory agencies, health technology assessment (HTA) bodies, patients, and prescribers
3) Recent developments with potential influence on conceptualization and assessment of value for new medicines
4) Recommendations for greater consideration of the concept of value early during the drug research and development process

*Value Creation in the Pharmaceutical Industry: The Critical Path to Innovation,* First Edition.
Edited by Alexander Schuhmacher, Markus Hinder, and Oliver Gassmann.

## 3.2
## Concepts and Definitions of Value

Merriam-Webster's dictionary defines "value" as (i) the amount of money that something is worth, (ii) something that can be bought for a low or fair price, or (iii) relative worth, utility, or importance. Similarly, for new medicines various stakeholders make value judgments about the worth of medicines and their usefulness, in making decisions regarding market authorization, formulary listing, reimbursement, prescription, and ultimately consumption. This involves evaluating the full range of consequences associated with new medicines, including clinical outcomes and effects on individuals (humanistic outcomes), the health-care system, and society at large:

1) Clinical outcomes include effects which are of clinical relevance, such as drug safety (measured by adverse events) and efficacy.
2) Humanistic outcomes include effects that people experience and care about, such as disease symptoms, health status, health-related quality of life (HRQoL), satisfaction with treatment, and adherence to treatment. HRQoL[1)] is multidimensional and is considered as the ultimate outcome of treatment (Bowling, 2001; Gill and Feinstein, 1994). Typically HRQoL is assessed using standardized generic health profiles (e.g., Short-Form 36 (SF-36)) or specific measures (i.e., specific to a disease or therapy such as Dermatology Life Quality Index). In addition, other types of HRQoL measures are those based on preferences (e.g., SF-6D and EQ-5D-5-L). Preference-based measures of HRQoL are relevant for calculating utilities, used in the calculation of quality-adjusted life years (QALYs).[2)]
3) Health-care system and societal consequences include the overall short- and long-term impacts associated with applying a new medicine/treatment in a specific patient population such as utilization of other health-care services, additional human resources and equipment needed/spent, need for organizational changes, and productivity losses, for example, due to absenteeism, with work, ethical, and political implications (O'Donnell *et al.*, 2009). These translate into a monetary cost.

Further consideration of value can be undertaken by making various types of comparative assessments, including assessing the benefit–risk profile, the therapeutic value, and economic value:

- *Benefit–risk balance*: Risks and benefits associated with a given medicine are conclusions based on subjective value judgment on the known treatment effects and adverse effects from various pieces of evidence such as clinical trials

1) Health-related QoL is a personal, evaluative statement summarizing the positivity or negativity of attributes that characterize one's psychological, physical, social, and spiritual well-being at a point in time when health, illness, and treatment conditions are relevant (Padilla *et al.*, 1996).
2) QALY is a generic outcome measure representing the length of affected time adjusted for the utility of the associated health status (Drummond *et al.*, 2005).

(Mussen, Salek, and Walker, 2009). This value judgment is made during the market authorization process of a new drug by clinicians and by patients.

- *Therapeutic value/benefit*: An evaluation of the added therapeutic benefit associated with a drug, in comparison with comparators in terms of medical outcomes, in the real-world clinical setting (Chandra, Jena, and Skinner, 2011). This mostly applies the concept of comparative effectiveness.
- *Economic value*: The balance between the costs and health outcomes associated with the use of a new medicine in a specific patient population and in a particular health-care setting. There are several variants depending on how health outcomes are measured, including (i) cost-effectiveness, (ii) cost–benefit ratio, and (iii) cost–utility.

## 3.3 Stakeholder's Perspectives on Value

### 3.3.1 Drug Regulatory Agencies

The task of evaluating new medicines for market approval involves reviewing results from phase III clinical studies to establish that a drug is safe and effective in a specific patient population. At the core of this process is a benefit–risk assessment of the drug, which involves a subjective judgment on evidence from clinical studies to establish whether the health benefits outweigh the risks, as understood. Other factors have an influence on regulatory decisions including the severity of the condition and unmet medical needs.

Currently, the benefit–risk assessment process at EMA and FDA is not systematic or formalized. Thus, there is little given, in the form of guidance from regulatory authorities to the pharmaceutical industry, regarding the basics and content of benefit–risk assessment process (Chuang-Stein, 2013). Nevertheless, both agencies are undertaking initiatives to develop frameworks for more formalized and systematic benefit–risk assessment and have raised concerns regarding making the review process overly quantitative (*see* Box 3.1 – systematic approaches to benefit–risk developments).

### 3.3.2 Health Technology Assessment

In general HTA bodies produce evidence-based reports summarizing the overall benefits of a new technology (clinical and economic) and some form of guidance decision on its usage and reimbursement (O'Donnell *et al.*, 2009). As a process, HTA typically involves stages of topic selection, assessment, and appraisal. Across Europe, HTA bodies vary in terms of (i) scope of their assessments, for example, whether an economic evaluation is included or not; (ii) the process of executing the HTA, for instance, whether the HTA body undertakes its own evaluation of evidence or whether it reviews evidence submitted by sponsors; and (iii) whether the result of the HTA process is utilized for making reimbursement or formulary listing decisions (Allen *et al.*, 2013; Eddy, 2009).

For example, in the HTA performed by the National Institute for Health and Care Excellence (NICE) in the United Kingdom, the assessment considers both clinical effectiveness and cost-effectiveness. The former is based on the synthesis of results from randomized clinical trials (RCTs), through systematic review or meta-analysis, whereas the latter is based on economic modeling, from the perspective of the National Health Service (NHS) and social services. The economic evaluation involves estimating the incremental cost-effectiveness ratio (ICER) associated with a technology (a new medicine).

During appraisal,

> A technology (new medicine) is considered clinically effective, if it confers a health benefit, taking into account any harmful effects and opportunity cost. A technology is considered cost-effective if its health benefits are greater than the opportunity costs of programmes displaced to fund the new technology, in the context of a fixed NHS budget
>
> *(NICE, 2013)*

Although NICE has not specified a threshold for reimbursement approval, ICER is expected to range from £20 000 to £30 000. Other factors may influence the appraisal decision regarding cost-effectiveness including uncertainty of results, quality of study design, innovative nature of technology,[3] and whether the technology is regarded as life-extending treatment at the end of life (NICE, 2013).

In Germany, HTA for new medicines is undertaken by the Institute for Quality and Efficiency in Healthcare (IQWIG) on behalf of the Federal Joint Committee (G-BA). Under a new framework in force since 2011, the assessment involves evaluating the additional benefit associated with a new treatment, in comparison with a relevant comparator, in terms of patient relevant outcomes. These include mortality, morbidity, and HRQoL. The final outcome of the assessment is a classification of additional benefit in any one of the following categories: (i) major benefit, (ii) significant benefit, (iii) marginal benefit, (iv) not quantifiable, (v) none, or (vi) less benefit than comparator (Schlette and Hess, 2013).

The assessment undertaken by the IQWIG takes into account the perspective/input of clinical experts and patients as well as current published evidence. The burden of disease (e.g., time and effort invested in managing a condition) and patient satisfaction with treatment are also considered.

The final decision in the appraisal stage, undertaken by the G-BA, may differ from recommendations from IQWIG and also takes into account findings from a public hearing.

3) If the innovation adds demonstrable and distinctive benefits of a substantial nature which may not have been adequately captured in the reference case QALY measure.

### 3.3.3 Patients

The patients' views of treatments go beyond clinical and pharmacologic concerns; of immediate importance are concerns about *how long and how much* a treatment of their condition alleviates their symptoms and affects their daily life and QoL and the sacrifices they may need to make in terms of symptoms and QoL as a result of the treatment (Lohr and Zebrack, 2009). Cancer patients on chemotherapy may ask themselves, "Can I go out with friends without the fear that I may vomit due to my chemotherapy?" (Salek and Kamudoni, 2013).

For example, Gelhorn *et al.* (2013a) looked at the preferences for oral medication attributes in patients with early and advanced type 2 diabetes mellitus (T2DB) in the United Kingdom using conjoint analysis methods. The attributes considered included efficacy, hypoglycemic events, weight change, gastrointestinal/nausea side effects, urinary tract infection and genital infection, blood pressure, and cardiovascular risk.

Assessing outcomes reported by the patient without any interpretation by a second party has become crucial to the understanding of the patient's perspective of therapeutic value of new medicines; these include symptoms, physical function, well-being, HRQoL, perceptions of treatment effect, and satisfaction with treatment. The FDA has published guidance on how patient-reported outcomes (PROs) may be reliably and validly measured in drug development programs (FDA, 2009). In order for the selected measure to be considered fit for the purpose, evidence of its psychometric properties must be assessed and evaluated, including reliability, content validity, construct validity, and responsiveness. Data supporting meaningful interpretation of scores should be available.

### 3.3.4 Prescribers/Clinicians

In one study in the United States, oncologists were more concerned with a treatment's effectiveness and the out-of-pocket costs borne by the patients and less with the cost-effectiveness (Nadler, Eckert, and Neumann, 2006). The benefit from a medicine is viewed in relative terms, with consideration of whether and how other possible doses of medicine may influence/benefit the drug (Spilker, 1994). Another consideration is whether the harms associated with the disease outweigh the risks associated with the medicine, with prescribers being more concerned about the latter (Edwards, Wiholm, and Martinez, 1996). This suggests that the therapeutic value of a drug is dependent on the individual patient.

Clinicians'/prescribers' views on value encompass issues such as (i) the position of a new medication in the treatment pathway and its clinical utility, (ii) existing unmet needs with current treatment choices (existing medications) and how the new medicine addresses these, and (iii) the potential impact of the new medicine on the clinical management of patients (e.g., poor safety profile may lead clinicians to be more conservative in their treatment approaches). Gelhorn *et al.* (2013b) explored the preferences of physicians for medication attributes in the prophylactic treatment of hemophilia A. The attributes considered included efficacy, costs, scientific evidence, dosing frequency, and administration time. The most important attributes were efficacy and scientific evidence, while treatment costs, dosing frequency, and administration time were less important.

## 3.4 Recent Developments Influencing the Definition and Assessment of Value

A number of changes are currently under way in the context and setting in which various stakeholders define and assess the value of new medicines. Recent technological advances have enabled the collection of large volumes of "big data" permitting various research questions to be explored. New models for providing scientific advice to pharmaceutical industry on their drug development programs are emerging, for example, parallel HTA–EMA scientific advice and collaboration across HTA agencies. There is a growing emphasis on capturing the perspective of patients in health care overall, particularly in the drug development process. These developments and their potential implications in how the concept of value is defined and assessed are described below:

1) Recent advances in technology have permitted/facilitated the establishment of large datasets of electronic clinical data, which are captured in routine clinical practice by clinicians and hospitals. Record linkages with other datasets (such as social services) and aggregation across facilities (e.g., across a nation) have created datasets with an unprecedented volume, complexity, and timeliness, now commonly referred to as *big data*. While previously the use of such data for commercial entities was prohibited (Statutory Health Insurance (SHI) prescription databank held by the Deutsche Arzneiprüfunginstitut e.V. (DAPI)), these data are now made available to the industry-funded clinical research (Box 3.2 – an example of "big data"). These are still early days in understanding the full potential of such "big data" in exploring the benefit–risk and benefit–cost profiles of new medicines. It is expected that impacts on HTA process might include (i) answering key questions not addressed by published studies, for example, detecting harms of treatment not captured in clinical studies; (ii) understanding heterogeneity of treatment effect; (iii) predicting the clinical impact of the uptake of a new treatment; and (iv) facilitating application of HTA results to local population (Jonathan, Erinoff, and Coates, 2013).
2) There is a renewed emphasis on the importance of the patient's perspective in health care, through their involvement in decision-making processes in the delivery of care and health-related research. This new role/activity, distinct from patients as subjects, has come to be known as *patient engagement*. A growing number of agencies which fund health research, such as the CIHR in Canada and the United Kingdom's NIHR, and the EU's Innovative Medicines

Initiative now formally require the active involvement of patients in the research process (Staniszewska *et al.*, 2012).

Under the Prescription Drug User Fee Act (PDUFA) V, the FDA is undertaking a "patient-focused drug development" initiative with the aim of obtaining the patient's perspective on certain disease areas, particularly on disease severity and unmet medical needs (FDA, 2014). This entails developing a systematic approach on how a patient's voice is included in the benefit–risk assessment by the FDA. However, the weight attached to the patient's perspective in the assessment process remains unclear. Similarly, HTA agencies such as NICE and IQWIG are beginning to "engage" patients in their technology appraisal processes through direct involvement in deliberations or by obtaining the patients' experience of living with the condition in written form.

3) An initiative is currently under way in the EU to streamline regulatory and HTA procedures to minimize the need for duplicated efforts in data collection to address the evidence needs of multiple stakeholders. Starting in 2010, the EMA has been offering a joint scientific advice process with HTA agencies to manufacturers/sponsors. This is intended to ensure that feedback is available to drug developers on the data required to establish the benefit–risk balance and value of drugs, from regulators' and HTA agencies' perspective, respectively, early during the drug development process. Inclusion of endpoints such as PROs (e.g., health status, symptoms) that meet the needs of multiple stakeholders is likely to be encouraged. Importantly, this also represents an opportunity for a common conceptualization and assessment of certain aspects of value, such as therapeutic value.
4) Following the financial crisis of 2008 and the public debt problem that ensued, most nations in Western Europe, including the United Kingdom, have been under austerity fiscal policy. This has created pressures on health budgets, including the resources available for paying for pharmaceuticals. If pharmaceutical budgets are capped, there is likely to be limited/no resources for financing future innovative medicines without displacing already existing treatments. Patients should also be expected to share an increasing portion of the cost of drugs. This entails a greater role for considerations of economic value (as at present) and affordability.

## 3.5 Recommendations: Implications for R&D

The foregoing discussion suggests that the notion of value runs throughout the lifecycle of a new medicine. Success in its uptake largely depends on the demonstration of value to all stakeholders including regulatory agencies, HTA bodies, prescribers, and patients. This calls for developers to explore and incorporate the various value perspectives early on during the drug development program. As much as possible, developers need to strive to make "value assessments" during the drug development process as those considered during the regulatory benefit–risk assessment, HTA, and the final use of the medicine. This approach would increase the opportunities for generating appropriate relevant data meeting subsequent value evidence required by various bodies. This should also ensure that appropriate and relevant information is provided to the end users on the therapeutic value of the medicines. With the growing enthusiasm for early engagement with pharmaceutical industry among regulatory agencies and HTA bodies, preliminary insights on value might also be useful during scientific advice meetings.

Various tools and techniques are available for making value considerations during the R&D process (Table 3.1):

1) *Preliminary technology assessment* may investigate the disease state, target population, epidemiological factors, and disease costs prior to clinical development of a new medicine. This would inform the target market uptake and cost-effectiveness of the new drug (Hartz and John, 2008).
2) *Early exploratory outcomes assessment,* such as at phase II, may be useful in various ways: it facilitates the identification of the most relevant patient-centered endpoints and validation of relevant instruments for their measurement, which would be useful during drug regulation and reimbursement stages.
3) *Early economic evaluation/modeling* may establish/identify parameters with the largest influence on cost-effectiveness, facilitating better and careful design and execution of clinical study, or collection of further data.
   a. Studies exploring the preferences of the target patient population for treatments with various characteristics may help to assign a monetary unit to the value of the treatment to the patients, for example, through willingness to pay (WTP).
   b. Such information might also be valuable in early formulation of a product's pricing strategy.
4) *Preliminary understanding of key outcomes and cost-effectiveness associated with a drug* would help to reduce reimbursement uncertainties, facilitating earlier identification of unsuccessful projects and enhancing efficiency/productivity (Hartz and John, 2008).

## 3.6 Discussion

Although the evaluation of value sends signals to the industry regarding which therapeutic areas/patient populations would be most rewarded, there are fears that the way in which some aspects of value such as HTA have been undertaken might hurt pharmaceutical innovation. The HTA, particularly evaluation of economic value, has come to be viewed as a cost-containment policy rather than an activity supporting and facilitating availability of new medicines to patients. Ultimately, as value judgment is subjective, potential friction in the perspectives of the different stakeholders may occur, although increasingly different

**Table 3.1** Strategies for early assessment of value during the drug research and development process.

| | Question/issue addressed | Phase | Technique | Application |
|---|---|---|---|---|
| 1 | What is the potential of a new medicine in this indication? | Preclinical phase | Disease–drug models | To characterize the disease and explore impacts of potential treatments |
| | | Preclinical phase | Health impact models | Potential consequences of developing a product with particular attributes can be explored |
| | | Preclinical phase | Burden of illness studies | Quantifying the degree of burden associated with a condition in humanistic and economic terms, including its epidemiology, impact on health status, lost productivity, and cost of treatment |
| | | Preclinical phase | Clinical trial simulation | To learn about the drug effectiveness and safety and to optimize study designs |
| 2 | Defining information needs to address value questions | Clinical phases I–III | Payer research, for example, interviews and scientific advice | To understand payer perspective on relevant endpoints and resources in demonstrating value |
| | | Clinical phases I–III | Qualitative interviews with patients | To identify major symptoms and disease impacts that are important to patients |
| | | Clinical phases I–III | DELPHI panels with clinicians | To understand the clinical utility of a new medicine and explore expected impact of medicine on clinical management strategy of patients |
| 3 | What is the potential marginal economic value of new product? | Preclinical phase | Cost-effectiveness gap analysis | Exploring minimum health benefits to be demonstrated by the new medicine in order for it to be effective |

(*continued overleaf*)

**Table 3.1** *Continued.*

| | Question/issue addressed | Phase | Technique | Application |
|---|---|---|---|---|
| | | Preclinical phase, clinical phases I–III | Preference studies, for example, conjoint analysis | Exploring patient's/clinician's preferences for various attributes of a medicine and understanding trade-offs between the attributes. Also useful for exploring willingness to pay |
| | | Clinical phases I–III | Patient engagement | To ensure that development plans and related study designs are adequate to capture medicines' value to patients |
| | | Clinical phases I–II | Value of information analysis | Useful in determining how much additional information will reduce the cost associated with uncertainty; helpful to identify which parameters require further research |
| | | Phase I | Early economic modeling | Preliminary estimates of cost-effectiveness based on synthesis of all available information about a product |
| 4 | Should we continue the clinical development program? | Preclinical, clinical phases I–III | Multiple-criteria decision analysis | To inform go/no-go decisions, taking into account all available information, priorities, and uncertainties at each stage of development |

Sources: IJzerman and Steuten (2011), Vallejo-Torres *et al.* (2011), Cosh *et al.* (2007), McAteer *et al.* (2007), and Basch (2013).

stakeholders are collaborating in establishing value. Notwithstanding, assessing whether the current value framework facilitates development of new medicines requires slightly complex study designs. However, in this chapter, we explore such issues by considering decisions made by regulatory authorities and HTA agencies.

Wang, Avorn, and Kesselheim (2013) performed a review of New Drug Applications (NDAs) and Biologic License Applications (BLAs) ($n = 52$) evaluated by the FDA from 2007 to 2009; 56% ($n = 30$) were not approved on the initial submission. The unapproved drugs included designations of priority review alone ($n = 4$, 13%), orphan drug status ($n = 3$, 10%), and both priority and orphan ($n = 4$, 13%). Reasons for non-approval included (i) excessive safety risk including request for additional safety data ($n = 13$, 48%), (ii) efficacy concerns ($n = 7$, 26%), and (iii) a combination of both excessive safety concerns and efficacy concerns ($n = 5$, 19%), labeling issues ($n = 4$, 15%), and manufacturing deficiencies ($n = 3$, 11%). The most important safety concerns included increased mortality risk, cardiovascular events, and hepatotoxicity. An additional 11 drugs (37%) were subsequently approved on second review. Of the 19 which remained unapproved, 7 drugs were approved in the EU with the same indication and 1 drug was approved in Canada and the EU with the same indication. This demonstrates the subjective nature of value judgments involved in the benefit–risk assessment and approval process.

Comparisons of decisions taken by HTA bodies across Europe also reflect variation in terms of share of technologies (new medicines) recommended, that is, new medicines judged to be adding value, as well as correlates of the decisions taken. Cerri, Knapp, and Fernandez (2014) reviewed the decisions taken by the HTA body in the Netherlands, *College voor Zorgverzekeringen* (CVZ), in the period 2004–2009, covering 256 technologies. Over half of the technologies (51%) were recommended, 33% were restricted, and 16% were not recommended for funding. The HTA decisions were associated with the use of active comparators, statistical superiority of the primary endpoint in clinical trials, pharmaceutical budget impact associated with technology, and prevalence of condition in target population. Dakin *et al.* (2015) reviewed 190 technology appraisals by NICE published in December 2011. Up to 82% of decisions by NICE were predicted by cost-effectiveness. Odds of favorable opinion seemed higher for musculoskeletal diseases and cancer technologies, but seemed lower for medicines treating respiratory conditions.

Similar observations have been made on the early benefit assessments performed by the IQWIG (on behalf of the G-BA) in Germany. In the first 18 months of the early benefit assessment procedure, out of the 27 medicines assessed, 50% were assessed to have some benefit (Ruof *et al.*, 2014). The G-BA in their final decision appraised 63% of medicines to have some benefit.

In general the proportion of medicines judged to have some value was comparable between the United Kingdom, the Netherlands, and Germany, although German HTA procedure had a higher proportion of medicines regarded to be beneficial. Although this may indeed suggest some social, cultural, and economic influences on value judgments, the HTA procedures differed across the countries. Economic evaluation is not included as part of the HTA procedures of the IQWIG/G-BA, while this is considered focal to the NICE (United Kingdom) and the CVZ (the Netherlands) procedures.

Perhaps more important than the hard numbers reflecting positive benefit–risk or benefit–cost balance, the perceptions of the industry on the transparency, predictability, and certainty of the process of assessing value have the most bearing on drug development-related decisions. The current emphasis on more systematic regulatory decision-making, and greater collaboration among stakeholders in subsequent value assessments stages, might indeed serve to incentivize the efficient development of new medicines.

## 3.7 Conclusion

The nature and type of information required to address value questions of the different stakeholders (drug regulatory agencies, HTA bodies, clinicians, and patients) throughout the lifecycle of a new medicine is different. There is a growing recognition of the need for, and benefit from, greater collaboration among the stakeholders, in meeting their information needs regarding the value of new medicines, with patients playing a more active and focal role. Moving forward, the pharmaceutical industry needs to adopt a holistic perspective on value and take into account the considerations of all stakeholders from early on (and throughout) in the drug development process. In this context, the development of new medicines that are of value to all stakeholders would seem to favor a collaborative and open regime of drug development.

**Box 3.1: An example of "big data": the clinical practice research datalink (CPRD)**

The clinical practice research datalink (CPRD) is a validated, anonymized longitudinal dataset of at least 5.4 living patients (with approximately 14 million overall) registered from 660 primary practices across England. The dataset contains coded diagnostic and clinical information including utilization of various health-care services. The CPRD is currently being used to address various types of research questions in pharmacovigilance, pharmacoepidemiology, and economic evaluation studies.

As part of the early scientific advice provided by NICE, developers are offered advice from the CPRD regarding the data available of relevance to a particular drug development program.

Sources: http://www.cprd.com, http://epi.grants.cancer.gov/pharm/pharmacoepi_db/cprd.html.

**Box 3.2: Systematic approaches to benefit–risk assessment: recent developments at the FDA**

Since 2009, and again re-emphasized in the PDUFA of 2010, the FDA has been undertaking an initiative to develop a structured approach to drug benefit–risk assessment. It is expected that this would serve as a template for product reviews and promote more transparency of the FDA's decisions. This is particularly crucial given the residual uncertainty associated with regulatory decisions and given that the information available about the drug during the review process does not address all questions and concerns. The framework under development is meant to fulfill the following roles/needs: (i) should be applicable throughout the lifecycle of drug, that is, pre- to postmarket, (ii) should identify areas of uncertainty and show how this influences/affects the decision, (iii) should be capable of weighing all relevant aspects/issues during deliberation, and (iv) should integrate existing work processes. This framework also incorporates the perspective of the patient on a condition and the currently available therapies. In short, the benefit–risk framework developed had the following structure:

1) Analysis of condition and current treatment options, representing the context of the decision.
2) Benefits and risks, based on the results of clinical trials and clinical meaning of primary and secondary endpoints, and subpopulation analyses. Severity and reversibility of harms (adverse events) are considered. Additional information such as clinical pharmacology and toxicology data, clinical pharmacology, chemistry, manufacturing, controls, and clinical microbiology is considered in evaluating risks and benefits.
3) Risk management and strategy/efforts that could help to mitigate identified safety issues.

Further, the framework includes the subjective evaluation of the above aspects comprising an interpretation of evidence, "*evidence and uncertainties*," and the ultimate conclusion reached, "*conclusions and reasons.*"

Sources: FDA (2014).

## References

Allen, N. *et al.* (2013) Development of archetypes for non-ranking classification and comparison of European National Health Technology Assessment systems. *Health Policy*, **113** (3), 305–312.

Basch, E. (2013) Toward patient-centered drug development in oncology. *N. Engl. J. Med.*, **369** (5), 397–400.

Bowling, A. (2001) *Measuring Disease: A Review of Disease-Specific Quality of Life Measurement Scales*, Open University Press.

Cerri, K.H., Knapp, M., and Fernandez, J.L. (2014) Public funding of pharmaceuticals in The Netherlands: investigating the effect of evidence, process and context on CVZ decision-making. *Eur. J. Health Econ.*, **15** (7), 681–695.

Chandra, A., Jena, A.B., and Skinner, J.S. (2011) *The Pragmatist's Guide to Comparative Effectiveness Research*, National Bureau of Economic Research.

Chuang-Stein, C. (2013) Quantitative Risk/Benefit Assessment: Where Are We? *Proceedings of the Fourth Seattle Symposium in Biostatistics: Clinical Trials Volume 1205 of the series Lecture Notes in Statistics*, pp. 119–135.

Cosh, E. *et al.* (2007) Investing in new medical technologies: a decision framework. *J. Commer. Biotechnol.*, **13** (4), 263–271.

Dakin, H. *et al.* (2015) The influence of cost-effectiveness and other factors on nice decisions. *Health Econ.*, **24** (10), 1256–1271.

Eddy, D. (2009) Health technology assessment and evidence-based medicine: what are we talking about? *Value Health*, **12**, S6–S7.

Edwards, I.R., Wiholm, B.-E., and Martinez, C. (1996) Concepts in risk-benefit assessment. *Drug Saf.*, **15** (1), 1–7.

Drummond, M. *et al.* (2005) *Methods for the economic evaluation of health care programmes*, Oxford University Press, Oxford.

FDA (2009) Guidance for industry. Patient-reported outcome measures–use in medical product development to support labelling claims. *Fed. Regist.*, **235**, 65132–65133.

FDA (2014) Prescription drug user fee act patient-focused drug development. *Fed. Regist.*, **79**, 60857–60859.

Gelhorn, H. *et al.* (2013a) Physician preferences for medication attributes for the prophylactic treatment of patients with severe haemophilia A with inhibitors to factor VIII. *Haemophilia*, **19** (1), 119–125.

Gelhorn, H.L. *et al.* (2013b) Preferences for medication attributes among patients with type 2 diabetes mellitus in the UK. *Diabetes Obes. Metab.*, **15** (9), 802–809.

Gill, T.M. and Feinstein, A.R. (1994) A critical appraisal of the quality of quality-of-life measurements. *JAMA*, **272** (8), 619–626.

Hartz, S. and John, J. (2008) Contribution of economic evaluation to decision making in early phases of product development: a methodological and empirical review. *Int. J. Technol. Assess. Health Care*, **24** (4), 465–472.

IJzerman, M.J. and Steuten, L.M. (2011) Early assessment of medical technologies to inform product development and market access. *Appl. Health Econ. Health Policy*, **9** (5), 331–347.

Treadwell, J.R., Erinoff, E., and Coates, V. (2013) How electronic clinical data can improve health technology assessment. *eGEMs* (Generating Evidence Methods to Improve Patient Outcomes), **1** (2), 4.

Lohr, K.N. and Zebrack, B.J. (2009) Using patient-reported outcomes in clinical practice: challenges and opportunities. *Qual. Life Res.*, **18** (1), 99–107.

McAteer, H. *et al.* (2007) Cost-effectiveness analysis at the development phase of a potential health technology: examples based on tissue engineering of bladder and urethra. *J. Tissue Eng. Regener. Med.*, **1** (5), 343–349.

Mussen, F., Salek, S., and Walker, S. (2009) *Benefit-Risk Appraisal of Medicines: A Systematic Approach to Decision-Making*, John Wiley & Sons, Ltd.

Nadler, E., Eckert, B., and Neumann, P. (2006) Do oncologists believe new cancer drugs offer good value? *Oncologist*, **11** (2), 90–95.

NICE (2013) *Guide to Methods of Technology Appraisal*, National Institute for Health and Care Excellence (NICE), London.

O'Donnell, J.C. *et al.* (2009) Health technology assessment: lessons learned from around the world—an overview. *Value Health*, **12** (Suppl. 2), S1–S5.

Padilla, G.V. *et al.* (1996) Quality of Life – Cancer, in *Quality of life and pharmacoeconomics in clinical Trials* (ed B. Spilker, *et al.*), 2nd ed., Lippincott-Raven, Philadelphia, USA, pp. 301–308.

Ruof, J., Schwartz, F.W., Schulenburg, J.-M., and Dintsios, C.-M. (2014) Early benefit assessment (EBA) in Germany: analysing decisions 18 months after introducing the new AMNOG legislation. *Eur. J. Health Econ.*, **15** (6), 577–589.

Salek, M. and Kamudoni, P. (2013) Quality of life measurement in dermatology consultation: impact on patient reported outcomes. *G. Ital. Dermatol. Venereol.*, **148** (3), 263–275.

Schlette, S. and Hess, R. (2013) Early benefit assessment for pharmaceuticals in Germany: lessons for policymakers. *Issue Brief (Commonw. Fund)*, **29**, 1–9.

Spilker, B. (1994) Incorporating benefit-to-risk determinations in medicine development. *Drug News Perspect.*, **7**, 53–59.

Staniszewska, S., Haywood, K.L., Brett, J., and Tutton, L. (2012) Patient and public involvement in patient-reported outcome measures. *Patient*, **5** (2), 79–87.

Vallejo-Torres, L. *et al.* (2011) Integrating health economics into the product development cycle a case study of absorbable pins for treating hallux valgus. *Med. Decis. Making*, **31** (4), 596–610.

Wang, B., Avorn, J., and Kesselheim, A.S. (2013) Clinical and regulatory features of drugs not initially approved by the FDA. *Clin. Pharmacol. Ther.*, **94** (6), 670–677.

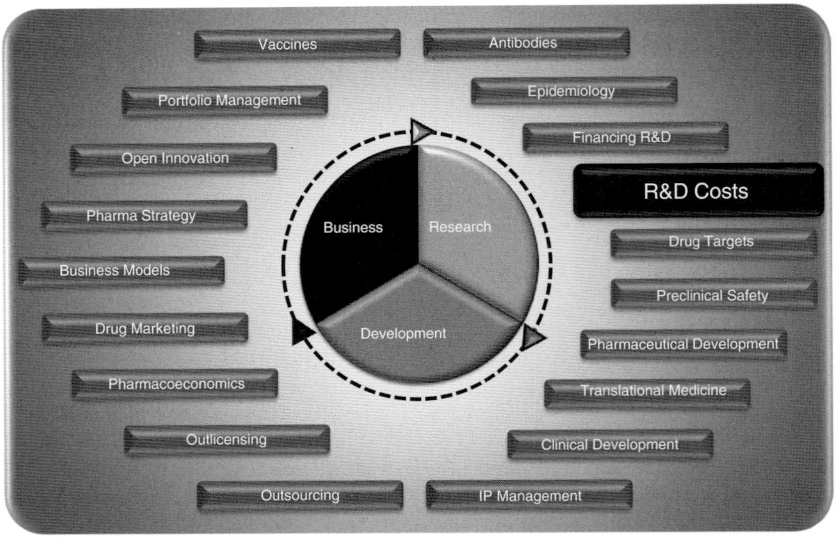

Vaccines
Antibodies
Portfolio Management
Epidemiology
Financing R&D
Open Innovation
R&D Costs
Pharma Strategy
Business
Research
Drug Targets
Business Models
Preclinical Safety
Drug Marketing
Development
Pharmaceutical Development
Pharmacoeconomics
Translational Medicine
Outlicensing
Clinical Development
Outsourcing
IP Management

# 4
# A Review of the Pharmaceutical R&D Efficiency: Costs, Timelines, and Probabilities

*Alexander Schuhmacher, Oliver Gassmann, and Markus Hinder*

## 4.1
## Introduction

The purpose of this chapter is to review the efficiency of pharmaceutical research and development (R&D). The key parameters in this context, namely, (i) cost, (ii) cycle times, (iii) probabilities of success (PoS), and (iv) number of new drugs launched, have been the subject of numerous research and review articles in the past years. In summary, it is known that the cost to develop a new drug increased over the past decades while the timelines to develop a new drug are increasing. Success rates are very low and the outcome of all R&D efforts is stagnating. In consequence, the R&D efficiency defined as the financial investment per drug has been steadily decreasing. As innovation is the major growth driver of the pharmaceutical industry, reliable data on R&D efficiency and new concepts to overcome these challenges are of great interest for R&D managers and the sustainability of the pharmaceutical industry as a whole.

With this chapter, we attempt to answer the following questions:

- What is the PoS of pharmaceutical R&D and what are the reasons for the relatively low success rates?
- What are the timelines of pharmaceutical R&D and what is the influence of the long timelines on R&D costs?
- What are the costs to discover and develop a new drug and what impacts these costs most?

Thus, we review in this chapter some R&D performance indicators over the past decades. Moreover, we summarize the PoS per phase, the total PoS of developing a new drug, and the factors influencing the success rates of pharmaceutical R&D. Moreover, we review the cycle times of pharmaceutical R&D, the timelines per phase, the changes in cycle times, and the factors impacting R&D cycle times most. In an additional step, we examine the total costs of drug R&D as well as the costs per development phase. Finally, we discuss the impact of these factors on the R&D efficiency and, to a smaller extent, on the productivity of the pharmaceutical industry.

*Value Creation in the Pharmaceutical Industry: The Critical Path to Innovation*, First Edition.
Edited by Alexander Schuhmacher, Markus Hinder, and Oliver Gassmann.

## 4.2 The Historical Perspective

From the 1950s to 2013, the American Food and Drug Administration (FDA) approved 1346 new molecular entities (NMEs). In this context, Munos (2009) reported that from 1950 to 2009 261 pharmaceutical companies existed that were successful in launching at least one NME. Of these only 12% were in the pharmaceutical business for the complete period of 60 years. And out of this group, 21 companies had launched 50% of all new drugs until today (Munos, 2009).

In the past 10 years, the research-based pharmaceutical companies have invested on average 15–20% of their total sales into R&D resulting for some of these companies in annual R&D costs of several USD billion. According to the European Commission, the pharmaceutical industry is among the top R&D investors globally and, more specifically, the one that is investing most in innovation (relatively and absolutely). According to the European Commission's 2013 European Union Industrial R&D Investment Scoreboard report (European Commission – Joint Research Centre), 15 of the 50 world's leading R&D investors come from the pharma sector, namely, Roche (6), Novartis (7), Merck & Co. (8), Johnson & Johnson (J&J, 9), Pfizer (10), Sanofi (15), GlaxoSmithKline (GSK, 20), Eli Lilly (26), AstraZeneca (33), Abbot Laboratories (35), Bayer (36), Amgen (37), Bristol-Myers Squibb (BMS, 40), Takeda (41), and Boehringer Ingelheim (42).

The importance of R&D as the major driver of growth in the pharmaceutical industry is most clear when reviewing the total expenditures for R&D across the whole industry. The International Federation of Pharmaceutical Manufacturers & Associations (IFPMA) reported an USD 137 billion investment in R&D in 2010 in the pharmaceutical and biotechnology industries (International Federation of Pharmaceutical Manufacturers and Associations). The Pharmaceutical Research and Manufacturers of America (PhRMA) published that its members have invested USD 48.5 billion in R&D in 2012 (PhRMA, 2013). Additionally, the EU Commission stated an R&D investment for the pharmaceutical sector of up to USD 100 billion worldwide in 2012 (European Commission – Joint Research Centre). Such high R&D expenditures both increase expectations for the return on investment (ROI) and call into question the long-term sustainability of the industry's R&D programs if the number of new drugs launched does not meet these heightened expectations.

In a 12-year period from 2001 to 2012, the most successful pharmaceutical companies received FDA approvals for 5–16 NMEs, namely, Novartis (16 NMEs), Pfizer (13) and GSK (12), BMS (9), Merck & Co. (8), Roche (7), Sanofi (7), Bayer (6), Amgen (5), Eli Lilly (5), and J&J (5). Within this group of 11 companies, an average 0.7 NMEs have been approved by the FDA per year and company. While some pharmaceutical companies benefit from the successes of their strong R&D organizations, other companies, such as Pfizer or Sanofi, only realized their R&D performance at least in part through successful mergers and acquisitions (M&A). Overall, for a portion of pharmaceutical companies, it will remain questionable if they can sustain the two to three NMEs per year that would be necessary to meet the companies' growth objectives exclusively through their own R&D productivity (Munos, 2009; Kola and Landis, 2004).

## 4.3 The R&D Phase Model

Figure 4.1 illustrates the typical R&D phase model for discovering and developing a new drug. Starting with applied research and followed by preclinical development, new drug candidates are tested on safety and efficacy in three subsequent phases of clinical development before a regulatory agency, such as the FDA, is approving the new drug candidate.

## 4.4 The Low R&D Success Rates

Pharmaceutical R&D has a low overall PoS. Only one out of more than 100 000 compounds that are screened in high-throughput screening (HTS) in drug

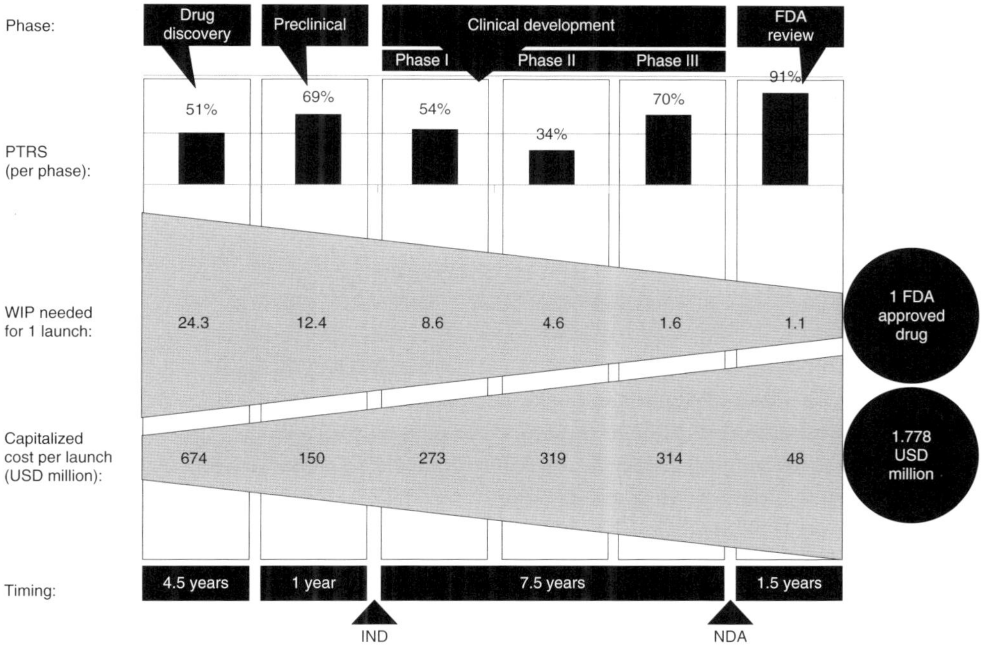


**Figure 4.1** The traditional R&D phase model; IND (Investigational New Drug), NDA (New Drug Application), FDA (Food and Drug Administration), PTRS (probability of technical and regulatory success), WIP (work in progress), USD (U.S. dollar). Data derived from Paul *et al.* (2010).

discovery and roughly 1 out of 100 compounds that are tested during preclinical development pass through drug development phases are registered as a new drug and make it to the market finally. Table 4.1 highlights selected research results and the reported probabilities per phase of drug R&D. In sum, the probability of discovering, developing, and registering one NME has been estimated to be around 4% (Paul *et al.*, 2010; Thomson Reuters, 2013).

The Center for Medicine Research (CMR) International reported in its Pharmaceutical R&D Factbook 2013 an average success rate of 4.1% from first toxicity dose to market approval and a change in success rates per phase over time (Thomson Reuters, 2013). While the PoS for phase III of clinical development improved slightly for compounds that entered phase III between 2006 and 2008 (PoS, 90%) compared to compounds that entered this phase earlier (PoS, 86%), the PoS for the previous phases in discovery research and preclinical development and clinical development decreased in the past years. More specifically, the CMR reported for the preclinical phase, phase I, and phase II of clinical development current success rates per phase of 67, 46, and 19%, respectively (Thomson Reuters, 2013).

The underlying causes of the high attrition rates are divers. It has been reported that adverse pharmacokinetics and bioavailability were the major reasons of attrition in 1991, while these factors were less relevant later and the lack of efficacy and safety were the main causes for the low PoS in clinical development in 2000 (Kola and Landis, 2004). In a review of the FDA approvals in 2012, it was reported that most failures in phases II and III still resulted from a lack of efficacy (56%), followed by safety (28%), changing strategies (7%), commercial reasons (5%), and operational challenges (5%) (Arrowsmith and Miller, 2013). In a recently published analysis of 142 drug R&D projects of AstraZeneca, it was confirmed that preclinical and phase I projects primarily failed for safety reasons while projects failing in phases II and III commonly lacked efficacy (Cook, 2014). In some areas like oncology and central nervous system (CNS), this might be due to the lack of predictive animal models in the discovery and preclinical testing phases (Bergren *et al.*, 2012).

A third reason for the high attrition rates may be found in the fact that the majority of today's drugs in development target novel mechanisms (Bergren *et al.*, 2012). The reason might be difficulties in target validation due to the concept of target-based drug discovery with the related advanced complexity of target selection, a competition for proprietary targets, and the complex process of target validation (Agarwal, 2013; Sams-Dodd, 2005). In an analysis of 259 drugs that were launched between 1999 and 2008, it was shown that the drug R&D approach of phenotypic screening exceeded the target-based approach. Most first-in-class drugs that have been evaluated were discovered by phenotypic screenings rather than by the target-based approaches (Swinney and Anthony, 2011). It was concluded that the hypothesis-driven target-based approach may contribute to higher attrition rates than the older phenotypic screening (Swinney and Anthony, 2011; Scannell *et al.*, 2012).

**Table 4.1** Success rates per phase of pharmaceutical R&D.

| Literature | Phase: PoS | Commentary |
|---|---|---|
| DiMasi (2001) | Approval success rates of NCEs (INDs filed between 1990 and 1992):<br>• Self-originated: 12.3%<br>• Acquired: 37.3%<br>• Anti-infective: 28.1%<br>• Antineoplastic: 15.8%<br>• Cardiovascular: 17.5%<br>• CNS: 14.5%<br>• Endocrine: 18.2%<br>• Gastrointestinal: 20.0%<br>• Respiratory: 12.0% | 671 NCEs analyzed with first INDs filed from 1981 to 1992 |
| DiMasi *et al.* (2003) | Probability for entering phase (%) starting with phase I:<br>• Phase I: 100.0%<br>• Phase II: 71.0%<br>• Phase III: 31.4% | 68 compounds, 61 NCEs, 7 NBEs or vaccines that were first tested in humans between 1983 and 1994 |
| DiMasi (2006) | Probability for entering phase (%) starting with phase I:<br>• Phase I: 100.0%<br>• Phase II: 75.0%<br>• Phase III: 36.2% | 93 NCEs tested that were first tested in humans during the period of 1970–1982 |
| DiMasi *et al.* (2010) | Probability for submitting a new drug:<br>• Phase I to submission (total): 19%<br>• Phase I to submission (biologics): 32%<br>• Phase I to submission (SMOLs): 13%<br>• Phase I to submission (self-originated): 16%<br>• Phase I to submission (in-licensed): 27% | 1 738 new drugs and biologics that entered clinical testing between 1993 and 2004 were analyzed |
| Paul *et al.* (2010) | Probability per phase:<br>• Preclinical to registration: 4.1%<br>• Target to hit: 80%<br>• Hit to lead: 75%<br>• Lead optimization: 85%<br>• Total discovery research: 51%<br>• Preclinical testing: 69%<br>• Phase I: 54%<br>• Phase II: 34%<br>• Phase III: 70%<br>• Submission to launch: 91% | Analysis based on data from 13 large pharmaceutical companies |

**Table 4.1** (*continued*).

| Literature | Phase: PoS | Commentary |
|---|---|---|
| Pammolli *et al.* (2011) | Average total success rates:<br>• PoS for acute diseases: 8.77%<br>• PoS for chronic diseases: 6.88%<br>• PoS for lethal diseases: 5.54%<br>• PoS for nonlethal diseases: 9.72%<br>• PoS of small organizations: 6.07%<br>• PoS of large organizations: 7.49%<br>• PoS of biotech: 5.14%<br>• PoS of nonbiotech: 7.86%<br>• PoS for rare diseases: 13.15% | 28 000 compounds from 1990 to 2010 have been investigated |
| Bergren *et al.* (2012) | Probability of clinical development (including review and launch):<br>• Phase I to launch (total): 8.3%<br>• Phase I to launch (SMOLs): 7%<br>• Phase I to launch (biologics): 12% | Success rates of NMEs approved between 2006 and 2010 |
| CMR International (2013) | Probability per phase:<br>• Preclinical: 67%<br>• Phase I: 46%<br>• Phase II: 19%<br>• Phase III: 77%<br>• Registration: 90 | Industry benchmark with active substances that entered phase I, II, III, or registration in the years of 2006–2008; total PoS (preclinical to registration): 4.1% |

SMOLs = small-molecule compounds, NCEs = new chemical entities, NBEs = new biological entities, IND = Investigational New Drug.

Further reasons for the high failure rates in R&D may relate to the increased complexity of clinical trials in combination with the higher demands of regulatory authorities. Finally, commercial restrictions may also negatively influence the probabilities of technical success of drug candidates, as the commercial demands defined in the target product profiles (TPPs) of drug candidates may still necessitate a revenue potential of more than USD 500 million. Despite meeting regulatory standards and the criteria for approvability, some drug candidates might need to be stopped because the TPP might not offer enough competitive advantage or differentiation in the market place. Thus, drug candidates can fail due to low market potential despite being scientifically and technically meaningful.

Differences in success rates of drug candidates depend on the drug class, the therapeutic area, the type of disease, the source of the drug candidate, and the size of the company. For example, it has been reported that biologics showed a higher PoS from phase I to submission than small molecules (DiMasi *et al.*, 2010). Likewise, a higher PoS for drugs that addressed acute diseases was published, when compared to drugs treating chronic diseases (Pammolli *et al.*, 2011). Furthermore, it could be shown that in-licensed drug candidates have a higher PoS for phase I to submission than self-originated drugs (DiMasi *et al.*, 2010). Moreover, the size of a company may also have an impact on attrition rates. While large organizations have a mean PoS of 7.86% from phase I to submission, small organizations have a PoS of 6.07% (Pammolli *et al.*, 2011). In the same context, biotechnology companies seem to have lower success rates in the clinical development phase than nonbiotechnology organizations (Pammolli *et al.*, 2011). Finally, in particular in oncology, predictive biomarkers are used in the design and execution of clinical development programs to increase the therapeutic window of efficacy to safety and thus to reduce attrition rates of the drug candidates (Simon and Maitnourim, 2004).

In summary, drug R&D has inherent uncertainties which result in high attrition rates in the drug discovery and development process.

## 4.5 The Long R&D Time Intervals

In a recent analysis by Paul *et al.* (2010), it was shown that discovery research, ranging from target identification to lead optimization, lasts on average 4.5 years, while preclinical testing takes 1 year and the three clinical development phases last 1.5, 2.5, and 2.5 years, respectively (see also Figure 4.1). The phase from submission to launch of a new drug lasts 18 months on average (Paul *et al.*, 2010). In short, the process of drug discovery and development as well as registration and approval of a new drug lasts on average up to 14 years (Paul *et al.*, 2010; Thomson Reuters, 2013; Pammolli *et al.*, 2011). Table 4.2 highlights selected results and review findings in respect to the R&D timing.

A central finding of the research performed on R&D timelines is the continuous increase in R&D cycle times in the past 25 years. Even in the last years, duration per phase did not decrease, although new high-throughput technologies and other tools to increase R&D efficiency have been installed. It has been reported that the average clinical development duration for drugs approved between 2005 and 2009 was 6.4 years (Deloitte and Thomson Reuters). According to the 2014 CMR Factbook, the composite median interval duration for ongoing development projects (2008–2011) was 9.1 years. In view of the data from the CMR, there is a trend toward longer preclinical development (+17%, 2004–2012) and phase I timelines (+58%, 2004–2012). At the same time, the average processing time for the FDA review and approval has decreased significantly since the enactment of the Prescription Drug User Fee Act (PDUFA) (Deloitte and Thomson Reuters; Kaitin and DiMasi, 2011; Dickson and Gagnon, 2004). It is also likely that new drugs in oncology and rare diseases benefited from the fast-track status or accelerated approvals and, therefore, contributed to the overall reduction in the review and approval timelines.

Differences in the timelines of clinical development phases of various therapeutic classes have been reported. New drugs addressing CNS last longest, needing 10

**Table 4.2** Average timelines of pharmaceutical R&D phases.

| Literature | Time/phase | Commentary |
|---|---|---|
| DiMasi (1991) | Average clinical phase lengths for approved NCEs:<br>• Phase I: 14 mo<br>• Phase II: 25.9 mo<br>• Phase III: 36.8 mo | 93 self-originated NCEs with first human dose between 1970 and 1982 |
| Reichert (2003) | Average time of phase I to approval: 6–8 yr<br>• From 1982 to 2001 the mean clinical development time increased from 61.6 to 63.9 mo<br>• The mean approval time was constant at a range of 23–37 mo (1982–2001)<br><br>Mean phase lengths of clinical development and approval (1982–2001):<br>• Anti-infective: 74.5 mo<br>• Antineoplastic: 116.0 mo<br>• Cardiovascular: 103.3 mo<br>• Endocrine: 115.3 mo<br>• Immunological: 100.2 mo | 554 drugs have been analyzed that were approved between 1980 and 2001 |
| DiMasi *et al.* (2003) | Time from start of clinical testing:<br>• Phase I to submission: 72.1 mo<br>• Phase I to marketing approval: 90.3 mo<br><br>Average phase times for investigational compounds (1985–2000):<br>• Phase I: 21.6 mo<br>• Phase II: 25.7 mo<br>• Phase III: 30.5 mo | 68 new drugs of these were 61 SMOLs and 7 biologics or vaccines |
| Paul *et al.* (2010) | Average time from preclinical to registration: 13.5 yr<br><br>Average time per phase:<br>• Target to hit: 12 mo<br>• Hit to lead: 18 mo<br>• Lead optimization: 24 mo<br>• Preclinical testing: 12 mo<br>• Phase I: 18 mo<br>• Phase II: 30 mo<br>• Phase III: 30 mo<br>• Submission to launch: 18 mo | Analysis based on data from 13 large pharmaceutical companies |

**Table 4.2** (*continued*).

| Literature | Time/phase | Commentary |
|---|---|---|
| Kaitin and DiMasi (2011) | Mean clinical phase times (phase I to submission) and mean approval times of NCEs and NBEs approved between 1980 and 2009:<br>• 1980–1984: 5.7 yr/2.8 yr<br>• 1985–1989: 5.8 yr/2.7 yr<br>• 1990–1994: 6.4 yr/2.4 yr<br>• 1995–1999: 6.5 yr/1.4 yr<br>• 2000–2004: 6.6 yr/1.5 yr<br>• 2005–2009: 6.4 yr/1.2 yr<br><br>Mean clinical phase time (phase I to submission) and mean approval time for NCEs and NBEs approved between 2005 and 2009 grouped by therapeutic class:<br>• CNS: 8.1 yr/1.9 yr<br>• Antineoplastic: 6.9 yr/0.7 yr<br>• Endocrine: 6.5 yr/1.2 yr<br>• Cardiovascular: 6.5 yr/1.3 yr<br>• Immunologic: 6.4 yr/1.0 yr<br>• Gastrointestinal: 5.8 yr/2.4 yr<br>• Anti-infective: 5.4 yr/1.2 yr<br>• Anesthetic/analgesic: 5.3 yr/0.8 yr<br>• AIDS antiviral: 4.6 yr/0.5 yr | Durations of clinical and approval phases for new drugs from 1980 to 2009 |
| Pammolli *et al.* (2011) | Average time for clinical development to submission increased from 9.7 yr for new drugs launched in the 1990s to 13.9 yr for new drugs launched between 2000 and 2008 | 28 000 compounds from 1990 to 2010 have been investigated |
| CMR International (2013) | Average time from preclinical to registration: 12.3 yr | Overall time for drug research and development decline from 15 yr (2010) to 12.3 yr (2012) |
| Remnant *et al.* (2013) | The total time for drug R&D: 14 yr | Analysis on a cohort of 12 companies |

years, while drugs for the treatment of AIDS antiviral had the shortest timelines, needing 4.9 years on average (Deloitte and Thomson Reuters).

Taking the information from the papers discussed herein and as outlined before, drug R&D lasts at least for 14 years. The timelines for basic research and postapproval times were not included in these studies. Assuming that basic research, such as the provision of knowledge in respect to new technologies or new drug targets, lasts for several years before enough information is available to rationalize and enable applied research of pharmaceutical R&D and assuming that the postapproval phase IV trials continue for years, one can conclude that the entire process of pharmaceutical R&D can last for two decades or even longer.

The four major risks related to long cycle times are the following:

- The capitalization of the costs
- The industry rivalry
- The generic competition
- The regulations and the public healthcare system

First, because costs associated with R&D were invested many years ago, they need to be capitalized until the date of ROI for the respective drug. The capitalization of drug R&D costs results in an enormous increase in overall R&D expenditure. As examined by Paul *et al.* (2010), the total out-of-pocket costs for new drug development cumulate on average to USD 873 million, while the total capitalized costs for one new drug are estimated to be on average USD 1.778 billion (Paul *et al.*, 2010). Consequently, any prolongation of the R&D timelines and any increase in the interest rate have a negative impact on R&D costs. For example, the average timelines for clinical development (phases I–III) have been reported to be 63.9 months in 2003 and newer data from 2010 showed an increase to a mean time of 78 months. Thus, and assuming these data are comparable, costs to develop a drug need to be capitalized on average for a longer period of 14.1 months (Paul *et al.*, 2010; Carpenter *et al.*, 2003).

Second, as companies work in the same therapeutic areas, address the same diseases by the same mechanisms and via the same drug targets, and as small molecules are (at least to some extent) identified by using nonproprietary compound libraries, most of the research-based pharmaceutical companies compete to be first in class or first to market to have a competitive edge over other market actors. Thus, the longer the timelines of drug discovery and development, the higher the likelihood that a competitor will be faster and first to the market, and the lesser the probability to be profitable. Longer drug R&D timelines increase the risk of competition and reduce the market potential and success.

Third, the effective date of generic competition influences the ROI of a new drug. In view of the first-to-file principal in international patent law, a patent application claiming the drug substance is usually filed promptly after inventing the optimized lead, which happens at the end of drug discovery. Starting with the date of filing of the patent application, the patent term is 20 years resulting in a market exclusivity of 8–10 years. Adding a potential 5-year patent term extension (PTE) or the supplementary protection certificate (SPC), this cumulates to a maximum exclusivity period of 15 years after launching the drug successfully. Any delay in drug development that occurs after the filing of the patent application claiming the drug substance results in a shortage of the patent term and, thus, in a reduction of the time between launching the product and the arrival of generic competition.

For example, assuming a patent application claiming the drug substance is filed during lead optimization, the subsequent drug development lasts on average for 12.5 years. Assuming also that the new drug is approved and the PTE is filed in time after the drug is approved, the total period of exclusivity is 12.5 years before generic competition arises. In addition, assuming that the therapeutic indication that has been tested first is not the indication that provides clinical proof of concept and the drug substance need to show its proof in a second therapeutic indication, the timelines of clinical development may easily extend by several years, thus reducing the overall time of exclusivity to less than 10 years despite the filing of a PTE or SPC.

Fourth, the pharmaceutical market is mainly driven by regulative forces on several levels, namely, the R&D regulations, product registrations requirements, price regulations, demands from national healthcare systems, and IP (Intellectual Property) rights. Legal R&D barriers have increased in most countries in the past years, partly due to decreasing public acceptance of new technologies such as biopharmaceutical, genetic, and other new technologies and partly due to restrictive regulations for experiments such as on animals and stem cell research. Drug prices are in most countries regulated by federal authorities, directly or indirectly. Sometimes the social costs of the society or the innovativeness of drugs, as, for example, measured by the number of patents in Brazil, are influencing the price and therefore the business model of a new drug. In consequence, the longer the time horizon of drug R&D, the more difficult is a business plan due to these legal and regulatory changes and uncertainties that may happen in between.

## 4.6 The High Cost of Pharmaceutical R&D

The long timelines and the high failure rate of pharmaceutical R&D result in enormously high costs for pharmaceutical innovation. Table 4.3 summarizes the development of costs of drug R&D in the past years, although the results are not directly comparable.

Munos (2009) reported in his analysis of the past 60 years that R&D costs have doubled every 8.5 years since 1950. The annual increase in capitalized costs per NME has been calculated as 12.3%, and in view of an average inflation rate in the past 60 years of 3.7% annually, the remaining inflation-adjusted increase in R&D costs was 8.6% per year (Munos, 2009).

Before the 1990s, costs per new drug were calculated to be less than USD 250 million (DiMasi, 2003). In 2003, the average out-of-pocket costs were already USD 403 million, and the capitalized costs had been calculated at USD 802

**Table 4.3** Costs of pharmaceutical R&D and costs per phase of R&D.

| Literature | R&D costs | Commentary |
|---|---|---|
| DiMasi (1991) | USD 231 million (in 1987 dollars) | Full cost estimation: out-of-pocket costs USD 114 million, 9% discount rate |
| DiMasi (2003) | USD 802 million (in 2000 dollars) | Full cost estimation: out-of-pocket costs USD 403 million and discount rate of 11% |
| Adams and Brantner (2006) | USD 868 million (in 2000 dollars) | Full cost estimation and discount rate of 11%<br>Drugs entering human clinical trials for the first time between 1989 and 2002 have been analyzed<br>Full costs in preclinical phases: USD 381 million<br>Full costs in clinical phases: USD 487 million |
| Adams and Brantner (2007) | USD >1 billion | Average expenditure in human clinical trials per year: USD 25 million |
| DiMasi and Grabowski (2007) | USD 1318 million | Capitalized costs per biologic (USD 1241 million) are comparable to capitalized costs per NCE (USD 1318 million) |
| Paul *et al.* (2010) | USD 1778 million | Full cost estimation: out-of-pocket costs USD 873 million; 9%<br>Capitalized costs per discount rate launch: USD 1778 million<br>Capitalized costs per launch per phase:<br>• Target to hit: USD 94 million<br>• Hit to lead: USD 166 million<br>• Lead optimization: USD 414 million<br>• Preclinical testing: USD 150 million<br>• Phase I: USD 273 million<br>• Phase II: USD 319 million<br>• Phase III: 314 million<br>• Submission to launch: USD 48 million |
| Remnant *et al.* (2013) | USD 1219 million | The capitalized costs of drug development per new asset increased from USD 1019 million (2010) to USD 1219 million (2013) |

USD = US dollars.

million (DiMasi, 1991). It has been stated that the increase was primarily related to increasing costs in clinical development (+350% from 1991 to 2003) (DiMasi, 1991). In 2010, the total out-of-pocket costs for drug R&D were calculated as USD 873 million, while the total capitalized costs are USD 1.778 billion. In particular the clinical development phases from phase I to submission were the major growth drivers as they accounted for 63% of these total R&D costs (Paul *et al.*, 2010).

Principally, the high costs of drug R&D arise from the high failure rates, the capitalization of costs, the implementation of new and expensive technologies in drug research, the target-based approach, the increased regulatory demands, and, more recently, the need for additional comparative data for reimbursement purposes.

While the high capitalized costs are due to the long R&D timelines, most of the out-of-pocket costs are associated with the low PoS and, thus, with the costs of failed research projects and development compounds (Paul *et al.*, 2010; DiMasi *et al.*, 2010). Increasing R&D costs may also be related to new technologies in drug research that have been implemented to increase the throughput in research, such as combinatorial chemistry, DNA sequencing, HTS, and computational drug design. They may also be associated with poor predictive preclinical animal models, increasing regulatory demands, larger clinical trial size, and higher costs for clinical infrastructure. Finally, a greater complexity of clinical trials conducted for drugs to treat chronic diseases and a higher number of R&D personnel may be additional factors (DiMasi, 1991; DiMasi, 2003).

The cost calculations and assumptions listed in Table 4.3 may not be complete, as they do not include all costs for basic research, costs related to phase IV trials, costs for regulatory approvals in non-US markets, or costs for developing the same drug in new indications. A hint for the "extra" costs that have not been included in the previous reports may be given by the data of the 2014 CMR Factbook, wherein the total R&D costs of 11 pharmaceutical companies have been analyzed. Within this peer group, 17.5% of all R&D costs were allocated to research; 26.7% to preclinical development, phase I, and phase II; 24.4% to phase III; and 5.7% to the submission phase, while 25.7% were dedicated to the international rollout and line extensions (Figure 4.2).

In summary, today's total capitalized costs of drug R&D are likely to amount to USD 1.3–1.8 billion per new drug. Together with the extended timelines for drug development, the increasing costs put pressure on the ROI per new drug.

## 4.7 The Reduced R&D Efficiency

R&D efficiency can be measured by the rendering of an input, such as a monetary investment, into an output, such as the first launch of an NME in one of the major pharmaceutical markets (Paul *et al.*, 2010). Thus and in the context of this book, the dimensions of the R&D efficiency are the costs per NME approved by the FDA. Scannell *et al.* (2012) have analyzed the decline of the pharmaceutical

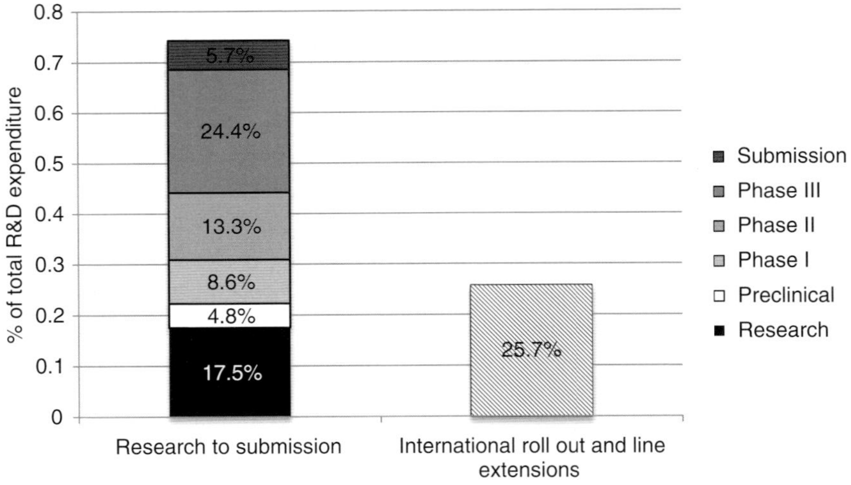


**Figure 4.2** Proportion of total R&D expenditures based on CMR 2014 data of 11 research-based pharmaceutical companies.

R&D efficiency in a period from 1950 to 2010 and concluded that the number of new drugs approved per USD 1 billion halved nearly every 9 years in the past 60 years (Munos, 2009; Scannell *et al.*, 2012). This trend is the result of a development in the pharmaceutical industry, whereby the number of new drugs launched by the industry was constant while the costs per new drug increased steadily. Today, the capitalized costs per new drug have been calculated to be around USD 1.8 billion, although it could be assumed that the actual costs of drug R&D are even higher (Paul *et al.*, 2010).

In a publication of PricewaterhouseCoopers (PWC), the total costs per launch of a new drug were analyzed as the ratio of total R&D costs of the industry to the total number of new drugs approved by the FDA (Cohen, 2005). Based on this, the total costs per NME in the years of 2002–2011 were between USD 2.2 and 4.6 billion. In a series of three *Forbes* articles, the costs of developing and launching new drugs were analyzed, highlighting that the average costs of drug development of top pharmaceutical companies are between USD 3 and 12 billion (http://www.forbes.com/sites/matthewherper/2012/02/10/the-truly-staggering-cost-of-inventing-new-drugs/). It has also been concluded that smaller pharmaceutical companies need less money to launch a new drug. This may relate to the fact that only successful small companies have been considered in the statistic and failed companies were disregarded (http://www.forbes.com/sites/matthewherper/2013/08/11/how-the-staggering-cost-of-inventing-new-drugs-is-shaping-the-future-of-medicine/). Finally, it was investigated that the top pharmaceutical companies, those that have launched more than four NMEs in the 10 years from 2002 to 2011, invested more than USD 5 billion per new drug (http://www.forbes.com/sites/matthewherper/2013/08/11/the-cost-of-inventing-a-new-drug-98-companies-ranked).

**Table 4.4** R&D efficiencies (2001–2012) of multinational pharmaceutical companies.

| | Total number NMEs approved by the FDA (2001–2012) | Total R&D costs (USD million, 2001–2012) | R&D efficiency (costs per launch) |
|---|---|---|---|
| AstraZeneca | 3 | 55 959 | 18 653 |
| Roche. | 7 | 83 888 | 11 984 |
| Takeda[a)] | 2 | 23 376 | 11 688 |
| Sanofi[b)] | 4 | 38 912 | 9 728 |
| Eli Lilly | 5 | 47 949 | 9 590 |
| Pfizer[c)] | 12 | 91 367 | 7 614 |
| GSK | 11 | 76 538 | 6 958 |
| Boehringer Ingelheim[d)] | 3 | 20 727 | 6 909 |
| Amgen | 5 | 34 119 | 6 824 |
| Novartis | 16 | 82 004 | 5 125 |
| Bristol-Myers Squibb | 9 | 40 292 | 4 477 |

NME = new molecular entities; FDA = Food and Drug Administration; R&D = research and development; GSK = GlaxoSmithKline; all data are derived from annual company reports.
a) Data of Takeda from 2006 to 2012.
b) Data of Sanofi from 2005 to 2012.
c) Data of Pfizer from 2002 to 2012.
d) Data of Boehringer Ingelheim from 2004 to 2012.

Table 4.4 summarizes an analysis of 11 multinational pharmaceutical companies, including their R&D costs, the number of NMEs approved by the FDA between 2001 and 2012, and the resulting R&D efficiencies.

From these data, it is apparent that the pharmaceutical companies listed had total R&D costs of USD 4.5–18.6 billion per new drug approved by the FDA in the past 10 years. On average pharmaceutical companies invested USD 9 billion per NME (median USD 7.6 billion), an amount that is significantly higher than the figures that have been calculated in previous publications based on analyzed out-of-pocket costs. In the period from 2002 to 2012, the most efficient pharmaceutical companies analyzed were Bristol-Myers Squibb (BMS) and Novartis with R&D efficiencies of USD 4.5 billion and USD 5.1 billion per NME, respectively.

Possible reasons for the low R&D efficiency have been discussed previously in the context of the low PoS and the high costs for pharmaceutical innovation. Furthermore, an insufficient number of projects in preclinical and early clinical phases may have negatively impacted the R&D efficiency (Paul *et al.*, 2010). Additionally, the increasing number of approved drugs raised the hurdle for approval and reimbursement of new drugs (Scannell *et al.*, 2012). In the same way, a lower-risk tolerance of drug regulators may have increased both the challenges for launching new drugs and the development-associated costs (Scannell *et al.*, 2012). An increasing number of M&As might also have influenced the growth of pharmaceutical companies and their R&D efficiencies negatively (PWC; Danzon,

2007; LaMattina, 2011). Comanor and Scherer (2013) discussed a more indirect negative impact of M&As on the output of pharmaceutical R&D, namely, that the decreasing number of well-established pharmaceutical companies is reducing the number of investors that will take the risk to progress drug candidates through the R&D process (Ornaghi, 2009). Moreover, DiMasi (2014) reported that half of the 256 drugs (46.9%) that were analyzed in their study were interrupted in their clinical development phases by licensing (25.4%), codevelopment (15.6%), M&A (5.5%), or joint venture (0.4%) and had median clinical development and approval phases that were 14.8 months longer than uninterrupted programs (Comanor and Scherer, 2013). Consequently, drug development projects that are interrupted have longer clinical development durations of approximately 20% and, thus, higher costs, hence reducing R&D efficiency. And, finally, it has been said that the low-hanging fruits have already been picked, resulting in technically more complex investigations for new drug targets and related preclinical and clinical studies (Scannell *et al.*, 2012).

## 4.8
## Can an Increase in R&D Value Compensate the Reduced R&D Efficiency?

In a monetary reflection of the R&D efficiency, the ROI of pharmaceutical R&D since the 1990s has been analyzed. In a first investigation of the profitability of R&D of new chemical entities (NCEs) and based on historical data, Joglekar and Paterson (1986) described a net present value (NPV) of USD 76 million (in 1976 dollars) for an average NCE with a break-even point at 12 years after marketing the product (DiMasi, 2014). Grabowski and Vernon (1990) reported that in a group of NCEs launched in the period of 1970–1979, only the top 30% of these drugs generated total sales that exceeded the average costs of drug R&D (Joglekar and Paterson, 1986). In another study, they reported that the mean NPV of new drugs launched in the United States between 1980 and 1984 was USD 22 million (Grabowski and Vernon, 1990). In addition, they presented that for NCEs launched between 1988 and 1992, the top 10% of the drugs accounted for more than 50% of the sales of all drugs in the study (Grabowski and Vernon, 1994). In several publications, the R&D productivity of the pharmaceutical industry has been analyzed (Grabowski and Vernon, 2000; Tollman *et al.*, 2010, Schulze *et al.*, (2015)). As the R&D spending of multinational pharmaceutical companies grew at a compound annual growth rate (CAGR) of 9% between 1990 and 2008, the authors concluded that a productivity ratio of 0.25–0.35 is required annually to meet its cost of capital of 9%. In fact, only seven multinational research-based companies achieved this figure, namely, Roche, J&J, GSK, Biogen Idec, Astellas, and Bayer, while other companies failed to achieve this goal.

Now, if the efficiency of R&D is reduced, its return on investment can still be maintained by increasing the value per drug launched to the market. It has been highlighted that the value of one NME, measured at the 5-year postlaunch sales, grew in the time period from 2010 (USD 10 billion) to 2012 (USD 16 billion) (Schulze, 2014). In contrast, it is expected that the average peak sales per NME will decline from USD 900 million (2012) to USD 600 million (2015). This phenomenon could be explained by the increased difficulty of offering benefits over existing treatments in light of the increasing price pressure and/or the development of new drugs for smaller and more targeted populations with higher response rates (Bergren *et al.*, 2012). In the mature markets of Europe and the United States, products face stronger competition and, thus, need to be developed for better profiled patient populations and to be launched to smaller market segments. In parallel, they face high cost pressure from the public and payors. In the same context, the projected revenues of all NMEs launched between 2012 and 2016 are expected to be USD 58 billion, whereas losses by patent expirations between 2013 and 2016 are forecasted to be USD 123 billion, showing that the new revenues based on NMEs will not compensate for the revenue lost by patent expirations in the industry (Bergren *et al.*, 2012; EvaluatePharma).

As a result, players in the pharmaceutical field have to think about new ways how to manage R&D efficiency in better ways and how to further create value and capture parts of the value. There are several strategic alternatives, which can be taken in order to circumvent the low R&D efficiency and increase the PoS:

1) Use better technologies in the early R&D process. For example, the consequent utilization of throughput technologies such as HTS may increase the probability to fail earlier and succeed sooner.
2) Cut experimental cycle times in drug discovery with combinatorial chemistry.
3) Improve the use of informatics along the whole R&D process. For example, bioinformatics has already been used in the past years in drug discovery but will be one of the most upcoming trends, as pure algorithm companies such as IBM are joining this segment and are in the meantime in the top 10 of the most patenting companies in the biopharmaceutical sector.
4) Open up the innovation process by increasing collaborations and strategic alliances. Out-licensing of promising substances with the option to buy back is an alternative. Moreover, biotechnology companies are embraced by the multinational pharmaceutical companies for their innovative power, combined with their inability to commercialize their inventions in the last steps through the FDA and push the drugs into the global markets.
5) Crowdsourcing approaches such as the Eli Lilly spin-off InnoCentive might create effective platforms to place problems and ask the whole scientific community for solutions. Today, most pharmaceutical companies have their own platforms where the crowd of external researchers can position their inventions, solutions, and research activities.
6) The full potential of personalized medicine needs to be realized. To this end, next steps to establish the use of genomics and other diagnostic tools need to be customized and made available for use in large populations and realize the benefits of personalized medicine.
7) As the healthcare systems continue to change, new business models are needed. New players from the consumer electronic area are entering the

scene today with smart devices and diagnostic and monitoring systems, which are more lifestyle products rather than medical devices. However, innovation research has taught us in the past that simple and cheap solutions mostly replace the sophisticated, slow, and expensive solutions. It is expected that software-based lifestyle companies are entering the business of the healthcare sector. Pharmaceutical companies should be ready to broaden their business models and look to provide service beyond the exclusive provision of the "pill."

In conclusion, the R&D efficiency of the pharmaceutical industry declined steadily since the 1950s. In view of the annual total R&D costs of pharmaceutical companies and the NMEs approved by the FDA per year and company, it can be assumed that today's average R&D efficiency is in the range of USD 1.8 billion per NME and may even hit more than USD 5 billion for some companies. Thus, the extraordinarily high costs per NME make it increasingly difficult for some pharmaceutical companies to generate enough ROI out of R&D and make it necessary to reexamine the overall R&D process, models for innovation and collaboration, portfolio management, and culture as discussed in later chapters in this book.

## References


Agarwal, C.P. and Agarwal, V.V. (2006) *Health Aff.*, **25**, 420–428.

Agarwal, C.P. and Agarwal, V.V. (2007) *Health Econ.*, **19**, 130–141.

Agarwal, P. (2013) *Nat. Rev. Drug Discovery*, **12**, 575–576.

Arrowsmith, J. and Miller, P. (2013) *Nat. Rev. Drug Discovery*, **12**, 569.

Bergren, R. *et al.* (2012) *Nat. Rev. Drug Discovery*, **11**, 435–436.

Carpenter, D. *et al.* (2003) *Health Aff.*, **W3**, 618–624.

CMR International (2013) CMR International Pharmaceutical R&D Factbook, http://cmr.thomsonreuters.com/services/factbook/ (accessed 25 August 2015).

Cohen, F.J. (2005) *Nat. Rev. Drug Discovery*, **4**, 78.

Comanor, W.S. and Scherer, F.M. (2013) *J. Health Econ.*, **32**, 106–113.

Cook, D. (2014) *Nat. Rev. Drug Discovery*, **13**, 419–431.

Danzon, P.M. (2007) *Manage. Decis. Econ.*, **28**, 307–328.

Deloitte & Thomson Reuters (2013) Measuring the Return from Pharmaceutical Innovation 2013.

Dickson, M. and Gagnon, J.P. (2004) *Nat. Rev. Drug Discovery*, **3**, 417–429.

DiMasi, J.A. (1991) *J. Health Econ.*, **10**, 107–142.

DiMasi, J.A. (2001) *Clin. Pharmacol. Ther.*, **69**, 297–307.

DiMasi, J.A. *et al.* (2003) *J. Health Econ.*, **22**, 151–185.

DiMasi, J.A. (2003) *J. Health Econ.*, **23**, 151–185.

DiMasi, J.A. (2006) *J. Health Econ.*, **10**, 107–142.

DiMasi, J.A. (2014) *Ther. Innov. Regul. Sci.*, **48**, 482–487.

DiMasi, J.A. and Grabowski, H.G. (2007) *Manage. Decis. Econ.*, **28**, 469–479.

DiMasi, J.A. *et al.* (2010) *Clin. Pharmacol. Ther.*, **87** (3), 272–277.

European Commission – Joint Research Centre (2013) The 2013 EU Industrial R&D Investment Scoreboard.

EvaluatePharma (2013) Biotech and Pharma 2012 Year in Review.

Grabowski, H.G. and Vernon, J.M. (1990) *Manage. Sci.*, **36**, 804–821.

Grabowski, H.G. and Vernon, J.M. (1994) *J. Health Econ.*, **13**, 383–406.

Grabowski, H.G. and Vernon, J.M. (2000) *Pharmacoeconomics*, **18**, 21–32.

International Federation of Pharmaceutical Manufacturers & Associations The Pharmaceutical Industry and Global Health – Facts and Figures 2014.

Joglekar, P. and Paterson, M. (1986) *J. Health Econ.*, **5**, 153–177.

Kaitin, K.I. and DiMasi, J.A. (2011) *Clin. Pharmacol. Ther.*, **89** (2), 183–188.

Kola, I. and Landis, J. (2004) *Nat. Rev. Drug Discovery*, **3**, 711–716.

LaMattina, J.L. (2011) *Nat. Rev. Drug Discovery*, **10**, 559–560.

Munos, B. (2009) *Nat. Rev. Drug Discovery*, **8**, 959–968.

Ornaghi, C. (2009) *Int. J. Ind. Organ.*, **27**, 70–79.

Pammolli, F. *et al.* (2011) *Nat. Rev. Drug Discovery*, **10**, 428–438.

Paul, S.M. *et al.* (2010) *Nat. Rev. Drug Discovery*, **9**, 203–214.

PhRMA (2013) Profile Biopharmaceutical Research Industry, http://www.phrma.org/sites/default/files/pdf/PhRMA%20Profile%202013.pdf (accessed 01 November 2014).

PWC (2012) From Vision to Decision Pharma 2020,

Reichert, J.M. (2003) *Nat. Rev. Drug Discovery*, **2**, 695–702.

Remnant, J. *et al.* (2013) Measuring the Return from Pharmaceutical Innovation 2013, http://thomsonreuters.com/business-unit/science/subsector/pdf/uk-manufacturing-measuring-the-return-from-pharmaceutical-innovation-2013.pdf (accessed 25 August 2015).

Sams-Dodd, F. (2005) *Drug Discovery Today*, **10**, 139–147.

Scannell, J.W. *et al.* (2012) *Nat. Rev. Drug Discovery*, **11**, 191–200.

Schulze, U. (2014) *Nat. Rev. Drug Discovery*, **13**, 331–332.

Schulze, U., Bädeker, M., Chen, Y.T., and Greber, D. (2015) BCG Perspectives, R&D Production 2014: A Breakthrough year for Biopharma.

Simon, R. and Maitnourim, A. (2004) *Clin. Cancer Res.*, **10**, 6763–6759.

Swinney, D.C. and Anthony, J. (2011) *Nat. Rev. Drug Discovery*, **11**, 507–519.

Thomson Reuters (2013) CMR International Pharmaceutical R&D Factbook, http://cmr.thomsonreuters.com/pdf/fb-exec-2013.pdf (accessed 07 November 2014).

Tollman, P. *et al.* (2010) *Nat. Rev. Drug Discovery*, **10**, 653–654.

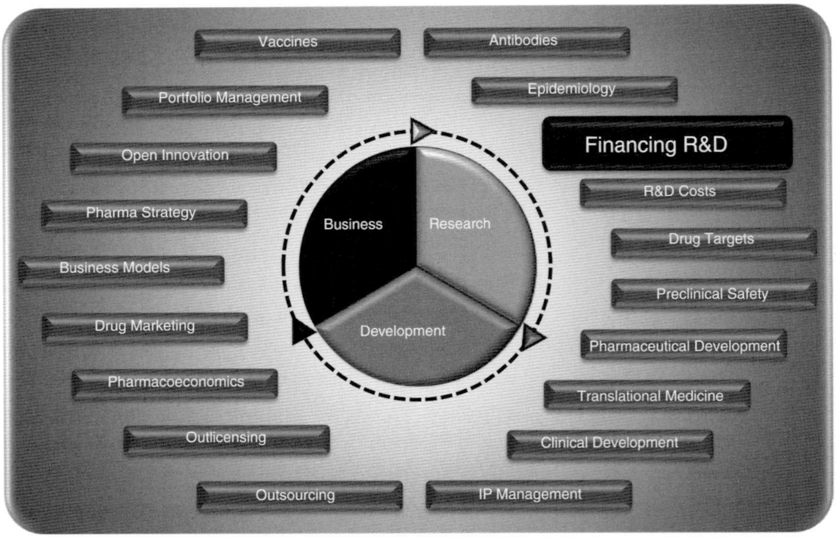
Vaccines
Antibodies
Portfolio Management
Epidemiology
Financing R&D
Open Innovation
R&D Costs
Pharma Strategy
Business
Research
Drug Targets
Business Models
Preclinical Safety
Drug Marketing
Development
Pharmaceutical Development
Pharmacoeconomics
Translational Medicine
Outlicensing
Clinical Development
Outsourcing
IP Management

# 5 Financing Pharmaceutical Innovation

*Sviataslau Sivagrakau*

## 5.1 Introduction

Finance is the denominator of Eroom's Law, which states that the number of new drugs launched per dollar spent has halved every 9 years since the 1950s (Scannell *et al.*, 2012). Therefore, finance is, by definition, at least equally important for the discussion of reduced research and development (R&D) efficiency as an output-related factor that enters the nominator. Eroom's law,[1] of course, evidences a substantial increase in gross R&D spending on the background of near-constant output. However, at a more granular level, pharmaceutical finance has changed not only in total volume but also in nature. It has moved across borders: from West Germany – global market leader in the 1960s – to the United States, which currently dominates worldwide drug R&D. It has also changed the public–private composition of its funding base, as some governments have significantly increased budget allocations for biomedical research, while others have not. In the private drug development segment, the industry, too, has experienced a transformation: smaller companies have gradually become a major source of innovation, while large pharmaceutical firms have consolidated – through waves of mergers and acquisitions – to form a concentrated industry landscape. These are the trends that will be further elaborated upon in this chapter.

The purpose of this chapter is to examine the economics of pharmaceutical R&D and to review the funding sources for innovation in drug development. With this goal in mind, it is instructive to start with the definitions that characterize the output of the pharmaceutical industry – new medicines admitted to the market. The first section, therefore, summarizes regulatory categories that assess innovativeness of new pharmaceutical products. Designations such as priority-review new molecular entities (pNMEs) and standard-review new molecular entities (sNMEs) will serve to differentiate between origins and funding sources of more and less innovative drugs. Generic pharmaceuticals, though a large industry on its own, are outside the scope of this chapter.

1) It is the reverse of Moore's Law in the computing hardware industry – a prediction that the number of transistors per silicon chip doubles every year.

*Value Creation in the Pharmaceutical Industry: The Critical Path to Innovation*, First Edition.
Edited by Alexander Schuhmacher, Markus Hinder, and Oliver Gassmann.

While much of the terminologies will follow the definitions of the US regulator – the Food and Drug Administration (FDA) – the overarching aim is to provide a comprehensive global overview. Thus, the focus will lie on the United States, Europe, and Japan since the development of new medicines outside these jurisdictions is currently insignificant. Drawing from national and international data sources and assessment studies, this chapter will provide both project-level and industry-level estimates of R&D expenditures and development timeframes.

To set the scene, an overview of the current drug universe and more detailed productivity statistics of the past decades are provided as a historical background. The total global figures of R&D expenditures are then dissected in several ways to better describe complex financing ecosystem for drug development worldwide. Specifically, pharmaceutical industry will be reviewed within three dimensions:

- Geography – Where in the world are the new medicines developed?
- Source – What are the roles of public and private sectors in the drug development market?
- Size – Within the private industry, how do large pharmaceutical firms fare in comparison to small- and medium-sized enterprises (SMEs)?

While each of these themes could deserve an entire book on its own, the objective of this chapter is to bring all of these important aspects together, giving the readers a brief, yet holistic and data-dense overview of drug development finance, with a specific focus on innovative medicines. In each of the three dimensions, the discussion of finance flows will finish with the reflection on how the money spent relates to the outcomes. This information, put into infoboxes at the end of the sections, anchors the text to the common storyline: how much novelty is generated on what spending and where? The chapter will conclude with a discussion on the cost of financing for pharmaceutical R&D, alternative financing schemes that may gain significance for drug development funding tomorrow, and relation of pharmaceutical finance to the broader field of public health economics.

For a summary, Box 5.1 below gathers the key figures of present-day pharmaceutical economics into a bulleted list.

**Box 5.1: Pharmaceutical industry and its economics in figures: a summary**

- From 1827 to 2013 a total of **1453** therapeutics had been authorized for use in the United States (Kinch *et al.*, 2014); currently there are 1088 patented drug products in the US market (Patented Medicine Prices Review Board, 2013).
- Still, out of the approximately **4000** diseases with known molecular cause, only about **250** have treatments available (Graybeal, 2012).

- More than **90%** of new drugs originate from the United States, Europe, and Japan (Kneller, 2010). Likewise, these three markets contribute to **90%** of global biomedical R&D expenditure (Chakma *et al.*, 2014).
- The regulators of the three aforementioned jurisdictions authorize **20–30** new active substances (NASs) annually since the 1980s. There is no clear trend of growth for this rate (Bujar and McAuslane, 2014; U.S. Food and Drug Administration, 2013; Munos, 2009).
- According to the recent figures from the FDA, about **40%** of medicines approved in the 2010s were first-in-class, that is, "drugs that use a new and unique mechanism of action for treating a medical condition" (U.S. Food and Drug Administration, 2011–2015).
- Only a small share of new drugs – about **1/10th** – show statistically significant superiority in efficacy compared to currently available medicines (Patented Medicine Prices Review Board, 2009; van Luijn *et al.,* 2010). Since July 2012, the FDA has designated such drugs as "breakthrough" (16.5% of new approvals in 2013–2014).
- Costs of drug development have been estimated to exceed **USD 1.5 billion** by several assessment studies (Mestre-Ferrandiz *et al.*, 2012; Adams and Brantner, 2010; Tufts Center for the Study of Drug Development, 2014; Herper, 2013).
- About **half** of drug development costs are attributed to cost of capital during **one to two decades** of R&D process.
- In the 1960s nearly one-third of all drugs were invented by pharmaceutical firms headquartered in Germany (Daemmrich, 2009). Today the United States is an unchallenged leader, originating about **50%** of all drugs and about **60%** of novel medicines (Kneller, 2010).
- US public spending for biomedical research is at least **twice** as big as in Europe, while the economy of the European Union marginally exceeds that of the United States in terms of gross domestic product (GDP) (Organisation for Economic Co-operation and Development, 2009; Berghmans *et al.*, 2011; Chakma *et al.*, 2014).
- In the United States, **5634** new products are estimated to be currently under development; in Europe, **2743** products (leading countries include the United Kingdom, Switzerland, Germany, and France) (EY, 2015).
- About **USD 266 billion** was spent on biomedical R&D globally in 2012 (Chakma *et al.,* 2014) – this roughly equals the money spent on buying cars of a single brand: Volkswagen (2014).
- The Japanese pharmaceutical industry collectively spent an equivalent of **0.6%** of the Japanese GDP on R&D – the highest proportion of all countries. On the other hand, the sharpest increase in R&D spending (public and private) from 2007 to 2012 – **313%** – happened in China (Chakma *et al.*, 2014).
- Globally **55–60%** of biomedical R&D is funded by the industry, **30–35%** from public funds, and about **10%** from private nonprofit organizations (Chakma *et al.*, 2014; Røttingen *et al.*, 2013).

- For all FDA priority-reviewed drugs in 1998–2005, **65%** of patents cited either a public-sector patent or a government publication; for standard-review drugs, only 36% (Sampat and Lichtenberg, 2011).
- More than **two-thirds** of all new molecular entities (NMEs) are controlled by the top 10 pharmaceutical companies (Kinch *et al.*, 2014).
- In 2002 38% of phase III compounds in the pharmaceutical industry's pipelines had been sourced externally. In 2010 this increased to **62%**, evidencing externalization of R&D to smaller biotechnology firms. The trend is not mirrored by early-stage assets (change in phase I compounds that originated externally was 34–38%) (McKinsey and Company, 2012).
- From 2006 to 2013 venture capital (VC) financing in biotechnology changed from \$3.9 to **\$4.3 billion** in the United States and from \$2.1 to **\$1.5 billion** in Europe. For comparison, Roche – the biggest player in biotechnology space – spent **\$9.8 billion** on R&D in 2013 alone (EY, 2015; Roche, 2014).

## 5.2
## Measuring Innovation: Categories of New Drugs

Not all drugs are created equal, and gross figures of annual market authorizations may be a poor gauge of pharmaceutical innovation. Looking at EMA's Annual Report 2013, a total of 81 medicines were recommended for marketing authorization, including five opinions for conditional market authorization. However, out of these 81, only 37 medicines contained a NAS. The rest were modifications of existing drugs: new dosage, new combination, or new indication, for instance (EMA, 2014). In the FDA's terminology, a term corresponding to EMA's NAS is NME (Branch and Agranat, 2014); NME's equivalent for new therapeutic biologics is Biologics License Application (BLA). Together NMEs and BLAs compose the segment of novel drugs. Nevertheless, even this definition comes with a caveat: per the FDA, "in some cases, while categorized as novel for technical and/or administrative purposes, a particular NME may not necessarily offer unique clinical advantages over existing therapies" (U.S. Food and Drug Administration, 2015). In order to measure the scientific impact of new medicines more precisely, the FDA has introduced several more stringent categories:

- *First-in-class* – drugs that use a new and unique mechanism of action for treating a medical condition. About 40% of FDA approvals since 2010 have been first-in-class. Still, even this classification does not ensure statistically significant superiority in efficacy compared to currently available medicines. The latter is signified by a breakthrough status, introduced by the FDA in 2012 with the aim to help shorten the development time of a promising new therapy. In 2013–2014 only 16.5% of all approvals were breakthrough (U.S. Food and Drug Administration, 2014–2015).

- Likewise innovative are the medicines targeting diseases that have no treatments available. These are assigned a fast track status – drugs with a potential to address unmet medical needs.
- *Priority review* – drugs with a potential to provide a significant advance in medical care.
- *Accelerated approval* – drugs for serious or life-threatening illnesses that offer a benefit over current treatments.

Novel drugs may fall into several of these categories simultaneously. For instance, Zydelig (a blood cancer drug approved in 2014) was submitted with two indications, of which one of the indications was granted a Breakthrough Therapy, Fast Track, and Priority Review and the other was granted Accelerated Approval.

While innovativeness of new medicines addressing unmet medical needs is not in doubt, most of the drugs entering into competitive segments are follow-on or me-too drugs, which do not provide statistically significant difference in efficacy. They may provide advances on properties such as safety, applicability, or convenience of administration, yet these are incremental improvements, not pharmaceutical innovation in a strict sense. Defining what constitutes innovation is essential not only for evaluating the performance of the pharmaceutical industry but also for wider public health economic issues, such as pricing of medicines. In Canada the public body mandated to ensure that prices at which patentees sell their patented medicines are not excessive and to report on pharmaceutical trends of all medicines is the Canadian Patented Medicine Prices Review Board (PMPRB). In fulfillment of its mandate, the PMPRB audits and appraises new medicines admitted to the Canadian market. In its review of all NASs, which gained market authorization in 2001–2009, the PMPRB has reported that only 10% were classified as Category 2, or new chemical entities that are a breakthrough or provide substantial improvement over comparable existing drugs. The rest were designated as Category 3, or new chemical entities that provide moderate, little, or no therapeutic advantage (Category 1 are line extensions) (Patented Medicine Prices Review Board, 2009).[2] Similar investigation of the European market during 1999–2005 by Dutch academic researchers also arrived at the 10% share for drugs with proven advantage in efficacy (van Luijn *et al.*, 2010).

NMEs nevertheless remain the standard unit of account when measuring the pharmaceutical industry's output. Smaller sample sizes and shorter time series of data for scientifically novel drugs make it difficult to answer questions such as does it cost more to develop a breakthrough therapy than a follow-on drug, and if yes, by how much? As discussed further in the chapter, R&D cost estimation studies do not differentiate between NMEs. That said, some attribution analysis

2) The new PMPRB Guidelines implemented in 2010 replace the three-category system with even more granular categorization into Slight or No Improvement; Moderate Improvement – Secondary; Moderate Improvement – Primary; Substantial Improvement; and Breakthrough drugs.

of priority-reviewed drugs by the type of discovering organization and geography is available, and this chapter shall draw on it further.

## 5.3 Productivity of Pharmaceutical Industry throughout Time

The Centre for Innovation in Regulatory Science, a subsidiary of the information vendor giant Thomson Reuters, tracked the approvals of NASs – chemical, biological, biotechnology, or radiopharmaceutical substances that have not been previously available for therapeutic use in humans and are destined to be made available as "prescription only medicines" – by three major global regulators: US FDA, European Medicines Agency (EMA), and Pharmaceuticals and Medical Devices Agency, Japan (PMDA), from 2004 to 2013 (Bujar and McAuslane, 2014). Annual numbers of NAS approvals are predominantly within a range of 20–30, without a clear growth trend.

London-based EMA was established in 1995 to harmonize marketing authorization in the European market, whereas PMDA was established in its current form only back in 2004. The FDA's continuous operations since 1906, however, have provided us with a common statistical ground to review the pharmaceutical industry's output on a longer timeframe. And the results display stalling performance: the rate of approval for the FDA's defined NMEs in the 2000s slowed down to the levels of the 1980s, while the first 5 years of the 2010s are on par with the 1990s (Figure 5.1).

Due to the varying regulatory procedures in sovereign jurisdictions, the introduction of new medicines is asynchronous around the globe. Still, within advanced economies the average lag is a matter of months, and the total annual NAS approval statistics are therefore similar. The data for the United Kingdom, for instance, mirrors the subdued growth rate dynamics of new drug introductions. Apart from the period of a statistically significant downward trend during the decade from 1997 to 2006, the number of new drug launches annually has been rising slowly.

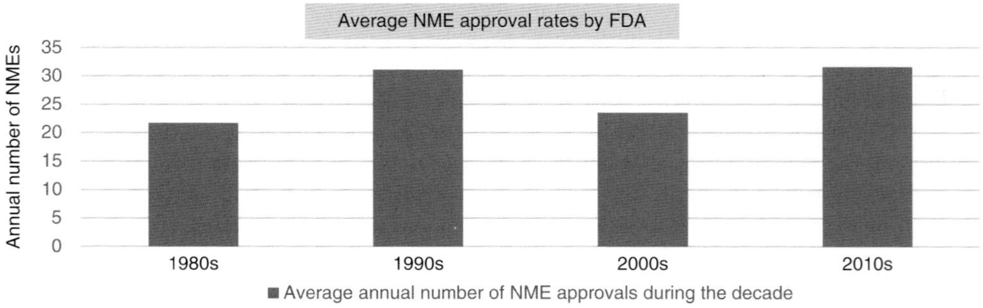


**Figure 5.1** Annual NME approval rates by decade. Source: U.S. Food and Drug Administration (2013).

Indeed, a longer timeframe evaluation shows the overall approval rate growing by only 0.16 new medicines per year over a 40-year period from 1971 to 2011 (Ward, 2013) – and all that despite sharply increased investments, strong performance within basic research on many fronts (from genome sequencing to stem cells to target therapies for cancer), and the progress in pharmaceutical technology, such as high-throughput screening and advances in bioinformatics and combinatorial chemistry.

Since the introduction of morphine in 1827, a total of 1453 therapeutics had been authorized for use in the United States by the end of 2013 (Kinch *et al.*, 2014). Most approvals in a single year came in 1996, when the FDA issued 51 market authorizations. That year, however, can be considered an outlier as the enactment of Prescription Drug User Fee Act (PDUFA) 4 years prior enabled the FDA to increase their personnel and ultimately process a backlog of applications by 1996 (Munos, 2009). More recently, 2013 was a stagnant year with only 27 approvals, whereas 2014 yielded 41 new treatments – highest in 18 years. According to the Canadian PMPRB, the total number of patented drug products in the market as of 2013 ranged from 760 and 919 in France and Germany to 1088 and 1306 in the United States and Canada (Patented Medicine Prices Review Board, 2013). The small total drugs universe (compare it, for instance, to the number of exchange-listed companies globally: 47 000) reflects the scientific challenge of bringing new treatments to patients and the fact that pharmaceutical R&D is one of the most regulated commercial activities performed in the modern economy. What it does not reflect is the demand for medical innovation: out of the approximately 4000 diseases with known molecular cause, only about 250 have treatments available (Graybeal, 2012).

## 5.4 Measuring the Cost of Developing New Medicines

Notwithstanding the vast market potential, pharmaceutical R&D is deterred by the lengthy timeframes commanded by the scientific and regulatory requirements. For all 113 first-in-class drugs approved by the FDA from 1999 to 2013, the median time from the first publication of the therapeutic concept, target, or chemotype to FDA approval was 22 years, while the initiation of drug discovery activities may have occurred several years before any such publication (Eder *et al.*, 2014). The period from patent publication to FDA approval differed significantly by drug discovery strategy: 8 years for target-based approaches and double that for systems-based approaches.

The long path of R&D to fill unmet medical needs does not necessarily end in profit for the innovator. In the 1990s, only the top 30% of the drugs were profitable, according to the calculations by Grabowski, Vernon, and DiMasi on a sample of 118 new chemical entities introduced into the United States between 1990 and 1994 (Grabowski *et al.*, 2002). The distribution was highly skewed: the top decile alone accounted for about 52% of the total net present value (NPV) generated by

**Table 5.1** Estimates of drug development costs.

| Publication year | Authors | Main affiliation | Data source | Sample | Total capitalized cost |
|---|---|---|---|---|---|
| 2010 | Adams and Brantner | Federal Trade Commission | Pharmaprojects dataset compiled by Citeline (a data vendor unit of UK-based publishing group Informa) | 183 firms over 12-year period: 1989–2001 | 1.214 billion (1999 USD) (1.7 billion in 2013 USD) |
| 2012 | Mestre-Ferrandiz, Sussex Towse | Office of Health Economics | CMRI database (Thomson Reuters) | 16 firms with at least one clinical development phase during 1997–1999 | 1.506 billion (2011 USD) |
| 2014 | DiMasi, Grabowski and Hansen | Tufts Center for the Study of Drug Development (CSDD) | Proprietary CSDD database of investigational compounds | 10 firms over 13-year period (1995–2007) | 2.558 billion (2013 USD) |
| 2013 | Matthew Herper (based on the data by Bernard Munos) | Forbes | Innothink Center for Research in Biomedical Innovation | 100 firms over 15-year period: 1998–2012 | 3.621 billion (2012 USD) |
| 2014 | EvaluatePharma | Evaluate Ltd. | Proprietary database | 35 NMEs in 2013 | 3.9 billion (2013 USD) |
| (3.7 billion when adjusted by a 3-year lag) | | | | | |

Sources: Mestre-Ferrandiz *et al.* (2012), Tufts Center for the Study of Drug Development (2014), Herper (2013), and EvaluatePharma (2014).

the entire sample, while the bottom seven deciles had NPV of less than the average development cost, which at that time stood at USD 480 million (this translates to USD 650 million in 2013 dollars). The most recent attempt of DiMasi, Grabowski, and Hansen at the Tufts Center for the Study of Drug Development to update this estimate resulted in a USD 2.6 billion figure, meaning that the R&D costs of developing new medicines have quadrupled in real terms (Tufts Center for the Study of Drug Development, 2014). The most recent update on profitability distribution shows even more skewed results: only one in five marketed drugs return revenues that match or exceed R&D costs (Vernon *et al.*, 2010). Despite methodological weaknesses of the Tufts cost study (particularly a small sample size of only 10 firms), the results lie in the range of estimates by other studies, all of which report that the cost of developing new medicines exceeds USD 1.5 billion.

A popular methodological choice for evaluating R&D efficiency is to impute an industry average based on total or per-phase R&D project costs using confidential survey data from a presumably representative sample of pharmaceutical companies. This method results in a 1.6–2.6 billion average expressed in 2013 USD, as detailed in Table 5.1.

An alternative to this approach is to compare the ultimate output – drugs launched to the market – to retrospective R&D expenditures. This approach shows that major pharmaceutical companies have been expanding their R&D budgets recently, with the cumulative expenditure over a decade ranging from USD 10 to 70+ billion per company among the top spenders. At the same time the number of FDA approvals gained per company during this period remained largely in single digits. The global top performer, Johnson & Johnson, brought 13 medicines to the market in the decade ending in 2012. Therefore a top-down approach of dividing the output by R&D dollars invested results in higher estimated average costs than alternative bottom-up calculation methodologies. It also shows substantially higher ranges for the cohort of traditional large pharmaceutical companies. Those firms that had developed more than four new drugs bore staggering median costs in excess of USD 5 billion (Herper, 2013).[3] Yet at the other end of distribution biopharmaceutical SMEs brought new medicines to the market at a small fraction of that expenditure – as low as several dozen USD million, though this calculation is based exclusively on success stories that lived up to market approval. Many more have failed, and those need to be accounted for in order to obtain accurate industry expenditure ranges. If not arriving at the same number, all studies invariably point to the same skewed distribution of incurred R&D costs, which evidences that the business of drug development is one of high-risk/high-reward. Blockbuster drugs are to compensate the costs of many failures.

3) The numbers are little affected by introducing a lag, which controls for the time difference between making the investment and the eventual market approval. EvaluatePharma, for instance, uses a 3-year lag: the R&D expense in 2010 divided by the number of NMEs in 2013, resulting in USD 3.7 billion, compared to USD 3.9 billion when R&D expense in 2013 is used.

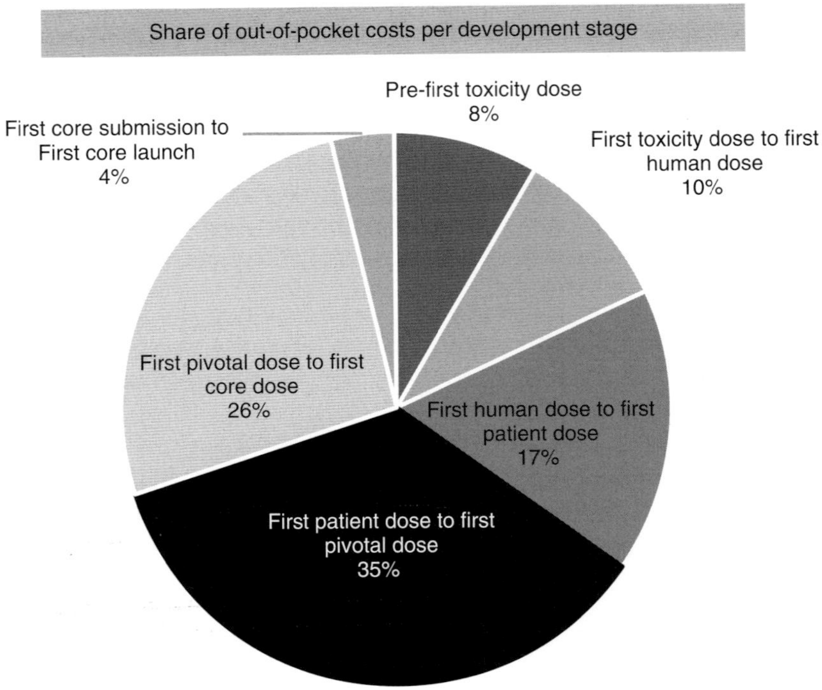


**Figure 5.2** Breakdown of out-of-pocket (before accounting for the cost of capital) costs by drug development stage. The milestone "first toxicity dose" is reached when the first dose is given in the first animal toxicity study required to support administration to a human. The "first patient dose" milestone is reached when the active substance for the relevant project is administered to patients for a specific indication with the intention of treating for that indication. The date for the "first pivotal dose" is the date when the first dose is given to the first patient in the first pivotal safety and efficacy trial – the large-scale clinical study necessary to support registration in one of the core markets. Source: Mestre-Ferrandiz *et al.* (2012) and Adams and Brantner (2010).

The final question "where do the R&D investments go?" is answered with the help of Figure 5.2, which is based on the dataset of the Office of Health Economics. Later stages of drug development, when the drug candidates are administered to patients, are the costliest. This is the period of a safety and efficacy trial and a subsequent large-scale clinical study on a wider patient population – called *phase II* and *phase III studies,* respectively. Naturally, a project failure further in development corresponds to larger losses. Accordingly, the largest share of the pie chart, from the first patient dose to the first pivotal safety and efficacy trial, is explained by the fact that the failure rate at this stage is still very high (31%), and so out of three projects that receive funding for this stage, only one can be expected to be a success. The risk – and therefore the cost – of failure accumulates during the development process, resulting in unyielding statistics for the pharmaceutical business model: only about 7% of compounds entering initial toxicity studies in

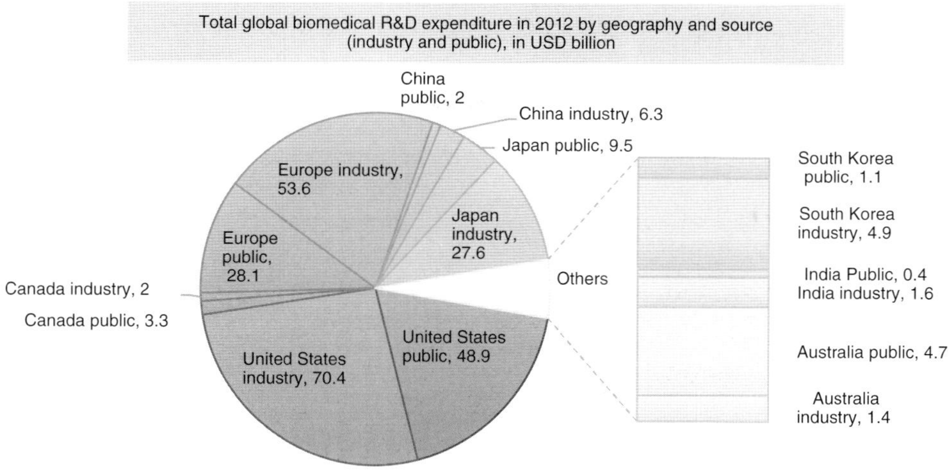


**Figure 5.3** Global public and private biomedical R&D expenditure. Additionally there were USD 2.4 billion expenditures coming from other Asia–Pacific countries. These are not included in the treemap as they cannot be allocated to the individual countries in the region. Since this contribution is less than 1% of the total global R&D expenditure, the effect of this omission on the global overview is immaterial.Based on the data from Chakma *et al.* (2014).

animals will eventually transition into the market (Mestre-Ferrandiz *et al.*, 2012), and even for those that do, many will not recover R&D expenses.

## 5.5 Funding Drug Development: a Global Endeavor

So where do we find the primary financiers of the expensive and failure-prone drug development enterprise? In 2012 the world on a whole spent close to USD 270 billion of private and public funds for developing new medicines. This is almost exactly the same amount that was spent in 2012 on buying cars of a single brand: Volkswagen (to be precise, Volkswagen's 2012 revenue was USD 254 billion). As the advanced economies are transitioning from their industrial past into knowledge-based bioeconomies, pharmaceutical industry gains significance in the GDP composition. The progress on this road, however, is not even among the developed countries, while emerging economies are only starting to catch up from very low levels. Figure 5.3 dissects the global expenditure into public and private contributions in the main R&D locations, while Figure 5.4 puts these expenditures in relation to the size of the domestic economy. In the case of Europe, it is the sum of EU's and EFTA's[4] GDPs.

4) European Free Trade Association, of which Iceland, Liechtenstein, Norway, and Switzerland are members.

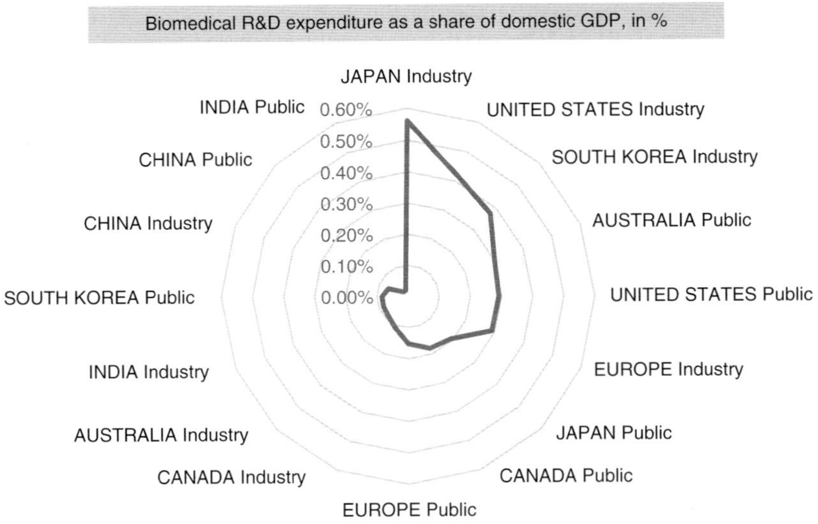


**Figure 5.4** Size of the public and private biomedical R&D expenditures relative to the domestic GDP. Author's calculations based on the data from Chakma *et al.* (2014).

The Japanese pharmaceutical industry collectively spent an equivalent of 0.6% of the Japanese GDP on R&D. In proportion to the economy's size, this is the highest share worldwide. It can be speculated that the Japanese industry invests heavily into the future healthcare in response to domestic demographic trends: Japan scores highest in the average life expectancy and median age globally. The US pharmaceutical industry provides the largest single source of funding volume – USD 70.4 billion, or about a quarter of total global funding. Nevertheless, the US industry lags behind Japan when adjusted by GDP. China's share is small: Chinese overall biomedical R&D investment is less than a quarter of Japanese, while the economy on the whole is twice the size. China, however, has witnessed the sharpest increase in R&D spending (public and private) from 2007 to 2012 – 313%. It still remains to be seen how efficiently the recently invested capital will translate into new treatments coming from Chinese developers. Chinese growth, however, provides a sharp contrast to decreasing private spending in the United States and Europe, where the decline during the same 2007–2012 period has been 15 and 4% correspondingly.

According to the Global Forum for Health Research[5] in 2005, only two countries – Iceland and Switzerland – spent in total (both industry and state) just over 1% of their GDP on investments into health R&D. Sweden, Denmark, and the United States followed with over 0.6%; the rest of Europe and Japan fell within 0.2–0.5% range; large developing countries such as China, Brazil, and Russia all lagged substantially at below 0.1% level (Landriault and Matlin, 2009). Figure 5.5 presents the data from an alternative dataset of the Organisation

5) It merged in 2010 with Geneva-based NGO Council on Health Research for Development.

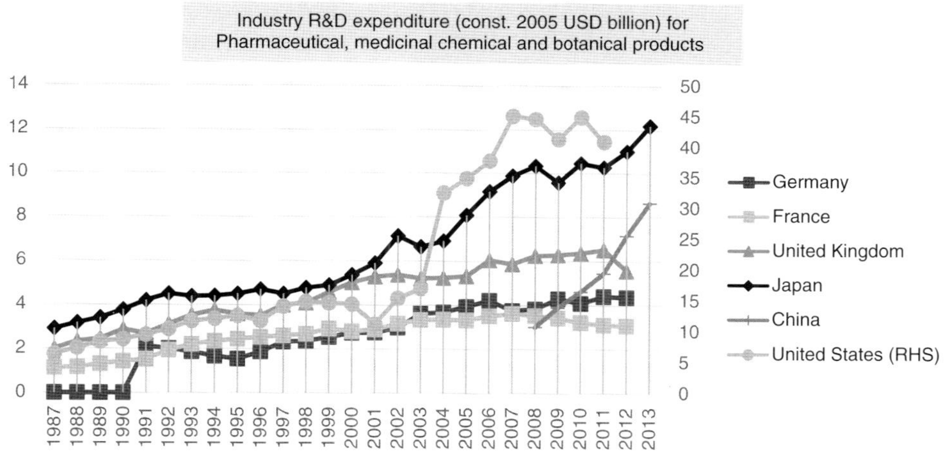


**Figure 5.5** Private sector expenditures in biomedical R&D during 1987–2013 for selected countries. The time series for the United States uses a secondary scale axis on the right hand side. Source: OECD ANalytical Business Enterprise Research and Development (ANBERD) database.

for Economic Co-operation and Development (OECD) in order to track the development of private sector R&D expenditures. The data show a sharp increase in the United States in the period from 2001 until the global financial crisis of 2007–2008, after which they plateaued. In Japan private investments in health R&D grew steadily before and after the crisis, effectively doubling in the past decade, while continental Europe (Germany and Switzerland) exhibited smaller increases in the corresponding period. The time series for China started much later in 2008, but the rate of increase mirrored the previously mentioned statistics that placed China on the top place globally in terms of R&D spending growth. Box 5.2 in turn shows the geographical breakdown for the outcomes of biomedical R&D, measured by the number of new *innovative* medicines originated.

**Box 5.2: Outcomes of biomedical R&D by geography**

Outcomes of R&D are even more geographically uneven than expenditures. Kneller (2010) traced the discovering institutions of nearly all drugs approved by the FDA in the decade from 1998 to 2007. Among them were 135 novel medicines: 98 pNMEs and 37 new therapeutic biologics (NTBs). Nearly two-thirds of these originated in the United States. Among the biologics the rate is in excess of 80%. This positions the United States firmly at the vanguard of pharmaceutical innovation. In the past decade the country had sweepingly assumed majority market share of the global pharma business. Back in the 1960s nearly one-third of all drugs were invented by pharmaceutical firms headquartered in Germany (Daemmrich, 2009). By now this share has dropped into the low single digits and, as Figure 5.6 demonstrates, is dwarfed by that of the United States.

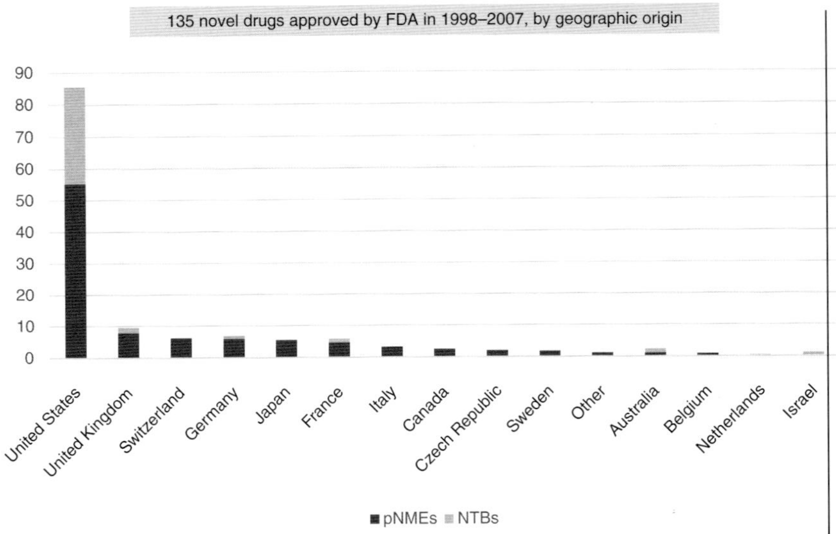


**Figure 5.6** Geographic origins of novel drugs. Based on the data from Kneller (2010).

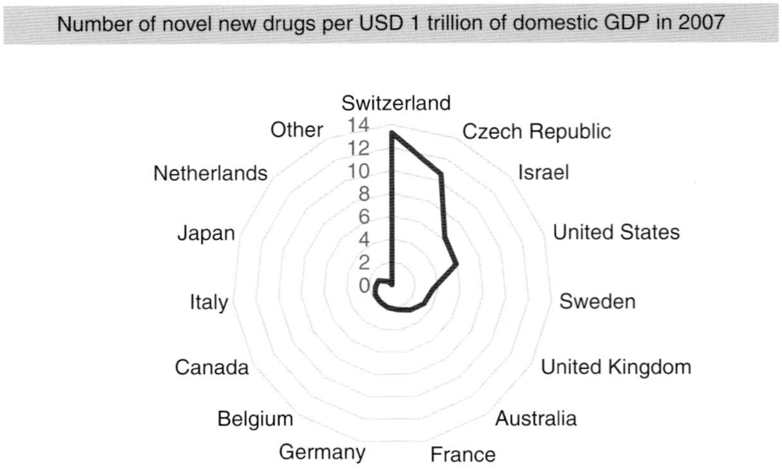


**Figure 5.7** Efficiencies of domestic pharmaceutical industries. Author's calculations based on the data from Kneller (2010).

The European performance, on the other hand, is uneven from country to country. Figure 5.7 shows the output relative to country's GDP. By this metric, Switzerland, a small country with about USD 0.5 trillion of GDP in

2007, originated 6.25 priority-reviewed NMEs.[6] This equals to about 12.5 medicines per USD 1 trillion of GDP; the corresponding figure in the United States is 3.8. Notably all of Switzerland's novel medicines were small molecule drugs, while a sizeable share of the US output came in the form of therapeutic biologics. It is equally conspicuous how the global landscape of pharmaceutical R&D lies entirely within the borders of the most advanced of developed nations – "innovation-driven" economies, in terms of the World Economic Forum (Schwab, 2014). While many parts of the drug development process chain have been, and continue to be, outsourced into cheaper emerging markets, R&D activities, which are innately intensive on human capital, are highly concentrated in the major innovation hubs: California and Massachusetts in the United States; Germany, France, the United Kingdom, and Switzerland in Europe; and Japan.

## 5.6 Public and Private Funds: Complementary Finance for Drug Development

Despite the popular image of economic policies across the Atlantic being labeled as pro-market in the United States and semi-socialist in Europe, it is perhaps surprising that the pharmaceutical sector evidences nothing of this sort. In fact, according to various estimates, the United States dedicates at least two times as much public funding for health research as European Union (Organisation for Economic Co-operation and Development, 2009; Berghmans *et al.*, 2011; Chakma *et al.*, 2014). Nevertheless, the budgets of the US National Institutes of Health, which is the largest public US institution financing biomedical R&D, have declined by more than 20% in real dollars since 2003 (Johnson, 2013). In Europe, although the total volume of public funding for biomedical R&D is still half that of the United States, there has been substantial growth. As the chart in Figure 5.8 shows, from 1998 to 2005, European public spending doubled in nominal terms: from about USD 10 billion in 2001 to just over USD 20 billion in 2005. However, not all European countries contributed equally to this increase: the United Kingdom, in particular, has been the major driving force behind it. Summing up across countries, a significant share – 35–40% of the total biomedical R&D globally – comes from public funds. This compares to about 55–60% of investments contributed by the industry globally and about 10% by private nonprofit organizations. GDP-adjusted global leaders of public R&D investment are the Nordics: Denmark, Sweden, and Norway are spending an equivalent of 0.25% of these countries' GDP on health R&D investments (Røttingen *et al.*, 2013).

6) The number is not exact because some of the drugs are a result of cross-border R&D efforts. In such cases each country is assigned a share of contribution.

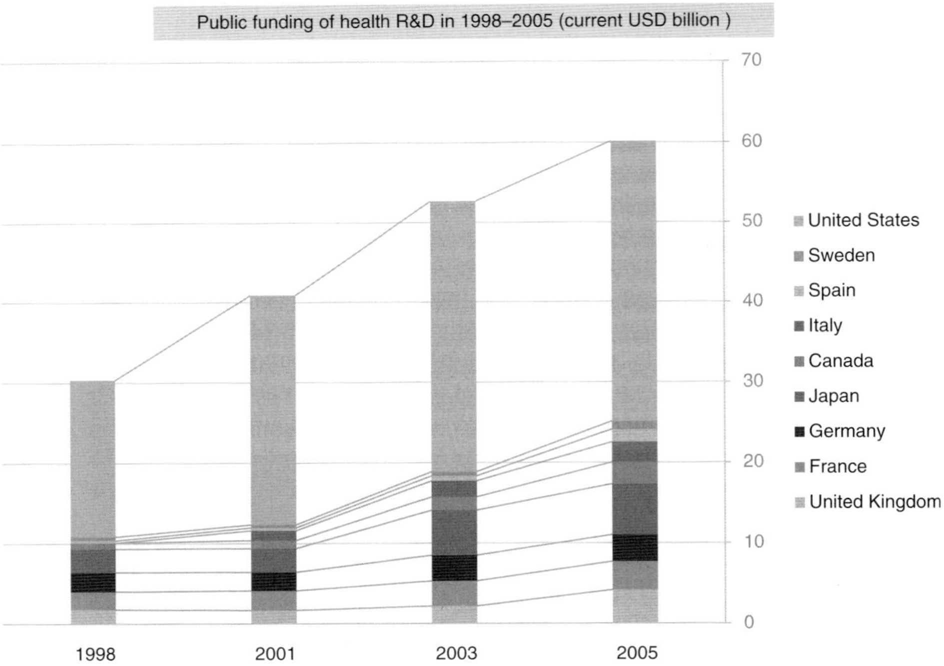


**Figure 5.8** Public funding of health R&D in 1998–2005, selected countries. Source: Burke and Matlin, Global Forum for Health Research (2008).

State funding comes via multiple avenues, the most straightforward of which are public research institutions and universities. Together with that, the state can act as an active investor beyond the funding of basic science, too. For instance, according to the EY's Global Life Sciences Center transaction-by-transaction data on VC funding for drug development in Germany, state VC is a major financier in this country. Out of the active VC funds, which had participated in more than two deals from 2011 to 2013, half of the deals can be attributed to state funds. Such funds include High-Tech Gründerfonds (HTGF – a private–public partnership, which is 87% funded by the German state and German state-owned development bank KfW), funds of KfW itself or regional Investitionsbank Berlin. The private VC market is much deeper in the United States. The total VC funds raised and invested in the biomedical industry there are almost three times higher than in Europe. Consequently the state funding in Europe has gained higher significance within the overall VC landscape. The European Private Equity and Venture Capital Association estimates that government agencies refinanced 38% of the entire European VC industry, which, in turn, allocated 23% of its investments into life sciences[7] companies (EVCA Research, 2014). Box 5.3 explores the link between public investment and innovation.

7) Funding for drug development is further diluted with investments into less-regulated biotechnology sectors of devices and research tools, software, and diagnostics.

**Box 5.3: Importance of public biomedical R&D investments for innovation and strength of domestic pharmaceutical industry**

Is public biomedical R&D investment paying off? For all priority-reviewed new drugs approved by the FDA in 1988–2005, 65% of patents cited either a public-sector patent or a government publication (Sampat and Lichtenberg, 2011), indicating the effectiveness of direct and indirect public funding support. For standard-review drugs, the percentage is much lower (36%), suggesting that public sector support was more pronounced for innovative drugs. Larger public spending is also associated with stronger domestic (bio)pharmaceutical industry. For instance, in 2005 the United Kingdom ranked third in the world in terms of public R&D expenditure, behind only by the United States and Japan. The United Kingdom is both the second largest jurisdiction for NME development worldwide and the second largest country of origin for priority-reviewed drugs and biologics.

## 5.7 How Commercial Drug Development Projects Are Financed Today: Big Firms, Small Firms, and Their Cooperation

In 2014 traditional large pharmaceutical companies brought 18 new drugs to the market (a share of 44%) – in line with its performance since the 2010s (Munos, 2015). British GlaxoSmithKline PLC (GSK) – a representative, large pharmaceutical firm with over USD 6 billion in R&D expenditure in 2010 – had a total of 147 projects across 13 therapeutic areas spread across different stages of development. It had 34 projects in phase I, 56 projects in phase II, 36 projects in phase III, 10 projects under application for approval, and 11 projects approved for launch (Ding *et al.*, 2014). As of 2013 the company's USD 27 billion capital structure includes almost USD 17 billion of net debt. GSK's access to debt financing rests on its ability to generate strong and consistent cash flows – USD 6.3 billion in 2013. In the absence of such cashflows, most of the smaller drug developers are unable to raise capital via bond issuance and therefore rely on higher-cost financing, such as VC.

VC financing in the 2006–2013 period changed from USD 3.9 to USD 4.3 billion in the United States and from USD 2.1 to USD 1.5 billion in Europe EY (2015). For comparison, Roche – the biggest player in biotechnology space – spent USD 9.8 billion on R&D in 2013 alone (Roche, 2014). It is true, of course, that Roche's figures include expenditures for phase III clinical trials, which are the most capital-intensive link in the drug development chain. Indeed, so costly these are that biotechs rarely undertake them internally, preferring instead to pass them on to a big pharmaceutical firm in one way or another. For large companies, therefore, this is an important source for replenishing own pipelines. Based on the statistics in the EvaluatePharma database, McKinsey reports that as of 2010, 62% of pipeline compounds in the large pharmaceutical companies originated externally. This share grew from just 38% in 2002. On the other hand, externally sourced phase I compounds have not exhibited such strong growth during the same period: the change was from 34% to 38%, indicating that big pharmaceutical firms focus on sourcing de-risked late-stage opportunities (McKinsey and Company, 2012). Box 5.4 highlights the growing importance of smaller firms for pharmaceutical innovation.

It is therefore a distinct feature of drug development business that in light of an exceptionally long R&D effort, which can span one to two decades, the owner–financier of the pipeline asset can change several times. The compound can be, for instance, discovered by a university, transferred to a biotechnology company, which is funded by a VC, and then sold to a large pharmaceutical company after reaching a key R&D milestone. As described above, a common financing strategy for smaller companies is to partner with larger ones, particularly for the costly late stages of drug development process. Still before phase III trials can commence, there are on average 8 years of development (Mestre-Ferrandiz *et al.*, 2012), and the further it is from an exit opportunity, the scarcer the funding becomes. Accessing finance is particularly challenging during the translational phase – a stage between basic research and clinical development (de Man *et al.*, 2013). As biomedical SMEs do not have access to capital markets, they face a complicated financing chain in the beginning of their R&D efforts. This may involve private funds, university funds, VC, state VC and grants, alliances, private–public partnerships, corporate VC, and acquisitions at various stages. Initial public offerings (IPOs) usually become a financing alternative only in the later stages of development. Though it sometimes happens, it is in general uncommon for a company with only a preclinical or phase I asset to raise capital via shares issuance. More often, this happens at phase II or phase III. In the United States 52% of the biotech capital raised via IPOs in 2013 was allocated to firms with phase II assets (EY, 2015). Post-IPO finance in the equity markets is more abundant. By this stage the pipeline assets have been substantially de-risked and are much closer to commercialization. Of course, some compounds will not pass the last leg of clinical trials successfully, yet the overall risk profile on the industry level is suitable for capital allocations from a wider investor base. Moreover, the recent market conditions have been very favorable for drug developers. The NASDAQ Biotechnology Index containing 150 biotechnology or pharmaceutical firms had reached its all time high on 20 March 2015. During 5 years ending in March 2015, the index gained 290%, strongly outperforming the broader NASDAQ Composite Index.

**Box 5.4: Importance of smaller firms for pharmaceutical innovation**

Access to finance for smaller firms, however, is vastly important for the performance of the entire industry, for these firms are major contributors to innovation in the sector. In the United States, Germany, France, Canada, and Australia, one-half to two-thirds of scientifically novel medicines originate outside of traditional large pharmaceutical firms. Instead they came from small pharmaceutical companies or biotechnology companies or were university-discovered and then transferred to biotechnology companies. This is the result of a reorganization that the pharmaceutical industry had undergone in the past decades. From the 1950s to the 1980s, large pharmaceutical companies originated about 75% of all NMEs. Starting from the 1980s, this share gradually declined to about 35%. The rest of the market share had been assumed by smaller companies. These smaller companies grew in their total number, doubling from 78 to 145 during the 1980s and 1990s, and also became more productive. The mean annual NME output of small companies has increased from ~0.04 to ~0.12 since 1995 (Munos, 2009). Interestingly, some important locations for pharmaceutical R&D can be outliers to this size transition trend. In Switzerland and the United Kingdom, most of the recent scientifically novel drugs originated from large pharmaceutical firms (Kneller, 2010). In the biologics segment the dominance of smaller firms is even stronger: in the United States only about 10% of all therapeutic biologics came from large pharmaceutical companies.

## 5.8 Public Health Economics and Financing Pharmaceutical Innovation

It is instructive to consider the pharmaceutical industry and its finance in the wider context of public health and economic burden of disease. It is estimated, for instance, that cancer alone costs European Union €126 billion per year, while only 27% of this expenditure attributed to medicines (Luengo-Fernandez *et al.*, 2013). The main cost contribution by far – 56% – came from inpatient care. In general, the burden of disease calculations is based on difficult-to-value concepts such as cost of life, cost of additional years gained, or informal costs of care provided by family and friends of a patient. Alternatively, instead of quantifying this into euros or dollars, we can turn to a more objective statistic of years gained due to pharmaceutical innovation. Using the data from IMS Health, World Health Organization (WHO), World Bank, and OECD, Lichtenberg (2014) estimated that 73% of the increase in life expectancy from 2000 to 2009 can be explained by better medicines (1.27 years out of the total 1.74 years increase during the period) (Lichtenberg, 2014).

The value of pharmaceutical innovation to society is incontrovertible, so an optimal capital allocation so as to increase the efficiency of the industry's output is a pressing challenge for the modern economy. Today the status quo is that most of the drug development is equity financed. For comparison, the ratio of debt in the whole capital structure ranges from about 4% in the Internet software industry to 50% in the power industry; drug development ranks at the lower end – 11% for pharmaceuticals and 7% for biotechnology (Damodaran, 2015). This is important because the cost of debt is about half that of equity: 5% versus 9.5% annually. Long investment horizons of drug development project turn the time costs into a significant component of total R&D expenditure. Indeed, USD 2.6 billion estimate of the Tufts Center for the Study of Drug Development is further decomposed into $1.4

billion in out-of-pocket cost and USD 1.2 billion in cost of capital. These are the expected returns that investors forego elsewhere while a drug is in development.

Naturally, the prevalence of equity finance reflects the risk associated with the business of developing new medicines. 7% chance of a preclinical compound to become a marketed drug, clearly, is not an attractive risk/reward profile for a bank or an institutional investor. Using financial engineering to transform this risk/reward profile was proposed by Fernandez *et al.* (2012). They have modeled a megafund that can invest into 40–200 compounds and refinance itself partly via various debt securities. However, further simulations for drug candidates to treat Alzheimer's disease have shown that such a project is unlikely to take off without government's participation to de-risk the entire scheme. This can be done, for instance, by providing guarantees or extending patent life of future drugs legislatively (Lo *et al.*, 2014). Likewise, Mazzucato (2011) argues for a more proactive role for the government in providing patient long-term capital for innovation, drawing parallels from the computer industry, where private electronics firms have moved into the segment only after the key technologies, such as Internet, World Wide Web, GPS, and touch screen displays, had been developed – all directly funded by governments (Mazzucato, 2011). Alternatively, Hensley and LaMattina (2015) propose to extend patent life of prescribed drugs for a year and channel the proceeds to support federally and industrially driven R&D efforts (Hensley and LaMattina, 2015). Amanda Christini, on the other hand, advocates for universities to become direct funders for research commercialization by setting up university venture funds (Christini, 2012). Indeed, there are already several examples of these, such as Imperial Innovations of Imperial College London and Karolinska Development initiated by Karolinska Institutet.

On the private industry side, big pharmaceutical companies have established corporate VC units in order to directly fund the development of earlier-stage external compounds. Such funds already provide a fifth of all VC financing (von Krogh, 2012). Together with that the number of big players in the industry is shrinking in a continuing consolidation wave. At the moment the top 10 pharmaceutical firms control more than two thirds of all NMEs in the market (Kinch *et al.*, 2014). This eventually limits the pool of potential corporate backers that smaller firms can turn to. Finally, there is an entire segment of preclinical assets, which are of no interest to private industry: nonpatentable compounds. These are either compounds based on information already released to public domain or existing medicines with expired patents, which can be tried for alternative indications. In fortunate cases, the development of such assets goes on with alternative funding. For instance, Cancer Research UK, a British charity, is funding a clinical trial of aspirin use after standard cancer therapy to improve recurrence and to improve survival of patients with solid tumors (Langley *et al.*, 2014). In Sweden, Uppsala University is funding a clinical trial for neuroendocrine tumors from a £2 billion Oncolytic Virus Fund, 1/10th of which was collected using a crowdfunding platform and the rest from a single wealthy individual (Uppsala University, 2015); but in other cases, nonpatentable academic research is put on hold due to the lack of financing.

Development of drugs and vaccines for tropical diseases that disproportionately affect populations in lower-income countries is likewise difficult to monetize. In this field charities and private foundations become important economic actors. The 2013 grant support of the world's largest charity – Bill & Melinda Gates Foundation – stands at USD 3.64 billion (Bill & Melinda Gates Foundation, 2014). This almost matches the size of the USD 3.96 billion biennial budget of the WHO for 2012–2013. It is true, of course, that most of the funding from both organizations is directed toward global health issues, with only a smaller share allocated to research for drug development. Further still, according to the WHO Programme Budget, the direct funding by the UNICEF/UNDP/World Bank/WHO Special Programme for Research and Training in Tropical Diseases for drug discovery for tropical diseases is to be wound down due to fiscal austerity in member countries and promising project leads transitioned to other partners (World Health Organization, 2011). On this background, funding from private nonprofits gains particular significance for R&D in the field of neglected and tropical diseases.

## 5.9 Conclusion

Compared to other forms of economic activity, drug development is an exceptionally lengthy and risky enterprise. Until recently it has been undertaken almost exclusively by the most advanced knowledge-driven economies that have access to human and financial capital commensurate with the task. More recently, emerging Asian economies, particularly China, have been rapidly increasing their biomedical R&D investment in a bid to join the global pharmaceutical industry. This happens on the background of slowing industry spending in the major traditional locations – the United States and Europe. There strong investment growth by commercial drug developers has not resulted in proportionate increase of the output, signaling a reduced R&D efficiency on an industry scale.

At the same time finance has been slowly adapting to a new innovation environment in the field, where most of the new medicines are discovered and originated outside of large pharmaceutical companies. Big firms continue to refinance themselves in equity and debt capital markets. The difference is that these firms are now responsible for originating about one-third of new medicines, not three quarters, as it had been the case in the 1980s. Financial technology has responded to this in a multitude of solutions designed to better serve smaller pharmaceutical and biotechnology companies, which are now the major sources of innovation. Corporate, state, and university-led venture funds have grown in importance, as well as different forms of partnering with large traditional pharmaceutical firms. Still, access to finance is fragmented: it is abundant in the capital markets for publicly listed firms and scarce for early-stage developers without consistent cashflows.

Naturally, this reflects the degree of risk aversion of the major investment base and the mismatch between the need for patient long-term capital on the developer's side and short-term investment horizons of capital providers. Financing shortage becomes most dire in the translational phase – a stage between basic research and clinical development. To counteract this problem, several national and supranational initiatives have been established, such as the National Center for Advancing Translational Sciences at the National Institutes of Health (NIH-NCATS, United States), Centre for Drug Research and Development (CDRD, Canada), and European Infrastructure for Translational Medicine (EATRIS) (European Union). The public sector in general has been and continues to be a major complementary funding source for drug development activities, with a sizeable share of about a third of the total global R&D expenditure. In contrast to centrally planned innovation management, modern digital infrastructure has allowed funding schemes as exotic as crowdfunding to enter into the pharmaceutical finance scene. While these remain insignificant, it is yet another manifestation of financial technology experimenting and evolving so as to better address the challenge of developing new medicines.

#### Acknowledgment

The author thanks Dr Ralf Kemkemer of the Max Planck Institute for Intelligent Systems for his advice and critical reading of the manuscript.

#### References

Adams, A. and Brantner, V. (2010) Spending on new drug development. *Health Econ.*, **19** (2), 130–141.

Berghmans, S. *et al.* (2011) A stronger biomedical research for a better European future. European Medical Research Councils (EMRC) White Paper II.

Bill & Melinda Gates Foundation (2014) Annual Report 2013, Bill & Melinda Gates Foundation, Seattle, WA.

Branch, S. and Agranat, I. (2014) "New drug" designations for new therapeutic entities: new active substance, new chemical entity, new biological entity, new molecular entity. *J. Med. Chem.*, **57** (21), 8729–8765.

Bujar, M. and McAuslane, N. (2014) New Drug Approvals in ICH Countries 2004–2013, Centre for Innovation in Regulatory Science R&D Briefing 54.

Burke, M. and Matlin, S. (eds) (2008) *Monitoring Financial Flows for Health Research*, Global Forum for Health Research, Geneva.

Chakma, J. *et al.* (2014) Asia's ascent — global trends in biomedical R&D expenditures. *N. Engl. J. Med.*, **370**, 3–6.

Christini, A. (2012) Why universities should step up in venture investing. *Nat. Biotechnol.*, **30**, 933–936.

Daemmrich, A. (2009) Where is the pharmacy to the world? International regulatory variation and pharmaceutical industry location. Harvard Business School working paper 09–118.

Damodaran, A. (2015) Costs of Capital by Industry Sector, http://pages.stern.nyu.edu/˜adamodar/New_Home_Page/datacurrent.html (accessed 25 March 2015).

Ding, M. *et al.* (2014) Portfolio management in new drug development, Chapter 3, in *Innovation and Marketing in the Pharmaceutical Industry*, Springer, New York.

Eder, J. *et al.* (2014) The discovery of first-in-class drugs: origins and evolution. *Nat. Rev. Drug Discovery*, **13**, 577–587.

EMA (2014) Annual Report 2013, European Medicines Agency, London.

EvaluatePharma (2014) *World Preview 2014, Outlook to 2020*, Evaluate Ltd, London.

EVCA Research (2014) *European Private Equity Activity Data 2007–2013*, European Private Equity and Venture Capital Association.

EY (2015) Biotechnology Industry Report 2014: Beyond Borders: Unlocking Value, Ernst & Young Global Limited.

Fernandez, J.-M. *et al.* (2012) Commercializing biomedical research through securitization techniques. *Nat. Biotechnol.*, **30**, 964–975.

Grabowski, H. *et al.* (2002) Returns on research and development for 1990s new drug introductions. *Pharmacoeconomics*, **20** (Suppl. 3), 11–29.

Graybeal, C. (2012) Collins: how can we do better? *The NIH Catal.*, **20** (3), 14.

Hensley, P. and LaMattina, J. (2015) A plan for sustainable funding for US biomedical research. *Nat. Rev. Drug Discovery*, **14**, 147–149.

Herper, M. (2013) How Much Does Pharmaceutical Innovation Cost? A Look At 100 Companies. Forbes, http://www.forbes.com/sites/matthewherper/2013/08/11/the-cost-of-inventing-a-new-drug-98-companies-ranked/ (accessed 25 March 2015).

Johnson, J. (2013) Brief History of NIH Funding: Fact Sheet. Congressional Research Service Report R43341, Congressional Research Service.

Kinch, M. *et al.* (2014) An overview of FDA-approved new molecular entities: 1827–2013. *Drug Discovery Today*, **19**, 1033–1039.

Kneller, R. (2010) The importance of new companies for drug discovery: origins of a decade of new drugs. *Nat. Rev. Drug Discovery*, **9**, 867–882.

von Krogh, G. (2012) The Changing Face of Corporate Venturing in Biotechnology. Bioentrepreneur, http://www.nature.com/bioent/2012/121001/full/bioe.2012.9.html#B1 (accessed 25 March 2015).

Landriault, E. and Matlin, S. (eds) (2009) *Monitoring Financial Flows for Health Research*, Global Forum for Health Research, Geneva.

Langley *et al.* (2014) *Add-Aspirin trial: A phase III, double blind, placebo-controlled, randomized trial assessing the effects of aspirin on disease recurrence and survival after primary therapy in common nonmetastatic solid tumors, Journal of Clinical Oncology*, 2014 ASCO Annual Meeting Abstracts, **32**, No 15_suppl (May 20 Supplement)

Lichtenberg, F. (2014) Pharmaceutical innovation and longevity growth in 30 developing and high-income countries, 2000–2009. *Health Policy Technol.*, **3** (1), 36–58.

Lo, A. *et al.* (2014) Parallel discovery of Alzheimer's therapeutic. *Sci. Transl. Med.*, **6** (241), 241cm5.

Luengo-Fernandez, R. *et al.* (2013) Economic burden of cancer across the European Union: a population-based cost analysis. *Lancet Oncol.*, **14** (2), 1165–1174.

van Luijn, J. *et al.* (2010) Superior efficacy of new medicines? *Eur. J. Clin. Pharmacol. Spec. Artic.*, **66**, 445–448.

de Man, F. *et al.* (2013) The current translational research landscape, in *Inaugural EATRIS Conference Highlights Report 2013*, EATRIS (European Infrastructure for Translational Medicine), Amsterdam.

Mazzucato, M. (2011) *The Entrepreneurial State*, Demos, London.

McKinsey & Company (2012) Evolution or Revolution. McKinsey Perspectives on Drug and Device R&D 2012, McKinsey & Company Pharmaceutical and Medical Products R&D Practice.

Mestre-Ferrandiz, J. *et al.* (2012) *R&D Cost of a New Medicine*, Office of Health Economics, London.

Munos, B. (2009) Lessons from 60 years of pharmaceutical innovation. *Nat. Rev. Drug Discovery*, **8**, 959–968.

Munos, B. (2015) 2014 New Drug Approvals Hit 18-Year High. Forbes, http://www.forbes.com/sites/bernardmunos/2015/01/02/the-fda-approvals-of-2014/ (accessed 25 March 2015).

Organisation for Economic Co-operation and Development (2009) *OECD Science, Technology and Industry Scoreboard 2009*, Organisation for Economic Co-operation and Development, Paris.

Patented Medicine Prices Review Board (2009) Annual Report 2009, The Patented Medicine Prices Review Board, Ottawa.
Patented Medicine Prices Review Board (2013) Annual Report 2013, The Patented Medicine Prices Review Board, Ottawa.
Roche (2014) *Annual Report 2013*, F. Hoffmann-La Roche Ltd., Basel.
Røttingen, J.-A. *et al.* (2013) Mapping of available health research and development data: what's there, what's missing, and what role is there for a global observatory? *Lancet*, **382** (9900), 1286–1307.
Sampat, B. and Lichtenberg, F. (2011) What are the respective roles of the public and private sectors in pharmaceutical innovation? *Health Aff.*, **30** (2), 332–339.
Scannell, J. *et al.* (2012) Diagnosing the decline in pharmaceutical R&D efficiency. *Nat. Rev. Drug Discovery*, **11**, 191–200.
Schwab, K. (ed) (2014) *The Global Competitiveness Report 2014–2015*, World Economic Forum, Geneva.
Tufts Center for the Study of Drug Development (2014) Cost to develop and win marketing approval for a new drug is $2.6 million. Press release, http://csdd.tufts.edu/news/complete_story/pr_tufts_csdd_2014_cost_study (accessed 25 March 2015)
Uppsala University (2015) The Oncolytic Virus Fund, http://www.uu.se/en/support/oncolytic/ (accessed 25 March 2015).
U.S. Food and Drug Administration (2011–2015) *Novel New Drugs Summary*, U.S. Food and Drug Administration, Silver Spring, MD.
U.S. Food and Drug Administration (2013) Summary of NDA Approvals & Receipts, 1938 to the Present, http://www.fda.gov/AboutFDA/WhatWeDo/History/ProductRegulation/SummaryofNDAApprovalsReceipts1938tothepresent/default.htm (accessed 25 March 2015).
U.S. Food and Drug Administration (2014–2015) *Novel New Drugs Summary*, U.S. Food and Drug Administration, Silver Spring.
U.S. Food and Drug Administration (2015) *Novel New Drugs 2014 Summary*, U.S. Food and Drug Administration, Silver Spring.
Vernon, A. *et al.* (2010) Drug development costs when financial risk is measured using the Fama-French three-factor model. *Health Econ.*, **19** (8), 1002–1005.
Volkswagen (2014) *Annual Report 2013*, Volkswagen Group, Wolfsburg.
Ward, D. (2013) Decline in new drug launches: myth or reality? Retrospective observational study using 30 years of data from the UK. *BMJ Open*, **3** (2), e002088.
World Health Organization (2011) *Programme Budget 2009*, World Health Organization, Geneva.

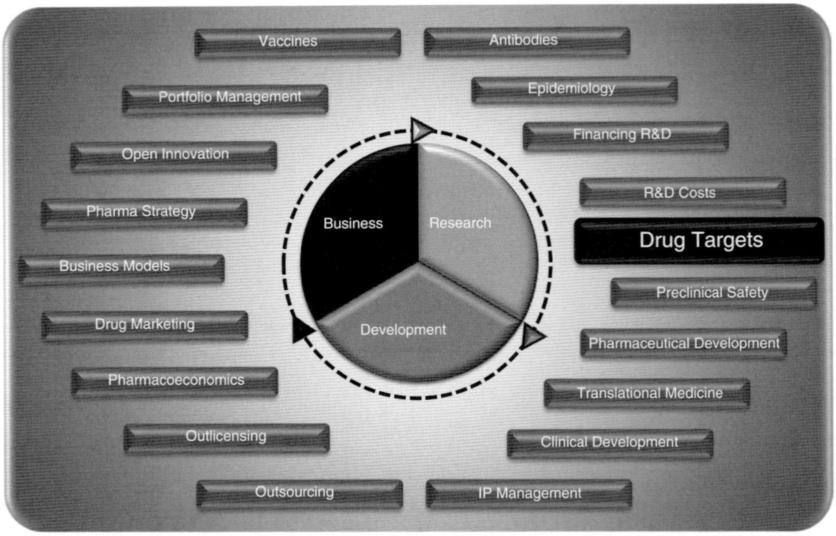
Vaccines
Antibodies
Portfolio Management
Epidemiology
Financing R&D
Open Innovation
R&D Costs
Pharma Strategy
Business
Research
Drug Targets
Business Models
Preclinical Safety
Drug Marketing
Development
Pharmaceutical Development
Pharmacoeconomics
Translational Medicine
Outlicensing
Clinical Development
Outsourcing
IP Management

# 6
# Challenges and Options for Drug Discovery

*Werner Kramer*

## 6.1
## Introduction

The purpose of this book chapter is to carve out the determining factors, circumstances, and parameters to be responsible for and underlying the dramatic changes in pharmaceutical discovery research with regard to its stagnation in medical innovation and decline of productivity. Pharmaceutical innovation and the finding of novel drugs of high medical need and value are becoming more and more difficult due to various challenges:

- Increasing scientific and regulatory requirements and challenges
- Increasing demands regarding safety of novel drugs
- Still a high attrition rate of drug candidates in phase II and phase III clinical trials
- Strongly increasing costs for research and development (R&D costs) to successfully launch a novel drug
- Increasing competition by generics and biosimilars
- Increasing competition between companies for an attractive emerging drug target with a shown proof of principle (PoP) in humans fostering huge investments into a few most promising drug approaches (an actual example is the development of monoclonal antibodies against proprotein convertase subtilisin/kexin type 9 (PCSK9) as a novel very effective way to lower LDL (low-density lipoprotein) cholesterol levels where currently involving only for the three most advanced drug candidates around 70 000 patients in phase II and phase III studies)
- Increasing pressure from public health claiming demonstration of significant improvement compared to existing therapies with a positive impact on disease modification and progression
- Increasing cost pressure of health organizations and healthcare providers
- Long development times combined with a quite short period of patent protection in order to regain the R&D expenditure costs and to achieve commercial success
- Public concerns and mistrust regarding the integrity and transparency of the pharmaceutical industry

*Value Creation in the Pharmaceutical Industry: The Critical Path to Innovation*, First Edition.
Edited by Alexander Schuhmacher, Markus Hinder, and Oliver Gassmann.

It is evident that the drug discovery and development processes being implemented and practiced to this day have to be ameliorated and adapted to the fundamentally changed landscape of the pharmaceutical industry and healthcare systems having occurred during the last 10–15 years in order to overcome the productivity and innovation challenges for biomedical progress. The following questions are explored in this book chapter:

- What are the major factors and circumstances underlying the productivity issue in pharmaceutical drug discovery?
- Is there an innovation gap in biomedical research?
- What are the reasons for the high failure rate of drug candidates in (early) clinical development?
- What are the implications from the "lessons learnt" to improve innovation and productivity of drug discovery research in the future?

We will start with a short consideration of the mindset changes and organizational consequences of the R&D departments in large pharmaceutical companies, followed then by an in-depth analysis of the various influence factors being relevant for a stringent scientific elucidation of putative promising novel target approaches and their translation into innovative new medicines with a clear and sustained benefit for disease treatments of the patient.

## 6.2
## Paradigm Shifts of R&D Organizations

Overall, the aforementioned and other inherent and environmental influence factors have induced a cultural change in the pharmaceutical industry. The coincidence of the appearance of the production gap innovation gap with the attempts to improve the R&D process by connecting it to market demands and blockbuster mentality and the introduction of management techniques suited to optimize well-known industrial processes in production and manufacturing as well as financial resource planning have turned the scientific R&D process more and more into a mere assessment business (Munos and Chin, 2011). This cultural change resembles a little bit to the attempt inspired by the successful moon landings that it should be possible to eradicate the cancer problem scientifically by just putting enough resources into a well-organized and structured national cancer program. This managerial approach, however, failed due to its noncompatibility to biomedical problems of unknown underlying biology with linear analytical approaches. It is obvious that the pharmaceutical industry has to be a commercially attractive business but the collectivity of the aforementioned factors has led to a certain bias in the drug discovery and development process characterized by a significant degree of risk avoidance regarding novel drug discovery approaches.

Pedro Cuatrecasas – former President of the Pharmaceutical Research Division of Warner–Lambert Company – worked more than 30 years in the pharmaceutical industry. Being involved in the discovery, development, and marketing registration of more than 40 drugs, he considers the decreasing productivity of the pharmaceutical R&D process as the outcome of a primary focus on business and markets rather than on medical science: "The basic scientific discovery and innovation are slightly incompatible with the current organizational structure, management systems and the context of R&D by marketing staff rather than the scientific staff" (Cuatrecasas, 2006). This limits the commitment for mid- to long-term (5–10 years) projects exploring the potential of novel targets and biochemical pathways possibly involved in a certain disease. As one consequence, big pharmaceutical companies bundle and concentrate a significant portion of their R&D investments into a powerful development machinery. As a result, many highly innovative but speculative and therefore risky approaches do not obtain the necessary support to be progressed to the decision point to investigate whether this particular approach could be of clinical benefit for the patient.

Over the last decade, it became obvious that the hitherto practiced R&D models will no longer be able to ensure a continuous flow of novel drugs serving a high medical need combined with a high commercial success. There is no single solution or recipe to overcome these obstacles, but options for improvement of the drug discovery and development process can emerge from two major angles:

1) Organizational changes of the R&D process including but not restricted to the following aspects:
   a. Open innovation
   b. Avoidance of redundant and repetitive research in numerous academic and industry labs
   c. Increasing precompetitive collaborations with academic groups, small and medium-sized enterprises (SMEs), and other pharmaceutical companies despite a strong competition on the commercial side
   d. Develop a network of specialized partners and turning the R&D organizations from a fully integrated pharmaceutical company into a fully integrated pharmaceutical network

For the specific aspects of R&D organization and collaboration models, the reader is referred to excellent reviews regarding this topic which will as well be handled in a different chapter of this book (Cuatrecasas, 2006; Paul *et al.*, 2010; Munos, 2009; Firehouse, 2011; Arrowsmith, 2012). The second pillar focuses on the improvement of the R&D process at the different stages of the value chain from understanding of disease taxonomy and heterogeneity to planning and execution of optimized clinical trials. This pillar will be highlighted in the following paragraphs of this book chapter.

## 6.3 Productivity of Drug Discovery

The productivity of pharmaceutical research measured as the approval of novel drugs remained over the last two decades relatively constant despite enormous breakthroughs in technologies (genetics, omic technologies, high-throughput screening and synthesis methods, big data management, etc.). Between 1993 and end of 2013, 541 new chemical entities (NCEs) have been approved by the US Food and Drug Administration (FDA) with an average annual approval rate of 25–26 NCEs per year. An annual high occurred with 53 drugs in 1996 and an annual low of 15 compounds in 2010; in 2014, 41 NCEs including 11 biologics were approved indicating a 50% increase compared to 27 drugs in 2013 (Mullard, 2015). In parallel the costs for R&D increased exponentially: drug companies have spent about $1.1 trillion in R&D over the last 10 years and around $135 billion are currently spent per year in the pharmaceutical industry to generate 25–30 novel drugs per year (Munos, 2013). In contrast, however, the market value of the current drug pipelines is estimated to be just $293 billion (Munos, 2013, http://www.evaluatepharma.com/worldpreview2018.aspx). To make things even worse, it was found that the commercial success of the new drug launches is often inversely related to the increasing R&D costs: mean peak sales of novel drugs launched decreased from $816 million in 2010 to $466 million in 2013 (Mullard, 2015). It is obvious that this increasing gap between R&D productivity on the one side and exploding costs and decreasing commercial success on the other side has to be overcome.

What are the reasons and obstacles underlying this discrepancy?

Quantum leaps in science and technology have occurred in the last 10–15 years such as:

- Elucidation of the human genome and the genomes of animal species being used in pharmaceutical research
- High-throughput technologies for synthesis and screening of chemical and natural product libraries involving hundreds of thousand or million compounds
- Information technologies to handle huge data sets and high-speed and capacity computer to model complex physiological networks and biochemical relationships

These fundamental improvements in science, molecular understanding of diseases, sophisticated technologies, and increases of R&D budgets did unfortunately not significantly improve the productivity of drug discovery and the translation into innovative medicines for patients. Quite the contrary, by an analysis of the phase II success rates for new molecular entities (NMEs) of 16 leading drug companies spending around 60% of global R&D, it was found that the success rate has fallen from 28% between 2006 and 2007 to meager 18% in the period 2008–2009 with no clear differences between therapeutic areas or between small molecules versus biologics (Arrowsmith, 2011) raising fundamental questions regarding the drug discovery strategies in the pharmaceutical industry. A recent analysis of 842 molecules having been in development between 2002 and 2011 with a known outcome for up to 18 different attributes identified the following correlations to success or failure (Ringel *et al.*, 2013):

- 205 drugs received approval and 637 failed in phase II trials or later.
- Size of the company is not an indicator of success for R&D productivity.

- Scientific track record of a company (publications, patents, citations per publication) correlates with the success of a molecule.
- A strong indicator of good judgment and higher success rate is seen in companies with an R&D leader with a long tenure.
- No significant impact on R&D success was found for attributes like R&D budget, indicative market size, therapeutic area (with the exception of neuroscience with negative and anti-infectives with positive impact on success rate), or target family.

Even if these analyses are incomplete, they indicate a stringent correlation between success rate for a drug candidate with a clear commitment to hold scientific expertise and thoroughness dear.

## 6.4 Is There an Innovation Gap in Biomedical Research?

It is often argued that a major reason for the low success rates of pharmaceutical research with regard to novel therapies is a high degree of conformism that most companies work on the same small number of targets rather than approaching a medical field broadly with multiple targets in a biochemical pathway, thereby limiting the potential probability of success (POS). A recent example comes from the field of incretins for diabetes treatment with a controversial discussion which one of the approaches – small-molecule inhibitors of dipeptidylpeptidase IV inhibitors or analogs of glucagon-like peptide 1 (GLP-1)/exendin 4 – would be the superior attempt. Both approaches have their strengths and weaknesses, and today with at least four drug entities of each class introduced into clinical practice, both approaches have become important pillars for effective diabetes treatment today. It is therefore often not possible to make an upfront clear decision. The argumentation has to be thoroughly experimentally elaborated in order to be able to outweigh benefits versus downsides of any approach.

Regarding the question of conformity or competitiveness of drug discovery and analysis of 1251 human proteins, 1027 have been identified as putative pharmacological targets; only 26% of these were investigated by more than five companies, 58% by more than one company, and 42% by only one company (Agarwal, Sanseau, and Cardon, 2013). Out of 712 novel targets with not yet proven efficacy for human diseases, 54% were followed by only one company, 82% by less than three companies, and for only 13% of these targets more than five companies are currently working (Agarwal, Sanseau, and Cardon, 2013). From these data, it is obvious that a lack of diversity regarding the different approaches by pharmaceutical companies cannot explain the apparent innovation gap. A consequence of the high percentage of novel target approaches in the early drug discovery pipeline of the pharmaceutical companies, however, is that the validity of these targets for the human disease, their limitations, and mechanism-based potential adverse side effect profile has to be intensively investigated to reduce the failure rate of drug candidates for these targets in early clinical trials.

### 6.4.1 To Go for First in Class or Best in Class?

Owing to the enormous costs to develop and launch a new drug as well as by the cost pressure of healthcare organizations, drug companies are under enormous force to be the first or among the first to market. As a consequence, further development of a considerable number of potential high-value drugs is prematurely terminated at early stages (phases I and II) due to a commercial risk evaluation if according to the actual assessment a "placement on the podium" seems unlikely, and thus to be first in class with a new drug approach remains a major priority for drug companies. A recently published analysis of 53 drugs introduced into the market between 1990 and 2010 belonging to 15 different modes of actions, however, revealed a balanced outcome with a slight advantage of first in class (Schulze and Ringel, 2013). For the commercial success of fast followers, the entry into market should not be later than 2 years to the frontrunner. In any case, follower drugs have to show significant advantages such as:

- Favorable differentiation characteristics in a subgroup of personalized precision medicine. As an example, the DPP-IV (Dipeptidylpeptidase-IV) inhibitor linagliptin being introduced as fourth drug of this class into practice can be prescribed to diabetic patients with kidney disease, a frequent complication of diabetes because the drug is predominantly eliminated from the body by the liver with biliary excretion.
- Breadth of indications broadening the drug's market potential such as aripiprazole as the sixth approved drug to treat schizophrenia; however, because of its efficacy for a broad spectrum of psychiatric indications, it has become a commercially successful drug.

### 6.4.2 How We Define Medical Innovation?

Medical innovation cannot be predominantly defined from the perspective of scientific novelty but the impact of a novel drug or treatment option for the benefit of the patient. A scientific breakthrough such as the identification of a novel target related to a disease process may be usefulness from the medical perspective if the science cannot be translated into a treatment option and does not improve the health-related quality of life and survival. The judgment of a medical innovation should take the following aspects into consideration (Aronson, Ferner, and Hughes, 2012):

- No existing therapy
- Higher efficacy of the novel approach
- Better safety of the new approach, particularly for drug–drug interactions
- More convenient application possible (i.e., oral vs. parenteral, once daily application, etc.)
- Stratification of patient population having benefit from the approach (e.g., B-Raf inhibitors for the treatment of cancers being driven by B-Raf mutations such

as vemurafenib and dabrafenib for the treatment of malignant melanoma with B-Raf V600E mutations)
- More cost-effective treatment possible

A survey covering 15 different medical areas involving 180 expert physicians in 30 leading academic medical centers addressed the questions on what defines an innovative drug and what are the success factors being far most transformative (Kesselheim and Avorn, 2012). The major outcomes and recommendations from this analysis were as follows:

- Effectiveness of drugs in managing the disease and their superiority to existing therapies
- A strong demand for more focus on breakthrough innovation with substantial advantages over existing therapy options as determined by hard clinical outcomes rather than efficacy parameters characteristic for the respective target (Munos and Chin, 2011; US Food and Drug Administration (FDA), 2011)
- A better benefit–risk ratio of novel drug approaches with regard to better efficacy, better safety, and the potential to be disease modifying
- Drugs addressing unmet clinical needs for which so far no treatment option was available (e.g., like sofosbuvir to treat and cure from hepatitis C)
- Drugs leading to incremental but significant benefit for the patient, particularly for chronic diseases with lifetime treatment (as an example, the development of long-acting insulins for once daily injection like insulin glargine significantly improved the treatment and metabolic control as well as the quality of life of diabetic patients)
- Drugs with an impact on practical use allowing a paradigm shift in treatment as was the case with the identification of *Helicobacter pylori* as the causative agent to develop peptic ulcer allowing eradication of the disease by drug treatment rather than surgery

A view on these expectations from clinical opinion leaders on novel drugs highlights the necessity to investigate the function, role, and physiology of novel targets in health and disease as well as their potential inherent mechanism-based side effects. Only by a deep understanding of the (patho)physiology of a novel potential drug target one will be in the position to rationally develop efficacious drugs with an acceptable safety profile.

## 6.5
## Why Did Drug Candidates Fail?

An analysis of a large database – the Pharmaceutical Industry Database – collecting information from R&D projects covering around 28 000 compounds investigated since 1990 and 200 000 patent applications since the early 1970s revealed clear trends regarding success or attrition rates of drug candidates (Pammolli, Magazzini, and Riccaboni, 2011):

- The probability that a compound reaches the next stage in the development process has declined. The attrition rate strongly increased at all phases of the drug development process between 1990 and 2004, with highest increases in the attrition in clinical phases II and III (preclinical, 65 > 75%; Ph I, 32 > 50%; Ph II, 45 > 70%; Ph III, 20 > 50%; registration, 15 > 20%). Interestingly, the success rate from initiation of a project to phase I trials also went down despite great progress in science and technology, from 10% between 2002 and 2004 to 5% between 2006 and 2008 (Anonymous, 2012) raising the question whether the biological validity of the respective targets was scientifically sound when the discovery programs were started.
- POS for the development of drugs for chronic diseases (such as diabetes, obesity, Alzheimer's disease, depression, multiple sclerosis, rheumatoid arthritis, or chronic obstructive pulmonary disease) showed a lower POS compared to drugs for acute diseases (6.88% vs. 8.77%).
- Average development times of drugs in the United States and 15 European countries increased from 9.7 years in the 1990s to 13.9 years for drugs developed after year 2000.
- POS strongly depended on the respective disease, drugs for the indication cancer, central nervous system, or metabolic diseases showing the lowest POS (1.2–4.1%), whereas success rates for antiviral, cardiovascular, dermatology, or genito-urinary drugs were found to be significantly higher ranging from 6.64 to 11.75%.

Analysis of many drug failures revealed one or more of the following aspects as major reason for the dropout of the drug candidate (Paul *et al.*, 2010):

- A doubtful nonproven validity of the target in the disease pathophysiology
- Inadequate compound characteristics being the cause of continuous issues regarding efficacy and safety during the development process such as the ADMET profile (i.e., absorption, distribution, metabolism, excretion, toxicity)
- Safety issues particularly in drug candidates which demand a chronic or lifetime treatment

A survey performed in 2009 analyzing the fates of drug discovery and development project from 14 major pharmaceutical companies revealed the following reasons for termination of the respective programs in the different stages from preclinical to clinical phase III (CMR International, 2010):

- *Strategic decisions*: 19–25%
- *Lacking resources*: 1–2%
- *Lack of efficacy*: preclinical, 4%; Ph I, 18%; Ph II, 53%; Ph III, 57%
- *Safety reasons*: preclinical, 58%; Ph I, 39%; Ph II, 22%; Ph III, 17%

The analysis of drugs approved or rejected by the European Medicines Agency in 2009 shows a similar pattern: from 48 new active substances (NAS) submitted, 29 (60%) were recommended for market authorization and 19 (40%) rejected. The major reason for rejection was an inadequate demonstration of efficacy in humans and/or safety concerns (Eichler *et al.*, 2010).

The fact that on the one hand the development of a drug candidate in the clinical phases I–III accounts for around 63% of the costs to launch one NME compared to 32% of costs for the entire preclinical phase from target selection to preclinical development (Paul *et al.*, 2010) and that on the other hand the attrition rate in the drug development cycle is by far highest in clinical phases II and III due to lack of efficacy and/or safety reasons puts the selection and validity of the respective target for the disease in question.

### 6.5.1 Why Is the Dropout Rate So High in Early Clinical Development?

The dramatic improvements in technologies (human genome elucidation, "omics" technologies, High Throughput Screening (HTS), computer sciences, etc.) in the 1990s have led to a paradigm shift in drug research away from low-throughput pharmacology with *in vivo* animal pharmacology and screening to *in vitro* high-throughput screening of chemical libraries on the protein or cellular level. Unfortunately, these important technical improvements indeed strongly increased the throughput in the R&D process but not really the output as measured by innovative drugs reaching clinical application. The drug discovery process was "industrialized" and a predominant focus was put on the achievement of compounds with as high as possible affinity/potency to a particular target, that is, to some extent there was a paradigm shift from quality to quantity. However, this phase in the pharmaceutical industry was characterized by significant degree of "biomedical naivety" and reductionism regarding the complexity of a living organism.

#### 6.5.1.1 Drug Behavior *In Vivo*: Role of Transport Proteins

Due to the focus on the potency of a drug candidate, often the ADMET characteristics of a drug in a living organism were not adequately addressed. As an example, the transport of a drug across biological membranes was (and is often still) considered as a biophysical problem to be solved by making the drug more lipophilic. However, there is more and more evidence that the majority of drugs cross biological membranes because they accidentally are substrates of membrane transport proteins developed during evolution for endogenous substrates. Therefore, membrane transport proteins determine and regulate the transport of drugs across cellular membranes and are thus major determinants of the pharmacokinetics and pharmacodynamics of a drug, thereby strongly influencing their efficacy and safety *in vivo*:

- Intestinal absorption or exclusion of a drug from systemic circulation.
- Renal and hepatic excretion.
- Organotropism of a drug determining its cell and organ specificity or selectivity.
- Interaction of a drug with the natural ligands for a certain transporter can be the cause of drug–food interaction or drug–drug interactions.

It is evident that the knowledge about all transport proteins of the genome (in human as well as animal species used in pharmaceutical research) and the natural substrates of the transporters as well as an interaction map of all known drug molecules with the transporterome would strongly improve drug discovery and development if this knowledge is a selection criterium for a drug candidate to be further investigated in human clinical trials. The methodological breakthroughs in cloning of transport proteins and technologies for HTS-transport measurements *in vitro* have led to a significant recognition of transport biology in the drug discovery and development process due to their fundamental role for the pharmacokinetic and pharmacodynamic characteristics of a drug (Dobson and Kell, 2008; Huang and Woodcock, 2010; Kramer, 2011; Giacomini *et al.*, 2013). An International Transporter Consortium was established to develop guidelines for the investigation of drug transporter interactions (International Transporter Consortium, 2010), and the systematic evaluation of some transporters has become mandatory for drug approval (Zhang *et al.*, 2008). A systematic evaluation performed by Pfizer investigating the circumstances of 44 internal drug programs reaching clinical phase II between 2005 and 2009 underlines the importance of thorough (molecular) pharmacokinetic characterization for success in drug development (Morgan *et al.*, 2012). The prominent cause of failure was lack of efficacy in humans in 43% of the cases. As a major reason therefore was found an inadequate attention to and consideration of the fundamental pharmacokinetic and pharmacodynamic principles which determine the behavior and fate of a drug *in vivo* and which are greatly determined by the interaction of the drug with transport proteins. The demonstration of exposure at the site of action over time, binding of the drug to its anticipated pharmacological target according to the expected mode of action, and showing of the desired and expected pharmacological activity are the three conditions ("three pillars of survival") which, if concomitantly being fulfilled, have a high predictability for the likelihood of survival of a drug candidate in phase II and its progression into phase III clinical trials.

#### 6.5.1.2
#### Hypes and Lack of Scientific Thoroughness

With the enormous progress in molecular and cell biology, protein purification and crystallography, automation, and computer sciences, the drug discovery process could be accelerated by orders of magnitude with the possibility to screen chemical or biological libraries with millions of individual entities or to synthesize large libraries of small chemical molecules or antibodies within days or months. In the 1990s, the drug discovery process shifted from a (patho)physiological paradigm to a technology-driven high-throughput paradigm focusing on the identification of ligands with high affinity or potency to the respective target primarily using isolated or recombinant proteins and cell lines expressing the respective targets ideally with an easily measurable and high-throughput-compatible readout signal. Fundamental biological principles such as membrane permeability, ADMET characteristics, or the involvement of the respective target in a biological pathway and network were considered as "technical problems" to be solved later on in the further development of the drug candidate. In essence, however, this paradigm shift in the drug discovery process was basically a shift from quality to quantity with a high-throughput but low output with regard to novel drug-like clinical candidates (Scannell *et al.*, 2012). on disease biology with the desire to find "the philosopher's stone" for pharmaceutical drug research with attempts such as outlined in the following chapters.

#### 6.5.1.2.1
#### Blockbuster Mentality

A great focus in the drug discovery attempts was laid on the search for "magic bullet" drugs being able to cure even very complex diseases such as cancer or diabetes with one drug only. This was driven by the euphoria during the human genome era when thousands of novel druggable targets have been expected. Whereas in the year 2002 482 molecular targets were known to be addressed by all marketed drugs at that time (Drews, 1996; Drews and Ryser, 1997) after sequencing of the human genome, the number of novel targets of pharmacological interest was estimated to around 8000, of which 5000 should be druggable by small-molecule approaches, 2400 by antibodies, and 800 as potential biopharmaceuticals (Burgess and Golden, 2002). "One drug fits all" in conjunction with a blockbuster mentality has led to an extreme simplification attitude of human biology according to the motto "one gene > one target > one disease."

#### 6.5.1.2.2
#### Belief That Technology Will Fix It

In the last two decades, drug discovery has faced several doctrines on how the productivity of drug discovery with regard to finding novel key targets and approaches for effective disease-modifying drugs can be fundamentally improved. This included the areas of high-throughput screening, high-throughput synthesis of chemical libraries, monoclonal antibodies, the human genome project, and nucleic acid-based drugs.

All of these approaches are important pillars of high value in drug research. However, with the appearance of the respective technology, a certain hype and belief emerged that by a strong focus and high input of resources into a selected technology the productivity challenges can be overcome. By the high extent of focusing on a certain technology, often a paradigm shift occurred, thereby neglecting a systematic scientific examination of inherent challenges of the respective approach and adequate consideration of the biological and physiological processes in a complex mammalian organism.

**Monoclonal Antibodies** After publication of the principal methodology to produce monoclonal antibodies (mAbs) highly specific for defined epitopes by Köhler and Milstein (1975), expectations were high that within a couple of years mAb-derived drugs of hitherto unknown specificity and safety would be available for most diseases, particularly also for those where at that time no therapy existed. However, the early approaches were not very successful, and it turned out that the biology of mAbs was much more complex than originally thought. The early used mAbs originated from mouse cells and were recognized as antigens by the human immune system resulting in an immune response against these mouse mAbs.

Furthermore, the interplay of antibodies with the different cell types of the immune system in the individual recipient was incompletely investigated and understood. With the capability to humanize the mouse mAbs by elimination of the mouse sequences and substitution by human sequences, these problems could be solved and in 1998 with rituximab the first humanized mAb was approved. It took around 20 years for mAbs from bench to bedside, and today around 42 mAbs are introduced in Europe and the United States in clinical practice.[1] They account with around $44.6 billion turnover in 2011 for about one-third of drug sales.

**Nucleic Acid-Based Drugs** Whereas conventional drugs modulate the function of a certain target protein, nucleic acid-based drugs (antisense oligonucleotides, siRNA therapeutics) aim to interfere at the level of the genetic information, either by inhibition of the formation of the target protein by blocking the translation of its sequence into a polypeptide chain (antisense oligonucleotides) or by silencing of the coding RNA (siRNA therapeutics). The strict complementarity of pairing DNA or RNA molecules thereby mediating the highest possible degree of biological specificity makes therefore such approaches in principle highly attractive as very effective and specific drugs, and consequently already in the 1990s antisense and triplex DNA drug approaches were en vogue in the pharmaceutical industry with similar high expectations as with the mAbs (Cohen and Hogan, 1994). The high hopes, however, did not materialize into drugs, and it became quiet in the field of nucleic acid drugs. The discovery of RNA interference by Fire and Mello and the finding by Tuschl that single genes in mammalian cells could be switched off with small RNA fragments led to a not yet seen race to become the leader in the field of RNAi therapeutics due to the theoretical advantages of this class of drugs:

- Specific switching off of genes early in the pathophysiological cascade of the disease process
- Less adverse side effects due to the “absolute” specificity for selected target gene
- Applicable to each target irrespective of its cellular location
- Applicable to targets difficult to approach with small molecules or antibodies

Due to these promising perspectives, big pharmaceutical companies like Roche, Pfizer, Abbott, Merck, and others invested billions of dollars into this field and pushed several drug candidates into clinical trials. Not surprisingly, the early clinical trials failed to fulfill the sky-high expectations – either due to insufficient efficacy or a strong immune response by the innate immune system (Bender, 2014). As a consequence, Roche shut down its RNAi activities in 2011 after having invested around $500 mio. After the approval of the first antisense

1) www.antibodysociety.org/news/approved_mabs.php

drug for systemic application in 2013 – mipomersen for the treatment of severe hypercholesterolemia – Roche reentered the field by spending at least $250 mio for the acquisition of Santaris (Roche, 2014). This phase of nucleic acid drug research was called *an era of irrational exuberance* (Bender, 2014). How could that happen? Oligonucleotides are very polar molecules, and as polyanions they cannot penetrate cell membranes by unspecific pathways like diffusion. These characteristics are known since decades, but already in the early phase of oligonucleotide drugs, this fundamental biological hurdle was not sufficiently investigated to find options for drug delivery to the cell interior: by chemical strategies like introduction of phosphorothioate, morpholino phosphorodiamidate, or locked nucleic acid building blocks, the biological stability and potency of oligonucleotides as well as their unspecific interaction with proteins could be improved, but the molecule-inherent membrane barrier hurdle persists. Today more emphasis is put on molecular delivery science either by packaging of the drugs into lipid nanoparticles or tailor-made synthesis of oligonucleotide–ligand candidates to make use of cell-type-specific uptake pathways via receptor-mediated endocytosis (Hausseker, 2014). By systematic and scientifically sound research regarding uptake pathways and the mechanisms underlying the stimulatory effect of oligonucleotides on the immune system mediated by Toll receptors, nucleic acid-based drug candidates may achieve the transition to clinical practice. Currently around 32 clinical trials with oligonucleotide drug candidates are reported in ClinicalTrials.com.

#### 6.5.1.2.3
#### The Irreproducibility Epidemic

In recent years, it became evident that a vast body of biomedical research being published in high-ranked scientific journals and often being considered as "breakthrough findings" with a high impact and importance for biomedical research and patient treatment is irreproducible (Mullard, 2011). Researchers at Bayer performed an analysis of their early in-house projects (target identification and target validation) of 67 different targets taken up from publications in the field of oncology, cardiovascular diseases, and women's health (Prinz, Schlange, and Asadullah, 2011). Only in 20–25% of the projects the relevant published data could completely be reproduced in their labs; a 100% reproduction rate was found in meager 7% of the investigated targets. In two-thirds of the projects, significant inconsistencies between published results and internal findings occurred leading despite experimental adaptions in most cases to termination of the project because the evidence for a role of the particular target/approach in the pathophysiology of the disease was insufficient. Similarly, Begley and Ellis at Amgen found that only 6 out of 53 selected so-called "landmark" studies in oncology could be reproduced by scientists in their company indicating a confirmation rate of the published findings in just 11% of the cases (Begley and Ellis, 2012). End of 2014 further devastating analyses regarding the quality of biomedical research results were published: a survey among 364 translational scientists working exclusively at academic or nonprofit institutions revealed that only 22% of them could reproduce published results from other labs. The Novartis Institutes for BioMedical Research reported that in its oncology division only 23% of 79 published claims could be reproduced (Dolgin, 2014). A thorough analysis of the factors underlying this poor performance of biomedical research laid open that there was obvious not a major single reason but instead a mèlange of influences of quite different characteristics (Mullard, 2011; Dolgin, 2014; Chalmers and Glasziou, 2009):

- *Inappropriate statistical analysis*: Particularly in preclinical research, a lack of awareness of the necessity for solid statistical power and potentially adequate education in statistical methods with the discovery researchers was spotted.
- *Lack of working under industry standards* (Mullard, 2011): Different criteria and lack of in-depth evaluation of safety and toxicity results in animal studies performed in academic labs for making assessments regarding biomedical findings lead to overinterpreted conclusions in many publications from academic labs such as "the compound was well tolerated and safe." Too little emphasis to include sufficient and reliable control experiments and a lack of skepticism versus the own generated data were also identified as contributing factors.
- *Quality of research tools* (Chalmers and Glasziou, 2009): Interestingly, in many cases, the quality of reagents delivered by vendors was questionable lacking adequate quality control documentation of the reagent. Particularly, reference compounds/drugs often had a poor quality with regard to purity and enantiomeric purity not fulfilling analytical quality standards (Dolgin, 2014). Even worse was the analysis regarding the quality of biological reagents. A typical practice among academic labs is the exchange of cell lines. However, very often these cell lines are delivered from one lab to the next without careful documentation regarding number of cell passages and changing phenotypes during passaging, contaminations by other cell lines, and so on. This results in turn often in research, that is, research with cells not really knowing capabilities and characteristics compared to the originator cell *in vivo*.
- *The incentive system in academia "publish or perish"*: The current incentive system in academia with a high pressure on scientists to receive recognition and reputation by publishing novel results first in high-impact journals fosters a frivolous handling of data and findings such as how to handle data outliers, leaving out findings not consistent with a consistent story line according to in-built (conscious or unconscious) biases. Additionally, there is today no culture and arena to publish so-called negative results to the scientific community to allow conclusions from these lessons learnt. There is as well a tendency in the academic scientific community to publish preferably molecular mechanistic studies rather than whole-body physiological examinations. There are increasing concerns and doubts among scientists including several Nobel Prize winners regarding our current system assessing the quality of scientific research. Initiated by a provocative viewpoint article called *Avoidable waste in the production and reporting of research evidence* by Chalmers and Glaziou (2009) concluded that as much as 85% of research investment was wasted. Lancet published end of 2014 a series of articles analyzing the underlying factors and recommending

options to improve (Ioannidis *et al.*, 2014; see also related articles published in Lancet 383, issue 9913 of 18–24 January 2014). It is evident that such a poor yield of evidence for the validity and robustness of a novel target is insufficient as a starting point for drug discovery attempts with enormous resources necessary. Taking into account the low reproducibility of biomedical results, the low success rate of novel targets and drug approaches despite quantum leaps in technology and speed and increases in R&D investment is not surprising but remains unsatisfactorily and inefficient.

#### 6.5.1.2.4
#### Static View on Targets

With the focus on rapid identification of high-affinity ligands for a particular target using high-throughput technologies, the assay systems had to be accordingly adapted by generation of "artificial" cell lines expressing the respective protein target with highly sophisticated readout systems such as fluorescence resonance energy transfer (FRET). These cell lines often express the respective target proteins at much higher concentrations than in the natural cell. Undoubtedly such cell lines allow the identification of target-specific high-affinity binding ligands. For example, in G-protein-coupled receptors (GPCRs) occurs a structural reorganization of the after binding of the ligand which initiates the downstream signaling cascade, thereby determining the physiological or pharmacological response. Thus, the dynamics of the interaction of a ligand with its receptor is a major factor for its biological activity, not just its high-affinity binding (Jonker *et al.*, 2005). Therefore, the "real" pharmacological target may not be the protein screened for ligand binding, but the interaction and orchestration of the target protein with other proteins is the actual determinant for a pharmacological response (Apic *et al.*, 2005). Additionally, the drug effect could as well be indirect because the interaction of the drug with its primary target may lead to changes in the microenvironment of cells, for example, by accumulation of metabolites which then trigger the pharmacological effect. As an example, the anti-Parkinson drug entacapone inhibits the enzyme catechol-*O*-methyltransferase (COMT), thereby preventing the degradation of dopamine and its precursor L-DOPA to inactive metabolites. The true pharmacologically active component is dopamine whose concentration in the (remaining) neurons of the substantia nigra is kept high due to prevention of its degradation by entacapone.

#### 6.5.1.2.5
#### Desire for Generally Applicable Standard Operating Procedures for Each Target

The drug discovery and particularly the drug development process has to be transparent and reproducible and fulfilling high-quality standards. Therefore, for the procedures and investigations relevant for final approval of a drug by health authorities, standard operating procedures for many investigations in development and production of a drug candidate for use in humans are mandatory, necessary, and meaningful. However, in the early drug discovery phase, sometimes such an inflexible rule-driven mentality may prevent the progression of innovative concepts. An example the author is aware of is the discussion regarding bioavailability. In pharmacokinetics, bioavailability is defined as the fraction of an orally applied administered dose of an unchanged drug that reaches systemic circulation and is a very valuable parameter for the development of a drug because by measurement of drug levels in peripheral blood peak levels, duration, and time dependence of drug appearance can easily be measured. However, for numerous approaches, the determination of bioavailability according to this definition is completely misleading and irrelevant:

- *Drugs acting on targets in the liver*: If the primary target for a pharmacological intervention is located within the liver, the ideal drug should exert its pharmacological activity in the liver and nowhere else in order to avoid undesired side effects in other organs. Since an orally applied drug is intestinally absorbed and reaches the liver with portal blood, a high first-pass effect for a hepatotropic drug is desirable. For drug development, however, this feature may be considered as a problem because the so-called bioavailability would be very low but for biomedical evaluation of such a drug irrelevant. However, for the development of such classes of drugs, more sophisticated methodologies are necessary to allow following the route of the drug *in vivo*.
- *Nonsystemic drugs acting in the lumen of the intestine*: An attractive approach for some indications is the development of drugs which exert their pharmacological activity not in the systemic circulation but at the level of the intestine or the intestinal lumen. Nonsystemically acting drugs have the principal advantage of no or very low systemic toxicity and adverse side effects because they do not reach ectopic sites in the body: examples are inhibitors of bile acid reabsorption (Kramer and Glombik, 2006) or cholesterol absorption, sequestrants for bile acids or phosphate or pancreatic lipase, as well as nonsystemically antibiotics, for example, against *C. difficile*. For these approaches, the drugs should not at all be intestinally absorbed. Therefore, novel and creative tailor-made methodologies are needed to monitor the pharmacokinetics of this kind of drugs during their clinical development. A certain misunderstanding between the inherent biological fundamentals of such a drug target, the need to monitor its fate *in vivo* and the desire to use standard methodologies "one size fits all" in drug development has hindered the progression of several innovative drugs into practice and is still a matter of controversial discussions in pharmaceutical companies and the reason to terminate innovate drug candidates. As mentioned earlier, an analysis performed by Pfizer had identified the inadequate consideration of the fundamental pharmacokinetic and pharmacodynamic principles of the behavior of a drug *in vivo* as the prominent cause of drug failure in human trials (Morgan *et al.*, 2012). Another example is the early determination of the drug formulation. Due to the ambitious timelines and the need to perform extensive stability investigations for drug approval, the drug developers want to define the final formulation of a drug very early in the development process, already in phase I or phase IIa trials where usually one has not yet an unequivocal answer regarding the efficacy of a novel drug candidate in humans. The author is aware

of several drug projects which failed because too early a decision was made regarding the final formulation prior to have valid data regarding its efficacy in humans. Particularly with drug candidates of low water solubility, a tablet formulation could lead to an early no go because the disappointing clinical efficacy was based on a limited solubility of the compound rather than a lack of biological activity. To reduce (false-positive) dropouts of drug candidates, I would advocate for an unequivocal demonstration of its pharmacological activity prior to selection and development of the most appropriate drug formulation.

## 6.6 Implications from the "Lessons Learnt" for Future Drug Discovery Research

In a nutshell, the thorough analysis of factors, circumstances, and paradigm shifts in the pharmaceutical industry within the last two decades indicate that the poor productivity and outcome of significant breakthroughs in drug therapy cannot be ascribed to a single specific reason, but is the resultant sum of an accumulation of shortcomings, lack of scientific rigor, misconception in the understanding of the rules and laws of science, and a certain haughtiness to the complexity of whole-body (human) physiology. There is no and probably cannot be one general recipe or philosopher's stone on how to build a successful and efficient R&D organization in a sort of a "blockbuster machine" which has just to be fed with enough information, resources, and money resulting in the continuous output of innovative drug candidates of high medical need and value.

However, a return and renaissance to scientific depth and scrutiny in the understanding of human and animal physiology in health and disease making use of the new available tools and knowledge in biology, chemistry, and information technology (IT) is an indispensable prerequisite for an efficient and rationale-based drug discovery process rather than a one guided by scientific fashions, hypes, hopes, and unrealistic expectations neglecting fundamental principles of science and human biology. The conclusions from numerous publications analyzing the factors, reasons, and mechanisms underlying the productivity issues in pharmaceutical R&D and the personal experience of more than 35 years of biomedical research in academia and pharmaceutical industry can be evolved into a great many of recommendations on how the R&D process in biomedical research can be improved as outlined in the following chapters. These suggestions do not claim a right to completeness but represent the personal opinion of the author of this chapter.

### 6.6.1 Organization of Drug Discovery and Development

In the past, drug discovery research on (a) novel target(s) was performed parallel in many pharmaceutical companies secretly in strict isolation without scientific exchange among them regarding results, issues, and failures driven by the opinion to ensure maximal intellectual property (IP) rights for a particular target/approach and with that strategy the commercial success for the company. However, regarding novel drugs, it turned out that finally IP can be protected predominantly by patents regarding the final drug candidate but only to a minor extent for claims on targets *per se*. Therefore, there is much more room for collaborative, precompetitive research than assumed in the past. As a consequence of the attempts to foster open innovation to make better and more efficient use of the scientific expertise available worldwide, many collaborative initiatives have been started between academic institutions, small biotech companies, and large pharmaceutical companies which made new forms necessary on how to organize research in such multipartner groups. There are many ways to organize open innovation, and I would like to refer to some recent excellent publications regarding this approach (Woodcock *et al.*, 2014; Said and Zerhouni, 2014; Goldman, Seigneuret, and Eichler, 2015; Hunter and Stephens, 2010; Hughes, 2009). How a truly collaborative precompetitive consortium can substantially contribute to biomedical progress can be illustrated by outcomes from the Innovative Medicines Initiative (IMI) project "Improving bet-cell function and identification of diagnostic biomarkers for treatment monitoring in diabetes" (IMIDIA), a public–private partnership between 8 leading pharmaceutical companies, one biotech company, and 12 leading academic partners in the field of β-cell medicine (IMIDIA, 2010).

Since decades diabetes researchers and pharmaceutical companies were looking for an insulin-secreting human β-cell line exerting the physiological characteristics of a β-cell in the human endocrine pancreas *in vivo*. Due to strong differences in the biology of pancreatic β-cells in rodents and humans, the availability of a "true" human β-cell line is a prerequisite to search for drug candidates which can correct β-cell dysfunction in diabetes and restore normal β-cell function or stimulate neoformation of functional β-cells from pancreatic precursor cells. Despite enormous efforts performed parallel in many labs, no real progress was made; millions of euros were spent by the pharmaceutical industry without the option to share their experiences and findings from failures. Within the IMIDIA consortium, academic researchers succeeded in the generation of a first human β-cell line (Ravassard *et al.*, 2011; Weir and Bonner-Weir, 2011). By an intensive collaboration between different academic research groups and three pharma companies, the findings could be reproduced and confirmed, the protocols were "standardized," and researchers from the different institutions and companies closely worked together sharing all their results and even mutual visits in the labs to ensure reproducibility and reliability of the cell line results. Together a second-generation β-cell line was developed (Scharfmann *et al.*, 2014) allowing now for the first time a rational search for drug candidates improving β-cell function by high-throughput screening of large chemical libraries using a robust and reproducible protocol and a stable well-characterized and quality-controlled human β-cell line. The benefits from such collaborative research networks can be manyfold:

- Availability and access to expertise of critical mass and tools to tackle major scientific challenges one partner alone would not able to do, both due to lack of the necessary network and resources as well as risk considerations, the latter being today a major decision-influencing factor to perform biomedical projects of high risk.

- A collaborative consortium allows a balanced focus between the necessary scientific depth and the demand toward application of research results for disease treatment.
- The quality of science with regard to reproducibility of results, "standardization" and harmonization of protocols, and elucidation of the value of scientific results for therapeutic options can significantly be imp by a strong collaboration and mutual "cross-check" of results by partners to overcome the irreproducibility epidemic.
- The collaboration ensures the availability and access of reliable tools, cell lines, materials, and knowledge.
- A true and intensive collaboration reduces the high degree of redundant work and helps to avoid to follow failed attempts and learnings from earlier failures. As a result, with the same level of resources, more drug discovery approaches can be evaluated in parallel, and thus the probability to have success is increased.

### 6.6.2 Elucidation of the Physiological Validity of a Target for the Human Disease

The recent analyses of the discrepancies between published supposed breakthrough findings in biomedical research and the low degree of robust reproducibility (Mullard, 2011; Prinz, Schlange, and Asadullah, 2011; Begley and Ellis, 2012; Dolgin, 2014) – the latter being a precondition to initiate a drug discovery program with significant binding of resources – revealed that the scientific starting position to initiate a drug discovery program is often quite preliminary and tenable necessitating to generate more convincing evidence for the (patho)physiological relevance of a target in a disease process prior to a full-blown drug discovery program.

#### 6.6.2.1 Extensive Inquiry of (All) Published Data of a Target or Pathway

As a first step to initiate a drug discovery program, the knowledge available in databases and literature should systematically and extensively be analyzed. Often projects are started after a sketchy literature search. The author is aware of numerous projects and approaches as well as of grants, both in academic institutions and pharmaceutical companies which have been approved and initiated despite the fact that the majority of results and conclusions attempted to be newly investigated in these proposals had already been published and being publicly available. This lack of thorough elucidation of the knowledge base available in the public domain is a huge waste of money and resources. Particularly, there seems to be an era prior to the Internet where scientific results were published in written form in journals and being available in (physical) libraries, whereas today with the World Wide Web scientific results and publications are available in electronic media with the consequence that more and more the scientific libraries with the printed journals disappear. Unfortunately, many of the already published results in printed media are only partially available via electronic search machines leading to the phenomenon that these "old" information is not used, forgotten, and not connected to actual science, and in many scientific projects the wheel is reinvented. The current academic incentive system with a certain hunt "to be the first" in scientific publication of new findings fosters this waste of resources.

A recent analysis confirms this observation that the exploding number of scientific papers – particularly in life sciences – makes it increasingly difficult for researchers to keep track of all the publications relevant for their work (Briotta Parolo *et al.*, 2015). A survey of millions of papers published in the fields of clinical medicine, molecular biology, chemistry, and physics has revealed that after a peak of citations a few years after publication the number of citations goes down relatively fast, this decay getting increasingly faster with the outcome that important and relevant papers and fundamental knowledge for the respective science field is forgotten more easily than in the past. This selective oblivion of scientific knowledge not only produces numberless redundant results but also can hinder or block biomedical progress into effective new drug approaches. An example is the development of inhibitors of the sodium-dependent D-glucose transporter SGLT2 in the kidney as a new therapeutic principle to treat diabetes. In the kidney SGLT2 is responsible for the reabsorption and "recycling" of up to 180 g of glucose being filtered in the glomerulus. Already published in 1987 that the natural product phlorizin can correct hyperglycemia in diabetic rats (Rossetti *et al.*, 1987), it took another 12 years until inhibition of SGLT2 was suggested as a novel approach to treat diabetes (Oku *et al.*, 1999). In 2012 then finally dapagliflozin was approved by the European Medicines Agency (EMA) in Europe as the first representative of this class of novel antidiabetic drugs. Today six SGLT2 inhibitors are approved and nearly another dozen is in clinical development (Haas *et al.*, 2014).

The story in the development of SGLT inhibitors is as well an illustrative and daunting example how prejudice, biases, and opinions rather than scientific facts can hamper medical innovation. Repeated caveats in the management as well as in R&D organizations of pharmaceutical companies but as well strong statements by outspoken and self-reliant medical opinion leaders prevented the realization of this approach for a long time by repeatedly stressing arguments that this mode of action would lead to polyuria with severe dehydration of the patients with disturbances of electrolyte balance and permanent thirst as well as relapsing urinary infections. However, a thorough look into the science literature would have allowed a more rational and evidence-based judgment of this approach. It was published that individuals with functional renal glucosuria (FRG) are generally asymptomatic (Santer *et al.*, 2003), except for unusual cases with complete absence of renal glucose reabsorption where polyuria and hypoglycemia have been described (Oemar, Byrd, and Brodhel, 1987). Therefore, a valuable risk assessment with regard to inherent adverse side effects caused by the mode of action would have been possible at the time where SGLT2 inhibition was suggested for the first time as a potential novel therapy for diabetes (Oku *et al.*, 1999).

#### 6.6.2.2
#### Integrative Knowledge Management

A strongly underdeveloped area in many biomedical R&D organizations is the lack of an integrated database with the results and data of early research projects and particularly of drug development compounds which have been terminated and did not reach approval as a rich source to analyze the (scientific) reasons leading to termination as valuable information for lessons-learnt conclusions. In most cases to my knowledge the information from such projects and compounds is not systematically analyzed and made available to the R&D scientists. Instead, very often data and information of drug candidates become forgotten, lost, or even destroyed! This valuable information is often closely associated to individual scientists involved in the project. With them leaving the company, due to retirement or other reasons, this scientific key knowledge might be irreversibly lost. As a consequence, it may happen that targets which have failed are taken up repeatedly in a pharmaceutical company because nobody knows that and why this approach failed or was given up earlier for particular reasons. Therefore, a clear and accessible documentation of all research projects and drug candidates is an essential factor to avoid redundant work and waste of resources and an important source of valuable information.

The explosion in the generation of huge amounts of data catalyzed by biomedical progress such as next generation sequencing or imaging (and its documentation) on the one side and improvements in the speed and storage capacities of IT bring new challenges to biomedical R&D with regard to integrating this information into databases and allowing access and integration of information of quite different characteristics. This complexity is further intensified by the increased collaborations of different partners in biomedical projects in drug discovery with regard to compatibility, linkage, and standardization of databases. Research organizations are aware of this challenge, and initiatives involving many partners were established to harmonize and standardize the data and knowledge management of biomedical R&D data (Marti-Solano *et al.*, 2014); the following needs should be implemented as soon as possible in a pharmaceutical R&D organization:

- Protocol design and data collection of the most relevant scientific questions necessary for decision making (prioritization of key challenges at each stage of the research value chain)
- Establish commonly agreed standards for characterization of data, methods, and models to improve compatibility and reproducibility of results
- Establish formats for data and models allowing easy access and critical assessment
- Establish a broad basis of data sharing to reduce redundancy and improve quality and comparability of results
- Establish sustainability models to ensure permanent and enduring use of the integrated data to combat oblivion of important and reliable information

#### 6.6.2.3
#### Demonstration of the Involvement of a Target in Human Disease

Current drug discovery projects usually start with a selected target with an anticipated role in the respective pathogenesis cascade mostly taken up from literature followed by the construction of molecular and cellular assays with a robust readout method allowing HTS formats for fast identification of lead compounds interacting with the target of interest followed by chemical optimization into a drug-like molecule. For cellular assays in most cases designed cell lines are used which may have a poor association to the respective cell type in the disease state. As a result, preclinical studies performed with proliferating (human) cell lines show a low predictability for drug behavior and efficacy in human, the latter being a major reason for failure in phase II clinical trials, because the target–disease relations elaborated in cell lines and animal models often do not translate in similar efficacy in the clinical testing (Edwards *et al.*, 2015). Therefore, the use of disease models based on human tissue samples is mandatory for an understanding of the (patho)physiology and the role of a particular new target in human disease. Because of the ethical and logistical challenges to obtain human cells in sufficient amounts and the logistical challenges in handling of natural cells to be isolated from human tissue, usually HTS approaches for lead identification using human cells are not possible and practicable.

##### 6.6.2.3.1
##### Pharmacogenetic Evaluation of Target Role

Human genetics, however, can be a very valuable approach to obtain relevant information regarding the involvement of a target in the (patho)physiology of a disease including the following options:

- Analysis of large patient cohorts can deliver insight into the heterogeneity of a disease and the disease population.
- Tissue samples from healthy and diseased individuals can unveil relevant differences in the polymorphisms, transcriptome, epigenome, proteome, or secretome, and gain-of-function or loss-of-function mutations in a particular cell type assumed to be involved in the disease biology.

A systematic pharmacogenetic approach can be suited to scrutinize the molecular changes being associated with human pathophysiology (Zerhouni, 2014) to yield convincing confidence into a novel target potentially being interesting for pharmacological modulation prior to the starting of the drug discovery machinery and should be applied whenever feasible.

A prominent example where a pharmacogenetic route worked extremely well with regard to the discovery of a novel breakthrough drug treatment option is the discovery of drugs inhibiting the action of PCSK9 as the most efficient approach to lower elevated LDL cholesterol levels. The finding that around 3% of the Afro-American populations are heterozygous for loss-of-function mutations of PCSK9 associated with very low plasma LDL cholesterol levels and a reduction of cardiovascular events by 80–90% (Abifadel *et al.*, 2003) has made PCSK9 to one of the most investigated novel drug targets to treat hypercholesterolemia. Less than 12 years after the publication (June 2003) that mutations in PCSK9 cause autosomal dominant hypercholesterolemia, the dossiers for the mAbs alirocumab and evalocumab were submitted for drug approval by authorities early 2015.

#### 6.6.2.3.2
#### Reverse Genetics

A possibility to receive evidence regarding the potential role of a certain target in physiology or pathophysiology is the systematic knockout or overexpression of the respective gene to obtain a phenotype potentially indicating the suitability of this target for pharmacological intervention. A retrospective evaluation of the knockout phenotypes of the targets for the 100 best-selling drugs in the year 2002 revealed a good correlation of the phenotypes with the known drug efficacy in human (Zambrovicz and Sands, 2003). A program to knock out the mouse orthologs of the 5000 most interesting putative genes from the human genome project was performed at Lexicon Genetics. In an interim after having analyzed 750 of these genes, they forecasted the number of new high-quality targets suited for pharmacological modulation out of these 5000 genes to 100–150 (Zambrovicz and Sands, 2003). A congruency between pharmacological intervention and the knockout phenotype of novel genes of hitherto unknown function was reported for angiopoietin-like 4 indicating that (in this case) the knockout phenotype reflects both the physiological function of the gene and the pharmacological consequences of its inhibition (Desai *et al.*, 2007). It cannot, however, be generalized that a systematic knockout approach will always work but it seems worth prior to initiate a high-effort drug discovery program, since a high degree of confidence into the biological role of a novel target involved in human disease was a strong predictor of clinically successful drug projects as was published in an analysis performed at Pfizer (Cook *et al.*, 2014): projects with evidence of a genetic link of the target to the disease survived to 73% clinical phase II trials compared to 57% of projects lacking such evidence of genetic validation.

#### 6.6.2.3.3
#### Time Course Analysis of Disease and Progression Marker

One important aspect to be considered is the time-dependent changes of the involvement of targets and pathways in the progression of the disease, that is, answering the question, "What causes a disease to resolve?" The factors and triggers initiating a pathogenic cascade can be different from those which maintain or progress the disease (Lesterhuis, Bosco, and Lake, 2014). Particularly for chronic diseases with a time-dependent continuum of biochemical changes in pathway involvements over time like cancer, cardiovascular diseases, or metabolic disorders like diabetes, these circumstances strongly influence the efficacy of a distinct drug target approach. Which factors determine that 10–25% of individuals with fatty liver progress to nonalcoholic steatohepatitis (NASH), for the 25–35% conversion to liver fibrosis, 9–20% to liver cirrhosis, and the 10% of cirrhotic patients finally coming down with hepatocellular cancer? This example illustrates that at different stages of the disease different targets may be dominating players in the disease development and accordingly quite different molecular targets probably can be the right choice for the treatment of gastrointestinal tumors positive for the mast/stem cell growth factor receptor kit (CD117).

Therefore, an integral part of a drug discovery program should be the parallel evaluation and search of biomarkers indicating or predicting the progression or transition of the disease into its next stage. Such indicators are of severalfold value for a drug discovery program:

- For monitoring the progression of the disease and monitoring of the efficacy of a drug candidate interfering with the disease process.
- To allow stratification of patient populations with regard to disease stage enabling small clinical trials with clearly defined and characterized patient subpopulations. By such trials the subpopulations of patients can be identified which would have a clear benefit from a distinct approach avoiding treatment of patients with a low expectation of therapy success.

Additionally, pharmacogenetic tests and predictive biomarker associated with the disease stage transition or progression are particularly essential for all attempts to develop "personalized medicines." For novel drug targets the parallel elucidation and availability of surrogate or biomarkers is key to obtain early in the clinical development process an unequivocal answer whether the drug reaches its anticipated target and whether this interaction is associated to the efficacy of the drug with regard to the desired pharmacological activity. In the course of personalized medicine, the availability of a diagnostic test will become increasingly important for the selection of patients having a clear clinical benefit from a treatment, and therefore, for drugs being effective in defined disease subpopulations, the parallel development of companion diagnostics will become part of the regulatory approval process as well as the clinical use of such a personalized drug (Fridlyand *et al.*, 2013). In oncology this concept is due to the heterogeneity of the disease already well established, just to mention the tyrosine kinase inhibitor imatinib for the treatment of gastrointestinal tumors being positive for the mast/stem cell growth factor receptor kit (CD117) or vemurafenib for the treatment of malignant melanoma in patients carrying the V600E mutation in the B-Raf protein being present in about 50% of human malign melanomas. In other chronic diseases with strongly heterogeneous patient populations such as diabetes with a continuous disease spectrum from asymptomatic prediabetics to severe diabetic complications, such a population-defined and stage-dependent treatment will be necessary for future drug therapy.

#### 6.6.2.3.4
#### Integrative Physiology and Translational Medical Research

As outlined earlier, the majority of drug candidates fail in phase II clinical trials with lack of efficacy and safety issues being the major reasons. In the past, the reason for failure due to safety issues was frequently caused by off-target effects of the drug candidate, whereas today termination often occurs due to mechanism-based undesired or toxic side effects (Kirschner, 2011). In the selection process of a drug candidate, usually the respective remaining molecules from lead optimization were screened for their interaction with more than hundred or even hundreds of other targets and a quite comprehensive target interaction profile elaborated allowing the final selection of the best suited molecule for further development. A prominent example of such a mechanism-based practical limitation of the efficacy of a drug target was shown with the development of inhibitors of the microsomal lipid transfer protein MTTP: MTTP (Microsomal Triglyceride Transfer Protein) catalyzes the assembly of very-low-density lipoprotein (VLDL) particles from apolipoprotein B and lipids in the endoplasmic reticulum of hepatocytes and enterocytes. Logically, by inhibition of MTTP, the secretion of VLDL particles by the liver and chylomicrons by the small intestine will decrease and concomitantly the concentration of LDLs and remnant particles as metabolic products of the secreted apolipoprotein-containing atherogenic lipoprotein particles. MTTP inhibitors are very effective in lowering of LDL cholesterol and triglyceride levels, but the tolerable dose is limited by the mechanism-based side effect of lipid storage in enterocytes and hepatocytes, the latter leading to hepatic steatosis and its toxic sequelae. However, for patients with homozygous familial hypercholesterolemia type IIb who are unresponsive to any other lipid therapy other than the invasive plasmapheresis, MTTP inhibitors are a new and important drug treatment option. Therefore, in order to estimate the potential benefit–risk ratio of a novel drug target, the elucidation of a drug candidate directed to this target in its physiological context *in vivo* is necessary to be able to make predictions regarding the long-term overall pharmacological outcome. A precondition to that goal is a detailed knowledge of the role of genes and gene products in the context of biological pathways and networks, but unfortunately our knowledge in systems biology is still in its infancy. Examples illustrating the difficulties for prediction of overall long-term effects of drug candidates *in vivo* are the poor outcome so far with many attempts for agonists for the peroxisome proliferator-activated receptors (PPARs) where sequestration of cofactor proteins subtly influences and guides the efficacy of the drug and the benefit–adverse side effect ratio, and we do not really know the rules for the necessary orchestration of these protein complexes. The anticancer drug imatinib is an example to underline the importance of a scientifically sound characterization of a drug candidate. Imatinib was claimed as a selective inhibitor of poly-ADP ribose polymerase (PARP) inhibitor for the treatment of breast cancer but failed in phase III clinical trials due to lack of efficacy. In independent investigations after its clinical failure it was demonstrated that imatinib and its metabolites are not all functional PARP inhibitors, but a nonselective modifier of cysteine residue-containing proteins in tumor cells (Liu *et al.*, 2011). In order to improve the success rate in drug discovery, a significant investment into research of integrative physiology and systems biology is worth to overcome the "black box" approach.

#### 6.6.2.3.5
#### Phenotypic Screening

The identification and optimization of a drug candidate in early discovery mostly occurs by an HTS model using either recombinant proteins or constructed cell lines expressing the target of interest with a smart readout system. Since these cell line constructs often do not reflect functionality of the corresponding cell type *in vivo*, it may be questioned whether a target-based approach or a functional test model would deliver better results. The target-based approach is the classical way to find a hit compound which subsequently can chemically be optimized to a drug molecule by measuring its activity on the respective target. In the functional approach in a so-called phenotypic screen, chemical libraries are tested in (natural) cells or cell lines to generate a physiological readout such as secretion of a metabolite. Since it is rather unlikely that a hit compound found by phenotypic screening fulfills all criteria for a drug candidate, this hit has to be chemically optimized to a drug-like molecule without knowing the mode of action whereby the hit compound generates the desired physiological readout being a major disadvantage or challenge in phenotypic approaches. Furthermore, the availability of "natural" human cells or cell lines limits the general applicability of this method. For example, it took decades to find a human insulin-secreting β-cell line behaving like a human β-cell *in vivo*. Now, with the availability of human β-cell lines (Weir and Bonner-Weir, 2011) for the first time, the possibility is there to identify compounds improving the functionality of β-cells by phenotypic screening.

Despite the challenges associated with phenotypic screening, this methodology should be envisaged as an option whenever possible because of the high success rate of drugs reaching medical practice which have originally been discovered by phenotypic screening (Moffat, Rudolph, and Bailey, 2014). Among the 183 small-molecule drugs approved between 1999 and 2008, 58 (32%) were discovered by a phenotypic assay. Additionally, 28 (56%) of the 50 first-in-class drugs came from phenotypic screening compared to only 17 (34%) from target-based approaches (Swinney and Anthony, 2011). Novel technology developments as the recently reported activity-based protein profiling (ABPP) are necessary to speed up the target deconvolution of hit compounds identified by phenotypic screens (Dominguez *et al.*, 2014). Dominguez and coworkers have identified with this technology an inhibitor of the lipid-metabolizing enzyme serine hydrolase carboxylesterase 3 (Ces3) which was able to improve the metabolic status of diabetes and obesity mouse models (Dominguez *et al.*, 2014).

### 6.6.2.4
### A Stringent and Comprehensive Test Sequence

In order to achieve better predictability for the efficacy of a drug *in vivo*, both in animal species and humans, particularly for novel drug targets, drug researchers are well advised to follow a stringent cascade of test systems starting from efficacy at the primary target to assay systems of increasing biological complexity finally to *in vivo* efficacy regarding the anticipated pharmacological effect. Insufficient robustness of preclinical studies as the starting point for drug development program is a critical reason for failure in clinical human trials (Woodcock *et al.*, 2014).

A general stringent decision tree should contain the following stepwise elucidation of the anticipated mode of action and efficacy of the drug candidate. Depending on the respective target and availability of the tools at each step of course, an individual adaption of the decision tree will be necessary. According to the long experience of the author, this systematic approach of getting unequivocal answers to the following scientific questions is extremely helpful to avoid dropouts and failures:

- What is the potency of the drug at the anticipated target?
- Does the drug produce a downstream effect after binding to the target?
- Is the drug able to pass the cell membrane(s) of the target cell?
- Does the drug induce a biologically relevant response after binding to its target in the target cell?
- Does the drug reach its target *in vivo* in the relevant target organ with changes of the parameters indicating drug action at the target?
- Is the drug intestinally absorbed and does it show a sufficient bioavailability? (Thorough evaluation of the pharmacokinetics and PK/PD (pharmacokinetics/pharmacodynamics) relationship)
- Does the drug have a clear beneficial effect on disease-specific and disease-characteristic parameters and syndromes in predictive animal disease models?
- What are potential mechanism-based adverse side effects of the approach?

Therefore, for any drug discovery and development approaches, a stringent and scientifically sound decision tree should be put in place. This will allow finding clear answers to the above questions and the identification of potential issues, roadblocks, and challenges which have to be solved in order to progress a drug candidate to an experimental safe and efficacious medicine for the treatment of human diseases. The numbering of the steps does not indicate a strict sequential elaboration of the respective questions but addresses the scientific challenges and tasks for what a clear answer is necessary in order to be able for a rational and clear judgment of the findings.

Step 1: Efficacy at the primary target (isolated protein)
  Outcome: Demonstration of binding of the drug to its anticipated target and efficacy
Step 2: Efficacy at the cellular level (cell line with expressed protein)
  Outcome: Demonstration of efficacy at the cellular level and indication for the anticipated pharmacological effect (depending on the reporter system downstream of the target; for intracellular targets demonstration of membrane permeability)
Step 3: Efficacy at the natural target cell (*ex vivo* cells or cultured cells/cell lines)
  Outcome: Demonstration of the anticipated pharmacological effect at the target cell (physiologically relevant readout parameters)

Step 4: Efficacy at the target organ (isolated organs, tissue slices)
Outcome: Demonstration of the pharmacological effect at the target organ considering cell–cell interaction and communication, paracrine effects, and (potential) metabolic conversion (activation, deactivation) in the target organ
Step 5: Efficacy *in vivo* after parenteral or local drug administration to healthy animals (preclinical PoP) (for both drug candidates for parenteral and for oral use)
Outcome: Demonstration of the anticipated pharmacological effect *in vivo* without potential influences by intestinal absorption or hepatic first-pass effects. Demonstration that interference of the drug with its molecular target produces the expected changes in the physiologically relevant parameters
Step 6: Efficacy *in vivo* after oral drug application to healthy animals (preclinical PoP)
Outcome: Demonstration of anticipated pharmacological effect after oral application. Demonstration of intestinal absorption and bioavailability
Step 7: Efficacy of drug candidate in relevant animal disease models (preclinical proof of concept (PoC))
Outcome: Demonstration of pharmacological efficacy to improve/correct disease-specific symptoms and outcome, not just the biochemical/physiological surrogate parameters
Step 8: Elucidation of the potential mechanism-based adverse side effect of the respective approach
Outcome: Enabling of a first judgment of the benefit–risk ratio of the respective approach

The latter step seems very important to avoid progression of an approach into a full-blown clinical development which in the end might not be very successful because initial beneficial effects may be outbalanced by detrimental metabolic changes inherent to the mode of action becoming manifest only after long-term treatment. Very often these "devil's advocate" approaches are neglected and not liked by the drug researchers because they have the potential to terminate an otherwise prospering and successful project prematurely. However, one cannot fool nature, and therefore the author strongly advocates for a consequent and scientifically sound approach in any drug discovery project focusing concomitantly both on the focus, the medical efficacy and the potential mechanism-based undesired effects. This is particularly of importance for approaches in chronic diseases which make a long-term or even lifetime treatment necessary. A good example on how the inherent mode of action may become dominant over time thereby limiting the efficacy and potency of an approach is the development of MTTP inhibitors. MTTP catalyzes the assembly of VLDL particles from apolipoprotein B and lipids in the endoplasmic reticulum of hepatocytes and enterocytes. Consequently, inhibition of MTTP would lead to a reduced secretion of VLDL particles by the liver and chylomicrons by the small intestine and due to the reduced number of VLDL and chylomicron particles during their catabolism as well less LDL and remnant particles would be formed resulting in a strong decrease of plasma LDL cholesterol and triglyceride levels, two major contributors to atherosclerosis. This occurs by inhibition of MTTP, but unfortunately over time the triglycerides are deposited in the hepatocytes and enterocytes as fat leading to hepatic steatosis with its sequelae, that is, the mechanism limits the clinical efficacy because of limiting the tolerable dose. As a consequence, the full potential of this target to treat dyslipidemia cannot be realized, but one MTP inhibitor – lomitapide – has received approval to treat patients with severe homozygous familial hypercholesterolemia for which so far only the option of invasive therapy with plasmapheresis existed.

#### 6.6.2.5
#### Translational Clinical Trials

The importance of a novel target involved in the human pathophysiology of a disease can finally only be confirmed by a clinical investigation in humans with a drug candidate specifically interacting with this target. Because of the limited predictability of animal experimentation for the human clinical situation, the limited predictability from findings of drug interaction with the target to the biochemical networks in whole-body physiology, and the heterogeneity of patient populations today, large clinical trials are necessary with a high number of participants in order to obtain statistically significant results regarding the efficacy of the respective drugs in humans. As a consequence, usually in phase I clinical trials, the tolerability and safety of a drug candidate after single and repeated dose are investigated followed by PoC/PoP studies in phase IIa and phase IIb to obtain first information regarding the efficacy of the drug candidate. In large long-lasting phase III clinical trials with hundreds or thousands of patients, finally the full target product profile and characteristics of the drug are investigated. Thus, the final unequivocal answer whether a novel drug approach is effective, safe, and better than existing therapy occurs quite late in the development process.

In order to increase the POS in clinical development, particularly in phase II with the highest failure rates, a PoP/PoC trial should be performed prior to progress the drug candidate into large phase III clinical trials. Smaller methodological studies with better defined and characterized patient populations could be an efficient and faster alternative to evaluate the potential of a novel drug candidate for therapy (Trusheim, Berndt, and Douglas, 2007). Furthermore, by testing a compound in clearly defined stratified subpopulations, it can earlier be found which patient populations would benefit from a novel approach. A key essential for stratified clinical trials for PoC/PoP is the availability of markers indicating an efficacy of the drug candidate at the respective target. The coincidence of demonstrating that the drug candidate has *in vivo* reached its anticipated target – for example, by measuring of a downstream metabolite of that target – and the expected pharmacological activity is a good indicator for the efficacy of the drug in human by the anticipated pathway and not by off-target effects.

The limited size of stratified trials in phase II and a clear demonstration of efficacy in a stratified population can be a cost-effective way to the definition of the right target population for that drug approach. An early and clear demonstration of PoC/PoP also can be seminal to judge the validity of a novel target approach for a human disease and avoid repeated attempts for the same target over years because its validity or nonvalidity was never shown with clear scientific scrutiny.

### Acknowledgment

The author cordially thanks Prof. Markus Hinder for his continuous support and critical reading of this manuscript.

### References

Abifadel, M., Varret, M., Rabés, J.P. *et al.* (2003) Mutations in PCSK9 cause dominant hypercholesterolemia. *Nat. Genet.*, **34**, 154–156.

Agarwal, P., Sanseau, P., and Cardon, L.R. (2013) Novelty in the target landscape of the pharmaceutical industry. *Nat. Rev. Drug Discovery*, **12**, 575–576.

Anonymous (2012) A decade in drug discovery. *Nat. Rev. Drug Discovery*, **11**, 3.

Apic, G., Ignjatovic, T., Boyer, S., and Russell, R.B. (2005) Illuminating drug discovery with biological pathways. *FEBS Lett.*, **579**, 1872–1877.

Aronson, J.K., Ferner, R., and Hughes, D.A. (2012) Defining rewardable innovation in drug therapy. *Nat. Rev. Drug Discovery*, **11**, 253–254.

Arrowsmith, J. (2011) Trial watch: phase II failures: 2008–2010. *Nat. Rev. Drug Discovery*, **10**, 328–329.

Arrowsmith, J. (2012) A decade of change. *Nat. Rev. Drug Discovery*, **11**, 17–18.

Begley, C.G. and Ellis, L.M. (2012) Drug development: raise standards for preclinical cancer research. *Nature*, **483**, 531–533.

Bender, E. (2014) The Second Coming of RNAi. The Scientist (Sep. 1).

Briotta Parolo, P.D., Pan, R.K., Gosh, R., Huberman, B.A., Kaski, K., and Fortunato, S. (2015) Attention Decay in Science. arXiv:1503.01881v1 [physics.soc-ph] 6 Mar 2015 http://arxiv.org/pdf/1503.01881v1.pdf (accessed 10 August 2015).

Burgess, J. and Golden, J. (2002) Cracking the Druggable Genome, http://www.bioitworld.com/archive/100902/firstbase.html (accessed 10 August 2015).

Chalmers, I. and Glasziou, P. (2009) Avoidable waste in the production and reporting of research evidence. *Lancet*, **374**, 86–89.

CMR International (2010) R&D General Metrics Study.

Cohen, J.S. and Hogan, M.E. (1994) The new genetic medicines. *Sci. Am.*, **271**, 76–82.

Cook, D., Brown, D., Alexander, R., March, R., Morgan, P., Satterthwaite, G., and Pangalos, M.N. (2014) Lessons learnt from the fate of AstraZeneca's drug pipeline: a five-dimensional framework. *Nat. Rev. Drug Discovery*, **13**, 419–431.

Cuatrecasas, P. (2006) Drug discovery in jeopardy. *J. Clin. Invest.*, **116**, 2837–2842.

Desai, U., Lee, E.C., Chung, K. *et al.* (2007) Lipid-lowering effects of anti-angiopoietin-like 4 antibody recapitulate the lipid phenotype found in angiopoietin-like 4 knockout mice. *Proc. Natl. Acad. Sci. U.S.A.*, **104**, 11766–11771.

Dobson, P.D. and Kell, D.B. (2008) Carrier-mediated uptake of pharmaceutical drugs: an exception or the rule? *Nat. Rev. Drug Discovery*, **7**, 205–220.

Dolgin, E. (2014) Drug discoverers chart path to tackling data irreproducibility. *Nat. Rev. Drug Discovery*, **13**, 875–876.

Dominguez, E., Galmozzi, A., Chang, J.W. *et al.* (2014) Integrated phenotypic and activity-based profiling links Ces3 to obesity and diabetes. *Nat. Chem. Biol.*, **10**, 113–121.

Drews, J. (1996) Genomic sciences and the medicine of tomorrow. *Nat. Biotechnol.*, **14**, 1516–1518.

Drews, J. and Ryser, S. (1997) The role of innovation in drug development. *Nat. Biotechnol.*, **15**, 1318–1319.

Edwards, A.M., Arrowsmith, C.H., Bountra, C. *et al.* (2015) Preclinical target validation using patient-derived cells. *Nat. Rev. Drug Discovery*, **14**, 149–150.

Eichler, H.-G., Aronsson, B., Abadie, E., and Salmonson, T. (2010) New drug approval success rate in Europe in 2009. *Nat. Rev. Drug Discovery*, **9**, 355–356.

Firehouse, R.A. (2011) Lessons from 54 years of pharmaceutical research. *Nat. Rev. Drug Discovery*, **10**, 963.

Fridlyand, J., Simon, R.M., Walrath, J.C., Roach, N., Buller, R., Schenkein, D.P., Flaherty, K.T., Allen, J.D., Sigal, E.V., and Scher, H.I. (2013) Considerations for the successful co-development of targeted cancer therapies and companion diagnostics. *Nat. Rev. Drug Discovery*, **12**, 743–755.

Giacomini, K.M., Balimane, P.V., Cho, S.K., Eadon, M., Edeki, T., Hillgren, K.M., Huang, S.M., Sugiyama, Y., Weitz, D., Wen, Y., Xia, C.Q., Yee, S.W., Zimdahl, H., Niemi, M., and International Transporter Consortium (2013) International Transporter Consortium commentary on clinically important transporter polymorphisms. *Clin. Pharmacol. Ther.*, **94**, 23–26.

Goldman, M., Seigneuret, N., and Eichler, H.-G. (2015) The innovative medicines: an engine for regulatory science. *Nat. Rev. Drug Discovery*, **14**, 1.

Haas, B., Eckstein, N., Pfeifer, V., Mayer, P., and Hass, M.D.S. (2014) Efficacy, safety and regulatory status of SGLT2 inhibitors: focus on canagliflozin. *Nutr. Diabetes*, **4**, e143. doi: 10.1038/nutd.2014.40

Hausseker, D. (2014) Current issues of RNAi therapeutics delivery and development. *J. Controlled Release*, **195**, 49–54.

Huang, S.M. and Woodcock, J. (2010) Transporters in drug development: advancing on the critical path. *Nat. Rev. Drug Discovery*, **9**, 175–176.

Hughes, B. (2009) Harnessing open innovation. *Nat. Rev. Drug Discovery*, **8**, 344–345.

Hunter, J. and Stephens, S. (2010) Is open innovation the way forward for big pharma? *Nat. Rev. Drug Discovery*, **9**, 87–88.

IMIDIA (0000) IMIDIA – Innovative Medicines Initiative for Diabetes: Improving Beta-Cell Function and Identification of Diagnostic Biomarkers for Treatment Monitoring in Diabetes, www.IMIDIA.org (accessed 10 August 2015).

International Transporter Consortium (2010) Membrane transporters in drug development. *Nat. Rev. Drug Discovery*, **9**, 215–236.

Ioannidis, J.P., Greenland, S., Hlatky, M.A., Khoury, M.J., Macleod, M.R., Moher, D., Schulz, K.F., and Tibshirani, R. (2014) Increasing value and reducing waste in research design, conduct, and analysis. *Lancet*, **383**, 166–175.

Jonker, D.M., Visser, S.A.G., van der Graaf, P.H., Voskuyl, R.A., and Danhof, M. (2005) Towards a mechanism-based analysis of pharmacodynamic drug-drug interactions *in vivo*. *Pharmacol. Ther.*, **106**, 1–18.

Kesselheim, A.S. and Avorn, J. (2012) The most transformative drugs of the past 25 years: a survey of physicians. *Nat. Rev. Drug Discovery*, **12**, 425–431.

Kirschner, M. (2011) An audience with Marc Kirschner. *Nat. Rev. Drug Discovery*, **10**, 894.

Köhler, G. and Milstein, C. (1975) Continuous cultures of fused cells secreting antibody of predefined specificity. *Nature*, **256**, 495–497.

Kramer, W. (2011) Transporters, Trojan horses and therapeutics: suitability of bile acid and peptide transporters for drug delivery. *Biol. Chem.*, **392**, 77–94.

Kramer, W. and Glombik, H. (2006) Bile acid reabsorption inhibitors (BARI): novel hypolipidemic drugs. *Curr. Med. Chem.*, **13**, 997–1016.

Lesterhuis, W.J., Bosco, A., and Lake, R.A. (2014) Comment on "drug discovery: turning the titanic". *Sci. Transl. Med.*, **6**, 229le2. doi: 10.1126/scitranslmed.3008763.

Liu, X., Shi, Y., Maag, D.X. *et al.* (2011) Iniparib nonselectively modifies cysteine-containing proteins in tumor cells and is not a *bona fide* PARP inhibitor. *Clin. Cancer Res.*, **18**, 510–523.

Marti-Solano, M., Birney, E., Bril, A., Della Pasqua, O., Kitano, H., Mons, B., Xenarios, I., and Sanz, F. (2014) Integrative knowledge management to enhance pharmaceutical R&D. *Nat. Rev. Drug Discovery*, **13**, 239–240.

Moffat, J.G., Rudolph, J., and Bailey, D. (2014) Phenotypic screening in cancer drug discovery – past, present and future. *Nat. Rev. Drug Discovery*, **13**, 588–602.

Morgan, P., Van de Graaf, P.H., Arrowsmith, J., Feltner, D.E., Drummond, K.S., Wegner, C.D., and Street, S.D.A. (2012) Can the flow of medicines be improved? Fundamental pharmacokinetic and pharmacological principles towards improving phase II survival. *Drug Discovery Today*, **17**, 419–424.

Mullard, A. (2011) Reliability of 'new drug target' claims called into question. *Nat. Rev. Drug Discovery*, **10**, 643–644.

Mullard, A. (2015) 2014 FDA drug approvals. *Nat. Rev. Drug Discovery*, **14**, 77–81.

Munos, B. (2009) Lessons from 60 years of pharmaceutical innovation. *Nat. Rev. Drug Discovery*, **8**, 959–963.

Munos, B.H. (2013) Pharmaceutical innovation gets a little help from new friends. *Sci. Transl. Med.*, **5** (168), 168ed1. doi: 10.1126/scitranslmed.3005281.

Munos, B.H. and Chin, W.W. (2011) How to revive breakthrough innovation in the pharmaceutical industry. *Sci. Transl. Med.*, **3** (89), 89cm16. doi: 10.1126/scitranslmed.3002273.

Oemar, B.S., Byrd, D.J., and Brodhel, J. (1987) Complete absence of tubular glucose reabsorption: a new type of renal glucosuria (type 0). *Clin. Nephrol.*, **27**, 156–160.

Oku, A., Ueta, K., Arakawa, K., Ishihara, T., Nawano, M., Kuronuma, Y., Matsumoto, M., Saito, A., Tsujihara, K., Anai, M., Asano, T., Kanai, Y., and Endou, H. (1999) T-1095, an inhibitor of renal Na+-glucose cotransporters, may provide a novel approach to treating diabetes. *Diabetes*, **48**, 1794–1800.

Pammolli, F., Magazzini, L., and Riccaboni, M. (2011) The productivity crisis in pharmaceutical R&D. *Nat. Rev. Drug Discovery*, **10**, 428–438.

Paul, S.M., Mytelka, D.S., Dunwiddie, C.T., Persinger, C.C., Munos, B.H., Lindborg, S.R., and Schacht, A.L. (2010) How to improve R&D productivity: the pharmaceutical industry's grand challenge. *Nat. Rev. Drug Discovery*, **9**, 203–214.

Prinz, F., Schlange, T., and Asadullah, K. (2011) Believe it or not: how much can we rely on published data on potential drug targets? *Nat. Rev. Drug Discovery*, **10**, 712–713.

Ravassard, P., Hazhouz, Y., Pechberty, S., Bricout-Neveu, E., Armanet, M., Czernichow, P., and Scharfmann, R. (2011) A genetically engineered human pancreatic ß-cell line exhibiting glucose-inducible insulin secretion. *J. Clin. Invest.*, **121**, 3589–3597.

Ringel, M., Tollman, P., Hersch, G., and Schulze, U. (2013) Does size matter in R&D productivity? If not, what does? *Nat. Rev. Drug Discovery*, **12**, 901–902.

Roche (2014) Roche to acquire Santaris Pharma to expand discovery and development of RNA-targeting medicines. Press release, August 4, 2014.

Rossetti, L., Smith, D., Shulman, G.I., Papochristou, D., and DeFronzo, R.A. (1987) Correction of hyperglycemia with phlorizin normalizes tissue sensitivity to insulin in diabetic rats. *J. Clin. Invest.*, **79**, 1510–1515.

Said, M. and Zerhouni, E. (2014) The role of public-private partnerships in addressing the biomedical innovation challenge. *Nat. Rev. Drug Discovery*, **13**, 789–790.

Santer, R., Kinner, M., Lassen, C.L., Schneppenheim, R., Eggert, P., Bald, M., Brodehl, J., Daschner, M., Ehrich, J.H., Kemper, M., Li Volti, S., Neuhaus, T., Skovby, P.G., Schaub, J., and Klaerke, D. (2003) Molecular analysis of the SGLT2 gene in patients with renal glucosuria. *J. Am. Soc. Nephrol.*, **14**, 2873–2882.

Scannell, J.W., Blanckley, A., Boldon, H., and Warrington, B. (2012) Diagnosing the decline in pharmaceutical R&D efficiency. *Nat. Rev. Drug Discovery*, **11**, 191–200.

Scharfmann, R., Pechberty, S., Hazhouz, Y., von Bülow, M., Bricout-Neveu, E., Grenier-Godard, M., Guez, F., Rachdi, L., Lohmann, M., Czenrichow, P., and Ravassard, P. (2014) Development of a conditionally immortalized human pancreatic ß-cell line. *J. Clin. Invest.*, **124**, 2087–2098.

Schulze, U. and Ringel, M. (2013) What matters most in commercial success: first-in-class or best-in-class? *Nat. Rev. Drug Discovery*, **12**, 419–420.

Swinney, D.C. and Anthony, J. (2011) How are new medicines discovered? *Nat. Rev. Drug Discovery*, **10**, 507–519.

Trusheim, M.R., Berndt, E.R., and Douglas, F.L. (2007) Stratified medicine: strategic and economic implications of combining drugs and clinical biomarkers. *Nat. Rev. Drug Discovery*, **6**, 287–293.

US Food and Drug Administration (FDA) (2011) Driving Biomedical Innovation: Initiatives to Improve Products for Patients, http://www.fda.gov/downloads/AboutFDAReportsManualsForms/Reports/UCM274464.pdf (accessed 10 August 2015).

Weir, G.C. and Bonner-Weir, S. (2011) Finally! A human pancreatic ß cell line. *J. Clin. Invest.*, **121**, 3395–3397.

Woodcock, J., Brumfield, M., Gill, D., and Zerhouni, E. (2014) The driving role on the critical path to innovative therapies. *Nat. Rev. Drug Discovery*, **13**, 781.

Zambrovicz, B.P. and Sands, A.T. (2003) Knockouts model the 100 best-selling drugs – Will they model the next 100? *Nat. Rev. Drug Discovery*, **2**, 38–51.

Zerhouni, E.A. (2014) Author response to comment on "drug discovery: turning the titanic". *Sci. Transl. Med.*, **6**, 229lr2. doi: 10.1126/scitranslmed.3008817.

Zhang, L., Zhang, Y.D., Strong, J.M., Reynolds, K.S., and Huang, S.M. (2008) A regulatory viewpoint on transporter-based drug interactions. *Xenobiotica*, **38**, 709–724.

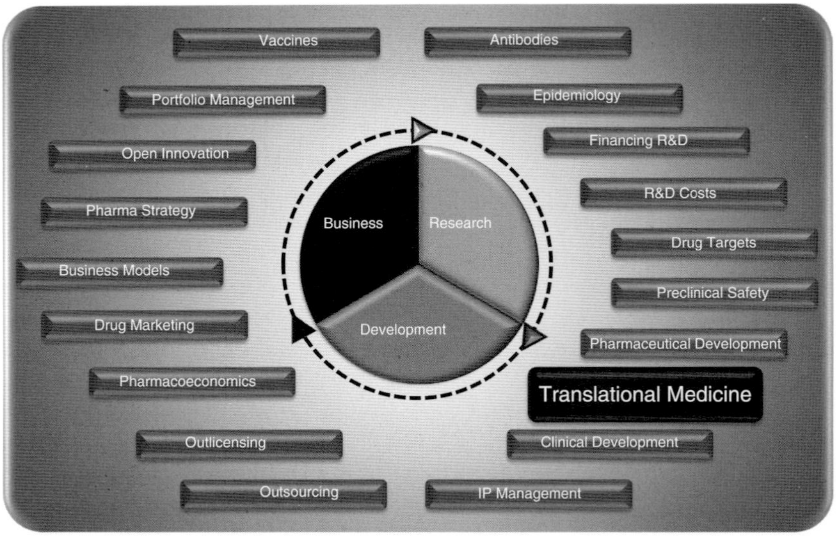
Vaccines
Antibodies
Portfolio Management
Epidemiology
Financing R&D
Open Innovation
R&D Costs
Pharma Strategy
Business
Research
Drug Targets
Business Models
Preclinical Safety
Drug Marketing
Development
Pharmaceutical Development
Pharmacoeconomics
Translational Medicine
Outlicensing
Clinical Development
Outsourcing
IP Management

# 7
# Translational Medicine: Enabling the Proof of Concepts

*Gezim Lahu and John Darbyshire*

## 7.1
## Introduction

The purpose of this book chapter is to review the basic concepts of translational medicine and how this translational discipline is providing value in the overall R&D process, how it is reducing development risk, and how maximum learning from the forward and back translation of research findings can be generated.

The key challenge in translational medicine has traditionally been the transition of preclinical molecules to the first administration to humans. Other key components of the translational phase are the establishment of target engagement and the detection of first hints for clinical efficacy. Because of these features, all pharmaceutical and biotech companies consider the translational piece as one of the decision-making processes to trigger further investments. Only recently has the value of back translation of early clinical findings been recognized. Learning from the clinical phase can help design better backup compounds or design molecules, which differentiate from existing therapeutic options or can derisk a program with alternative indications and added value to the asset.

This chapter will discuss the following topics:

- The role of translational medicine in knowledge generation in drug discovery and development.
- The relevance of the animal model to human disease and the quality of the dose selection. What are the best approaches in estimating the starting and efficacious dose?
- Model-based drug development (MBDD) concepts and implications in the translational space.
- Application of biomarker strategies and modeling & simulation (M&S) in designing a successful proof of concept (PoC).

Translational Medicine (TM) has rightly been identified as one of the main pillars in drug development and plays a significant role in providing the necessary input not only in product discovery, development, and differentiation but also in reinventions, life cycle management, and positioning activities. The term *translational medicine* although as a concept has always been applied and present

*Value Creation in the Pharmaceutical Industry: The Critical Path to Innovation,* First Edition.
Edited by Alexander Schuhmacher, Markus Hinder, and Oliver Gassmann.

in drug discovery and development; albeit under a different guise, the term is now being used to broadly describe, lead, and integrate activities from even basic discovery research to PoC and beyond.

Translational medicine apart from championing the forward translation from *in vitro* and nonhuman models to PoC rational designs has also championed the backward translation, which has played a significant part in rethinking the validity of the targets. Back translation has served as the basis for improved quality of thought to allow new hypothesis generation and has stimulated further the exploration of the disease and compound space (Venkatakrishnan *et al.*, 2015; Szalma *et al.*, 2010).

The consequence of the improved quality of thought and better characterization of the models resulted in the birth of innovative animal and human models that may allow testing of some of the hypotheses (Szalma *et al.*, 2010). This in turn has allowed the generation and validation of the research and development (R&D) hypotheses with human disease information derived from patient characterization (Szalma *et al.*, 2010).

Translational medicine has been instrumental in developing the concepts and studies that enable drug development by translating a molecular and genetic understanding of disease biology and drug response into the clinical use of pharmacodynamic (PD) biomarkers of response/safety and patient enrichment/selection strategies and rational repurposing of drugs.

Translational medicine has focused on understanding the pathophysiology of disease, the PD biomarkers of target engagement, and the hypotheses to enrich for responsive patients. This in turn has required understanding the predictability of the animal models in terms of safety and efficacy. New frontiers in technologies that have allowed the acquisition of human disease tissue and have enabled the characterization of target modulation and its link to pharmacological and consequently to disease modification have required the development of biomarkers and executable biomarker strategies and have enabled informed and quantitated decision.

The desire to be the best in industry and deliver game-changing therapies has raised the awareness of pharmaceutical organizations to deliver first or best in class drugs that clearly provide exceptional value to patients. A way forward to achieving this goal was through translational research and the application of a personalized medicine approach to drug development.

In research, linking drug mechanism of action to disease and patient phenotype/genotype has been used to inform indication and patient selection beyond classical definitions of disease by clinical symptoms or histopathology. The integration into early phase clinical development of PD biomarkers measuring target engagement and downstream biological effects will support the determination of PoC, derisking early-stage programs by confirming drug activity and establishing biologically active doses in humans. Such knowledge improves the probability of success supporting late-stage development and associated financial commitment. The development, registration, and reimbursement of more effective and differentiated therapies require tailoring strategies tested through patient enrichment/stratification approaches in drug target identification, late phase clinical trials, and the co-development/approval of companion diagnostics (CDx).

A comprehensive understanding of disease pathobiology and drug mechanism of action as well as feedback of clinical results to research, supports better prediction of human responses and purposing/repurposing of drugs. Such generation of new knowledge characterizing preclinical and clinical samples, requires the identification and application of innovative technologies to appropriately collected and annotated samples. Most importantly, given the inadequacy of many preclinical models to represent human disease and drug response, advanced analysis of large dataset coupled with the characterization of human samples obtained in phase 0 in an iterative loop of preclinical and clinical activities.

Phase 0 trials can be first-in-human studies of limited duration using lower doses conducted prior to phase I safety and tolerability testing. Such trials allow the investigation of drug activity and characterization of patient responses early in drug development. Other phase 0 studies may not involve investigational drugs but provide data collected on patients currently being treated with the standard care. These are done to improve the understanding of disease natural history and therapy, validate new end points or outcome measures, discover new targets or biomarkers, and begin early development of a biomarker for clinical use. Patients characterized in a phase 0 study may also become a pool of prospective patients for subsequent investigational studies.

## 7.2 Translational Medicine and Its Role/Value in Early Development

Translational medicine produces momentum across a pharmaceutical company for products that deliver on the truly differentiated value proposition of tailored medicines. TM is one of the key levers to implement the biomarker and personalized medicine and CDx strategies. Indeed translational medicine focuses on translating the laboratory and other findings into how to develop treatments and treat patients.

The three pillars of translational medicine are knowledge generation, tools, and biomarker-tailored strategy. The concepts operate by providing translational strategy and biomarker technical expertise to ensure the most efficient (time and money) delivery of robust PoC decisions. Translational research will:

- Drive a personalized medicine approach to drug development by translating a molecular and genetic understanding of disease biology and drug response into the clinical use of PD biomarkers of response/safety and patient enrichment/selection strategies and rational repurposing of drugs.
- Support the development of CDx to enhance the value and differentiation of therapies for patients and in the marketplace.

- Enable biomarker-driven trials to enrich responder patients in order to increase the probability of success for PoC (early development) and registration/commercialization (late development).
- Enable quantitative decision making by building biological, computational, and pharmacometric models.
- Generate and validate R&D hypotheses with human disease information derived from patient characterization clinical trial data using innovative technologies.

## 7.3 Knowledge Generation

The generation of the knowledge and improvement of the quality of thought are key to innovation and also essential for good decision making. Knowledge generation is the product of intelligent assembly of data from analytic and biological tools that allow deduction of unexplored diseases, targets, and phenotypic responses. Typically knowledge is not generated from a single source. People and brainpower, quality of internal and external data, and methods to validate findings by informatics and experimental approaches (animal models or human experimental medicine models) are some of the principal key components of knowledge generation. The impact of applied knowledge generation can significantly influence product differentiation and life cycle management for approved products. It can propose and/or support drug repositioning; subject/patient stratification for higher probability of approval, better dosing, and clinical trial design; broader or deeper understanding of disease mechanisms; and identification of targets for unmet medical needs. It requires a strong collaboration, which demands a multidisciplinary approach. As knowledge generation is a huge topic, we will not discuss the aspects from where translational medicine fuels its ideas and uses the information generated to create knowledge but rather focus on the traditional translational medicine impact on the type of experiments and data needed to move from animal models and nonclinical research to human trial and early clinical development.

## 7.4 Types of Data, Experiments, and Tools Needed to Move from Basic Research to Early Clinical Development

One of the key roles of translational medicine is to integrate the knowledge generated in the nonclinical space into meaningful recommendations that allow the transfer of the experiments from animals to humans. In general the selection of a new chemical entity (NCE) as a suitable candidate for development is based on the pharmacology, physicochemical properties, metabolism, and pharmacokinetics and formulation aspect in general, which are called druggability properties. Following this stringent testing regime, the regulated experiments are tailored to facilitate efficient knowledge generation to fulfill the basic criteria before moving into the first-in-human trials. The purpose of the first-in-human trials is to show that the NCE is safe and well tolerated and to explore the pharmacokinetics in humans. At this stage in development, the general efficacy *per se* is less relevant, and the focus is mainly in assessing the translational exploratory biomarkers of the target engagement or in some instances the safety and efficacy markers known to be relevant in the target that is being pursued.

With this goal in mind, a tailored series of *in vivo* and/or *in vitro* experiments whenever applicable are consulted before one of the biggest challenges in the development chain, namely, the definition of the starting dose in humans. In the next paragraphs, the focus will be to describe this process the strategies and tools applied to achieve the goal. There are many publications and guidances used for strategies, design, and implementation of safety studies to mitigate the risk for first-in-human trials for the investigational medicinal product (EMA, 2007; Industry, 2005, 2010).

### 7.4.1 Dose Selection

Over the years, there has been and still is a continuous attempt to replace animal models with various methodologies and approaches. However getting into the mechanistic details and safety assessment has proven to be very difficult, and a minimum number of models will continue to be important in supporting the transition (Hester and Harrison, 2006; Lipinski and Hopkins, 2004; Rowland, 2006).

The current guidelines describe in detail the types of information needed for the decision of the starting dose. In general the initial step is generating data in safety pharmacology where the assessment of the vital functions such as the respiratory, central nervous system, and cardiovascular safety are performed. An abbreviated list and description of the studies that should be considered are listed in Table 7.1. Some programs may require a more specific tailored approach. In oncology, the nonclinical programs may be more abbreviated, shorter, and aggressive so the NCE that has been efficacious in various animal models and has the potential to make a difference can be moved swiftly to testing in patients with advanced cancer that have failed the existing therapy. For more details of these studies, the reader should refer to the FDA, EMA, and ICH guidelines.

In general the types of studies and experimental designs described in Table 7.1 are considered before the recommendation of the starting dose and approximate efficacious and upper expected tolerable doses are recommended. In the succeeding text, we will briefly describe the animal models and tools that are used for dose predictions and the generations of safety and efficacy packages to allow testing into humans. The integrated assessment of this available information will give the confidence to proceed with caution with the first-in-human studies.

**Table 7.1** Types of studies conducted and their outcome to mitigate the risk for first-in-human trials for the investigational medicinal product.

| Type of testing | Design | Outcome/impact |
|---|---|---|
| Single- (acute) and repeated-dose toxicity testing | Both conducted in rodent and nonrodent species | Defines the acute and chronic toxicity including the duration of the first clinical trial in man |
| Genotoxicity studies | Bacterial and mammalian | Reference to the ICH guideline |
| Safety pharmacology studies | Assessment of vital functions in animal models | Safety assessment of respiratory, cardiovascular, and central nervous system functions |
| Toxicokinetic studies | Assessment of exposure data from safety pharmacology and animal toxicology studies | Adequate exposure adhered to the doses given and evidence that safety and toxicology observations resulted from the exposure |
| Metabolism and pharmacokinetic studies | Assessment of pharmacokinetics from *in vitro* and *in vivo* systems from the species used in efficacy studies | Identification of enzymes involved in the disposition, assessment of the cytochrome P450 inhibitory potential, bioavailability, and so on |
| Nonclinical pharmacology studies (dose–exposure–response studies) | Assessment of the dose–exposure–response in the relevant animal models | Establishment of exposures that are efficacious and characterization of the dose–exposure–response curve |

### 7.4.2
### Animal Models

While the animal models for safety assessment are well characterized and a lot is known about the background pathology/pathophysiology, a greater challenge has been in developing disease animal models representative of the actual disease and with the potential to guide the selection of the candidates/doses, surrogate markers, and so on. The importance of these models is that in principle they allow the estimation of the expected therapeutic doses/exposure needed to be achieved in humans for potential benefit. Furthermore, in addition to the toxicology studies if appropriately designed, the animal dose–response studies for the efficacy may also provide safety relevant data, especially if the exaggerated pharmacology results in safety concerns. In general the translation of animal data to humans is based on the considerations that the NCE and the target protein/receptor interaction are similar across the species and that the basic pharmacological concept in animals and humans follows the same principles. Over the years many animal models have been successfully semi-validated as representing the pathophysiology of interest. Some of the animal models are well established and are very predictive of human efficacy and exposure ranges needed to achieve the desired efficacy. The best example of these predictive models are for erectile dysfunction, diabetes, cardiovascular disease, infectious disease, and other therapeutic areas (Gajbhiye *et al.*, 2015; Rees and Alcolado, 2005; Marsden and Zack, 2015). While there is sufficient confidence in the animal models about the power of their predictability, there are also areas where they have been less predictive.

Animal models as the tools for investigating compounds for inflammatory disease or diseases with inflammation as the chief suspect have been particularly difficult to mimic. Inflammatory diseases may manifest locally but are driven by systemic pathobiophysiology. The biological mechanisms responsible for disease induction and progression are mostly unknown and can change during the disease. Most inflammatory diseases are defined by classical clinical syndromes, resulting in heterogeneous patient populations, and this requires an improved understanding of disease pathophysiology to treat patients based on heterogeneity and thus improve the success of therapy.

With new knowledge of the molecular phenotypes of disease, there is a potential for tailored therapies and enrichment of the trials with the responding patients to maximize risk–benefit through personalized medicine.

Translational animal disease models from chemically induced models to transgenic and knockout animal models are continuously developed and used for guidance on stopping or moving forward with the program. A guiding principle for the general selection of translational animal models for different diseases is to start the testing with models that include lower organisms, from bacteria to drosophila to understand some basic mechanisms and genetics of the signaling pathway. Models that are based on chemical inductions are often used in cases where the molecule does not pertain to any pathway or process. The advantage of these models is often that the severity of the disease can be controlled based on the amount of the inducer/chemical agent given. Information about the animal models, their utility and quality of information that it provides for traditional indications are well described in the literature. There is a lot known about the predictability potential of these models from data supporting the indication, dosing, and targeted translatable markers. In general most of the animal models are based on the known target modulation and their downstream/upstream effect on relevant readouts from the animals. Some animal models are more specific for the indications, while some only address safety aspects. Overall there is no single model for a given indication that provides all the answers one needs to move ahead. Low and colleagues have recently described the animal models used in the development of treatments for ulcerative colitis (Low, Nguyen, and Mizoguchi, 2013). It became apparent that there was no one specific animal model with all the answers and that was essential to have the time to work on the quality of thought and define the specific question in mind to identify the model that is most likely to address scientific questions or test the hypothesis. Naturally animal models have been used earlier in development where scientists have tested their hypotheses and created information that gives the confidence to move ahead on the development ladder. Also the questions raised or the hypotheses generated from the knowledge from available clinical data either on the disease or the competitors, may come from the compounds that either will be developed or are already in the market. In this instance, these compounds can be tested in the disease-specific or indication-specific animal models for hypothesis testing before performing an experimental/PoC trial.

Often the decision to test these compounds in animal models before seeking alternative experimental medicine trials to test the hypothesis in the human setting is constrained by the lack of confidence or reluctance to invest in expensive human trial although safety may not be a concern.

Mathematical pharmacokinetic/pharmacodynamic (PK/PD) modeling of the animal data with respect to safety/efficacy also provides some guidance on dosing. Basic assumptions when using the data from these models is that the pharmacokinetics somehow translates to PD biomarker or some sort of the pharmacodynamics effect. Key to these assumptions this is that in the PK/PD model, the PD components do not significantly change across species and the drug receptor or target interaction are similar. These models are consequently humanized by changing the PK component and adapting it to predict human PK, either using the traditional allometric scaling methods or using the *in silico* PKPB models that recently have been heavily utilized and give very confident and accurate results. Various softwares like GastroPlus, PK-Sim, and Simcyp are very common armament in the translational medicine (Figure 7.1).

However the challenging part comes from the biomarkers and their translatability into the clinic. While this can be better defined with the targets that may potentially be validated, it becomes much more uncertain when targets have not been validated and there are no competitors exploring the target space of interest. In this instance, scientific rationale and any experimental data including that which may be relatively unstructured can be useful to support or even discard the hypothesis. New technologies and methods are being developed to measure all sorts of biomarkers and metabolomics. Imaging, PGx,

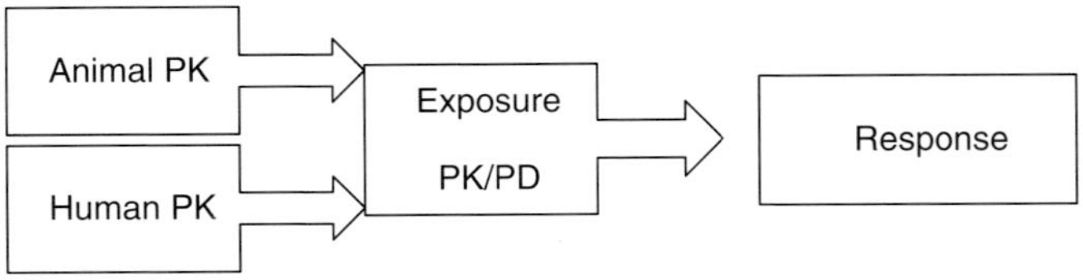


**Figure 7.1** The exposure at the site of action that is responsible to trigger a pharmacological action and humanization of the animal PK/PD models.

and gene technologies combined together are providing a huge impact on target, treatment, and patient selection with the best outcome on the risk–benefit assessment.

### 7.4.3 Fraction of NOAEL and Efficacious Dose

The no observed adverse effect level (NOAEL) is the concept that is used for guiding the safe and starting human dose. The selection of the NOAEL is discussed elsewhere in this book, and we will only briefly discuss the concepts here with respect to dose selection for first-in-human studies. When the dose–exposure recommendations are made in addition to the toxicology outcomes, the least responsive pharmacological models are also considered when selecting the starting dose. The NOAEL is established from the most sensitive species. The fraction of the NOAEL is generally considered as the safe starting dose, and this fraction is different in different therapeutic areas. As a general rule, we recommend the reader to look at the FDA/EMA guidance document. Principally the therapeutic area, known pharmacological class effects, and toxicity are key to the considerations of the fractions of the NOAEL as the starting dose. In oncology the fraction of the NOAEL used for the starting dose ranges depends on the species used, for example, the 6th of the NOAEL fraction from the monkey or the 10th from the rat is used for starting dose selection. For nononcology drugs, the fraction may vary and various approaches have been described in the literature. The FDA guidance 2005 (Industry, 2005) specifically highlights that the smaller fractions of the NOAEL should be considered if the dose toxicity–response curve is steep and there is high biological variability and also in cases where a new target is being explored and there is an irreversible toxicity and/or unexplained mortality. In contrast to unprecedented targets and well-characterized pharmacological class or where the NCE is an analog of the already established predecessor, higher fraction of the NOAEL may be used.

The guiding principle in selecting the starting dose for first-in-human healthy volunteers is to consider that for the selected doses planned to be tested, there is sufficient evidence that they are safe and can be tolerated. While in the nononcology space this is the norm, higher fractions of the NOAEL and consequently more aggressive escalation schemes are considered in the oncology area. In oncology, the first-in-human trial enrolls patients that have failed any therapy and are at the late and terminal stages of the disease. These patients are in general very heterogeneous, and there is a great complexity when designing and interpreting the data from these trials (Venkatakrishnan *et al.*, 2015).

Once the gathered evidence justifies the fraction of the NOAEL to be used as the starting dose, the next step is to project the doses that result in exposure explored in the animal PK/PD studies that resulted in the desired PD outcome or the exposure that other compounds in the same pharmacological class have resulted in the desired outcome. These exposures also define the escalation schemes and may further influence their refinement. Escalation schemes need to be designed in such a manner to create a quick but safe dose increase that will produce the exposure anticipated to produce the desired response. The escalation schemes may also follow the Bayesian approach although in general phase I studies are Bayesian by design. Toxicology studies and animal PK/PD experiments result in definitions of exposure that are safe and efficacious; however, they fail to estimate which doses in human may result in the desired exposures. Traditionally the doses selected have also followed the fractions of the dose from the NOAEL and were based on scaling the doses by following allometry (Boxenbaum and DiLea, 1995) and now increasingly the physiologically based pharmacokinetic (PBPK) modeling.

### 7.4.4 Allometric Scaling and PBPK

Some aspects of the allometric scaling have been discussed elsewhere in this book, and we will only briefly discuss the approaches that are used nowadays. Estimation of the starting dose has traditionally been done using various multipliers that are a composite of the interspecies difference in pharmacokinetics and pharmacodynamics and the uncertainty around them. Safely factors 10- to 100-fold below the NOAEL have often been used, without any real scientific evidence. In addition to the aforementioned methodology often employed by traditional toxicologists, there are exposure predictive softwares that are powerful in estimating the key PK parameters for dose estimation. Historically estimation of the absorption and bioavailability has been more difficult to perform, but the presence of software platforms such as Simcyp, GastroPlus, and PK-Sim has raised the bar and can predict the key parameters from the physicochemical properties alone. It is understandable that these predictions at the early stage are more prone to inaccuracies but are a good starting point and allow the experimenter to define and address the key questions in their experiments that will be used to feed its models and modules to achieve predictions of the desired quality. Often in-house developed software is available that can be used in conjunction with the commercially available software. These software platforms collectively known as *PBPK softwares* are expanding and incorporating even more of the physiological factors in their predictions. Their predictability is very powerful for the entities that undergo metabolism or are also cleared unchanged.

In general most of the pharmaceutical industry or at least large pharmaceutical companies use more than one approach for allometric scaling. The classical approach has been utilized to scale the clearance ($CL_i$) and volume of distribution ($Vd_i$) by using the well-established and frequently used power function:

$$CL_i \text{ or } Vd_i = a \times (\text{total body weight})_i^{\,b}$$

where $a$ and $b$ are the scaling factors. Boxenbaum and DiLea (1995) have found that in general the exponents of around 0.75 for clearance, around 1 for volume, and 0.25 for half-life are relatively predictable for human exposure. However it is very important to recognize that not all drugs behave like this; there have also been a number of cases where this concept did not work. Traditional allometry has been poor in predicting compounds that undergo extensive metabolism especially if the metabolism rate and enzyme activities do not scale. While allometry has been relatively poor in this dimension, recent PBPK models have been quite successful in addressing some of the gap in the predictability of allometry.

### 7.4.5 Physiologically Based Pharmacokinetic Models PBPK

PBPK modeling is an approach that uses mathematical modeling and physiology to predict the disposition of the compound in the body. The key component of PBPK is the mathematical construction of the physiology via the differential equations, the rates of the physiological processes and the physicochemical properties of the compound that are integrated to generate the key PK parameters that allow prediction and projections of the dose. PBPK outputs are also generated for different species and integrate information from various sources, including drug-dependent, physiological, and biological properties, as they vary with age in between species and patients. Several software packages, including Simcyp, PK-Sim and Gastro Plus have increased our understanding of the relationship between physiological and physiochemical properties and consequently dramatically increased the predictability of exposure (Jones *et al.*, 2015). The increase in confidence is not only within the pharmaceutical industry but also within the regulatory agencies where these types of analyses are frequently requested and also utilized for decision making. A more recent development in the field of PBPK is the integrations of PBPK models with the PD models to predict the impact of genotypic variability, in target binding capacity and target site drug concentrations on drug responses and variability, and also to predict the formulation differences on the aforementioned.

Furthermore, PBPK modeling has often been successfully applied in incorporating the knowledge on the pH-dependent solubility in informing the development of the final formulations for the clinical development (Chung *et al.*, 2015). The generation of the extensive PBPK databases in different species across compounds with various physicochemical properties is a continuous process and is key to increasing the confidence and predictability of PBPK models (Suenderhauf and Parrott, 2013).

### 7.4.6 Pharmacokinetic and Pharmacodynamic Modeling

PK and PD modeling as a key component of MBDD has, in the past several years, been applied extensively to integrate and use the information available for better decisions in designing and executing successful clinical trials.

MBDD is a paradigm and a mind-set that covers the whole spectrum of the drug development process, and it is not limited to a specific modeling technique or application area. The implementation of this paradigm requires pharmaceutical companies to foster innovation and make changes at many levels and specifically to establish mind-sets of the importance and use of MBDD, to align and if necessary set up the processes that are adaptive to the requirements of MBDD, and to create a closely collaborating organization by sharing the responsibility and institutionalization of the paradigm. Pharmaceutical companies that are able to embrace the changes MBDD poses will likely be able to improve their success rate in drug development, and the beneficiaries will ultimately be the patients in need.

The importance of integrating PK, PD, and toxicokinetic principles in drug development has been discussed in many conferences and consortia and also in publications that arose from these discussions. PK/PD M&S techniques as the core of MBDD have been very successful in integrating these principles in drug development. Several reports have criticized pharmaceutical companies for slowly adopting M&S as part of their development process. Consequently this criticism and raised awareness has resulted in regulatory agencies drafting various guidances advocating a more integral role of M&S in the drug development process and strongly encouraging pharmaceutical companies to increasingly utilize M&S (Report, 2004).

PK/PD models are extensively used to describe the temporal relationship between study drug administration, exposure, and response both for efficacy and important adverse events, following drug administration. Challenges, such as formulation, routes of administration, dosage regimens, differences between patient populations, and differences between drug analogs can also be addressed using these PK/PD models. The appropriate utilization of these models for the selection of the dose and for the patient.

Mixed-effects models are used to estimate the population parameters (fixed effects) of the PK, PK/PD, or disease models; to quantify between-subject variability; and to identify factors (covariates) that may influence these parameters.

Drug development is a continuous process during which a great amount of data is acquired from experiments in *in vitro* systems, animals, and humans. One of the biggest hurdles in the pharmaceutical industry is its inability to integrate the knowledge generated. Although the main purpose in generating this knowledge is linked to decision making in a more informative fashion, it appears that very often, these principles are completely ignored. Very often, the lessons learned from previous trials or experiments are not utilized to their full potential, and not considered when designing new studies or even to terminate the development program. There is a lack of continuous updates of the knowledge generated. PK/PD modeling activities are a continuum that starts as early as discovery and continue throughout the development phase and beyond. The appropriate integration of available data and translation of the animal PK and pharmacodynamics data via PK/PD modeling into the human situation is where PK/PD modeling can have a massive impact. Especially in establishing the starting doses and defining the dose escalation schemes.

As polypharmacy and multidrug regimens have become common practices in today's often complicated pharmacotherapy, the need to investigate the drug–drug interaction is increasing. In general these considerations have traditionally been a major concern of clinicians with regard to safety and tolerability. PK/PD modeling has been successfully used to identify these concerns by estimating the doses of the two interacting drugs that provide the acceptable risk–benefit associated with the use of both comedications on the regimens that are recommended for one or the other monotherapy. In general there are two types of interactions a) change in exposure (PK interaction) that may consequently result in pharmacodynamic interactions (PD) and b) PD interactions that may result from synergy, antagonism, competition at sites of action or from indirect alteration of physiology and can be quantified using PK/PD concept. Koch and coworkers have recently described the modeling concept of additive and synergistic effect using preclinical data to also distinguish synergistic and/or additive drug effect in detail (Koch *et al.*, 2009).

There are many examples in the literature where these concepts have been successfully applied either in future trial designs, decisions to continue with the development or in the decision to stop the development. Examples include gemcabene that was under development and intended to be given in combination with statins much like ezetimibe. The purpose of this meta-analysis was to assess the value of an integrated model of these NCEs with likely clinical profile in making data-driven decisions during the drug development process (Mandema *et al.*, 2005).

Acute stroke therapy by inhibition of neutrophils was another example of an adaptive phase II dose–response, proof-of-concept study to establish whether UK-279,276 improved recovery in acute ischemic stroke (Krams *et al.*, 2003). Many other examples of the MBDD have been reported and reviewed (Lalonde *et al.*, 2007). Either way utility of these concepts has been tremendous in savings and investments and also reducing the unnecessary exposure of patients. In general, these considerations traditionally have been a major concern of clinicians with regard to safety and tolerability.

## 7.5 FIM (Dose Escalation and MTD)

The confidence to move a NCE into first-in-human comes only after the successful completion of the nonclinical program minimizing the risk that the compound will not jeopardize the safety of volunteers in the trial. The toxicology and safety program, including the allometric scaling performed using different approaches, and the conservative selection of the fraction of the first-in-human dose and the precautionary principles considered, give the confidence to enter human trials. Various designs with different escalating schemes are considered. Furthermore well-defined inclusion and exclusion criteria are explained in various industry guidances (Industry, 2010). These schemes primarily consider safety and tolerability but also exposure. The general guiding principle is to consider an escalating scheme that allows a rapid uptitration to exposures where the desired effect may be anticipated and then apply less aggressive dose escalations by reducing the increment of the increased dose to fine-tune the desired doses. The most commonly used schemes are the generic geometric or algebraic series or concentration-guided designs where the exposure achieved in human trials approaches the exposure from the toxicokinetic data. The dose escalation continues until the minimum intolerated dose (MID) may be observed (Cutler *et al.*, 1997). The dose before the MID is called the maximum tolerated dose (MTD). Although the estimation of the MTD is very useful to know for later drug–drug interaction and safety studies, sometimes it is not pursued especially if the desired therapeutic range is much below the MID. In oncology trials, the concept of dose-limiting toxicity (DLT) is applied for the determination of the MTD. The definition of the MTD and DLT are very well defined in the guidances.

At this time, the single rising dose data generated would have been analyzed, and the basic human PK model would have had been built and used to humanize the animal PK/PD model built during the nonclinical development. Following the characterization of the single rising doses, the multiple rising doses are conducted to investigate safety and tolerability including pharmacokinetics of the multiple rising doses of the NCE. Simulations from the humanized PK/PD model which predict the desired exposure for efficacy are key when designing the multiple-dose escalation trials, although the primary objective of the multiple-dose escalation trial is to test the safety and tolerability of multiple doses of the NCE.

The doses and dosing interval selected for the multiple rising doses are based on the outcome of the single rising dose trial, and it is balanced with the simulations of the PK/PD model previously developed. The guiding principle for frequency, in a multiple-dose escalation trial, is that the area under the curve (AUC) and maximum concentration (Cmax) from the last day of the multiple-dose escalation trial are less than the AUC and Cmax from the highest safe and tolerable dose from the single dose.

Employment of various translational safety and efficacy biomarkers that would give an indication of the proof of presence or proof of engagement would be essential and key to increasing the confidence in moving to the next step.

During the single- and multiple-dose escalation trials, it is becoming very common that a food effect is also investigated although being only informative. The food effect on the safety and pharmacokinetics of the NCE is explored later during development; the initial assessment is a good guidance for the later decision making and also for PoC trials on the small cohort of patients.

## 7.6 Proof of Concept (PoC)

PoC is a state that may be a temporally spaced concept anywhere throughout discovery and development and suggests that there is sufficient evidence to suggest that the key principle on which the hypotheses were based and tested is present and valid. It also suggests that the key risk factors that could hamper the development are absent or identified and known. There are many definitions forged around the philosophical definition and concept of PoC, but in general, it really means the state where we are for the first time faced with the decision based on the presence or absence of the key parameters defined earlier, whether the body of evidence is sufficient to invest resources and effort in further developing the compound. In general considering the duration and the objectives of PoC trials in many therapeutic areas especially when developing the drug for the treatment of the disease, whose symptoms and clinical end points are not obvious within the first few weeks of treatment, it is essential that PoC trials are armed with various tools and measurements that may suggest sufficient evidence that the desired changes on the clinically relevant end points are likely.

One of the most difficult discussions to have, is regarding on the levels of confidence we are ready to accept for the decisions to be made. Of course, for such confidence to be sufficient, it is essential to have tools including biomarkers that ideally are translatable into the meaningful clinical end points and methodology such as PK/PD modeling that allow informed decisions to be made. The levels of confidence in general start from the availability of the information that is converted into knowledge. The tools to report that information also provide answers to many questions. Biomarkers as measurements, various pharmacometric approaches including the PK/PD modeling, statistical tests, and adaptive designs are tools mostly used in PoC. The biomarker selection strategies have been very different from company to company, but they are essentially developed to build that confidence ultimately that the target is present, that we are engaging the target, and that there is a dose–exposure–response relationship. Further they should also facilitate the assessment of any other off-target toxicity or other effects, which are also dose-related (Wagner, 2008).

Furthermore biomarkers have been very useful in selecting the patients that are most likely to respond (Chien *et al.*, 2005). PK/PD modeling and adaptive design have also been employed as tools to raise confidence. One approach successfully used in pharmaceutical industry is MBDD which utilizes quantitative pharmacology and pharmacometric methods that help to separate the unexplained variability from the real signal (Sheiner, 1997). Often such information is gained by pooling information across trials, doses, and more importantly compounds with the same mechanism of action in order to help us understand and separate the noise or uncertainty from the real effect of the new compound. For example, they are used to help identify subpopulations where dose adjustments may be required. For planning subsequent clinical trials, these models are used to predict the anticipated treatment effect size and as a basis for clinical trial simulations (CTSs) (Holford *et al.*, 2000). The choice of concepts, models, and approaches depends also on the underlying disease and mode of action of the drugs.

The level of confidence including the effect size should be set up before conducting of the trial. It is essential to understand the probability distribution of the effect size and also understand the threshold below which the effect size would not be desirable and not worth pursuing. The probability distribution and confidence intervals of the effect size would allow comparison of the threshold below which development becomes meaningless. In general the assessment of the effect size should not only focus on one measurement or outcome of the trial but should also include more parameters and consider the development of the clinical utility index (CUI). CUI is a composite parameter from the weighted individual readouts that can be of a safety and efficacy nature, merged into one value, which can then be used for further modeling. There are several examples describing the applicability of the CUI in Ouellet et al. (2009).

These aspects are generally captured in the target product profile or TPP and should serve as the reference for decision making and comparison. While these types of TPP where the target is validated or there is another compound being developed or already in the market are more straightforward because of the reference already established, the novel targets and new class of compounds with their TPPs may be more difficult and should be dynamic and adapt to the new body of evidence from not only the clinical outcomes but also the change in regulatory and commercial environment.

The key elements that should be considered in PoC stretch from safety and efficacy, to regulatory and commercial. The quantitative assessment of these can also be done by applying the concept of CUI where the health economic parameters maybe incorporated and weighted into the CUI.

### Summary

To summarize, we have attempted to shed some light on the translational space and have discussed the approaches and the tools that are used when moving from research to development and entering the human trial. We have briefly discussed the data needed to generate knowledge that allows us to move safely to the human trial. While no specifics were made, these approaches are very common and can serve as the bases and platforms for defining new strategies. Bench to bed and bed to bench translation combined with the availability of personalized medicine strategies, big data, and outcome research data will in the future play a massive role in chasing and designing a PoC. This with not only improved probability of success but also more importantly a firm corroboration of the hypothesis and confidence to move the NCE into the full clinical development. Although data analysis has improved substantially, it is still an area where we could do much more especially since there is a large amount of information waiting to be processed into knowledge. Supercomputers like IBM's Watson, that use artificial intelligence and provide answers rather than a selection of relevant documents listed according to relevance, will be reliable partners for scientists to make decisions and choices. Ultimately the accesses to the relevant technologies in biomarker research, outcome, and big data information in a form that can be quantified and analyzed, in tandem with the artificial intelligence and significant human component, will be the features of not only translational space but also drug discovery and development.

## References


Boxenbaum, H. and DiLea, C. (1995) First-time-in-human dose selection: allometric thoughts and perspectives. *J. Clin. Pharmacol.*, **35**, 957–966.

Chien, J.Y., Friedrich, S., Heathman, M.A., de Alwis, D.P., and Sinha, V. (2005) Pharmacokinetics/Pharmacodynamics and the stages of drug development: role of modeling and simulation. *AAPS J.*, **7**, E544–E559.

Chung, J., Alvarez-nunez, F., Chow, V., Daurio, D., Davis, J., Dodds, M., Emery, M., Litwiler, K., Paccaly, A., Peng, J., Rock, B., Wienkers, L., Yang, C., Yu, Z., and Wahlstrom, J. (2015) Utilizing physiologically based pharmacokinetic modeling to inform formulation and clinical development for a compound with pH-dependent solubility. *J. Pharm. Sci.*, **104**, 1522–1532.

Cutler, N.R., Sramek, J.J., Greenblatt, D.J., Chaikin, P., Ford, N., Lesko, L.J., Davis, B., and Williams, R.L. (1997) Defining the maximum tolerated dose: investigator, academic, industry and regulatory perspectives. *J. Clin. Pharmacol.*, **37**, 767–783.

EMA (2007) Clinical Efficacy and Safety: Clinical Pharmacology and Pharmacokinetics, Strategies to Identify and Mitigate Risks for First-in-Human Clinical Trials with Investigational Medicinal Products.

Gajbhiye, S.V., Jadhav, K.S., Marathe, P.A., and Pawar, D.B. (2015) Animal models of erectile dysfunction. *Indian J. Urol.*, **31**, 15–21.

Hester, R.E. and Harrison, R.M. (2006) *Alternatives to Animal Testing*, Issues in Environmental Science and Technology, Royal Society of Chemistry.

Holford, N.H., Kimko, H.C., Monteleone, J.P., and Peck, C.C. (2000) Simulation of clinical trials. *Annu. Rev. Pharmacol. Toxicol.*, **40**, 209–234.

Industry, F. G. F. (2005) Estimating the Maximum Safe Starting Dose in Initial Clinical Trials for Therapeutics in Adult Healthy Volunteers, U.S. Department of Health and Human Services, F. A. D. A., Center for Drug Evaluation and Research (Ed.).

Industry, F. G. F. (2010) FDA Guidance for Industry, M3(R2) Nonclinical Safety Studies for the Conduct of Human Clinical Trials and Marketing Authorization for Pharmaceuticals, U.S. Department of Health and Human Services, F. A. D. A. (Ed.).

Jones, H., Chen, Y., Gibson, C., Heimbach, T., Parrott, N., Peters, S., Snoeys, J., Upreti, V., Zheng, M., and Hall, S. (2015) Physiologically based pharmacokinetic modeling in drug discovery and development: a pharmaceutical industry perspective. *Clin. Pharmacol. Ther.*, **97**, 247–262.

Koch, G., Walz, A., Lahu, G., and Schropp, J. (2009) Modeling of tumor growth and anticancer effects of combination therapy. *J. Pharmacokinet. Pharmacodyn.*, **36**, 179–197.

Krams, M., Lees, K.R., Hacke, W., Grieve, A.P., Orgogozo, J.M., and Ford, G.A. (2003) Acute Stroke Therapy by Inhibition of Neutrophils (ASTIN): an adaptive dose-response study of UK-279,276 in acute ischemic stroke. *Stroke*, **34**, 2543–2548.

Lalonde, R.L., Kowalski, K.G., Hutmacher, M.M., Ewy, W., Nichols, D.J., Milligan, P.A., Corrigan, B.W., Lockwood, P.A., Marshall, S.A., Benincosa, L.J., Tensfeldt, T.G., Parivar, K., Amantea, M., Glue, P., Koide, H., and Miller, R. (2007) Model-based drug development. *Clin. Pharmacol. Ther.*, **82**, 21–32.

Lipinski, C. and Hopkins, A. (2004) Navigating chemical space for biology and medicine. *Nature*, **432**, 855–861.

Low, D., Nguyen, D.D., and Mizoguchi, E. (2013) Animal models of ulcerative colitis and their application in drug research. *Drug Des. Dev. Ther.*, **7**, 1341–1357.

Mandema, J.W., Hermann, D., Wang, W., Sheiner, T., Milad, M., Bakker-Arkema, R., and Hartman, D. (2005) Model-based development of gemcabene, a new lipid-altering agent. *AAPS J.*, **7**, E513–E522.

Marsden, M.D. and Zack, J.A. (2015) Studies of retroviral infection in humanized mice. *Virology*, **479–480**, 297–309.

Ouellet, D., Werth, J., Parekh, N., Feltner, D., Mccarthy, B., and Lalonde, R.L. (2009) The use of a clinical utility index to compare insomnia compounds: a quantitative basis for benefit-risk assessment. *Clin. Pharmacol. Ther.*, **85**, 277–282.

Rees, D.A. and Alcolado, J.C. (2005) Animal models of diabetes mellitus. *Diabet. Med.*, **22**, 359–370.

Report, F. C. A. O. (2004) Innovation or Stagnation: Challenge and Opportunity on the Critical Path to New Medical Products, Administration, F. A. D. (Ed.).

Rowland, M. (2006) Microdosing and the 3Rs, National Center for the Replacement, Refinement, and Reduction of Animals in Research (NC3Rs).

Sheiner, L.B. (1997) Learning versus confirming in clinical drug development. *Clin. Pharmacol. Ther.*, **61**, 275–291.

Suenderhauf, C. and Parrott, N. (2013) A physiologically based pharmacokinetic model of the minipig: data compilation and model implementation. *Pharm. Res.*, **30**, 1–15.

Szalma, S., Koka, V., Khasanova, T., and Perakslis, E.D. (2010) Effective knowledge management in translational medicine. *J. Transl. Med.*, **8**, 68.

Venkatakrishnan, K., Friberg, L., Ouellet, D., Mettetal, J., Stein, A., Troconiz, I., Bruno, R., Mehrotra, N., Gobburu, J., and Mould, D. (2015) Optimizing oncology therapeutics through quantitative translational and clinical pharmacology: challenges and opportunities. *Clin. Pharmacol. Ther.*, **97**, 37–54.

Wagner, J.A. (2008) Strategic approach to fit-for-purpose biomarkers in drug development. *Annu. Rev. Pharmacol. Toxicol.*, **48**, 631–651.

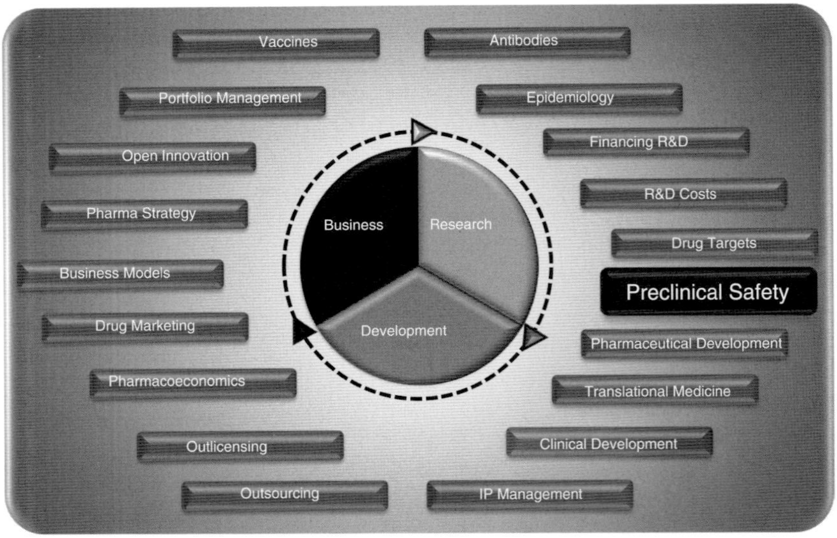

Vaccines
Antibodies
Portfolio Management
Epidemiology
Financing R&D
Open Innovation
R&D Costs
Pharma Strategy
Business
Research
Drug Targets
Business Models
Preclinical Safety
Drug Marketing
Development
Pharmaceutical Development
Pharmacoeconomics
Translational Medicine
Outlicensing
Clinical Development
Outsourcing
IP Management

# 8
# Preclinical Safety and Risk Assessment

*Paul Germann and Rob Caldwell*

## 8.1
## Introduction

Administering a new molecular entity (NME) to human patients or healthy volunteers is a decision based upon an assessment of unfavorable toxicity or side effects compared to the potential beneficial therapeutic effect for the patient to cure the disease or to improve the health status. The generation and analysis of an NME preclinical safety profile (hazard identification, hazard description, and exposure data) serves the need to generate a risk assessment (RA). The goals of a scientifically sound RA are to achieve regulatory acceptance and to inform human safety physicians and patients. The following chapter describes the various aspects of preclinical safety, their interdependencies and how they contribute to the RA. The RA can be divided into three main concepts. First, hazard identification is conducted through experimental approaches utilizing *in vitro* and *in vivo* models. Second, hazard characterization is conducted based upon these experimental results with an effort to predict human relevance. Finally, the weight of evidence of these accumulated data sets is utilized to evaluate the safety margins between observed toxicities and the predicted efficacious doses for humans.

This chapter should answer the following key questions:

1) What are the components of a preclinical safety assessment?
2) What are the relevant questions which need to be answered by scientists before a new drug can be applied in humans?
3) How can we manage the risk which is exposed to humans when taking a new chemical or biological entity?

## 8.2
## Test Systems

### 8.2.1
### *In Silico* Analysis

The concept of hazard identification and risk assessment of a new therapy depends on the generation of experimental data. These data sets are generated in various

*Value Creation in the Pharmaceutical Industry: The Critical Path to Innovation*, First Edition.
Edited by Alexander Schuhmacher, Markus Hinder, and Oliver Gassmann.

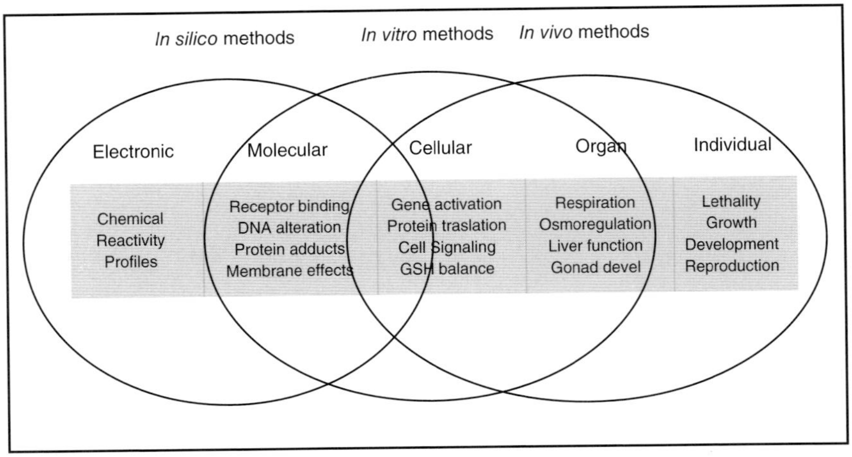


**Figure 8.1** Schematic screening cascade for selection of compounds through application of "*in silico*" (left), "*in vitro*" (center), and "*in vivo*" methods (right). From http://www.vet.utk.edu/reactivity/.

test systems during the detection and maturation of a new drug candidate (see Figure 8.1, left side). In the very early stages of its development, a compound is competing with a lot of other very close compound structures and the compound availability is limited. Therefore, a screening and selection filter cascade using *in silico* tools and programs is an efficient approach to narrow down the number of "good" candidates. Based on broadly, worldwide sourced *in silico* tools and prediction programs like M-Cat or Derek, compounds with a highly reactive chemical and toxicity profile are deselected in favor of less reactive compound classes. These computer-based assessments can also identify compounds with mutagenic or clastogenic potential, thereby further winnowing the pool of favorable compounds.

### 8.2.2
### *In Vitro* Experiments

*In vitro* experiments typically utilize cell-based assays including bacteria, immortalized cell lines, or primary cells derived from animals or humans. Cell-based assays lend themselves to high-throughput, repetitive, and cost-effective methods to generate pharmacologic and/or toxicologic data sets. This contrasts to the relatively much more rate and resource-limiting expenditures needed to conduct *in vivo* animal (and eventually human) assessments *In vitro* experiments are generating real experimental data, that serve as an important basis for hazard identification prior to ever exposing living organisms to the NME. Modern *in vitro* test systems have potential to further refine toxicity profiles before conducting *in vivo* experiments, thereby reducing the number of animals utilized for *in vivo* experimentation. Finally, the number of assays and the number of validated real data points generated by *in vitro* methods are much more cost effective to achieve than in any other test system.

## 8.3 Case Study: hERG Assay

Receptor binding studies are playing a crucial role to discriminate unwanted side effects (off-target effects) of molecules by binding to off-target receptors binding sites, which induces toxicity in a different organ.

One prominent example to detect such undesired off-target toxicity is the hERG (human ether-a-go-go-related gene) assay. The function of this hERG channel is the conduction of potassium ($K^+$) ions out of the muscle cells of the heart (cardiac myocytes). This out-flux ion channel current is critical in correctly timing the return to the resting state (repolarization) of the cell membrane during the cardiac action potential. Inhibition of the hERG channels may lead to short QT syndrome with the risk of ion channel dysfunction that can lead to the risk of potentially fatal cardiac arrhythmias (e.g., *torsades de pointes*; see Figure 8.2), due to repolarization disturbances of the cardiac action potential (Sanguinetti *et al.*, 1995; Moss *et al.*, 2002). This channel is sensitive to drug binding, as well as decreased extracellular potassium levels, both of which can result in decreased channel function and drug-induced (acquired) long QT syndrome. Among the drugs that can cause QT prolongation, the more common ones include antiarrhythmic, antipsychotic agents, and certain antibiotics (including quinolones and macrolides) (Sanguinetti and Tristani-Firouzi, 2006; Zhang *et al.*, 2010). Although there exist other potential targets for cardiac adverse effects, the vast majority of drugs associated with acquired QT prolongation are known to interact with the hERG potassium channel. Due to the awareness of the potential danger of such QT drugs, the regulatory authorities issued recommendations for the establishment of cardiac safety during preclinical drug development: ICH S7B, The nonclinical evaluation of the potential for delayed ventricular repolarization (QT interval prolongation)

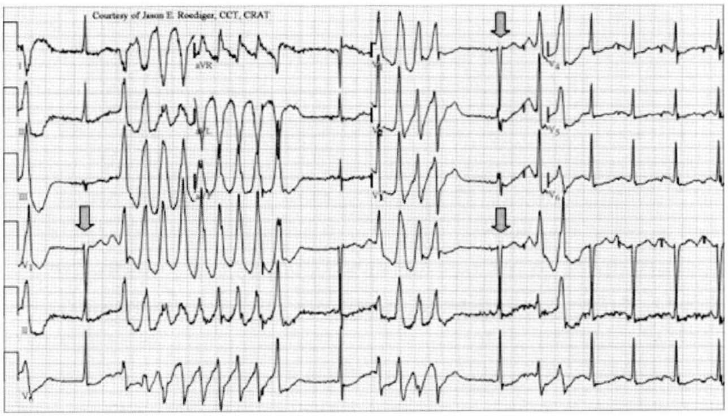


**Figure 8.2** Exemplary electrocardiogram with induction torsades de pointes (TdP). The arrows represent falling T-wave which is initiating the induction of the TdP. Modified from Wikipedia, courtesy of Jason E. Roediger, CCC, CRAT. 26 February 2014.

by human pharmaceuticals, issued as CHMP/ICH/423/02, adopted by CHMP in May 2005. Preclinical hERG studies should be accomplished in GLP environment (Good Laboratory Practice), mostly in a high-throughput *in vitro* test system.

Other general *in vitro* methods are depicted in Figure 8.1 in the center. Some of them are deemed mandatory by regulatory agencies, and some are preferred as more exploratory in nature for use in screening funnels. Even though a constant introduction of better and more predictive *in vitro* methods helps to improve the selection of new innovative drug candidates, the complexity of an integrated animal or human organism needs also the conduction of *in vivo* experiments.

### 8.3.1
### *In Vivo* Experiments

With the introduction of NME into experimental animals, researchers are trying to generate data to better understand the interactions between the new compound and the whole organism. *In vivo* experiments are tested on entire, living organisms, while *ex vivo* experiments are using organs or organ parts. *In vivo* testing is only employed after *in vitro* testing so as to reduce the total number of animals that must be utilized. The major goals of *in vivo* experiments with an NME in animals can be summarized as:

1) (De)validation of an experimental model for a certain scientific assessment
2) Describe the desired pharmacological activity in an adequate (disease) animal model
3) Exposure assessment
4) Dose – effect relationship and reversibility
5) Toxicity assessment
6) Determine the metabolic pattern, the compound fate, and excretion routes
7) Identify liabilities not identified in the preceding experimental tiers (i.e., *in silico*, *in vitro* experimentation)

To achieve these goals, a rational scientific design and specific species selection are of central importance. For the species selection, researchers have to balance between the selection of expected "most humanlike" species and the principles of reducing, replacing, and reusing animals as best possible. Preexisting data proposing a close proximity of the species to the human metabolism and physiology should lead to the choice from the scientific point of view. Sometimes the ethical situation drives the species selection to a species which is a little more "away," means less comparable with the human setting, but is still acceptable from the relevance of the expected scientific data. It is also important to note that after many decades of data collection using, for example, mice and rats, researchers have a broad database of reference range of normal variations in these species with regard to behavior, physiology, genetics variation, and overall tolerability. There is also a preference to utilize relatively lower-order (e.g., Rodentia) versus higher-order (e.g., nonhuman primates) phyla. Every effort should be made to reduce the total number of animals utilized while obtaining best translatable and accurate data sets from the applied *in vivo* models.

The more a new drug candidate progresses in its development, the more *in vivo* experiments for hazard identification are performed. These experiments are addressing the various aspects of hazard identification and their detailed description (see Figure 8.1, left part):

1) Physiologic impairment detection prior to overt moribund toxicity
2) Organ toxicity
3) Growth impairment
4) The reproductive capability
5) Developmental toxicity
6) Impairment of learning and memory
7) Effects on drug tolerance and drug dependency
8) Effects on drug distribution, metabolism, and pharmacokinetics

## 8.4 The Preclinical "Package" during the Development of an NME

The process of drug innovation and drug development is taking place in a highly regulated environment to secure the validity and reproducibility of the generated scientific data which are serving as the basis for the approval of an NME for the intended human therapeutic use. Therefore, most of the experimental designs for *in vitro* and *in vivo* experiments for hazard identification are described in general ICH guidelines (ICH: International Conference on Harmonization of Technical Requirements for Registration of Pharmaceuticals for Human Use). Figure 8.3 gives an overview about the key deliverables of preclinical safety.

The top bar in this illustration called *stage* describes the various stages of a pharmaceutical development cycle. Underneath the different teams which are supporting the NME project are listed. The exact composition of these experiments is constantly adapted to the various needs of the different phases of clinical development. These program or project phases are depicted in the following text starting from the research phase over the "preclinical development phase," entering into and progressing through the clinical phases I–IV. Important decision points related to investment decisions based on the available data set are also shown. Selection of the final lead NME candidate to enter clinical trials is a substantial financial investment compared to the NME optimization which has occurred up to that point. Typically only the lead most desirable NME among the pool of available compounds is selected for *in vivo* animal assessment in regulated animal toxicology studies. The next major financial investment occurs when a compound moves from phase I human safety and tolerability studies into phase II proof-of-pharmacology (or proof-of-concept (PoC)) studies. The final major resource investments occur upon execution of large global phase III clinical

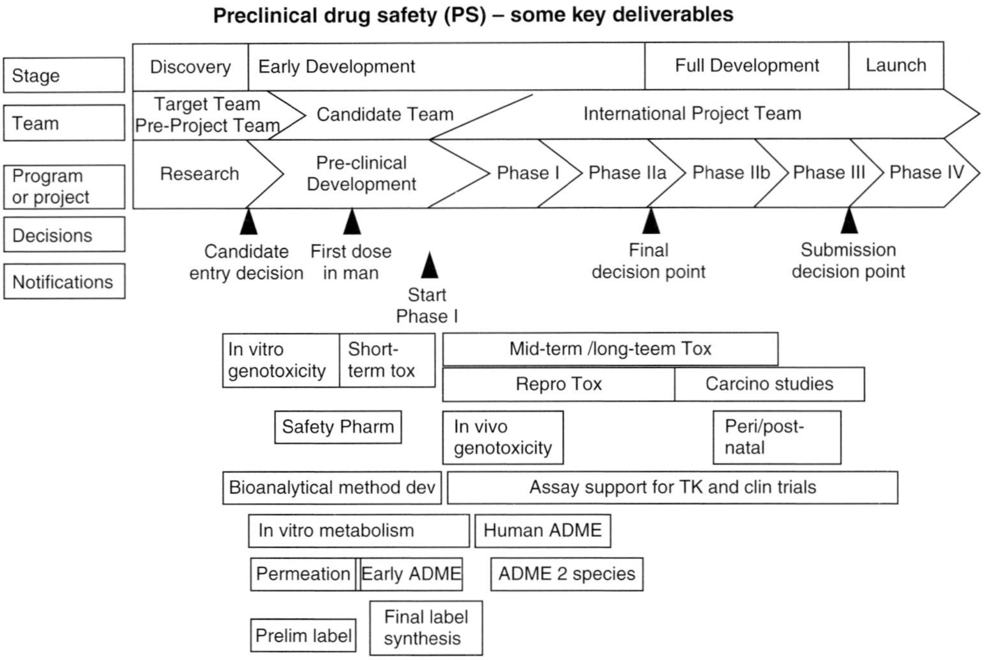


**Figure 8.3** The development of a compound and the preclinical data set.

trials. The decision to execute these sequentially more resource-consuming stages of clinical development are all based upon the continuous preclinical assessments that either (i) enable subsequent stages of clinical development, (ii) investigate the comparative safety of NME changes throughout the development, or (iii) evaluate the risks associated with new unintended findings only observed based upon the size and scale of late-stage clinical development trials.

## 8.5 Factors Influencing the Preclinical Data Set

There is not a fixed preclinical data package. The main factors can be summarized as follows:

- Timing and costs
- Intended clinical application route
- Treatment duration and treatment frequency
- Clinical indication
- Ongoing changes of the regulatory landscape
- New drug formats

### 8.5.1 Timing and Costs

The timing of these preclinical safety experiments reflects an attempt to streamline the development in terms of costs and efficacy. For example, relatively short-term animal studies (single dose to 4 weeks) enable short-term phase I human clinical trials of similar duration. Longer-term (e.g., 13 weeks) and more complex (embryofetal development) animal safety studies occur in parallel to phase I human clinical trials so that animal data can be accumulated to derisk pending phase II clinical trials. It is important that animal trials be conducted in judicious manner while as best possible not delaying human clinical trials simply on the basis of timing. Other examples are carcinogenicity assessments. In the current regulatory environment, 2-year rodent carcinogenicity studies are generally not initiated until a favorable human PoC study is completed. Delaying these resource-consuming 2-year rodent toxicology studies has the benefit of saving around USD 4 million and saving animal lives of around 1000 mice and/or rats if the compound does not exhibit the desired pharmacology in patients.

### 8.5.2 Intended Clinical Application Route

The design and size of a preclinical package are substantially influenced by various factors. The intended clinical application route like oral intake of tablets or fluids, dermal application, intravenous or subcutaneous injection, or inhalation should be reflected and reproduced in the preclinical safety package. This attempt to simulate the clinical situation leads to an adaptation of the preclinical package by adding or deleting specific experiments.

### 8.5.3 Treatment Duration and Treatment Frequency

Also the intended treatment duration and frequency (e.g., once daily, twice weekly, three-cycle infusion per 2 weeks) have an essential influence on the design of the preclinical package. The closer the clinical treatment is simulated in the preclinical package, the more predictive is the hazard identification and risk assessment.

### 8.5.4 Clinical Indication

The intended clinical indication influences the scope and details of the preclinical package. Treatment of a life-threatening disease like cancer leads to a higher accepted risk for unwanted side effects by the authorities, physicians, and also the patients. A fast track of the compound development in these indications may save patient lives for those with imminent poor prognosis. With this in mind, sometimes a limited preclinical package prior to approval is accepted by the authorities when it will be supplemented by additional postapproval commitments. The studies are then submitted during the launch or marketing phase to complete the preclinical data set later.

### 8.5.5 Ongoing Changes of the Regulatory Landscape

The scientifically sound generation of experimental data with state-of-the-art methods is the basis for acceptable hazard identification. The ICH is unique in bringing together the regulatory authorities and pharmaceutical industry of Europe, Japan, and the United States to discuss scientific and technical aspects of drug registration. Since its inception in 1990, the ICH has gradually evolved, to respond to the increasingly global face of drug development, so that the benefits of international harmonization for better global health can be realized worldwide. With this permanent ongoing process of improving the scientific standards in the experimental data generation, the preclinical data packages are constantly evolving to maintain regulatory acceptance.

### 8.5.6 New Drug Formats

The recent increase of therapeutic innovation is reflected by the introduction of a number of *new drug formats*. The requirements for an accurate preclinical safety profile (hazard identification) and risk assessment of these drug formats are addressed in different guidelines. For them, a broad variation of open and different safety questions needs to be answered. As various drug formats, the following list is an attempt for completeness:

- Small molecular entities (SMEs)
- Biologics (for details, see following text)
- Cell-derived therapies
- Medicinal devices
- RNA-derived therapies
- Gene therapies
- Vaccination
- Other new drug formats
- Combination therapies

As the details of the various aspects of hazard identification in the different drug formats would overload the scope of this chapter, the following detailed "Deep Dive Case Study" will give insight into one of these drug formats (biologics).

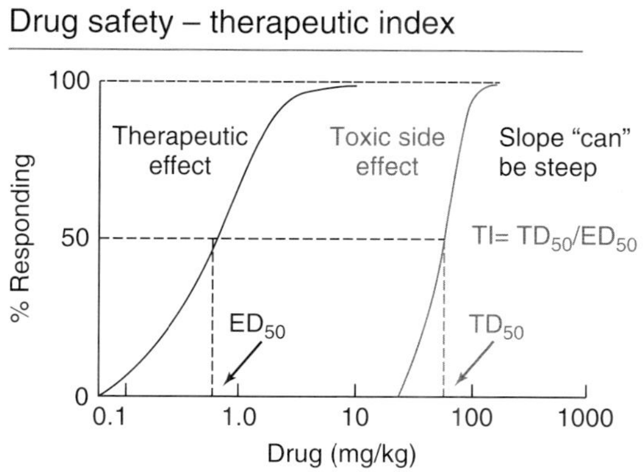


**Figure 8.4** Schematic figure describing the therapeutic window (or range, or index). From http://tmedweb.tulane.edu/pharmwiki/doku.php/drug_receptor_theory?s%5b%5d=therapeutic&s%5b%5d=window.

## 8.6 Translation into Humans: The "Therapeutic Window"

All generated experimental data during the hazard identification are serving as a basis to estimate or calculate a so-called therapeutic window or a therapeutic index (Figure 8.4). The therapeutic window of a drug is the range of drug dosages which can treat disease effectively while staying within a safe drug exposure range. In other words, it is the dosages of a medication between the amount that gives an effect (effective dose) and the amount that gives more adverse effects than desired effects. Medication with a small pharmaceutical or therapeutic window must be administered with care and control, frequently measuring blood concentration of the drug, since it easily gives adverse effects, including irreversible damage.

## 8.7 Influence of Intended Therapeutic Use on the Risk Assessment (RA)

As mentioned before, the intended clinical indication is influencing the scope and details of the preclinical package.

In addition, an RA is highly dependent on the severity of disease and the existence of alternative treatment options. Rare and severe diseases, where no curative therapies exist, have desperate patient populations who are hoping for new curative approaches. The patients are looking for new and probably also more risky treatment alternatives. Therefore, this high medical need is taking into account by the approving authorities accepting a risk assessment with a higher probability of undesired side effects.

In these situations, a good risk management with closer monitoring of the patients toward the toxic side effects of a compound and with supportive therapies which limiting the drug-related toxicity is of central importance. As an example, a cancer patient treated with a chemotherapeutic cytotoxic agent might be closely monitored for the occurrence of diarrhea and suppression of the hematopoiesis. If this monitoring for side effects shows emerging signals of this drug-related toxicity, a supportive blood transfusion and fluid replacement is the second level of risk management. As a third level of risk management, in some instances, the treatment scheme is stopped due to the toxic side effects.

## 8.8 Deep Dive Case Study: Safety Assessment of Biological Drug Formats

The majority of therapeutic drugs approved in the marketplace are often referred to as *small molecules*. These are chemically synthesized molecules (e.g., salicylic acid, acetaminophen, simvastatin) with molecular weights in the range of hundreds of daltons. By comparison, the class of therapeutic agents often referred to as *large molecules* (or biologics) is biosynthetically produced by engineered prokaryotic or eukaryotic cell lines. Many of these marketed biologics (e.g., infliximab, trastuzumab, ranibizumab) are either monoclonal antibodies (mAbs), other types of soluble receptors, or bioengineered molecules based upon these structural frameworks. The common feature of all of these new biological entities (NBEs) is that they are proteins formed from amino acids. They are engineered to bind specific amino acid sequences (or epitopes) in target proteins important in disease processes. These two factors (protein structure, amino acid target selectivity) can lead to unique challenges and opportunities for assessing preclinical safety of these types of drugs.

Because of the unique specificity of biologic drugs, researchers must justify that the chosen toxicology species is relevant for assessing safety. Put another way, the biologic must bind to the target epitope in the selected toxicology species in a manner comparable to the human epitope. For example, many biologics will bind to the cognate epitope in cynomolgus monkeys. This is perhaps not surprising given the phylogenetic proximity of humans to monkey. It is less common for biologics to bind to other common toxicology species such as mouse, rat, and dog. This species comparison was historically done utilizing *in vitro* immunohistochemical techniques (often referred to as *tissue cross-reactivity* ). A fluorescently labeled version of the therapeutic biologic was applied to a panel of tissues from each toxicology species and humans. The staining pattern and intensity of immunoreactivity of the labeled biologic were compared across species and tissues in order to determine epitope (and presumably pharmacologic) comparability. Access to experimental procedures for gene cloning and expression, surface plasmon resonance (Biacore) techniques, and generation of epitope-specific cellular reporter assays have made this a much more quantitative process. For example, a therapeutic mAb against interleukin-6 (IL-6) can be evaluated for respective binding affinity against recombinant IL-6 from mouse, rat, rabbit, dog,

**Table 8.1** Affinity ($K_D$, dissociation constant) of an IL-6 mAb against various preclinical species epitopes.

| Species | Affinity ($K_D$) |
|---|---|
| Human | 200 pM |
| African green monkey | 280 pM |
| Cynomolgus monkey | 280 pM |
| Dog | >1 μM |
| Mouse | >1 μM |
| Rabbit | >1 μM |
| Rat | >1 μM |

Recreated from http://www.google.com/patents/WO2014074905A1?cl=en.

and nonhuman primates (Table 8.1). In addition, the investigational mAb can also be placed into *in vitro* reporter experiments sensitive to soluble recombinant IL-6 concentrations. Increasing concentrations of the mAb will decrease output from the reporter assay for each species. The respective dose–response and resulting empirical inhibition constant (e.g., $IC_{50}$) across species can be utilized to assess relative potency (or efficacy) in the cellular reporting system. The relatively higher selectivity of biologics as compared to small molecules means that (theoretically) there will be less off-target toxicity, but it also requires that the biologic exhibits relevant binding in the utilized toxicology species. Species selection for both small and large molecules is enhanced if a known pharmacology can be identified in the chosen species.

Assuming a relevant toxicology species can be identified, the next unique challenge to biologic therapeutics is demonstrating the ability to maintain sustained systemic exposure of the biologic agent throughout the duration of the toxicology study. The nominally competent immune systems present in drug-naive toxicology animals will not infrequently respond to a parenterally administered biologic by recognizing it as a foreign nonself protein and generate an immune response to the biologic. One aspect of this immune response can be generation of antidrug antibodies (ADAs) which can bind to the therapeutic agent. These ADAs can act by neutralizing both the systemic circulation and the intended pharmacologic activity of the biologic. Therefore, relatively small-scale pilot *in vivo* pharmacokinetic studies can be conducted to assess the ability to maintain NBE exposures following utilizing the intended route of administration and dosage levels (both of which can affect relative immunogenicity) as the definitive pivotal regulatory toxicology studies. The results from these experiments can be used to demonstrate that although the NBE can bind to mouse, rat, rabbit, dog, and monkey species with comparable affinity and/or potency, only the monkey can be utilized as a reliable experimental species based upon a neutralizing ADA response evident in remaining species.

Final selection of relevant toxicology species allows the researcher to evaluate the most appropriate toxicology strategy based upon the relevance of species-specific physiology for human risk assessment and contingent upon regulatory agency expectations for a given stage of clinical trial involvement. For example, identification of rat, rabbit, and monkey as relevant species would allow a toxicology strategy comparable to small-molecule drug development as indicated in the following table. Alternatively, a toxicology program only enabled in cynomolgus monkey poses some unique challenges with respect to developmental and reproductive toxicology assessments. In particular, mAbs do not generally cross primate placenta during the period of fetal organogenesis. This precludes use of monkey as a species for embryofetal toxicology assessments. Regulatory guidance documents indicate that embryofetal toxicology studies should not routinely be conducted with monkeys. It should be noted however that these are only guidances and that scientific data indicating transplacental fetal exposure during organogenesis and/or an especially unique developmental or teratogenic pharmacology may warrant unique study designs in order to appropriately derisk human fetal exposures under therapeutic applications. The interpretation of all available developmental and reproductive toxicology assessments (or the absence thereof) can impact the approved label for the therapeutic agent.

Fertility assessments with a monkey-only enabled biologic also pose a unique challenge for the toxicology strategy. Small-molecule fertility studies generally utilize a schema where males and females are administered test compound and then allowed to mate and the resulting study endpoints (e.g., estrous cycle duration, number and distribution of corpora lutea/implantation sites, viable/nonviable embryo enumeration) are evaluated. It is generally recognized that the comparatively higher specificity of biologics and the impracticality of monkey breeding preclude routine conduct of monkey fertility studies. Therefore, for monkey-only enabled programs, sexually mature animals are specifically selected for use within at least one of the general toxicology studies. Reproductive tissues (among other routine tissues) are specifically evaluated from control and treated animals to ascertain if the NBE exhibited effects on organ weights, tissue anatomy and histology, and staging of gametogenesis in ovaries and testes. Additional endpoints may include NBE-dependent effects on estrous, reproductive hormones, and ejaculate constitution.

Genetic toxicology studies are generally not conducted based upon the inability of therapeutic biologics (particularly mAbs) to cross both the plasma and/or nuclear membrane and thereby interact with DNA. However, if the biologic did have potential to interact directly with DNA or alter chromosomal architecture following systemic exposure in target patient populations, routine *in vitro* and *in vivo* genetic toxicology studies may be warranted based upon a weight of evidence for a particular NBE. Interesting examples are antibody–drug conjugates (ADCs) wherein toxic small molecules are bound to an antibody. The antibody delivers the toxin to a specific cell type (e.g., tumor cell) where the toxin destroys the cell. In this case, the toxin portion of the ADC would undergo genetic toxicology assessments.

Safety pharmacology studies are generally not conducted unless weight of evidence suggests otherwise. Again, the ability to conduct standard *in vivo* safety pharmacology studies is based upon relevance of routine toxicology species. And again, if the NBE did have pharmacologic activity (as evidenced through pilot screening studies or relevant literature review of mechanism) which could alter vital organ system physiology (e.g., central nervous system, cardiovascular, pulmonary, gastrointestinal), then dedicated relevant *in vitro* and/or *in vivo* safety pharmacology studies may be warranted based upon a weight of evidence for a particular NBE.

Historically, routine metabolism studies were not conducted. Two examples of where some metabolic assessments are conducted are when the pharmacologic activity of the biologic can impact expression of metabolizing enzymes or the metabolism/distribution of ADC toxins. Recombinant cytokine therapies (e.g., IL-6) can induce detoxifying liver enzyme CYP gene expression, which can in turn alter the metabolic homeostasis of a patient resulting in pharmacokinetic changes to normal endogenous physiologic processes as well as concomitant NCE (New Chemical Entities) therapies for particular patient populations. Toxicologic assessment of ADCs poses a particular set of challenges in that the pharmacokinetics and metabolism of all components of this type of NBE must be evaluated in order to understand observed/potential toxicities. This necessitates developing analytical methods to different native antibody, native conjugate, perhaps linker, and intact ADC distribution, metabolism, and pharmacokinetics in each relevant toxicology species and human patient populations.

Preclinical carcinogenicity risk assessments for all NMEs (both NCE and NBE) should be based upon assessment of relevant toxicology species (metabolic pathways, exposure margins, and pharmacologic relevance), evaluation of literature-based pharmacology review as it relates to tumor/oncogenesis, and epidemiologic data from homologous therapeutics and/or congenital/acquired genetic conditions. Based upon this weight of evidence, the NME sponsor should propose an appropriate and relevant risk assessment strategy to regulatory agencies. The risk assessment strategy for NCEs frequently includes assessment of *in vivo* carcinogenic risk of lifetime exposure to the NCE in rodent species (e.g., 2-year rat carcinogenicity study). It is generally agreed that lifetime exposure studies are not warranted for large animal species including dog and monkey. What this means for NBE candidates is that these moieties frequently do not exhibit relevant pharmacologic properties in rodent species, due to the unique epitope specificity of these agents. However, this does not preclude the necessity to conduct and vet an integrated carcinogenicity risk assessment for NBEs. In some cases, a lifetime exposure study in rodent species may be feasible and beneficial to conduct. Two examples of how this might occur are exemplified by the abatacept and ustekinumab toxicology programs. Abatacept was utilized in a 2-year rat carcinogenicity study. Rats exhibited an increased incidence of mammary tumors due to the pharmacologic immunosuppression of the NBE. The chronic immunosuppression resulted in activation of recrudescent DNA virus and eventual increase in mammary tumors. Another strategy is exemplified by ustekinumab. The NBE was utilized in a lifetime exposure study in transgenic mice expressing the target antigen. The results of all of these experiments were controversial with regard to the translatability of the mechanism of tumorigenesis, the ultimate necessity of these resource-consuming *in vivo* studies as compared to *in vitro* alternatives, and the relevance to human risk assessment. Nonetheless, the results culminated in labeling requirements which must be integrated by physicians and their patients.

Juvenile toxicology risk assessments are conducted in a similar fashion for both NCE and NBE therapeutics. As indicated for other assessments earlier, the feasibility of NBE juvenile toxicology studies is impacted by the ability to maintain exposures (without a neutralizing antidrug immune response) in juvenile animals of biologically relevant species. In addition, the physiologic distribution of NBEs may impact feasibility of juvenile toxicology studies. If pilot experiments evidence that the NBE is not distributed to target organs (testes, eyes) or that the physiology of the relevant toxicology species (immune, CNS, reproductive systems) is markedly dissimilar for juvenile animals as compared to juvenile human patients, this must also be integrated into the toxicology strategy. All of these facets must be integrated into a pediatric investigational plan (PIP) and vetted with regulatory agencies. All of this is essential to inform physicians and patients of the risk–benefit profile of a potentially beneficial therapeutic to a particularly naive target patient population.

The following paragraphs present case studies regarding the unique challenges of conducting NBE toxicology studies in relevant toxicology species. Each of the case studies involves a mAb or mAb-based NBE (Table 8.2).

**Table 8.2** Comparison of various cross-species properties of biologics and their influence on the design of the safety package.

| | Small molecule | Biologic #1 | Biologic #2 | Biologic #3 | Biologic #4 |
|---|---|---|---|---|---|
| Relevant species | Mouse, rat, rabbit, dog, monkey | Mouse, rat, rabbit, dog, monkey | Rat, rabbit, monkey | Monkey | Transgenic mouse, surrogate mouse mAb |
| Placental transfer during organogenesis | Mouse, rat, rabbit | Mouse, rat, rabbit | Rat, rabbit | N/A | Mouse |
| Routine embryofetal toxicology | Mouse or rat, and rabbit | Mouse or rat, and rabbit | Rat and rabbit | N/A | Mouse |
| Nonroutine embryofetal toxicology | N/A | N/A | N/A | Monkey | N/A |
| Routine fertility toxicology | Mouse or rat | Mouse or rat | Rat | N/A | N/A |
| Nonroutine fertility toxicology | N/A | N/A | N/A | Monkey | N/A |

## 8.9 NBE Case Study 1

mAb1 was designed for lifetime therapeutic application for various pain indications. The target epitope was mainly expressed in neurologic, immune, and epidermal tissues, among others. mAb1 exhibited adequate binding to mouse, rat, rabbit, and monkey epitopes.

Repeat-dose toxicology studies of up to 6 months' duration were conducted in both rats and cynomolgus monkeys. No test item-related changes were observed in monkey. Test item-related skin lesions and secondary systemic inflammation and septicemia were observed in a significant enough number of rats to warrant early termination of study. These findings increased in severity with increasing systemic exposure. The findings were also consistent with the neutralization of target epitopes in skin, immune system, and neuronal tissues. Dose levels in rat which did not produce the skin lesions were also associated with ADA responses which neutralized the test item. Reproductive toxicology studies were conducted in rat and rabbit revealing no test item-related effects on fertility, embryofetal development, or postnatal development. No test item-dependent changes were observed among immune function tests. Dedicated neurobiology studies conducted in rats identified a test item-dependent change consistent with the pharmacologic activity of the compound. Carcinogenicity studies and juvenile toxicology studies were not conducted based upon the stage of clinical development for this compound.

This case study represents several features unique to NBE toxicology studies. It is uncommon for multiple species to be qualified for use for NBE toxicology studies. The use of rodent and rabbit species allowed for a rigorous (two species, >10/sex/group, extensive historical databases) reproductive biology risk assessment. The use of rodent species precluded the ability to maintain system exposures throughout chronic dosing periods due to production of test item-neutralizing ADAs. Only the rat was able to identify test item-dependent effects (skin lesions, neurologic changes) consistent with the pharmacological mode of action of the compound. Both the increased susceptibility to skin lesions (and potential skin infections) and the neurologic changes were provided to physicians, clinical trial volunteers, and regulatory agencies.

## 8.10 NBE Case Study 2

mAb2 was designed for lifetime therapeutic application for various immunological indications. The target epitope was primarily expressed by the immune system. mAb2 only exhibited adequate binding to cynomolgus monkey but not to any rodent or rabbit species. Therefore, the entire toxicology program utilized only cynomolgus monkeys.

Repeat-dose general toxicology studies were conducted in cynomolgus monkeys for up to 6 months' duration. Test item-related changes were limited to a fungal infection. This finding is consistent with the immunosuppressive pharmacology of the test item. Breeding studies are not routinely conducted in cynomolgus monkeys unless there is a specific cause for concern based upon weight of evidence from literature, *in vitro*, and/or *in vivo* studies. No such risks were identified for mAb2. Nonetheless, only animals deemed sexually mature based upon physiologic assessments (testis size, estrous cycle, age, and weight) were utilized in the 6-month study. This allowed for assessment of test item-related effects on the reproductive tissues in both male and female animals. No test item-related changes were observed among the reproductive tissues for male and female animals on this study. Based upon the results of histologic assessments in these sexually mature animals, no dedicated cynomolgus monkey breeding studies were conducted with this compound. Embryofetal development toxicology studies were not conducted based upon the stage of clinical development for this compound. Repeated administration of test item did cause neutralizing ADA production among multiple animals at multiple dose levels and dose routes (intravenous, subcutaneous). With continued dosing, animals would progress to producing acute postdose hypersensitivity reactions resulting in a range of clinical observations including recumbency, decreased blood pressure and heart rate, emesis, skin reddening, and in some cases mortality. The cause of these adverse clinical signs was ultimately attributed not to mAb2 binding to target epitope but rather to the severity of the ADA response. The ADA formed immune complexes with test item that triggered an anaphylactoid-like response including complement cascade activation and immune complex deposition in the microvasculature of lung, kidney, and liver (among other) tissues. Both the increased susceptibility to opportunistic infections and the heightened awareness of potential hypersensitivity reactions were relayed to physicians, clinical trial volunteers, and regulatory agencies.

## 8.11
## Carcinogenicity Risk Assessment for Marketed Drugs

In the following case study with pantoprazole, the dependencies of hazard identification and RA are described with particular focus on the carcinogenicity potential. Figure 8.5 depicts a possible decision tree for an in-depth risk assessment of carcinogenicity findings (Figure 8.6).

Pantoprazole is a marketed drug, several years on the global market for the treatment of gastric ulcers and gastroesophageal reflux disease (GERD). More than 1 billion patients have been successfully treated with this NCE. Pantoprazole irreversibly inhibits the $H^+/K^+$-ATPase enzyme of the parietal cells in the stomach. With this, the secretion of gastric acid is substantially reduced, leading to a very efficient pain relief and healing of the ulcerated gastric mucosa. The maximum intended treatment duration with pantoprazole is 8 weeks.

In the carcinogenicity experiments, rats and mice were treated for 2 years with a 100-fold of the intended human dose (high dose, approximately 560 × fold AUC,

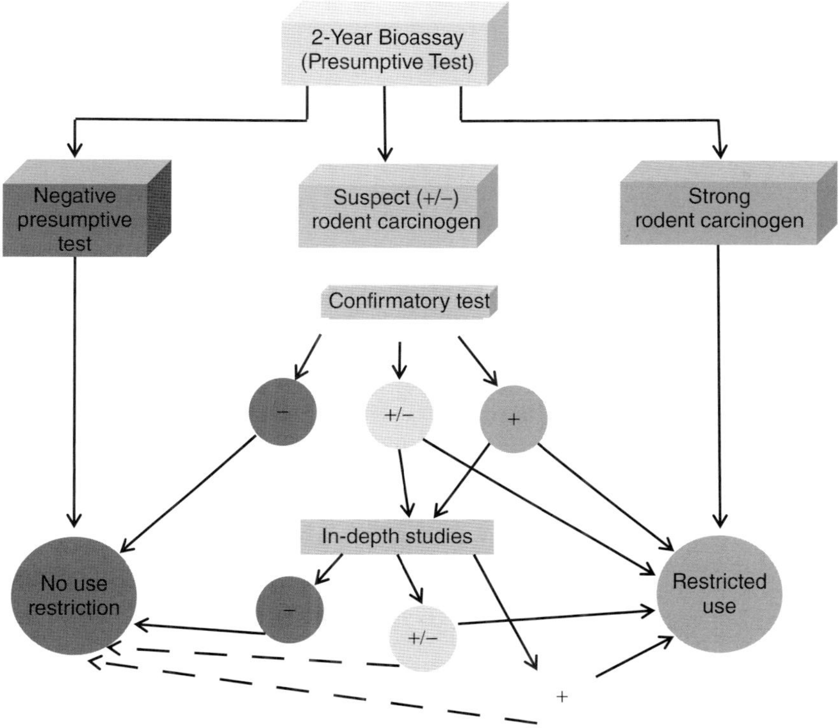


**Figure 8.5** Exemplary decision tree for carcinogenicity risk assessment.

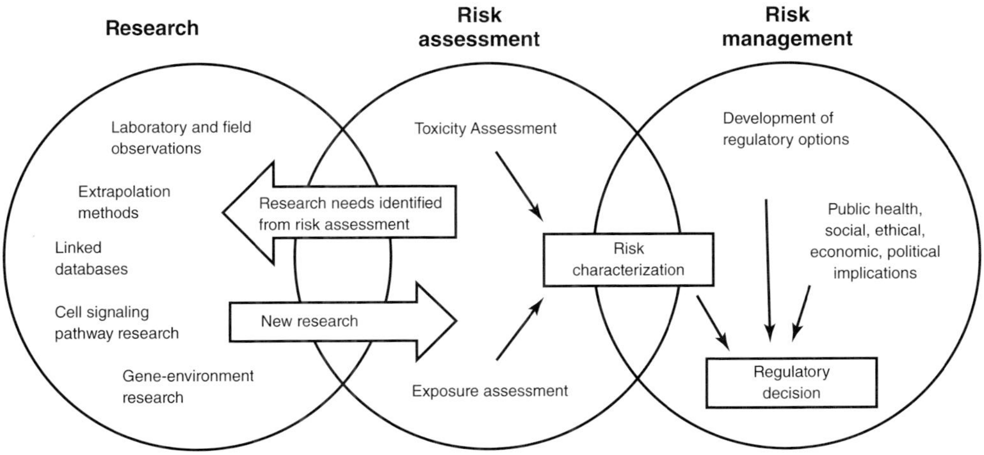


**Figure 8.6** The concept of hazard identification and risk assessment. From http://www.nap.edu/openbook.php?record_id=9871&page=27.

area under the curve of drug concentration in the plasma; mid dose, approximately 280 × fold). Individual rats of the high and mid dose developed under this treatment paradigm benign and malignant tumors of the glandular stomach. These enterochromaffin-like cell (ECL) tumors are hormone-producing endocrine cell in the stomach. Mice did not exhibit these tumors but showed the formation of a few benign hepatocellular liver tumors in high-dose group animals. The *in vitro* and *in vitro* mutagenicity experiments did not show any genotoxic potential of pantoprazole. To conduct an adequate RA of these carcinogenicity findings, the following two questions are of central importance: How high is the likelihood that humans treated with pantoprazole will develop tumors in the stomach or in the liver? How do these data extrapolate and predict their relevance for the intended clinical use?

## 8.12
## Treatment Duration

During the carcinogenicity study, the rats and mice are treated for their whole life long The maximum period of an 8-week treatment in humans would expose them to a time-correlated risk of drug exposure of 0.0026 compared to that exposure in the carcinogenicity experiment (based on a human life expectancy of 80 years, reduced by 20 years for childhood – 60 years).

*Secondly the AUC* (see preceding text) gives an additional safety factor (260×). Important here is the fact that the used rat strain exhibited the so-called chronic progressive nephropathy (CPN) of old rats, leading to a reduced elimination of pantoprazole through the urine and with this increasing the AUC of the compound. Very important is the fact that no initial *genotoxic potential* of pantoprazole was determined; otherwise the safety factor calculation and application as therefore mentioned would become difficult.

Finally, the elucidation of the *histogenesis and mechanism of the specific tumor* formation is very important. So a potential rodent carcinogen (see Figure 8.5), in this case pantoprazole, will go into further analysis with in-depth investigative mechanistic studies. For the benign liver tumor in mice, the comparison with control data incidences is important. While the "in study control incidences" in comparison with the treated groups exhibited an increase in the high-dose group mice. In the perspective of a broader control data bank of the same mice strain and breeder, the incidences were comparable to that of the high-dose incidence. Secondly a general observation in the *species comparison* is important: While liver cell adenomas (benign tumors) are frequent in mice, these tumors are very rare in humans. In mice it is easy to induce benign liver cell adenomas during the constant "drug overload" of the liver during a lifelong experiment as a signal of deregulation of the drug-induced metabolism and detoxification processes. This species difference for liver tumors is known and has also to be taken into account.

For the ECL tumors seen in the stomach of the rat, the situation has an additional component of hazard identification: in several investigative mechanistic studies in the various preclinical animal species (rats, mice, hamsters, monkeys, dogs, pigs), it was shown that the rat has firstly many more ECL cells per square centimeter mucosal area than the humans and also many more than all other species (*species sensitivity*). Regarding the mechanism of the tumor formation, the following hypothesis could be validated with data of several mechanistic experiments: when pantoprazole is constantly blocking the acid production in the rat stomach, the permanent increased pH is triggering a high production and secretion of an endocrine hormone called *gastrin*, produced by the G-cell in the last part of the rat stomach – the antrum. This increased gastrin level leads to a stimulatory effect on the cell proliferation of the ECL cell and ultimately to the tumor formation in the rat stomach. Important here is the fact that the first indicator of this mechanism, the cell proliferation of ECL cells, was only seen in rodents and here, namely, in rats as very high doses. All nonrodent species did not show this indicative signal even in long-term toxicity studies with high doses. In addition, as a confirmation of this species specificity of this lesion confined to the rat, gastric biopsies where taken in human clinical trials and also later in phase IV postmarketing trials. All these human follow-up studies for risk management did not exhibit any proliferative signal for the ECL cells in humans.

All these described factors were taken into account for generating the *risk assessment for pantoprazole* regarding the tumor formation. Based on the broad "therapeutic window" between the intended treatment duration and dose in humans and the doses which induced tumors in rats and mice, pantoprazole exhibits no risk for humans in the intended therapeutic indication. In addition, due to the proven species-specific sensitivity (rat, stomach) and low relevance for human situation (mouse, liver), this compound was registered without any restrictions. Of course, all these carcinogenicity findings are described in the so-called product label.

## 8.13 Conclusion – the "Art" of Preclinical Safety: Summarizing the Concept of Hazard Identification and Description, Risk Assessment, and Risk Management

Preclinical risk assessment of NMEs is an important part in the development process of new therapeutic agents. Based on an experimental data set, generated by *in silico*, *in vitro*, and *in vivo* experiments, the identification of potential hazards for humans is the first step of this process (see Figure 8.4). By adding further experiments and applying state-of-the-art available knowledge, the hazard characterization is done in close alignment with regulatory guidelines. This data profile is then assessed in the RA where the "art" of preclinical safety starts.

### Acknowledgment

The authors wish to thank Mrs. Marion Schoebel for her superior support in the edition of the graphical figures.

## Disclosures

This presentation was sponsored by AbbVie. AbbVie contributed to the design, research, and interpretation of data as well as writing, reviewing, and approving this presentation/publication. Paul Germann and Rob Caldwell are employees of AbbVie.

## References

Moss, A.J., Zareba, W., Kaufman, E.S., Gartman, E., Peterson, D.R., Benhorin, J., Towbin, J.A., Keating, M.T., Priori, S.G., Schwartz, P.J., Vincent, G.M., Robinson, J.L., Andrews, M.L., Feng, C., Hall, W.J., Medina, A., Zhang, L., and Wang, Z. (2002) Increased risk of arrhythmic events in long-QT syndrome with mutations in the pore region of the human ether-a-go-go-related gene potassium channel. *Circulation*, **105** (7), 794–799. doi: 10.1161/hc0702.105124. PMID 11854117.

Sanguinetti, M.C., Jiang, C., Curran, M.E., and Keating, M.T. (1995) A mechanistic link between an inherited and an acquired cardiac arrhythmia: HERG encodes the IKr potassium channel. *Cell*, **81** (2), 299–307. doi: 10.1016/0092-8674(95)90340-2. PMID 7736582.

Sanguinetti, M.C. and Tristani-Firouzi, M. (2006) hERG potassium channels and cardiac arrhythmia. *Nature*, **440** (7083), 463–469. doi: 10.1038/nature04710. PMID 16554806.

Zhang, X.M., Zhu, Z.H., Sun, X.L., Guo, J., Zhao, Z.Z., and Zhang, Z. (2010) Chinfloxacin hydrochloride inhibit HERG potassium channel at open state. *Acta Pharmacol. Sin.*, **45**, 1491–1496.

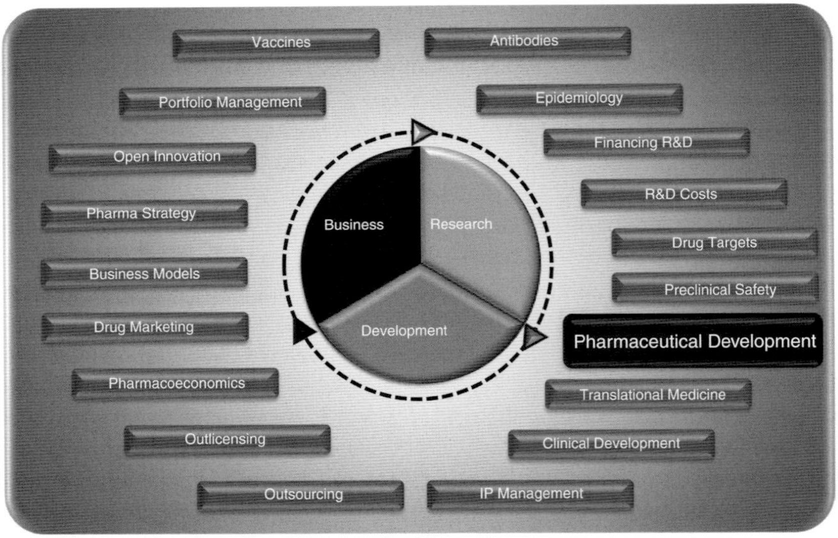
Vaccines
Antibodies
Portfolio Management
Epidemiology
Financing R&D
Open Innovation
R&D Costs
Pharma Strategy
Business
Research
Drug Targets
Business Models
Preclinical Safety
Drug Marketing
Development
Pharmaceutical Development
Pharmacoeconomics
Translational Medicine
Outlicensing
Clinical Development
Outsourcing
IP Management

# 9
# Developing Commercial Solutions for Therapeutic Proteins

*Galina Hesse*

## 9.1
## Introduction

The purpose of this book chapter is to provide an overview about several challenges observed in the pharmaceutical development of new molecules, especially therapeutic proteins, during the last decade and will require further innovation in the future.

New product innovation in the pharmaceutical industry is often interpreted as the invention of new molecules, leading to breakthrough therapies to treat or ideally cure diseases. While the invention of new molecule classes is and stays at the core of the industry and could be a "jump innovation" or "disruptive innovation," a lot of other innovations are necessary to turn a new pharmacological principle into a real product available to those who prescribe or take them. These innovations are often many "steps" or "stairs" to go and climb or "sustaining innovations" but mandatory to create value from the original innovation (Weidenauer and Maeder, 2010). It is important to understand that an invention only becomes valuable for patients, healthcare professionals, and pharmaceutical industry if it can be transformed into an affordable marketed product with good quality.

The following questions are explored in this book chapter:

- How has the *development of formulations and analytical tests* for the molecules changed that ensure the molecule arrives at the expected place of activity and being sufficiently stable during an acceptable storage period? What are the challenges in *development of containers* protecting the product from the environment until use? Are *delivery systems* required allowing an optimal application of the new products?
- Do new medicines require innovative thinking and *new approaches* regarding the manufacture of the final product, covering all steps of the value chain from the active ingredient, formulation, preparation of the drug presentation,

*Value Creation in the Pharmaceutical Industry: The Critical Path to Innovation*, First Edition.
Edited by Alexander Schuhmacher, Markus Hinder, and Oliver Gassmann.

packaging of the final product, and distribution to pharmacies, hospitals, and patients?
- Are there examples for *technical evolutions* in the area of production?

While in the 1980s and 1990s the major focus of these developments was on oral forms in order to ease the use of small-molecule chemical products, today the arrival of more and more drugs for parenteral application route like proteins, peptides, and especially monoclonal antibodies in the pipelines of the pharmaceutical companies as indicated in their annual reports requires further innovation focusing on injectable products or related technologies (Long and Works, 2013). Since in 1986 the first monoclonal antibody Orthoclone™ (anti-CD3) was approved in the United States, another 48 have been approved or are under review by early 2015 (FDA: Food and Drug Administration). Originally the monoclonal antibodies were developed for the acute treatment of life-threatening conditions like cancer and prevention of kidney transplant rejection (Beck *et al.*, 2010). More recent developments are also for chronic conditions like rheumatoid arthritis, psoriasis, and lately hypercholesterolemia (National Institute of Health .).

This chapter will focus on the challenges related to develop commercial solutions for therapeutic proteins giving some examples of recent innovations and evolvement of regulatory guidelines that mandate new ways of working taking into account all the aspects mentioned above.

## 9.2 Developing Commercial Solutions for Therapeutic Proteins

### 9.2.1 Defining a Target Product Profile

Many factors have to be taken into account when developing formulations, manufacturing processes, and application solutions of commercial products. While in research areas many experiments are performed with an exploratory mindset and the outcome of each experiment creates new hypothesis for the next, the development of a commercial product needs to be focused and well planned. This requires a typical project management approach, where as a first step the project goal is to be defined and thereafter a systematic development plan is executed, allowing for appropriate adjustment based on findings during the process or change of the environment impacting the project goal.

The definition of targeted product profile during an early stage of the development activities requires cross-functional discussions between various departments within a pharmaceutical company, a deep understanding of the disease against which the product is being developed, and an intensive market analysis in the countries where the product is likely to be used in the future. Some key parameters to be investigated and well selected during development are described in Figure 9.1.

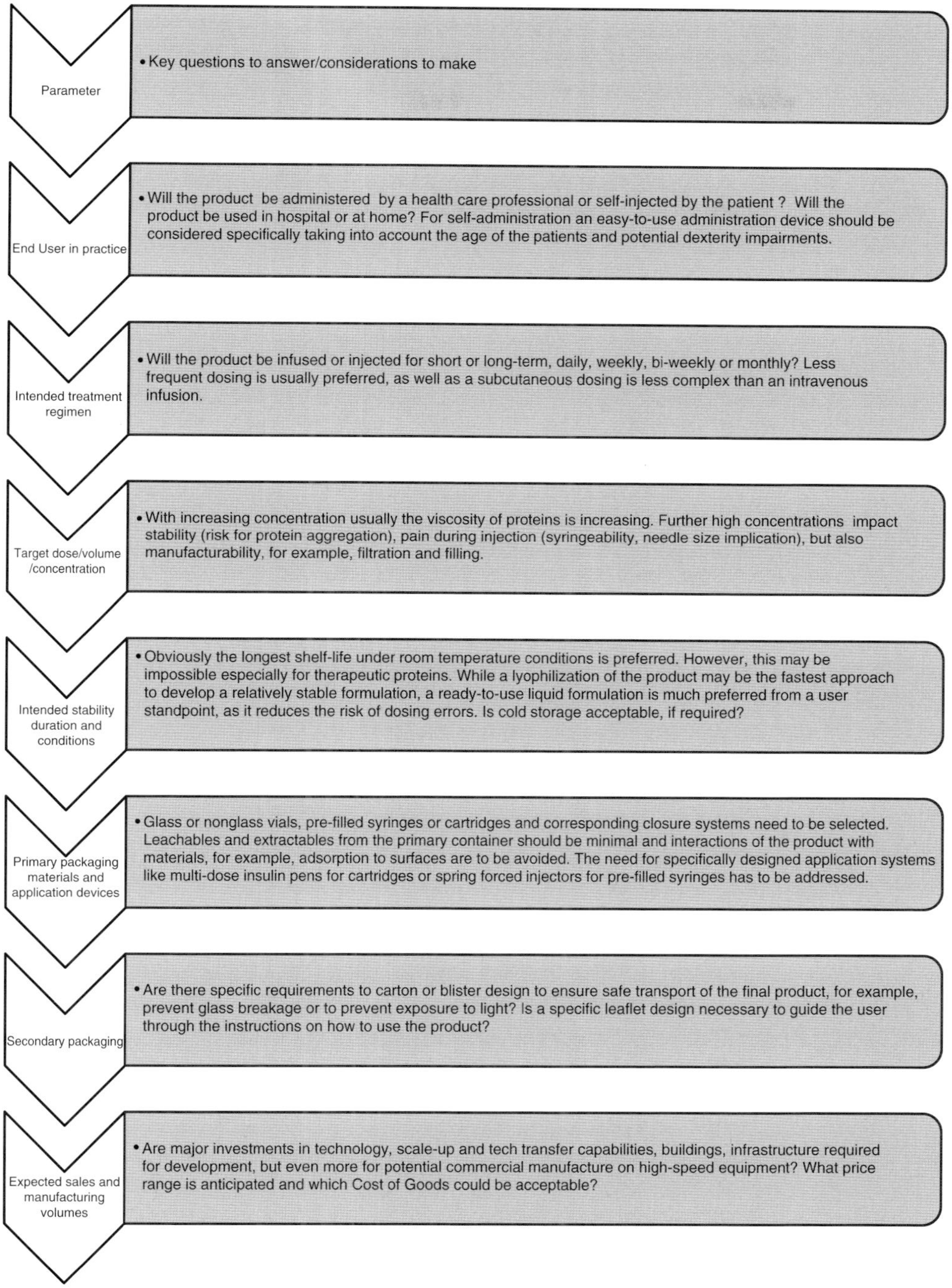


**Figure 9.1** Key parameters to be investigated and well selected during product/target product profile for an injectable peptide/protein product.

### 9.2.2 Developing Formulations for Therapeutic Proteins

Proteins are highly complex molecules. Besides chemical stability considerations, their activity is dependent on the correct folding, meaning their three-dimensional structure (Rathore and Rajan, 2008). The loss of conformational integrity leading to aggregation during manufacturing and storage led to loss of activity but has also raised potential safety concerns due to immunogenicity. This problem has become especially acute not only in terms of defining shelf life and ensuring proper administration, but it arises frequently as a comparability issue during the biopharmaceutical drug development process.

Several factors have been identified to impact proteins. Changes in temperature or pH can lead to aggregation, deamidation, and hydrolyses but also impact the structure integrity. Mechanical stress or adsorption to surfaces may lead to aggregation or changes in the 3-D structure (Wang and Roberts, 2010). Exposure to oxygen or light could induce oxidation of the proteins. Different buffer types and the concentration of ionic components also impact aggregation (Wang *et al.*, 2007). Generally a protein should have a stability shelf life of at least 18 months, preferably more, under specified storage conditions. This requires the 3-D structure of the protein to be intact as well as minimal protein aggregation at the time of manufacture but of course also during the product shelf life (Bye, Platts, and Falconer, 2014). The rate of aggregation can be minimized by selection of the "right" active ingredient but also by development of an appropriate formulation. Two main forms of protein formulations are usually commercialized: ready-to-use liquid forms and solid lyophilizates (FDA). The freeze-dried protein formulations show less protein modification; however, liquid formulations are preferred due to less risk of dosing errors, ease of handling, and lower manufacturing cost. Therefore, a liquid formulation always is the preferred option for a commercial product (Bye, Platts, and Falconer, 2014).

Several steps need to be performed during the development to come to the final product definition that should be as close as possible to the target product profile. The first step is to perform preformulation studies where formulation scientists gather all available information on the specific protein, as size, purity, structure, glycosylation state, and isoelectric point. Based on this, a variety of experiments are performed testing the impact of different pH and ionic strength on solubility and the protein status, especially aggregation and chemical stability (Weidenauer and Maeder, 2010). Excipients used for the formulation of therapeutic proteins need to be generally recognized as safe. They have to be toxicologically safe and fulfill the requirements of good manufacturing practice. The FDA has established a database with all Inactive Ingredient Approved Drug Products (FDA, n.d.). Any excipients to be used not yet listed in this database need to be fully qualified regarding the abovementioned prerequisites.

During the preformulation studies, buffers like phosphate, citrate, or acetate are used to regulate the pH at acceptable values (between pH 4 and 8, depending upon the product stability). Parenteral formulations should be isotonic with human plasma so as to avoid damage to the tissues. Therefore, the addition of tonicity-adjusting agents to the formulation is usually needed. The most commonly used tonicity agents are sugars like dextrose and sucrose; sugar alcohols like glycerol, mannitol, and xylitol; and sodium chloride (Rathore and Rajan, 2008). Solubilizing agents help in dissolving or increasing the drug solubility into the formulation. Surfactants increase the dissolution by reducing the surface tension of the drug substances. Frequently used surfactants are polysorbate 20 or polysorbate 80. Additionally studies could be performed on different shaking or freeze-drying conditions. Furthermore, exposure to air–liquid interfaces might be assessed at that stage (Wang and Roberts, 2010).

Based on the outcome of these experiments, the formulation and stability evaluation is planned and executed. If the protein candidate has shown tendency to aggregate during the initial tests, formulation studies need to be conducted to identify the minimal aggregation tendency. During these experiments, the amounts of excipients can be varied, but also the excipient types, ideally not only to test individual parameters but take into account the interaction and variability of the parameters (Wang *et al.*, 2007). Depending on the intended dose of the product, the concentration of the protein in the solution needs to be defined. High protein concentrations above 100 mg/ml have shown more protein aggregation and challenges during the scale-up of production processes (Shire, Shahrokh, and Liu, 2004). Initial short-term stability is evaluated. Stability studies are often conducted at several storage temperatures, for example, −40, 5, 15, 25, and 40 °C. High storage temperature helps to identify early tendencies of protein aggregation under tight development timelines. Unfortunately, these results are not always predictable for long term. Therefore, the impact of different temperatures on the product candidate needs to be carefully assessed to understand the predictability of the accelerated stability studies.

If no effective formulation can be found to successfully control protein aggregation in liquid state, preparation of a dry form and corresponding diluent are potential alternatives. For vacuum-drying, freeze-drying, or spray-drying processes, the addition of filling materials is necessary to prevent aggregation during lyophilization and in solid state. Saccharides like trehalose or sucrose have been used successfully in some cases, but also amino acids like glycine might be used. For the lyophilization in a first step, a dissolved substance is frozen in its final package (vial, syringe, or cartridge). During a primary drying step, the water or diluent is extracted via vacuum, resulting in a porous, dry "lyo cake." A final drying step removes residual unfrozen water molecules. The process parameters like temperature during process, duration of steps, and so on need to be investigated and the integrity of the protein confirmed. At the end of the process, the solid usually contains less than 10% of water. All of these processes might lead to protein aggregation, depending on the excipients used, temperature curves, speed of freezing or drying, and protein candidate (Rathore and Rajan, 2008).

### 9.2.3 Testing Formulations for Therapeutic Proteins

While the approach is in principle similar to a classical development of a chemical drug product, the challenges for biologic products are different. A chemical product can be well characterized by its physical properties and testing the attributes of products is considered to fully confirm the quality. Many tests are state of the art and described in pharmacopeias. This is only partly true for therapeutic proteins. Here both the physical and the biological characterization ("activity") need to be performed.

In physical protein characterization, many new approaches are available and being developed (Weidenauer and Maeder, 2010). Some few examples are peptide mapping with mass spectrometry and glycoform analysis applied for identification of product isoforms. Structures can be assessed by electron microscopic tomography of molecules or multidimensional nuclear magnetic resonance (NMR) spectrometry. Aggregation and size can be evaluated using analytical ultracentrifugation, field flow fractionation, dynamic light scattering, and other approaches (Wang and Roberts, 2010).

The biological characterization requires a detailed understanding of the biological activity, the pathways, and the development and use of product-specific assays. Models of biological activity can occur at many levels and may also be used in drug discovery and pharm–tox evaluations. These could cover binding to receptors, signal transduction for pathways, cell culture effects for cellular targets, and *in vivo* studies in appropriate animal models for the diseases to investigate the safety (Weinberg *et al.*, 2005). The selection of relevant models is crucial but also the understanding of what attributes really matter and at what level. Most of these models have an important intrinsic variability which needs to be taken into consideration when being used for assessing product characteristics. Some ensuring the safety and efficacy of the product are defined in ICH Q6B. A combination of several tests might be required. A specific challenge is the assays for high protein concentrations. Some analytical methods involve dilution to lower concentrations that are very different than the initial formulation composition. This may alter a protein's physical state in a way that the assay results are not representative of the initial conditions (Shire, Shahrokh, and Liu, 2004).

For the formulation development and even more for the stability programs, the appropriate interpretation of the test results is a must (FDA, 2013). As these activities are initiated early during the development, this is a major challenge for the developers as the full understanding often only comes with the clinical data. The validation of the assays is time consuming and expensive but should not be postponed to a late phase.

### 9.2.4 Development of Primary Containers

Whatever formulation development is selected finally, it also needs to consider the choice of the primary container used. In 1999, the FDA has published guidance on container closure systems for packaging human drugs and biologics. Four key pillars have to be considered:

- Protection of the product from light, solvent loss/leakage, microbial contamination, water vapor, or gas
- Compatibility of the product with the components of the container closure system
- Safety of the material
- Performance of the assembled packaging system

There are six basic primary container systems: ampoules, glass; vials, glass and plastic; prefilled syringes, glass and plastic; cartridges, glass and plastic; bottles, glass and plastic; and bags, plastic.

Vials have a market share of about 50% of small volume injectable packaging for both liquid and lyophilized products; prefilled syringes having today 25–30% but representing the fastest-growing market segment, with ampoules (more and more decreasing) and cartridges, fill the rest (Sacha *et al.*, 2010). A number of manufacturers are developing dual-chamber or multichamber syringes or cartridges to contain drugs that must be reconstituted at the point of delivery. Bottles and bags are the only packaging systems for large volume injectable products, which is not the focus of this chapter.

Prefilled syringes have some key advantages (Ingle and Agarwal, 2014):

- Less need for overfill of product compared to vials and other containers, which is especially of interest for expensive biological products.
- Reduction of dosage errors because syringes contain the exact amount of dose needed – reduction of medication errors and misidentification.
- Ease of administration – more convenient for healthcare professionals and end users, easier for home use, and easier in emergency situations.
- Sterility assurance is increased, because fewer manipulations are required.
- Lower injection costs – less preparation, fewer materials, and easy storage and disposal.

A potential disadvantage compared to, for example, ampoules is the higher potential reactivity between the drug product components and the packaging components resulting in the appearance of particulate matter, detection of extractables, evidence of protein aggregation, and other concerns (Sharma, 2007; Adler, 2012). Traditionally glass has been the preferred raw material for manufacturing as it is chemically inert, strong, stable, and transparent (Pareek and Khunteta, 2014). However, protein aggregation in prefilled syringes has been described. Two examples are given hereafter. Tungsten has been identified as a leachable in glass prefilled syringes (Jiang *et al.*, 2009). Tungsten pins are used to keep the fluid path open at the nozzle end of the syringe during the glass-syringe formation process. Upon cooling, a needle is staked in with adhesive to make a glass prefilled syringe with a staked needle. Residual tungsten can migrate into the drug product and cause some protein to form aggregates (Bee *et al.*, 2009). To enable good gliding of the rubber plungers within the glass tubes, silicone oil is being used as a lubricant (Badkar *et al.*, 2011). Protein aggregation and particle formation due to silicone oil has been reported with different proteins; however, the mechanisms are as of today not fully understood (Depaz *et al.*, 2014; Thirumangalathu *et al.*, 2009). Recent studies have shown that silicone oil is often distributed unevenly, thus leaving certain areas of the prefilled syringe surface with insufficient lubrication. Uneven silicone coating may increase travel forces and cause failures such as incomplete injections (Wen *et al.*, 2009).

Alternatively plastic syringes could be used. Over the last years syringes made of new polymers have been developed, for example, cyclic olefin polymers (Stout and Vilivalam, 2009). While their physical appearance is similar to glass, protein surface adsorption properties have been shown to be favorable. The above described issues with tungsten and silicone oil are being avoided (Jezek *et al.*, 2013). Lastly, there is a reduced risk for breakage of the container, which is a safety concern for the patients (West Pharmaceutical Services, .n.d.). As of today these systems are not being widely used as mostly prefilled syringes are available for large volumes (20 – 50 ml), but given the concerns with glass systems, this is likely to change with the further increase in protein developments (Sacha *et al.*, 2010).

In both glass and polymer syringes, rubber stoppers/plungers are inserted to close the containers but also to extract the product from the container. There are four general types of rubber interactions with the drug product: adsorption of the protein at the surface of the rubber, absorption of one or more formulation components into the rubber, permeation of a formulation component through the rubber, and leaching of rubber components into the drug product, especially critical for low concentrated compounds like preservatives or surfactants. In order to avoid these issues, coatings are being investigated (West Pharmaceutical services), which in return may lead to challenges regarding container closure integrity.

### 9.2.5 Development of Application Systems

While prefilled syringes are convenient, the needles can still pose a potential safety threat. Both the United States and EU have implemented specific regulations to prevent and protect healthcare workers against injuries and infections from medical sharps. In the United States, OSHA (Occupational Safety and Health Administration,.n.d.) has implemented the Bloodborne Pathogen Standards: Occupational Exposure to Bloodborne Pathogens (OSHA). The EU has established the Framework Agreement on Prevention from Sharp Injuries in the Hospital and Healthcare Sector which has been endorsed by HOSPEEM (European Hospital and Healthcare Employers' Association) and EPSU (European Public Services Union), the recognized European social partners in the hospital and healthcare sector. Various needlestick prevention systems are available, many of them being bulky systems attached to the prefilled syringes. Systems for single-dose injection of prefilled syringe content have been developed for monoclonal antibodies used in rheumatoid arthritis or heparin injection preventing needlestick injuries. Due to the size of the systems, the costs for storage and transport are high, especially for products to be cooled to 2 – 8 °C (Sacha *et al.*, 2010).

In addition to the safety of the user, another key dimension is the ease of use of the product presentation. Originally insulin and monoclonal antibodies have only been available in simple glass vials, either as liquids or as lyophilized products that need to be reconstituted by the user. The appropriate dose has to be extracted by the user with a syringe and then injected or further diluted for an infusion. Today more and more delivery devices are being developed and used as, for example, insulin injector systems composed of an insulin cartridge and a dial to measure the dose. Some pen systems use a replaceable insulin cartridge. When the insulin cartridge is empty, the empty cartridge is disposed of and a new one is inserted into the pen. Alternatively prefilled pens are entirely disposable. The pen comes prefilled with insulin, and when the insulin cartridge or reservoir is empty, the entire unit is discarded (Perfetti, 2012). Application systems for single-dose injections contain prefilled syringes. Patients perceive them to be easier to use and more convenient (Kivitz *et al.*, 2006).

The drug application systems fall in the United States generally under the medical device regulations of the Center for Devices and Radiological Health (CDRH), while the Center for Drug Evaluation and Research (CDER) and the Center for Biologics Evaluation and Research (CBER) have the primary jurisdiction over New Drug or New Biologics License applications including the review and approval of the manufacturing and control aspects. The prefilled systems (syringes with safety systems, pens, autoinjectors) are regulated as combined drug product systems under leadership of these offices with consultancy by the CDRH. The FDA has created in 2002 an Office of Combination Products to help streamline and clarify the process of obtaining market approval in the United States. In EU the regulatory pathway follows the medicinal product.

The Code of Federal Regulations 820.30 and various ISO standards exist describing the principles of system or design developments. Several steps have to be applied:

During the design input phase, the intended use of the device, including the needs of the user and patient, has to be analyzed as described in Figure 9.1, either by market survey or by preliminary analyses of human factors. Based on this, an existing system might be selected and adjusted specifically to the needs or a system developed completely from scratch. The user interfaces development to be supported by human factor and usability engineering analyses. Design output contains acceptance criteria, models, drawings, engineering analyses, and other documents and shall ensure that those design outputs that are essential for the proper functioning of the device are identified. The review of all design development and testing activities needs to be well documented and results finally in the design history file (DHF). Design verification tests allowing for an adequate evaluation of conformance to design input requirements have to be successfully conducted to confirm that the design output meets the design input requirements. In case of failure, further design changes have to be made, design output has to be adapted, and tests have to be repeated.

The device design needs to be translated into production specifications. Lastly, the design is validated by producing initial production units, lots, or batches performed under defined operating conditions. Furthermore, it needs to be proven in well-designed human factor studies that devices conform to defined user needs and intended uses under actual or simulated use conditions.

## 9.3 Quality by Design

Research and development departments were historically organized separately from production units, both having very different performance indicators, speed of development versus cost and reliability. Therefore, the industrial manufacturability was of limited importance to the developers, at least during early development stages. This system was also reflected in the regulatory documents required to get a new drug approved, where mostly information was requested about the final end product specifications (Shah, 2009). The underlying quality concept was that quality of a product was ensured, if intermediate and final product was within the specifications which should be selected as narrow as possible. Changes to processes or some equipment were potentially leading to long painful reregistration procedures, different for Europe, the United States, and Japan. This situation resulted in inflexible production systems hindering innovation after the initial approval of a product.

The International Conference of Harmonization (ICH) is an international organization comprised of scientists from regulatory agencies and the regulated industry in the United States, Europe, and Japan; it publishes documents on clinical safety and efficacy, manufacturing and product quality, and nonclinical pharmacology/toxicology to facilitate global drug development. The ICH agreed during the years 2003–2009 to a new quality paradigm: science is no longer isolated; it is living across the life cycle of the product/process within a quality management system. Their aim was to develop a harmonized pharmaceutical quality system applicable across the life cycle of the product emphasizing an integrated approach to quality risk management and science. At the core of this principle is the quality by design (QbD) principle.

The basic idea of QbD is that quality must be mainly built in by a science-driven formulation and process development and it will not only improve by additional testing and inspection. Better utilization of modern science including statistical methodology throughout product life cycle should enable identifying quality risks from materials or processes used to manufacture. This obviously requires an integrated, structured approach to development, manufacturing, and quality, both in industry and by the regulators. The concept is described in the ICH guideline Q8 as follows: The aim of pharmaceutical development is to design a quality product and its manufacturing process to consistently deliver the intended performance of the product. The information and knowledge gained from pharmaceutical development studies and manufacturing experience provide scientific understanding to support the establishment of the design space, specifications, and manufacturing controls. An overview of the key differences between classical ways of working and QbD is provided in Table 9.1.

**Table 9.1** Moving from a minimal approach to quality by design.

| Aspect | Minimal approaches | Enhanced, quality by design approaches |
|---|---|---|
| Overall Pharmaceutical development | • Mainly empirical<br>• Developmental research often conducted one variable at a time | • Systematic, relating mechanistic understanding of material attributes and process parameters to drug product critical quality attributes<br>• Multivariate experiments to understand product and process<br>• Establishment of design space<br>• Process analysis tools (PATs) utilized |
| Manufacturing Process | • Fixed<br>• Validation primarily based on initial full-scale batches<br>• Focus on optimization and reproducibility | • Adjustable within design space<br>• Life cycle approach to validation and, ideally, continuous process verification<br>• Focus on control strategy and robustness<br>• Use of statistical process control methods |
| Process Controls | • In-process tests primarily for go/no go decisions<br>• Off-line analysis | • PAT tools utilized with appropriate feed forward and feedback controls<br>• Process operations tracked and trended to support continual improvement efforts postapproval |
| Product specifications | • Primary means of control<br>• Based on batch data available at time of registration | • Part of the overall quality control strategy<br>• Based on desired product performance with relevant supportive data |
| Control Strategy | • Drug product quality controlled<br>• Primarily by intermediates (in-process materials) and end product testing | • Drug product quality ensured by risk-based control strategy for well-understood product and process<br>• Quality controls shifted upstream, with the possibility of real-time release testing or reduced end product testing |
| Life cycle management | • Reactive (i.e., problem solving and corrective action) | • Preventive action<br>• Continuous improvement facilitated |

Source: ICH Q8, R2.

For the development of a therapeutic protein commercial solution, each of the components described before – formulation, primary packaging, and drug delivery system – needs to be addressed not only individually but also as a fully integrated system. The impact of variability of the various parameters of the full system needs to be considered to define the design space (Rathore *et al.*, 2011). In most of the cases, the development of the delivery systems is performed after the initial development and approval of a new therapeutic solution as life cycle management activities throughout the full time a product is being manufactured and provided to patients (FDA). At that stage several design input requirements like dose, volume, and viscosity (Shire, Shahrokh, and Liu, 2004) of the liquid drug product are known and not subject to change any more. However, developing the application solutions already in parallel to the drug development is a major challenge as these factors still vary during the formulation and clinical development phase.

A well-designed process is expected to be robust and to enable predictable productivity and a product of consistent quality. The primary source of variability is the interaction of two or more variables. The term *design space* describes a multidimensional combination of several parameters with potential variability. This variability could be certain attributes of raw material components, used for the manufacture of a pharmaceutical product or of process parameters. There is a need for extensive use of design of experiments (DoEs) to map primary effects on the final product and interactions between variables during process characterization (robustness) studies.

The acceptable variability in product quality and process performance attributes is established based on clinical exposure of the product, knowledge from other similar products, and general scientific understanding about the molecule. For key and critical operational parameters, process characterization studies have to be performed to establish acceptable ranges (Shieu *et al.*, 2014). The combination of these acceptable ranges defines the so-called design space. Operating within this space provides the assurance of quality. The characterization studies should cover wide ranges for product quality and process performance attributes, extending beyond what will typically be the operating ranges.

While this concept is applicable generally in the pharmaceutical development, it is of even more importance for biologic products, for example, monoclonal antibodies due to lack of international reference standards, compendial test methods for potency, and potential immunogenicity (Sharma, 2007a).

## 9.4
## Examples for Innovations in Manufacture of Sterile Pharmaceutical Products

Biopharmaceutical products injected or infused need to be sterile filtered to prevent either infection by the patients due to microbial contamination or immunological reactions due to particles in the solution. First of all each starting material, for example, bulk product, excipients, and primary container components, should be either sterile or of an acceptable bioburden. Sterilization can be performed via thermal energy (standard conditions, e.g., autoclaving at 121 °C or dry heat sterilization at 160 °C or higher) or for solutions by a filtration through a filter with minimum 0.22 μm pore size. As biological products can rarely be terminally heat sterilized, the filling into the primary container needs to take place under aseptic conditions.

All handling of the sterilized product or equipment needs then to happen in clean room environment. Different classes of clean rooms are defined in the European Pharmacopeia (Ph. Eur.), the compendial quality documents developed by a Commission of the member states of the European Union, and the good manufacturing practice requirements from the World Health Organization (WHO) and United States Pharmacopeia (US Pharm.) also reflected in regulations published by the North American FDA Department. All of them require rigorous microbiological control and minimization of airborne particles. The higher the risk is for the final product, the more stringent are the requirements for the environmental conditions and monitoring. All aseptic preparation and filling of product need to take place in the most stringent environment, Class A. The limits defined for different classes of clean rooms for both particles and microbial findings are shown in Tables 9.2 and 9.3.

Clean rooms used to be constructed in standard ways by covering bricks with easy-to-clean material as, for example, epoxy or stainless steel or as softwall clean rooms (frame structure with clear vinyl curtain wall material). To control particles by appropriate airflow within the clean rooms, the key technologies to operate the facilities are fan filter units (FFUs), high-efficiency particulate air filters and systems, laminar airflow systems and biosafety cabinets, and air diffusers and showers. All material and personnel enter or leave through airlocks.

**Table 9.2** WHO: Maximum permitted airborne particle concentration (World Health Organization, 2011).

| | Maximum permitted number of particles per m$^3$ greater than or equal to the tabulated size | | | |
|---|---|---|---|---|
| | At rest[a)] | | In operation[b)] | |
| Grade name | 0.5 μm | 5.0 μm | 0.5 μm | 5.0 μm |
| A | 3 520 | 20 | 3 520 | 20 |
| B | 3 520 | 29 | 352 000 | 2 900 |
| C | 352 000 | 2 900 | 3 520 000 | 29 000 |
| D | 3 520 000 | 29 000 | Not defined | Not defined |

a) The "at rest" state is the condition where the installation is complete with equipment installed and operating in a manner agreed upon by the customer and supplier, but with no personnel present.

b) The "in operation" state is the condition where the installation is functioning in the defined operating mode and the specified number of personnel is present. The areas and their associated environmental control systems should be designed to achieve both the "at rest" and "in operation" states.

**Table 9.3** WHO: Recommended limits for microbial contamination[a] (World Health Organization, 2011).

| Grade | Air sample ($CFU/m^3$) | Settle plates (diameter 90 mm) (CFU/4 h)[b] | Contact plates (diameter 55 mm) (CFU/plate) | Glove print (five fingers) (CFU/glove) |
|---|---|---|---|---|
| A | <1 | <1 | <1 | <1 |
| B | 10 | 5 | 5 | 5 |
| C | 100 | 50 | 25 | — |
| D | 200 | 100 | 50 | — |

a) These are average values.
b) Individual settle plates may be exposed for less than 4 h.
CFUs, colony-forming units.

The key advancement of clean room technology was the use of modular hardwall clean rooms. The modules are constructed as ready-made walls and ceilings that are assembled on site. A large variety of these systems are available today, and clean rooms can be built much faster and allow higher flexibility. All of the systems need to be well maintained and controlled for appropriate operation. Furthermore, the environment within the clean room is rigorously controlled for particles and microorganisms against the described requirements (Whyte, 1999).

While there is long-lasting experience in clean room production and detailed good manufacturing practice guidance and regulations available from health authorities, the sterile manufacture remains a major risk factor for the industry. The number of warning letters from the FDA related to sterile manufacture increased steadily over the last years (2010, 12 WL; 2011, 12; 2012, 13; 2013, 16, 2014, 21) (FDA). Many of the well-known pharmaceutical companies and contract manufacturers have been under this enforcement of regulations. Many of the warning letter findings are related to inadequate processing of product or samples, often simply nonadherence to existing procedures by the personnel, for example, during gowning or exchange of gloves leading to microbial contaminations. Another concern regularly raised is insufficient capability to ensure the appropriate condition of clean room classification, for example, due to insufficient airflow or construction flaws.

One key area of innovation has been in the construction of room-in-room concepts. The first advancement is passive restricted access barrier systems (RABSs), where the clean room is separated in different rooms by walls/glass walls so that necessary manual operations can only be performed via built-in gloves. The most advanced technology is the so-called isolators that are built as totally closed systems with integrated air filtration and flow control and in-place cleaning or sanitization systems (Akers and Agalloco, 2011). This reflects the progressive demand for gapless protection of the sterile product by separation of operators. But the advancing technological development stages are also an expression of a higher degree of automation in the production, for example, in-process automatic controls, automatic loading/unloading systems for starting materials and end products, or integrated microbial air sampling equipment/particle measuring instruments.

The other area of innovation is in the field of consumables. The consumables used in clean rooms can be classified into two major categories, namely, safety consumables and cleaning consumables. The safety consumables comprise apparel, gloves, and other safety consumables, while cleaning consumables are categorized as vacuum systems, wipes, disinfectants, and other products to enable cleaning or sanitization of the rooms and equipment. Today vaporized hydrogen peroxide (VHP) is used as a decontamination method. The same highly effective method is applied in many hospitals in order to prevent the spread of methicillin-resistant *Staphylococcus aureus* (MRSA). A study with MRSA showed reduced effectiveness of the method as compared to *Geobacillus stearothermophilus* (Pottage *et al.*, 2012). Whether the broad usage will lead to the development of even further resistant microorganisms is speculation, but alternative methods should be investigated. Important other consumables are gloves usually made of polymers, polyethylene, butyl, latex rubber, or other materials. These gloves need to fulfill two important purposes: to ensure the integrity of the sterile environment and to protect the operators from potentially toxic substances handled. A special case is gloves built into isolator to enable manual activities within the isolator without opening the system. To determine which material would be best suited for the products to be handled, the sanitization method and the degree of necessary dexterity have to be considered (Whyte, 1999). New materials or coatings being developed for other purposes, for example, in aerospace environment, should systematically be screened for use in the pharmaceutical manufacturing context.

## 9.5 Summary

Development of modern medicines, for example, therapeutic proteins, has become much more complex than the development of small-molecule products. Figure 9.2 provides an overview on the various interlinked development steps to be performed during the journey from a targeted product profile to a commercial solution.

In order to address the complexity but also to ensure these new products can be well manufactured, good project management practices need to be applied. This starts with a clear description of the targeted product followed by a systematic way of execution. One of the key new execution processes combining science and goal-oriented ways of working is the approach of QbD. A strong collaboration between all involved people with different areas of expertise is required.

Technical innovations and automation could further improve the consistent manufacture of high-quality products. Continuous innovation in the field of

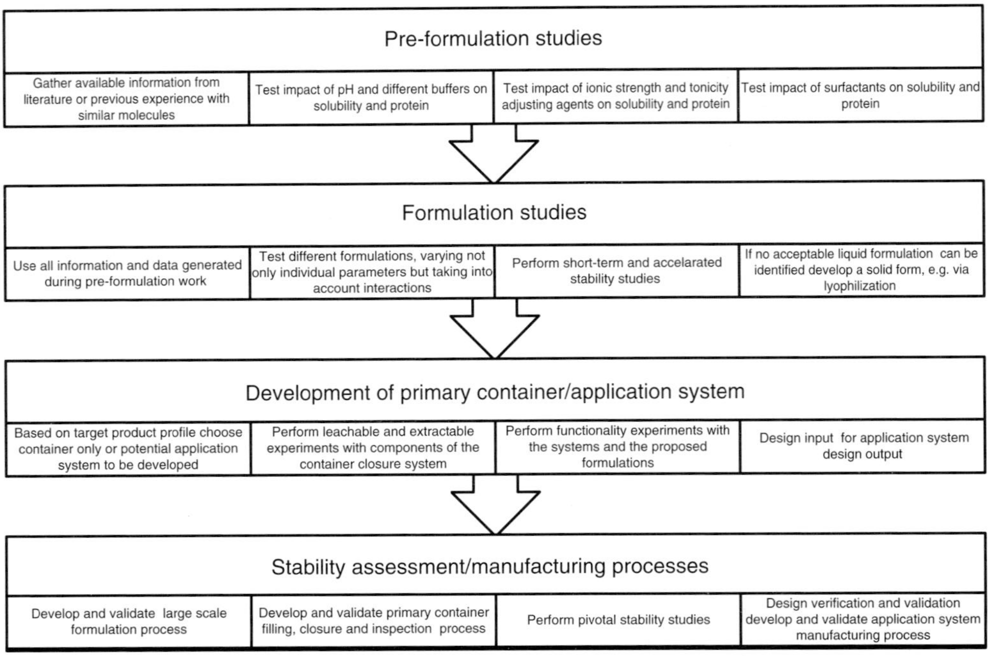


**Figure 9.2** Schematic high-level overview on development steps for a commercial solution.

engineering and material science should go hand in hand with the development of new products.

## List of FDA/ICH Guidances Referenced

FDA: Bioanalytical Method Validation, 9/2013
Quality Considerations in Demonstrating Biosimilarity to a Reference Protein Product, 2/2012
Container Closure Systems for Packaging Human Drugs and Biologics, 7/1999
Demonstration of Comparability of Human Biological Products Including Therapeutic Biotechnology Derived Products, 3/1996
Immunogenicity Assessment for Therapeutic Protein Products, 8/2014
Interpreting Sameness of Monoclonal Antibody Products Under the Orphan Drug Regulations, 4/2014
Residual Solvents in Drug Products Marketed in the United States, 11/2009
Allowable Excess Volume and Labeled Vial Fill Size in Injectable Drug and Biological Products, DRAFT, 3/2014
Analytical Procedures and Methods Validation for Drugs and Biologics, DRAFT, 2/2014
Assay Development for Immunogenicity Testing of Therapeutic Proteins (I), DRAFT, 12/2009

Submission Documentation for Sterilization Process Validation Applications for Human and Veterinary Drug Products (I), DRAFT, 11/1994
Glass Syringes for Delivering Drug and Biological Products: Technical Information to Supplement International Organization for Standardization (ISO) Standard 11040-4 – Draft Guidance for Industry and FDA Staff, 4/2013
Investigating Out of Specification (OOS) Test Results for Pharmaceutical Production, 10/2006
Process Analytical Technology – A Framework for Innovative Pharmaceutical Manufacturing and Quality Assurance, 10/2004
Process Validation: General Principles and Practices, 1/2011
Quality Systems Approach to Pharmaceutical Current Good Manufacturing Practice Regulations, 10/2006
Sterile Drug Products Produced by Aseptic Processing, 10/2004
Comparability Protocols – Protein Drug Products and Biological Products – Chemistry, Manufacturing, and Controls Information (I), DRAFT, 9/2003
Current Good Manufacturing Practices for Combination Products,10/2004
Q1A(R2) – Stability Testing of New Drug Substances and Products, 11/2003
Q3B(R) – Impurities in New Drug Products, 7/2006
Q5C – Quality of Biotechnological Products: Stability Testing of Biotechnology/Biological Products, 7/1996
Q8 (R2) – Pharmaceutical Development, 11/2009
Q9 – Quality Risk Management, 6/2006
Q10 – Pharmaceutical Quality System, 4/2009
Formal Meetings Between the FDA and Biosimilar Biological Product Sponsors or Applicants, 4/2013
Technical Considerations for Pen, Jet, and Related Injectors Intended for Use with Drugs and Biological Products, 6/2013
Glass Syringes for Delivering Drug and Biological Products: Technical Information to Supplement International Organization for Standardization (ISO) Standard 11040-4, 4/2013
Current Good Manufacturing Practice Requirements for Combination Products, DRAFT, 1/2015

## Disclaimer

The views and opinions expressed in this chapter are those by the author and do not necessarily represent the official policy or position of Sanofi.

## References

Adler, M. (2012) Challenges in the development of Pre-Filled Syringes for Biologics from a Formulation Scientist's Point of View, AmercianPharmaceuticalReview, http://www.americanpharmaceuticalreview.com/Featured-Articles/38372-Challenges-in-the-Development-of-Pre-filled-Syringes-for-Biologics-from-a-Formulation-Scientist-s-Point-of-View/ (accessed 11 December 2014).

Akers, J. and Agalloco, J.P. (2011) www.americanpharmaceuticalreview.com, http://www.americanpharmaceuticalreview.com/Featured-Articles/36878-Clean-Rooms-RABS-and-Isolators-Validation-and-Monitoring-in-the-Diverse-World-of-Aseptic-Processing/ (accessed 05 January 2015).

Badkar, A., Wolf, A., Bohack, L., and Kolhe, P. (2011) Development of biotechnology products in pre-filled syringes: technical considerations and approaches. *AAPS Pharm. Sci. Technol.*, **12** (2), 564–572. doi: 10.1208/s12249-011-9617-y

Beck, A., Wurch, T., Bailly, C., and Corbaia, N. (2010) Strategies and challenges for the next generation of therapeutic antibodies. *Nat. Rev. Immunol.*, **10** (5), 345–352. doi: 10.1038/nri2747

Bee, J., Nelson, S., Freund, E., Carpenter, J., and Randolph, T. (2009) Precipitation of a monoclonal antibody by soluble tungsten. *J. Pharm. Sci.*, **98** (9), 3290–3301.

Bye, J.W., Platts, L., and Falconer, R.J. (2014) Biopharmaceutical liquid formulation: a review of the science of protein stability and solubility in aqueous environments. *Biotechnol. Lett.*, **36**, 869–875. doi: 10.1007/s10529-013-1445-6

Depaz, R.A., Chevolleau, T., Jourrfay, S., Narwal, R., and Dimitrova, A.N. (2014) Cross-linked silicone coating: a novel prefilled syringe technology that reduces subvisible particles and maintains compatibility with biologics. *J. Pharm. Sci.*, **103**, 1384–1393.

FDA (2013) Bioanalytical Method Validation, September 2013.

FDA (n.d.-a) Inactive Ingredient Search for Approved Drug Products, http://www.accessdata.fda.gov/scripts/cder/iig/index.Cfm (accessed 11 August 2015).

FDA (n.d.-b) Warning Letters, http://www.fda.gov/ICECI/EnforcementActions/WarningLetters/ (accessed 29 December 2014).

FDA: Food and Drug Administration (n.d.) http://www.accessdata.fda.gov/scripts/cder/drugsatfda/index.cfm (accessed 24 December 2014).

Ingle, R.G. and Agarwal, A.S. (2014) Pre-Filled Syringe – A Ready-to-Use Drug Delivery System: A Review, http://informahealthcare.com/doi/abs/10.1517/17425247.2014.923400 (accessed 23 December 2014).

Jezek, J., Darton, N.J., Derham, B.K., Royle, N., and Simpson, I. (2013) Biopharmaceutical formulations for pre-filled delivery devices. *Expert Opin. Drug Delivery*, **10** (6), 811–828.

Jiang, Y., Nashed-Samual, Y., Li, C., Liu, W.P., Mallard, D., Wen, Z.-Q. *et al.* (2009) Tungsten-induced protein aggregation; solution behaviour. *J. Pharm. Sci.*, **98** (12), 4695–4710.

Kivitz, A., Cohen, S., Dowd, J., Edwards, W., Thakker, S., Wellborne, F. *et al.* (2006) Clinical assessment of pain, tolerability, and preference of an autoinjection pen versus a prefilled syringe for patient self-administration of the fully human, monoclonal antibody adalimumab: the TOUCH trial. *Clin. Ther.*, **28** (10), 1619–1629.

Long, G. and Works, J. (2013) *Innovation in the Biopharmaceutical Pipeline: A Multidimensional View*, Analysis Group.

National Institute of Health (n.d.) ClinicalTrails.gov, https://www.clinicaltrials.gov/ (accessed 24 December 2014).

Occupational Safety and Health Administration (n.d.) 1910.1030 – Bloodborne Pathogens, https://www.osha.gov/pls/oshaweb/owadisp.show_document?p_table=STANDARDS&p_id=10051 (accessed 11 August 2015).

Pareek, V. and Khunteta, A. (2014) Pharmaceutical packaging: current trends and future. *Int. J. Pharm. Pharm. Sci.*, **6** (6), 480–485.

Perfetti, R. (2012) Reusable and disposable insulin pens for the treatment of diabetes: understanding the global differences in user preference and an evaluation of inpatient insulin pen use. *Diabetes Technol. Ther.*, **12** (Suppl. 1), 79–85. doi: 10.1089/dia.2009.0179

Pottage, T., Macken, S., Walker, J.T., and Bennett, A.M. (2012) Meticillin-resistant Staphylococcus aureus is more resistant to vaporized hydrogen peroxide than commercial Geobacillus stearothermophilus biological indicators. *J. Hosp. Infect.*, **80**, 41–45.

Rathore, N., Pranay, P., Eu, B., Ji, W., and Walls, E. (2011) Variability on syringe components and its impact on functionality of delivery systems. *PDA J. Pharm. Sci. Technol.*, **65** (5), 468–480. doi: 10.5731/pdajpst.2011.00785

Rathore, N. and Rajan, R.S. (2008) Current perspectives on stability of protein drug products during formulation, fill and finish operations. *Biotechnol. Prog.*, **24**, 504–514.

Sacha, G.A., Saffell-Clemmer, W., Abram, K., and Akers, M.J. (2010) Practical fundamentals of glass, rubber and plastic sterile packaging systems. *Pharm. Dev. Technol.*, **15** (1), 6–34.

Shah, R.B. (2009) *Quality by Design in Pharmaceutical Manufacturing*, AAPS, Chicago, IL.

Sharma, B. (2007a) Immunogenicity of therapeutic proteins. Part 2: impact of container closures. *Biotechnol. Adv.*, **25**, 318–324.

Sharma, B. (2007b) Immunogenicity of therapeutic proteins. Part 3: impact of manufacturing changes. *Biotechnol. Adv.*, **25**, 325–331.

Shieu, W., Torhan, S.A., Chan, E., Hubbard, A., Gikanga, B., Stacu, O.B., and Maa, Y.F. (2014) Filling of high-concentration monoclonal antibody formulations into pre-filled syringes: filling parameter investigation and optimization. *PDA J. Pharm. Sci. Technol.*, **68** (2), 153–163. doi: 10.5731/pdajpst.2014.00973

Shire, S.J., Shahrokh, Z., and Liu, J. (2004) Challenges in the development of high protein concentration formulations. *J. Pharm. Sci.*, **93** (6), 1390–1402.

Stout, D. and Vilivalam, V. (2009) Plastic Prefilled Syringes: A Better Fit for Autoinjector Systems, Pharmtech.com, http://www.pharmtech.com/plastic-prefilled-syringes-better-fit-autoinjector-systems (accessed 05 January 2015).

Thirumangalathu, R., Krishnan, S., Speed Ricci, M., Brems, D., Randoph, T., and Carpenter, J. (2009) Silicone oil- and agitation induced aggregation of a monoclonal antibody in aqueous solution. *J. Pharm. Sci.*, **98** (9), 3167–3181.

Wang, W. and Roberts, C.J. (2010) *Aggregation of Therapeutic Proteins*, John Wiley & Sons, Inc., Hoboken, NJ.

Wang, W., Singh, S., Zeng, D.L., King, K., and Nema, S. (2007) Antibody structure, instability and formulation. *J. Pharm. Sci.*, **96**, 1–26.

Weidenauer, U. and Maeder, K. (2010) in *Innovative Arzneiformen* (eds M. Karsten and U. Weidenauer), Wissenschaftliche Verlagsgesellschaft GmbH, Stuttgart.

Weinberg, W.C., Frazier-Jessen, M.R., Wu, W.J., Weir, A., Hartsough, M., Keegan, P. *et al.* (2005) Development and regulation of monoclonal antibody products: challenges. *Cancer Metastasis Rev.*, **24**, 569–584.

Wen, Z.-Q., Vance, A., Vega, F., Cao, X., Eu, B., and Schulthesis, R. (2009) Distribution of silicone oil in prefilled glass syringes probed with optical and spectroscopic methods. *PDA J. Pharm. Sci. Technol.*, **63** (2), 149–158.

West Pharmaceutical Services (n.d.) http://www.westpharma.com/en/products/Pages/CrystalZenithRU.aspx (accessed 28 November 2014).

Whyte, W. (1999) *Front Matter and Index in Cleanroom Design*, 2nd edn, John Wiley & Sons, Ltd., Chichester.

World Health Organization (2011) WHO Good Manufacturing Practice for Sterile Pharmaceutical Products. WHO Technical Report Series 961, World Health Organization, New York.

West Pharmaceutical Services (n.d.) http://www.westpharma.com/en/products/Pages/Novapure.aspx (accessed 24 December 2014).

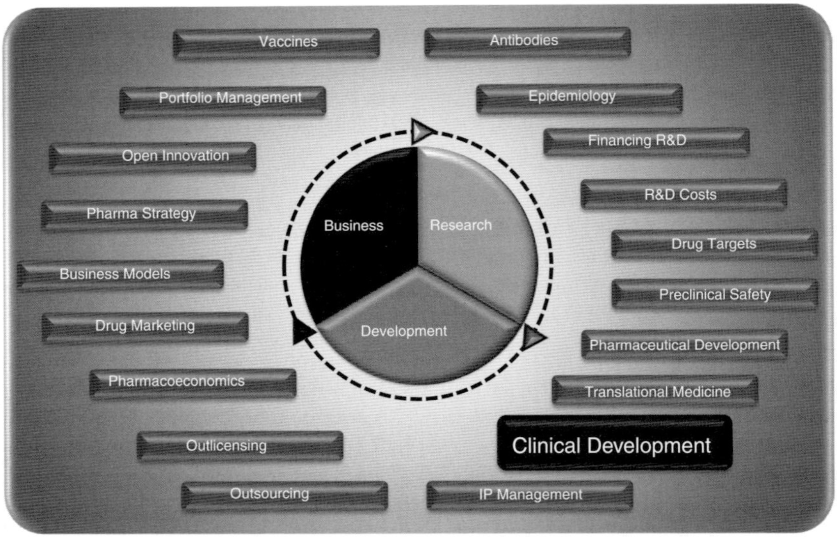
Vaccines
Antibodies
Portfolio Management
Epidemiology
Financing R&D
Open Innovation
R&D Costs
Pharma Strategy
Business
Research
Drug Targets
Business Models
Preclinical Safety
Drug Marketing
Development
Pharmaceutical Development
Pharmacoeconomics
Translational Medicine
Clinical Development
Outlicensing
Outsourcing
IP Management

# 10
# The Evolution of Clinical Development: From Technical Success to Clinical Value Creation

*Markus Hinder and Alexander Schuhmacher*

## 10.1
## Introduction

The purpose of this chapter is to review the function of clinical development (CD) and its evolution over the past years and to give an outlook on new trends in CD.

Within the R&D process, the clinical phases of drug development are traditionally denoted as phases 1–4. Whereas phases 1 and 2a are often defined as the "exploratory phase" of drug development, the phases 2b and 3 are called *confirmatory phase*, and phase 4 trials are conducted to support marketing of a new drug.

The two most important milestones during the exploratory phase are the transition into the "first-in-human" trial, namely, the first testing of a new molecule in human beings, and proof of concept. Proof of concept is supposed to answer the question whether the new molecule is able to confer beneficial effects to treat or cure a human disease.

The main goals of the exploratory phase are to:

- Test if the findings from preclinical pharmacology and toxicology are translatable to the human situation
- Provide information as to whether further a investment in the drug candidate is warranted to proceed its development in the confirmatory phase

In many of the research-based pharmaceutical companies, proof of concept (phase 2a) is followed by a formal decision to advance the new compound to full CD, namely, into the confirmatory phase. This part of the pharmaceutical R&D process usually comprises at least three elements: dose finding, one or more large confirmatory trials, and a comprehensive clinical pharmacology program. All three components are used to provide scientifically sound, reliable, and convincing data on the following questions:

1) Which patient populations benefit most from treatment with the drug candidate?
2) How long do these patients need to be treated?
3) What is the optimal dose and dose regimen for the treatment?

*Value Creation in the Pharmaceutical Industry: The Critical Path to Innovation,* First Edition.
Edited by Alexander Schuhmacher, Markus Hinder, and Oliver Gassmann.

4) Can the drug candidate be given to patients with concomitant diseases?
5) Can the drug candidate be given together with other drugs?
6) Do the benefits provided by the drug candidate outweigh the risks?
7) How effective is the drug candidate compared to already marketed therapies?

Subsequently, the data resulting from these questions form the intellectual basis for the marketing authorization of the new medicine.

As outlined in the book chapter on the R&D efficiency, the major challenges in CD are the still too high failure rates of late-stage R&D projects, the long development timelines, and the increasing cost related to drug development. Thus, CD is a central part of the R&D process for the following two reasons:

- First, it delivers the confirmatory data that are needed to support a marketing authorization application.
- Second, it is the most expensive and the longest part of pharmaceutical R&D (Sartori *et al.*, 2011; Orloff *et al.*, 2009; Orloff and Stanski, 2011).

In this chapter, we will review the principal changes that occurred in CD in the past years. We will also examine technical and commercial challenges of CD, elaborate on the principles of the evidence-based medicine (EbM), outline the potential of patient stratification and personalized medicine, discuss the potential of strategic outsourcing, and address the innovation that might come with the use of social media.

## 10.2
## CD: Changes and Challenges

CD is an activity and discipline responsible for the confirmatory phases of drug development. The primary and ultimate goal of CD is to generate convincing and persuasive data to provide proof of efficacy and safety of a drug candidate and, thus, to support the regulatory approval.

During the last few decades the way how a positive benefit–risk assessment is provided has undergone considerable changes. The main areas discussed in this chapter are:

1) Clinical endpoints in studies and subsequently study size
2) Determination and quantification of risks
3) Assessment of medical progress in context of available therapeutic options

### 10.2.1
### Clinical Endpoints: From Symptom-Oriented Endpoints to Hard and Predefined Endpoints

The standards based on which efficacy and safety need to be demonstrated have changed significantly in the past decades. From the early days of CD in the 1960–1980s, it was the state of the art to base assessments of clinical efficacy primarily on clinical symptom-oriented parameters. Usually, improvements on symptoms had to be demonstrated in at least two adequately powered and independent clinical trials. A typical historical example for the use of symptomatic endpoints comes from the use organic nitrates in ischemic heart disease. While there is compelling evidence that nitrates are potent antianginal and anti-ischemic drugs, data on modification of the disease and positive hard morbidity and mortality endpoints are lacking (Abrams, 1995; Wakai *et al.*, 2013; Darius, 1999).

One of the first trials which systematically investigated the effect of an intervention on mortality in a nondeadly disease was the University Group Diabetes Program (UGDP) (Meinert *et al.*, 1970). During the 1980s a number of important landmark clinical trials, especially in the cardiovascular area, shifted the clinical trial focus from mechanistic and intermediate endpoints to "hard" outcomes (The CONSENSUS Trial Study Group, 1987; Matthews and Tsapas, 2008; UK Prospective Diabetes Study Group, 1998; Eidelman, Hollar, and Hebert, 2014).

Even in areas where the quantification of morbidity and mortality was less common than in the therapeutic areas of cardiovascular or oncology, similar developments took place. In areas in which symptomatic treatment is still the gold standard, the use of accepted and standardized endpoints for confirmatory development programs is today the rule. Illustrative examples for this development are among other the (HAM-D) Hamilton D scale in depression and Visual analogue (VAS) Scale for the assessment of analgesics, clearly defined cure and response rate for anti-infectives, or total and pain-free walking distance for peripheral artery disease (Norgren *et al.*, 2007; Downing *et al.*, 2014; FDA, 2011; EMA, 2013, 2014).

In summary the shift from individually and symptomatically defined endpoints to hard or at least standardized clinical trial endpoints has resulted in an increased demand to deliver results for drug registration and in turn for CD programs.

### 10.2.2
### Determination and Quantification of Risks

It is state of the art since a long time that the sponsor of a clinical program needs to provide convincing and compelling evidence of efficacy of the drug candidate. Generally, an accepted methodology was the demonstration of a clinically relevant effect seconded by a meaningful $p$-value. This approach has more recently been complemented and in part replaced by the use of confidence intervals and point estimates (ICH E14, 2005; Nuzzo, 2014). Safety findings and endpoints have traditionally been analyzed in a "no difference" approach. This included the comparisons of continuous variables, such as transaminases, serum creatinine, or heart rate, and categorical variables, such as the number of adverse events and the number of patients with abnormal liver function tests by means of testing for statistical differences between groups using parametric or nonparametric tests (Brass *et al.*, 2006). This "no difference" approach has been criticized for being underpowered to detect signals and therefore as being too insensitive to convincingly exclude a risk (The PLoS Medicine Editors, 2011; Ledford, 2013). Brass *et al.* (2006) have recently published data on cilostazol and how an "exclusion of risk" approach could eliminate a certain critical incidence of undesirable events in a target population (Brass *et al.*, 2006). A more recent example for the increasing importance of drug safety for the approval status comes from the market withdrawal of rosiglitazone for safety concerns. This example illustrates that safety assessments need to be of equal importance as efficacy assessments in the future (Winterstein, 2009). Although logical and justified, this stipulation leads to the dilemma that due to their low(er) incidence, more focus on safety data prior to drug approval will increase the need for probably larger, longer, and more expensive clinical programs (Wright, 2012; Zerhouni, 2013).

### 10.2.3 Assessment of Medical Progress in Context of Available Therapeutic Options

The determination of the benefits of a drug versus the risk associated with its use is primarily a "stand-alone assessment" of the drug under investigation performed by the originator company and subsequently checked by the health authority. Pharmaceutical R&D has successfully provided in a number of therapeutic areas gold standard therapies, such as in arterial hypertension or cholesterol-lowering therapy. In view of already marketed drugs, a new drug candidate might pass all classical technical test criteria of a stand-alone assessment while it provides a negative benefit–risk evaluation in view of other available therapeutic alternatives. In the past years, some marketed drugs, such as mibefradil, gatifloxacin, or cerivastatin, were withdrawn, although these drugs passed the formal regulatory tests on quality, efficacy, and safety and were approved by health authorities. The identification of new or additional (mibefradil) or the increase frequency of known risks (gatifloxacin, cerivastatin) together with the availability of similarly effective drugs finally led to their market withdrawals (Food and Drug Administration, 2008; Meinertz, 2001; Sorelle, 1998; Furberg and Pitt, 2001).

In addition to the aforementioned three direct mechanisms, two indirect but meanwhile similarly important mechanisms shape the CD area, namely, the EbM, health economic and pharmacoeconomic considerations, and the fourth hurdle.

#### 10.2.3.1 EbM

David Sackett and colleagues defined EbM as "the conscientious, explicit, and judicious use of current best evidence in making decisions about the care of individual patients ... " (Sackett *et al.*, 1996). As a consequence any treatment which cannot (by means of accepted scientific methods (randomized clinical trials (RCTs) in conjunction with statistical methods)) prove its clinical benefit will not be regarded as evidence based. It will therefore usually not make its way into international and national treatment guidelines. The chances of such a treatment to be considered a new good standard (not to mention gold standard) are very low.

The stringency of the drug approval process today leads to drug labels that are exclusively reflective of the trials and populations studied and that follow the motto "what you study and what you have evidence for is what you get in your label." Eichler *et al.* (2011) reviewed the wording of the treatment indications in the labels of new active substances authorized in the year 2009 in the United States and the European Union (EU). Their findings support "the notion that the label is frequently restricted to what some may consider an artificial subset because, in most cases, the limited indication reflects the clinical trial scenarios. It is indeed biologically plausible that some of the more narrowly authorized drugs might be equally safe and efficacious in extended settings, such as first-line therapy or monotherapy. However, too much extrapolation in the wording of the label would violate the principles of evidence-based drug licensing" (Eichler *et al.*, 2011).

In addition to regulators a number of academic groups are involved in refining the methodology by which evidence-based recommendations are developed today (e.g., the Evidence-based Practice Center (EPC) program) (Owens *et al.*, 2010)).

The restriction of perspective by focusing exclusively on RCTs has recently more and more been criticized to be artificial and to exclude the classical aspects of patient acceptance and efficiency (Charlton, 2001; Smith and Pell, 2003; Penston, 2007). A model which explains the difficulties of getting the right medication at the right dose to the right patient has been proposed by Seidling *et al.* (2013). This paper summarizes 10 different processes which need to take place and which potentially influence the therapeutic success in a real-life setting. The thinking in strictly exclusive EbM categories excludes these additional factors contributing to the therapeutic success.

#### 10.2.3.2
#### Health Economics, Pharmacoeconomics, and the Fourth Hurdle

The importance of health economic considerations has steadily increased in recent years not least because of the increased demand driven by demographic change but also by the availability of treatment in disease areas where a few decades ago no treatment was available.

Due to the need for comparative data versus established medicines, placebo-controlled trials alone are often neither sufficient nor ethically justifiable. The need for actively controlled trials in turn leads to an increase in sample size and consequently trial size. Nevertheless, in many cases regulators may ask for placebo-controlled trial data in addition to active controls. For drugs which are not intended to provide a higher efficacy but better safety, compliance, or patient acceptance, this necessitates to run noninferiority trials which again due their statistical nature increase the sample size (Hung, Wang, and Neill, 2005; Downing *et al.*, 2014; Seidling *et al.*, 2013; Schumi and Wittes, 2011).

#### 10.2.3.3
#### Results of These Changes and Challenges

The remit of CD has been considerably expanded in recent years. The consequences of the aforementioned mechanisms are:

- A shift from symptomatic endpoints to hard endpoints
- An increase in the number, size, and duration of clinical trials to support registration and risk assessment
- A need to generate more comparative data to support the benefit–risk assessment in context of available therapeutic options

The core deliverable has still remained, namely, the provision of evidence for clinically relevant and statistically significant effects that demonstrate a positive benefit–risk profile. On the other hand, new and meanwhile equally important duties have been added, such as the provision of data on the specific place of a new therapy in the therapeutic armamentarium. This includes comparative data on efficacy and safety. Regarding the clinical safety, this can include dedicated studies in patients or healthy volunteers to actively rule out a known class effect or a theoretical risk (Gurbel *et al.*, 2009, 2010; Camm *et al.*, 2012). Studies dedicated to provide additional data on efficacy might include but are not limited to comparative investigations on onset of effect, switch ability, and compatibility with concomitant medication (Angiolillo *et al.*, 2014; Zhou *et al.*, 2014; Heise *et al.*, 2007; Lorenz *et al.*, 2013). In summary one can say that the larger the number of available therapeutic options, the more data on use and optimal application of the medication and the derived benefits are needed.

The increased demands toward more meaningful and hard outcomes have impacted the size and conduct of clinical trials as the core piece of CD with the need for bigger and longer studies (Reichert, 2003; Kola, 2008; DiMasi *et al.*, 2009). In summary, the aforementioned mechanisms have considerably increased the overall hurdles to develop drug candidates to provide approved and reimbursed products. Today, CD is thus a long-term, resource- and knowledge-intensive, and high-risk endeavor (Table 10.1).

## 10.3
## Technical Success and Clinical Value Creation in CD in the Future

The new challenges and additional tasks which have been added to the CD phase necessitate a rethinking of what is being done during this phase but also how CD needs to be conducted.

One of the key learnings in drug development in recent years was the fact that the biggest value for the patient can be achieved if one can identify the right medication for the right patient at the right dose and at the right time (e.g., during the disease process) (Elliott and Liu, 2010; Piquette-Miller and Grant, 2007; Lillie *et al.*, 2011; Hamburg and Collins, 2010). This has led to a more science-based approach in identifying and classifying disease and predicting and monitoring

**Table 10.1** Changes and their impact on the clinical development area.

| Changes and their impact in the clinical development area | |
|---|---|
| **Changes** | **Impact** |
| Shift from symptomatic and/or intermediate endpoints to hard endpoints | Increase in sample size<br>Larger trials |
| Increased demands on patient population characterization and proactive risk assessment | Increase in number size and duration of clinical trials to support registration and risk assessment |
| Need for comparative data to support the benefit–risk assessment in context of available therapeutic options | More complex trial designs including placebo and/or active comparators<br>Need for registries<br>Additional trial endpoints (PROs, QoL, resource utilization parameters)<br>Need for registries |

therapeutic success in human disease, an approach which has been coined personalized or stratified medicine (Trusheim, Berndt, and Douglas, 2007; Orloff *et al.*, 2009; Woodcock, 2007).

The traits of clinical programs which satisfy these increased demands have a number of characteristics in common:

- Clinical trials will have more than one objective.
- Endpoints are chosen to address the questions of the clinical, regulatory, and reimbursement positions.

Late-stage clinical trials today need to be multidimensional. They need to provide the basis for the registration and for the fourth hurdle. They need to provide compelling evidence for the efficacy claim and exclude safety risks. And last, they need to be cutting edge from a scientific perspective and need to provide evidence which can be intuitively understood by a broader public who is involved in the decision-making process (Breckenridge and Walley, 2008; Silva *et al.*, 2005; Moors, Cohen, and Schellekens, 2014).

Clinical trial objectives reflect the overall increased remits of the CD phase. The need to align several distinct objectives with a limited number of clinical trials makes most companies today strive to find a consensus on the trials and necessary outcomes for drug registration before starting the late-stage program. Regulatory feedback cannot be regarded as binding as it constitutes an important risk mitigation element (Kaitin, 2010; Sartori *et al.*, 2011).

In addition to standard efficacy objectives, the trial(s) might in addition be aimed to detect differences for an unacceptable safety risk which could not be eliminated during the exploratory phase, as, for example, due to the small number of patients exposed or the short treatment duration. For known rare risks, pharmaceutical companies have adapted a model which proactively assesses theoretical and potential risks to avoid market withdrawals. Examples for this approach are omapatrilat (Coats, 2002), torcetrapib (Editorial, 2011; Sofat *et al.*, 2010), and more recently dalcetrapib (Roche, 2012).

The situation is more challenging with R&D activities in areas where neither a standard treatment nor commonly accepted endpoints exist. This obviously affects the fields of rare and orphan disease or areas with a high medical need for which however no treatment exists (Nisticò, 2011; Committee for Orphan Medicinal Products and the European Medicines, 2011). As a consequence validated and relevant parameters to measure treatment effects need to be created in parallel to the drug development process, a task which is extremely challenging as this usually precedes the therapeutic advances and usually takes decades (Suckfüll *et al.*, 2011; Eng *et al.*, 2001; Food and Drug Administration, 2014; Breitenstein, Scheuer, and Holsboer, 2014).

### 10.3.1 Established and Novel Approaches to Determine the Dose–Exposure–Response Relationship

The true therapeutic potential of a drug needs to be seen in the context of its dose–exposure–effect relationship. Science-based dose finding would be the seamless elucidation of the relationship between dose, concentration, pharmacodynamics, desired therapeutic effect, and undesired adverse side effects. Optimally a "chain of evidence" ranging from dose to pharmacokinetics[1)], biomarkers of target engagement, clinical effects, and outcomes should be developed (Figure 10.1). This chain of evidence would then form a solid basis to justify the therapeutic doses but also the rationale for the positive benefit–risk relationship which is critical for the overall approval of the drug.

The easiest case of dose finding is to conduct a parallel-arm study in the target population and to analyze the targeted approval endpoint as clinical endpoint of the study. For drugs with an immediate effect and a simple efficacy readout, this approach has been successfully applied (e.g., development of analgesics) (Stegmann *et al.*, 2008; Kleinert *et al.*, 2008).

However, most of the indications studied today have a chronic pathophysiology, and many of the pharmacological targets do not display immediate effects. In these cases, dose finding is an iterative data integration process across several studies rather than a single study activity. The tools supporting this include physiologically based pharmacokinetic (PBPK) modeling, PK/PD modeling, mechanism-based PK/PD modeling, and statistical models (Danhof *et al.*, 2008; Csajka and Verotta, 2006; Gabrielsson and Green, 2009; Pinheiro and Bornkamp, 2014; Lon and Dubois, 2011; Bretz and Pinheiro, 2010).

Combinations of these methods in the decision-making process are frequent, and there is not a single best method. In contrast, it is rather conceivable to utilize

1) PKS

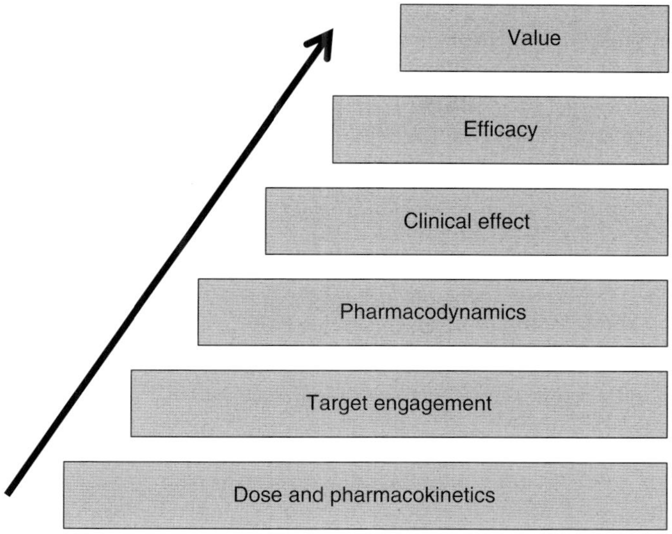


**Figure 10.1** Chain of evidence in clinical development.

the full range of data and analyze them with the different models. The variability and consistency of predictions as well as the pathophysiological and pharmacological plausibility should subsequently drive the decision-making process (Danhof *et al.*, 2008).

The "chain of evidence" approach ranks the different parameters and readouts from multiple experiments and studies according to their proximity from the pharmacological target versus the clinical endpoint for registration (or of interest in more general terms). A well-built model and decision process is characterized by consistent and plausible readouts along this chain of evidence (Paccaly *et al.*, 2006; Hinder *et al.*, 2006; Sabatine *et al.*, 2009).

A method which adds to the classical benefit and risk dimensions is the Clinical Utility Index (CUI) (Poland *et al.*, 2009; Korsan, Dykstra, and Pullman, 2005; Ouellet, 2010). This index integrates both internal data on the compound's preclinical and clinical efficacy and safety properties and puts these data in a competitive context by comparing it to already available or future therapeutic options. Despite its clear rationale and structure, this method has not widely been used yet.

In areas where phase 2 studies cannot be performed on the relevant clinical registration endpoints, dose finding needs to be done on a series of more or less validated and predictive biomarkers. The development of heart failure drugs is one example. Dose finding is often done on hemodynamic markers and/or biomarkers (Lapp *et al.*, 2009; Pang *et al.*, 2011; Erdmann *et al.*, 2013). The dose determined on these parameters does not reflect the final registration endpoint, such as cardiovascular mortality. Thus, there is often a need to carry forward several safe and efficacious doses into late-stage trials with the option to later drop unnecessary or ineffective treatment arms (Sabatine *et al.*, 2009; Ridker *et al.*, 2011; Steg *et al.*, 2012).

### 10.3.2 Comparators

In order to fully assess a new therapy's value, comparative data on efficacy and safety are needed. The answer to question as to whether a new treatment is superior, comparable, or inferior does have an enormous impact on the market access and the pricing of the new drug. In the future, it is likely that comparative data will even have consequences for the drug approval process (Sorenson, Drummond, and Burns, 2013; Sorenson *et al.*, 2011; Piccini *et al.*, 2009; Sanofi-Aventis, 2009).

The European Network for Health Technology Assessment (HTA) has issued a guideline which might help investigators and sponsors to select the best comparator(s). Numerous dilemmas might complicate this process (EUnetHTA – European network for Health Technology Assessment (2015))

- The best or right comparator from a scientific perspective might be ethically not acceptable; this is in many cases true for a placebo comparison.
- For some indications a true (scientific) gold standard might not have been established.
- From a HTA or health economics (HE) perspective, the most interesting comparison might come from data of drugs routinely used in clinical practice. The most routinely used drugs in clinical practice however might not conform to the criteria of EbM.

Interestingly, the choice of the comparators does not only support the efficacy comparison but also the rate of expected adverse events: reporting of adverse events which were expected for the comparator might be reported for the new drug just because of the expectedness of adverse events in this therapeutic area (Rosenthal effect) (Dinubile, 2008).

### 10.3.3 Patient Stratification to Increase Treatment Response and Benefit and Reduce Risk

By identifying patients who either benefit from treatment or patients who might experience harm, the benefit–risk balance for patients can be dramatically increased. This increase in benefit for the individual translates into an improved cost efficiency ratio, which is appreciated by payors and political decision makers (Hamburg and Collins, 2010). Patient identification and stratification often necessitate additional diagnostic testing. In most of the cases, this is an extra laboratory, bedside, or point-of-care test.

If the combination of therapy and companion diagnostic is intended for the marketplace, the drug candidate needs to be codeveloped together with the diagnostic. For example, for many diseases in oncology and psychiatry, the use of diagnostic or genetic tests to qualify patients for treatment is no longer an option but rather a prerequisite for some treatments (FDA, 2013). For diseases in which disease develops as a complex interplay of genetic disposition, environmental factors, and personal lifestyle, the (exclusive) use of genomic markers is usually less powerful. Nevertheless, a better understanding of the different factors has led to the consideration that the patient characterization in complex and multifactorial diseases can be achieved by the use of a multidimensional matrix. In addition to genetic or genomic information, this matrix utilizes factors from multiple sources, such as information arising from clinical biomarkers or imaging (Lee *et al.*, 2012). The FDA recently reported an increased review activity in the personalized medicine space, such as a 21% review activity for cardiovascular drugs, which has been a difficult area for personalized medicine traditionally. Recently published examples on optimized antiplatelet therapy illustrate that patient identification, stratification, and development of treatment recommendations can be based both on genetic and functional data (Trenk *et al.*, 2012; Trenk and Hochholzer, 2014).

To fully leverage its opportunities, personalized medicine needs to provide a more aligned collaboration between CD and diagnostics (Jorgensen, 2013; Sartori *et al.*, 2011). This includes considerations for biomarkers already with the start of drug discovery, markers for target engagement during the early CD phases, the "confirmatory" use of diagnostics during the CD phase, and the provision of drug-diagnostic tandems to the market. One central aspect of this approach is the fact that both the drug candidate and the companion diagnostic are developed in parallel (Woodcock, 2007; Ehmann *et al.*, 2013). In view of the stringent timelines of drug R&D, additional up-front investments are required to ensure the codevelopment as there is a need for an open and science-based supportive framework from the regulators (Ehmann *et al.*, 2013; Breitenstein, Scheuer, and Holsboer, 2014; Palmer, 2014; Fryburg *et al.*, 2014).

Palmer has recently reviewed the use of biomarkers in the area of drugs targeting the central nervous system (CNS). His framework for the use of biomarkers for patient selection, stratification, demonstration of target engagement, pharmacodynamics, and clinical effects like disease modification could be a useful structure to systematize and optimize the application of available markers (Palmer, 2014).

### 10.3.4 New Operational Tools to Succeed in Trials with Increased Complexity, Special Populations, or Large Size

The increased complexity in the CD programs can be partially counterbalanced by new models of trial conduct and by technologies, which recently have become available.

Involvement of more countries will facilitate patient recruitment, increase geographical diversity, and align data acquisition, which in turn should lead to larger acceptance of the data. From an operational point of view, digitization and synchronization of activities will become the standard. Examples of these new standards are specialized service providers who deliver online services for data review and interpretation, such as ECG or other safety data handling, storing, and reading. This way the advantages of integrating specialized services into a bigger context can be fully leveraged, an approach which under the term telemedicine has already found its way into clinical practice. Apple has recently proposed to use their "ResearchKit" to acquire and use data for personal health monitoring but also to inform patients about new potential treatments available or under investigation (https://www.apple.com/researchkit/).

Certain diseases, such as diabetes or Alzheimer's disease, show a geographical clustering based on environmental and genetic factors. Clinical studies will increasingly be conducted in areas of high disease prevalence and incidence rather than just focusing on the classical five key countries (EFPIA, 2013). It will be important to develop partnerships with patients, investigators, and authorities to ensure that all parties involved benefit therefrom (Van Meter *et al.*, 2008; Kukko *et al.*, 2004; WHO, 1999; Belluck, 2010; Kalaria *et al.*, 2008). As some of these partners have no experience in the conduct of clinical trials, the respective infrastructure and expertise need to be developed first (Dole *et al.*, 2014; Novartis, 2013; Novartis Institutes of Biomedical Research, 2013).

Willingness of eligible subjects to participate in clinical trials has been a long neglected matter and deserves more attention from the sponsor and the investigator. In a study conducted by Agoritsas, Deom, and Perneger (2011), safety, convenience, oversight, and open communication were attributes which were highly valued by patients and trial participants, whereas aspects which often are important for the scientific value and integrity of the trial, such as placebo controls, random allocation, or destruction of samples, were deemed less critical by patients (Agoritsas, Deom, and Perneger, 2011). Today, we have only started to understand why patients are willing or sometimes eager to participate in clinical trials. Two extreme examples of this phenomenon are on the one hand the "ACT UP movement" in the 1980s where HIV patients were fighting for their right for access to new AIDS medicines (http://www.actupny.org/) and on the other hand the recruitment of patients with rheumatoid arthritis into clinical trials which can be a huge challenge today. More research but also analysis of existing data will help conduct clinical trials faster and more efficient to finally reach patients who benefit most from their participation (Ding *et al.*, 2007; Agoritsas, Deom, and Perneger, 2011; Thoma *et al.*, 2010).

The art and science of how to optimize patient recruitment will finally serve both patients and pharmaceutical companies to develop drug candidates (Thoma *et al.*, 2010; Covington, 2013). Critical success factors range from pure planning, project management, and scientific standards of care to infrastructural components. Although each of them alone might not be overly difficult to manage, early planning and mock runs will reduce the likelihood of failures. In respect to communication, clinical trials require a dedicated effort from the investigators to be present and ready to provide answers to the numerous questions that may come from trial participants.

### 10.3.5 Collaboration and Outsourcing as Tools to Work in Networks

The impact of outsourcing on the conduct of clinical trials is examined in detail in the book chapter "Trends and Innovations in Pharmaceutical R&D Outsourcing."

The authors believe that the role of outsourcing will change from a third-party payment of externalized work packages to a true collaboration between partners. First moves in this direction were the preferred provider agreements by Merck & Co. and Merck Serono some years ago (Loftus, 2011; Wright, 2014a). The authors know from their own experiences that the establishment of a preferred provider constellation necessitates enormous up-front investments in terms of time and money before the partnership pays off. The primary benefit of such collaborations is related to the familiarity of culture, mindset, processes, and systems of both sides. On the other hand, the general spirit of "telling and being told" of fee for service is maintained, and the full skill set, motivation, and ownership of the contractors usually cannot be exploited. A last but important aspect in favor of outsourcing is the better risk and financial controls (Zerhouni, 2013; Wright, 2012).

The next level of collaboration, which puts more focus on complementary skill sets, has been coined as moving from tactical to strategic outsourcing. In this case, the collaboration is sought because the contracted partner has access to knowledge, skills, or infrastructure which is not available on the sponsor's side. Strategic outsourcing has been successfully applied in the past, but it has also been criticized as being an asymmetric and a deficit-oriented collaboration model (Wright, 2014b).

The third and probably most advanced model of collaboration between a sponsor and a service provider would be the true sharing of work and skill sets. In contrast to the former two models, strategic outsourcing is sharing the risks and the rewards that are related with the project among partners. One example is the Lead Discovery Center Dortmund, which has been formed as an international network of academic and industry partners for lead discovery. Thereby, the public side is represented by the Centre for Drug Research and Development (CDRD), Canada; the Scripps Research Institute, Florida, United States; the Centre for Drug Design and Discovery (CD3), KU Leuven, Belgium; the Medical Research Council Technology, United Kingdom; and Cancer Research Technology, United Kingdom. Partners from the industry comprise AstraZeneca, Daiichi Sankyo, Bayer, and Merck Serono. Although the authors are not aware of any similar models in the area of clinical trials or CD, the fact that in the complex area of lead discovery partners from industry, academia, and applied research collaborate to achieve common goals is encouraging for researches in other areas of pharmaceutical R&D to install similar models of collaboration.

The decisive difference to today's models of cooperation is to develop a common vision and mindset and agree on the general idea of truly sharing ideas to reach a common goal and exploit ideas in a shared fashion.

### 10.3.6
### Collaboration across Sectors and Industries to Boost the Next Wave of Innovation

The most obvious change in CD in the near future will be the aforementioned variations in working with service providers. The more revolutionary aspect will be in a rethinking of the business model of CD, in particular in view of the potential open innovation is providing.

Patients that participate in clinical trials today might well only be a small subpopulation of those individuals who if adequately approached would be willing to participate. Rather than bringing the patient into the trial, one could bring the clinical trial closer to the patients. The Walgreen–Novartis collaboration investigates presently the feasibility and acceptance of such a setting by placing the study center into a Walgreens pharmacy (Dole *et al.*, 2014). This approach offers several advantages for the parties involved: it can ease patient access to the study center, reduce time required for travel, and free up resources in the hospital for more complex tasks in the context of the study. As a consequence, a collaboration of nearby clinical service centers, providing services such as blood sampling or vital sign measurements, could be combined with more specialized clinical centers, such as academic university hospitals, which manage the more complex clinical tasks.

Allen and Marks have proposed a cross-industry collaborative model whereby several companies that work in the same therapeutic area or indication conduct their clinical trials jointly (Allen, Marks, and Arroyo, 2014). This would allow to potentially apply more effective trial designs and in turn enable the early comparison of several new investigational compounds and, at the same time, reduce the number of patients being treated with placebo or the gold standard. In that respect these trials would resemble large publicly funded trials rather than classical development studies. This approach seems to be attractive for a number of reasons. At the same time it will be crucial in the future to find ways to balance potentially competing interest with respect to alignment of trial designs and questions around investments and intellectual property (IP).

A meaningful impact on the public perception of R&D activities comes from the social media (Haslam and Kim, 2014). In general, social media facilitate the access to data and accelerate the dissemination of information. This has already led to an increased generation of personal data which are administered via Internet-based platforms and which comprise data that have not been available years ago. Of course their use is restricted today to a selected population which uses these wearables ("quantified self"). Nevertheless, it shows the large potential the new technologies have to monitor physiological human functions and provide in a second step counsel or advise for patients.

The second important impact on CD that social media will have in the future is the increase in visibility and open opportunities for patients to follow scientific efforts and trial progresses and, thus, become an active part of the trials and studies. Examples of this are already visible such as the patients' willingness to participate in trials for new experimental immunomodulatory therapies (e.g., PD-1, PD-1L) where information and knowledge are shared (https://www.smartpatients.com).

To a large extent this role has been filled in the past by disease societies and patient associations (Cattell, Chilukuri, and Levy, 2013). Their role will remain but will be complemented by social media. Not least the increased visibility and transparency of data and efforts by which the pharmaceutical industry works on

**Table 10.2** Factors driving technical success and value creation in the clinical development area.

*Trial objectives, design, and endpoints*

- Scientific and registrational endpoints
- Combined with patient-centric endpoints
- For certain diseases neither registrational nor patient-centric endpoints are available and need to be developed during clinical development
- Adaptive designs including combination of dose-finding part with confirmatory phase 3 part

*Enhanced dose–exposure–response information*

- Optimized benefit–risk ratio for individual and patient populations
- Reduced risks in patients with concomitant medications and diseases

*Patient stratification*

- Optimal and expanded patient disease characterization allows targeted therapeutic approaches
- Only patients with relevant molecular target will be treated
  - Increased responses in treated population
  - Avoidance of treatment in patients who will not benefit and/or bear a risk
  - Optimized resource utilization

*New operational approaches and tools*

- Digitization and synchronization of data acquisition and analysis
- Trials to come to the patient and active patient involvement
- Disease- or syndrome-oriented trials versus compound- or company-centric trials
- Buildup of trial infrastructure in regions with high medical need and disease prevalence

*Outsourcing versus collaboration*

- Distinguish outsourcing from collaboration
  - Active resource, budget, and risk management
  - Active collaboration between institutions to complement skill sets

*Collaboration beyond pharma – across other industries and service sectors*

- Involvement of disease and patient societies in addition to scientific societies
- Social media to facilitate a true two-way communication and ease access to trials and services
- Increase external communication to understand patients' needs and assessment of relevance of services
- Integration of new technologies into patient diagnosis/treatment pathways

improving patients' lives will help to restore trust in pharmaceutical companies (Hernandez *et al.*, 2014; Bauchner and Fontanarosa, 2014) (Table 10.2).

## Disclaimer

The views and opinions expressed in this chapter are those of the author and do not represent any official policy or position.

## References


Abrams, J. (1995) The role of nitrates in coronary heart disease. *Arch. Intern. Med.*, **155**, 357–364.

Agoritsas, T., Deom, M., and Perneger, T.V. (2011) Study design attributes influenced patients' willingness to participate in clinical research: a randomized vignette-based study. *J. Clin. Epidemiol.*, **64** (1), 107–115, http://www.ncbi.nlm.nih.gov/pubmed/20558036 (accessed 06 August 2014).

Allen, J., Marks, L., and Arroyo, S. (2014) Imagine running clinical trials together: what would this look like? Disruptive Innovations, The Fairmont Copley Plaza, September 11–12, 2014.

Angiolillo, D.J. *et al.* (2014) Pharmacodynamic evaluation of switching from ticagrelor to prasugrel in patients with stable coronary artery disease: results of the SWAP-2 Study (Switching Anti Platelet-2). *J. Am. Coll. Cardiol.*, **63** (15), 1500–1509, http://www.ncbi.nlm.nih.gov/pubmed/24333493 (accessed 01 June 2014).

Bauchner, H. and Fontanarosa, P.B. (2014) Restoring confidence in the pharmaceutical industry. *JAMA*, **309** (6), 2013–2015.

Belluck, P. (2010) The vanishing mind. Alzheimer's Stalks a Colombian Family. New York Times, p. A1.

Brass, E.P. *et al.* (2006) Risk assessment in drug development for symptomatic indications: a framework for the prospective exclusion of unacceptable cardiovascular risk. *Clin. Pharm. Ther.*, **79** (3), 165–172.

Breckenridge, A. and Walley, T. (2008) Risk sharing and payment by results. *Clin. Pharm. Ther.*, **83** (5), 666–667, http://onlinelibrary.wiley.com/doi/10.1038/clpt.2008.15/pdf.

Breitenstein, B., Scheuer, S., and Holsboer, F. (2014) Are there meaningful biomarkers of treatment response for depression? *Drug Discovery Today*, **19** (5), 539–561. doi: 10.1016/j.drudis.2014.02.002

Bretz, F. and Pinheiro, J.C. (2010) Practical considerations for optimal designs in clinical dose finding studies. *Stat. Med.*, **29**, 731–742.

Camm, a.J. *et al.* (2012) Ziprasidone and the corrected QT interval: a comprehensive summary of clinical data. *CNS Drugs*, **26** (4), 351–365, http://www.ncbi.nlm.nih.gov/pubmed/22452529 (accessed 01 June 2014).

Cattell, J., Chilukuri, S., and Levy, M. (2013) How Big Data Can Revolutionize Pharmaceutical R & D, April 2013.

Charlton, B.G. (2001) Fundamental deficiencies in the megatrial methodology. *Curr. Control. Trials Cardiovasc. Med.*, **2**, 2–7.

Coats, A.J.S. (2002) Omapatrilat- the story of overture and octave. *Int. J. Cardiol.*, **86**, 1–4.

Covington, M. (2013) Diversity in Clinical Trials – And Why It Matters Speaking of Sanofi, http://blog.sanofi.us/2013/05/23/diversity-in-clinical-trials/ (accessed 12 August 2015).

Csajka, C. and Verotta, D. (2006) Pharmacokinetic – pharmacodynamic modelling: history and perspectives. *J. Pharmacokinet. Pharmacodyn.*, **33** (3), 227–279.

Danhof, M. *et al.* (2008) Mechanism-based pharmacokinetic-pharmacodynamic (PK-PD) modeling in translational drug research. *Trends Pharmacol. Sci.*, **29** (4), 186–191.

Darius, H. (1999) Role of nitrates for the therapy of coronary artery disease patients in the years beyond 2000. *J. Cardiovasc. Pharmacol.*, **34** (S2), S15–S20.

DiMasi, J.A. *et al.* (2009) Trends in risks associated with new drug development: success rates for investigational drugs. *Clin. Pharm. Ther.*, **87** (3), 272–277. doi: 10.1038/clpt.2009.295

Ding, E.L., Powe, E.R., Manson, J.E., Sherber, N.S., and Braunstein, J.B. (2007) Sex differences in perceived risks, distrust, and willingness to participate in clinical trials: a randomized study of cardiovascular prevention trials. *Arch. Intern. Med.*, **167**, 903–912.

Dinubile, M.J. (2008) Double-blind active-control trials: beware the comparator you keep. *Clin. Infect. Dis.*, **47** (8), 1064–1067, http://www.ncbi.nlm.nih.gov/pubmed/18781878 (accessed 19 August 2014).

Dole, K., Orloff, J., Gleeson, G., Nelson, A., and Van Dam, J. (2014) Innovation in alternative clinical settings: the walgreens-novartis collaboration. Disruptive Innovations, The Fairmont Copley Plaza, September 11–12, 2014.

Downing, N.S. *et al.* (2014) Clinical trial evidence supporting FDA approval of novel therapeutic agents, 2005–2012. *JAMA*, **311** (4), 368, http://jama.jamanetwork.com/article.aspx?doi=10.1001/jama.2013.282034 (accessed 22 January 2014).

Editorial (2011) Learning lessons from Pfizer's 800 million $ failure. *Nat. Rev. Drug Discovery*, **10** (3), 163–164.

EFPIA (2013) The Pharmaceutical Industry in Figures. Key Data 2013, pp. 1–28.

Ehmann, F. *et al.* (2013) Gatekeepers and enablers: how drug regulators respond to a challenging and changing environment by moving toward a proactive attitude. *Clin. Pharmacol. Ther.*, **93** (5), 425–432, http://www.ncbi.nlm.nih.gov/pubmed/23549149 (accessed 18 June 2014).

Eichler, H. *et al.* (2011) Bridging the efficacy-effectiveness gap: a regulator's perspective on addressing variability of drug response. *Nat. Rev. Drug Discover*, **10** (7), 495–506. doi: 10.1038/nrd3501

Eidelman, R.S., Hollar, D., and Hebert, P.R. (2014) Randomized trials of vitamin E in the treatment and prevention of cardiovascular disease. *Arch. Intern. Med.*, **164**, 1552–1556.

Elliott, M. and Liu, Y. (2010) The nine rights if medication administration : an overview. *Br. J. Nurs.*, **19** (5), 300–305.

EMA (2013) Guideline on Clinical Investigation of Medicinal Products in the Treatment of Depression, May 2013, pp. 1–19.

EMA (2014) Clinical Efficacy and Safety: Anti-Infectives for Systemic Use, http://www.ema.europa.eu/ema/index.jsp?curl=pages/regulation/general/general_content_000401.jsp&mid=WC0b01ac0580034cf2 (accessed 12 August 2015).

Eng, C.M., Guffon, N., Wilcox, W.R., Germain, D.P., Lee, P., Waldek, S., Caplan, L., Linthorst, G.E., Desnick, R.J., and International Collaborative Fabry Disease Study Group (2001) Safety and efficacy of recombinant human alpha-galactosidase a–replacement therapy in Fabry's disease. *N. Engl. J. Med.*, **345** (1), 9–16.

Erdmann, E. *et al.* (2013) Cinaciguat, a soluble guanylate cyclase activator, unloads the heart but also causes hypotension in acute decompensated heart failure. *Eur. Heart J.*, **34** (1), 57–67.

EUnetHTA – European network for Health Technology Assessment (2015) COMPARATORS & COMPARISONS. Guideline on Criteria for the Choice of the Most Appropriate Comparator(s) Summary of Current Policies and Best Practice Recommendations, pp. 1–24. https://eunethta.fedimbo.belgium.be/sites/5026.fedimbo.belgium.be/files/news-attachments/2015-10-02_md_gl_sag_publiccons.pdf.

FDA (2011) Guidance for Industry and Food and Drug Administration Staff Clinical Investigations of Devices Indicated for the Treatment of Urinary Incontinence.

FDA (2013) Paving the Way for Personalized Medicine.

Food and Drug Administration (2008) Agency : action. *Fed. Regist.*, **73** (175), 97–98.

Food and Drug Administration (2014) Public Workshop – Complex Issues in Developing Drug and Biological Products for Rare Diseases, http://www.fda.gov/drugs/newsevents/ucm367820.htm (accessed 12 August 2015).

Fryburg, D.A. *et al.* (2014) Systems diagnostics: anticipating the next generation of diagnostic tests based on mechanistic insight into disease. *Drug Discovery Today*, **19** (2), 108–112. doi: 10.1016/j.drudis.2013.07.006

Furberg, C.D. and Pitt, B. (2001) Withdrawal of cerivastatin from the world market. *Curr. Control. Trials Cardiovasc. Med.*, **2**, 205–207.

Gabrielsson, J. and Green, A.R. (2009) Quantitative pharmacology or pharmacokinetic pharmacodynamic integration should be a vital component in integrative pharmacology. *J. Pharmacol. Exp. Ther.*, **331** (3), 767–774.

Gurbel, P.A. *et al.* (2009) Randomized double-blind assessment of the ONSET and OFFSET of the antiplatelet effects of ticagrelor versus clopidogrel in patients with stable coronary artery disease: the ONSET/OFFSET study. *Circulation*, **120** (25), 2577–2585, http://www.ncbi.nlm.nih.gov/pubmed/19923168 (accessed 03 June 2014).

Gurbel, P.A. *et al.* (2010) Response to ticagrelor in clopidogrel nonresponders and responders and effect of switching therapies: the RESPOND study. *Circulation*, **121** (10), 1188–1199, http://www.ncbi.nlm.nih.gov/pubmed/20194878 (accessed 12 June 2014).

Hamburg, M.A. and Collins, F.S. (2010) The path to personalized medicine. *N. Engl. J. Med.*, **363** (4), 301–304.

Haslam, J. and Kim, J. (2014) Social media revolution in clinical trials: when? Disruptive Innovations, The Fairmont Copley Plaza, September 11–12, 2014.

Heise, T. *et al.* (2007) Insulin glulisine: a faster onset of action compared with insulin lispro. *Diabetes Obes. Metab.*, **9** (5), 746–753, http://www.ncbi.nlm.nih.gov/pubmed/17593235 (accessed 05 June 2014).

Hernandez, J.F. *et al.* (2014) Restoring trust in the pharmaceutical sector on the basis of the SSRI case. *Drug Discovery Today*, **19** (5), 523–527, http://www.ncbi.nlm.nih.gov/pubmed/24295641 (accessed 28 July 2014).

Hinder, M. *et al.* (2006) Direct and rapid inhibition of factor Xa by otamixaban: a pharmacokinetic and pharmacodynamic investigation in patients with coronary artery disease. *Clin. Pharm. Ther.*, **80** (6), 691–702, http://www.ncbi.nlm.nih.gov/pubmed/17178269 (accessed 01 September 2014).

Hung, H.M.J., Wang, S., and Neill, R.O. (2005) A regulatory perspective on choice of margin and statistical inference issue in non-inferiority trials. *Biom. J.*, **47**, 28–36.

ICH E14 (2005) Guidance for Industry Interval Prolongation and Guidance for Industry. U.S. Department of Health and Human Services Food and Drug Administration Center for Drug Evaluation and Research (CDER) Center for Biologics Evaluation and Research (CBER), October 2005.

Jorgensen, J.T. (2013) Companion diagnostics in oncology – current status and future aspects. *Oncology*, **85**, 59–68.

Kaitin, K.I. (2010) Deconstructing the drug development process: the new face of innovation. *Clin. Pharmacol. Ther.*, **87** (3), 356–361.

Kalaria, R.N. *et al.* (2008) Alzheimer's disease and vascular dementia in developing countries: prevalence, management, and risk factors. *Lancet Neurol.*, **7** (9), 812–826, http://www.pubmedcentral.nih.gov/articlerender.fcgi?artid=2860610&tool=pmcentrez&rendertype=abstract (accessed 16 July 2014).

Kleinert, R., Lange, C., Steup, A., Black, P., Goldberg, J., and Desjardins, P. (2008) Single dose analgesic efficacy of tapentadol in postsurgical dental pain: the results of a randomized, double-blind, placebo-controlled study. *Anesth. Analg.*, **107** (6), 2048–2055.

Kola, I. (2008) The state of innovation in drug development. *Clin. Pharm. Ther.*, **83** (2), 227–230.

Korsan, B., Dykstra, K., and Pullman, W. (2005) A Clinical Utility Index (CUI) Openly Evaluates a Product's Attributes — and Chance. Pharmaceutical Executive, March 2005.

Kukko, M., Virtanen, S.M., Toivonen, A., Simell, S., Korhonen, S., Ilonen, J., Simel, O., and Knip, M. (2004) Geographical variation in risk HLA-DQB1 genotypes for type 1 diabetes and signs of beta-cell autoimmunity in a high-incidence country. *Diabetes Care*, **27** (3), 676–681.

Lapp, H. *et al.* (2009) Cinaciguat (BAY 58-2667) improves cardiopulmonary hemodynamics in patients with acute decompensated heart failure. *Circulation*, **119** (21), 2781–2788.

Ledford, H. (2013) Diabetes drugs ride a bumpy road. *Nature*, **504** (7479), 198, http://www.ncbi.nlm.nih.gov/pubmed/24336264 (accessed 10 August 2015).

Lee, M. *et al.* (2012) Personalized medicine in cardiovascular diseases. *Korean Circ. J.*, **42**, 583–591.

Lillie, E.O. *et al.* (2011) The n-of-1 clinical trial: the ultimate strategy for individualizing medicine? *Pers. Med.*, **8** (2), 161–173.

Loftus, P. (2011) Merck Teams Up with Parexel on Biosimilars. Wall Street Journal.

Lon, H. and Dubois, D.C. (2011) NIH public access. *Pharm. Res.*, **28** (7), 1622–1630.

Lorenz, M. *et al.* (2013) Effects of lixisenatide once daily on gastric emptying in type 2 diabetes--relationship to postprandial glycemia. *Regul. Pept.*, **185**, 1–8, http://www.ncbi.nlm.nih.gov/pubmed/23665027 (accessed 02 June 2014).

Matthews, D.R. and Tsapas, A. (2008) Four decades of uncertainty : landmark trials in glycaemic control and cardiovascular. *Diabetes Vasc. Dis. Res.*, **5**, 216–218.

Meinert, C.L. *et al.* (1970) A study of the effects of hypoglycemic agents on vascular complications in patients with adult-onset diabetes. II. Mortality results. *Diabetes*, **19** (Suppl.), 789–830.

Meinertz, T. (2001) Mibefradil — a drug which may enhance the propensity for the development of abnormal QT prolongation. *Eur. Heart J.*, **3**, 89–92.

Moors, E.H.M., Cohen, A.F., and Schellekens, H. (2014) Towards a sustainable system of drug development. *Drug Discovery Today*, **19** (11), 1711–1720, http://www.ncbi.nlm.nih.gov/pubmed/24657626 (accessed 25 May 2014).

Nisticò, G. (2011) Orphan drugs assessment in the centralised procedure. *Ann. Ist. Super. Sanita*, **47** (141), 98–99.

Norgren, L. *et al.* (2007) Inter-society consensus for the management of peripheral arterial disease (TASC II). *J. Vasc. Surg.*, **45** (Suppl. S), S5–S67, http://www.ncbi.nlm.nih.gov/pubmed/17223489 (accessed 26 May 2014).

Novartis (2013) The Novartis Malaria Initiative in Zambia Engagement in Zambia across Novartis divisions, August.

Novartis Institutes of Biomedical Research (2013) Advancing Global Health.

Nuzzo, R. (2014) Statistical errors. *Nature*, **506**, 150–152.

Orloff, J.J. and Stanski, D. (2011) Innovative approaches to clinical development and trial design. *Ann. Ist. Super. Sanita*, **47**, 8–13.

Orloff, J. *et al.* (2009) The future of drug development: advancing clinical trial design. *Nat. Rev. Drug Discovery*, **8** (12), 949–957, http://www.ncbi.nlm.nih.gov/pubmed/19816458 (accessed 10 August 2015).

Ouellet, D. (2010) Benefit – risk assessment: the use of clinical utility index. *Expert Opin. Drug Saf.*, **9** (2), 289–300.

Owens, D.K. *et al.* (2010) Grading the strength of a body of evidence when comparing medical interventions--agency for healthcare research and quality and the effective health-care program. *J. Clin. Epidemiol.*, **63** (5), 513–523. doi: 10.1016/j.jclinepi.2009.03.009

Paccaly, A. *et al.* (2006) Pharmacokinetic/pharmacodynamic relationships for otamixaban, a direct factor Xa inhibitor, in healthy subjects. *J. Clin. Pharmacol.*, **46** (1), 45–51, http://www.ncbi.nlm.nih.gov/pubmed/16397283 (accessed 01 September 2014).

Palmer, A.M. (2014) The utility of biomarkers in CNS drug development. *Drug Discovery Today*, **19** (3), 201–203, http://www.cerebroscience.com/The utility of biomarkers in CNS drug development.pdf (accessed 10 August 2015).

Pang, P.S. *et al.* (2011) Rationale, design, and results from RENO-DEFEND 1: a randomized, dose-finding study of the selective A1 adenosine antagonist SLV320 in patients hospitalized with acute heart failure. *Am. Heart J.*, **161** (6), 1012–1023. doi: 10.1016/j.ahj.2011.03.004

Penston, J. (2007) Patients' preferences shed light on the murky world of guideline-based medicine. *J. Eval. Clin. Pract.*, **13**, 154–159.

Piccini, J.P. *et al.* (2009) Comparative efficacy of dronedarone and amiodarone for the maintenance of sinus rhythm in patients with atrial fibrillation. *J. Am. Coll. Cardiol.*, **54** (12), 1089–1095. doi: 10.1016/j.jacc.2009.04.085

Pinheiro, J. and Bornkamp, B. (2014) Model-based dose finding under model uncertainty using general parametric models. *Stat. Med.*, **33**, 1646–1661.

Piquette-Miller, M. and Grant, D.M. (2007) The art and science of personalized medicine. *Clin. Pharm. Ther.*, **81** (3), 311–315.

Poland, B. *et al.* (2009) The clinical utility index as a practical multiattribute approach to drug development decisions. *Clin. Pharm. Ther.*, **89** (1), 105–108.

Reichert, J.M. (2003) Trends in development and approval times for new therapeutics in the United States. *Nat. Rev. Drug Discovery*, **2** (9), 695–702.

Ridker, P.M. *et al.* (2011) Interleukin-1 β inhibition and the prevention of recurrent cardiovascular events : rationale and design of the Canakinumab Anti-inflammatory Thrombosis Outcomes Study (CANTOS). *Am. Heart J.*, **162** (4), 597–605. doi: 10.1016/j.ahj.2011.06.012.

Roche (2012) Roche provides update on Phase III study of dalcetrapib. Media Release, May 07.

Sabatine, M.S. *et al.* (2009) Otamixaban for the treatment of patients with non-ST-elevation acute coronary syndromes (SEPIA-ACS1 TIMI 42): a randomised, double-blind, active-controlled, phase 2 trial. *Lancet*, **374** (9692), 787–795. doi: 10.1016/S0140-6736(09)61454-9.

Sackett, D.L., Rosenberg, W.M., Gray, J.A.M., Haynes, R.B., and Richardson, W.S. (1996) Evidence based medicine: what it is and what it isn't. *Br. Med. J.*, **312** (7023), 71–72.

Sanofi-Aventis (2009) MULTAQ® (DRONEDARONE) Briefing Document Advisory Committee Meeting of the Cardiovascular and Renal Drugs Division of the US Food and Drug Administration, FDA, March 18, 2009, http://www.fda.gov/ohrms/dockets/ac/09/briefing/2009-4417b1-02-Sanofi_Aventis.pdf (accessed 14 November 2015).

Sartori, V. *et al.* (2011) Value-driven drug development — unlocking the value of your pipeline, in *Pharmaceutical and Medical Products Practice*, McKinsey and Company.

Schumi, J. and Wittes, J.T. (2011) Through the looking glass: understanding non-inferiority. *Trials*, **12** (1), 106, http://www.trialsjournal.com/content/12/1/106 (accessed 10 August 2015).

Committee for Orphan Medicinal Products and the European Medicines (2011) European regulation on orphan medicinal products: 10 years of experience and future perspectives. *Nat. Rev. Drug Discovery*, **10** (5), 341–349. doi: 10.1038/nrd3445.

Seidling, H.M. *et al.* (2013) Safeguarding the process of drug administration with an emphasis on electronic support tools. *Br. J. Clin. Pharmacol.*, **76**, 25–36.

Silva, S.G. *et al.* (2005) Reviews and overviews the case for practical clinical trials in psychiatry. *Am. J. Psychiatry*, **162** (5), 836–846.

Smith, G.C.S. and Pell, J.P. (2003) Parachute use to prevent death and major trauma related to gravitational challenge: systematic review of randomised controlled trials. *Br. Med. J.*, **19** (2), 126–128, http://www.ncbi.nlm.nih.gov/pubmed/16602356 (accessed 10 August 2015).

Sofat, R. *et al.* (2010) Separating the mechanism-based and off-target actions of cholesteryl ester transfer protein inhibitors with CETP gene polymorphisms. *Circulation*, **121**, 52–62.

Sorelle, R. (1998) Withdrawal of posicor from market. *Circulation*, **98**, 831–832.

Sorenson, C. *et al.* (2011) Evidence of comparative efficacy should have a formal. *Br. Med. J.*, **343**, 1–6.

Sorenson, C., Drummond, M., and Burns, L.R. (2013) Evolving reimbursement and pricing policies for devices in Europe and the United States should encourage greater value. *Health Aff.*, **32** (4), 788–796.

Steg, P.G. *et al.* (2012) Design and rationale of the treatment of acute coronary syndromes with otamixaban trial: a double-blind triple-dummy 2-stage randomized trial comparing otamixaban to unfractionated heparin and eptifibatide in non-ST-segment elevation acute coronary syndromes with a planned early invasive strategy. *Am.*

*Heart J.*, **164** (6), 817–824.e13. doi: 10.1016/j.ahj.2012.10.001.

Stegmann, J.U., Weber, H., Steup, A., Okamoto, A., Upmalis, D., and Daniels, S. (2008) The efficacy and tolerability of multiple-dose tapentadol immediate release for the relief of acute pain following orthopedic (bunionectomy) surgery. *Curr. Med. Res. Opin.*, **24** (11), 3185–3196.

Suckfüll, M. *et al.* (2011) A randomized, double-blind, placebo-controlled clinical trial to evaluate the efficacy and safety of neramexane in patients with moderate to severe subjective tinnitus. *BMC Ear Nose Throat Disord.*, **11** (1), 1–10, http://www.biomedcentral.com/1472-6815/11/1 (accessed 10 August 2015).

The CONSENSUS Trial Study Group (1987) Effects of enalapril on mortality in severe congestive heart failure. Results of the Cooperative North Scandinavian Enalapril Survival Study (CONSENSUS). The CONSENSUS Trial Study Group. *N. Engl. J. Med.*, **316** (23), 1429–1435.

The PLoS Medicine Editors (2011) Why drug safety should not take a back seat to efficacy. *PLoS Med.*, **8** (9), 8–9.

Thoma, A., Farrokhyar, F., McKnight, L., and Bhandari, M. (2010) How to optimize patient recruitment. *Can. J. Surg.*, **53** (3), 205–210.

Trenk, D. and Hochholzer, W. (2014) Genetics of platelet inhibitor treatment. *Br. J. Clin. Pharmacol.*, **77** (4), 642–653, http://www.ncbi.nlm.nih.gov/pubmed/23981082 (accessed 29 July 2014).

Trenk, D. *et al.* (2012) Personalizing antiplatelet therapy with clopidogrel. *Clin. Pharmacol. Ther.*, **92** (4), 476–485, http://www.ncbi.nlm.nih.gov/pubmed/22948893 (accessed 29 July 2014).

Trusheim, M.R., Berndt, E.R., and Douglas, F.L. (2007) Stratified medicine: strategic and economic implications of combining drugs and clinical biomarkers. *Nat. Rev. Drug Discovery*, **6** (4), 287–293, http://www.ncbi.nlm.nih.gov/pubmed/17380152 (accessed 10 August 2015).

UK Prospective Diabetes Study Group (1998) Intensive blood-glucose control with sulphonylureas or insulin compared with conventional treatment and risk of complications in patients with type 2 diabetes (UKPDS 33). *Lancet*, **352** (9131), 837–853.

Van Meter, K.C. *et al.* (2008) A procedure to characterize geographic distributions of rare disorders in cohorts. *Int. J. Health Geographics*, **7**, 26, http://www.pubmedcentral.nih.gov/articlerender.fcgi?artid=2430550&tool=pmcentrez&rendertype=abstract (accessed 16 July 2014).

Wakai, A. *et al.* (2013) Nitrates for acute heart failure syndromes (review). *Cochrane Database Syst. Rev.*, (8) (Art. No.:CD005151)). doi: 10.1002/14651858.CD005151.pub2.

WHO (1999) Genetics and Diabetes, pp. 1–15, http://www.who.int/genomics/about/Diabetis-fin.pdf (accessed 12 August 2015).

Winterstein, A.G. (2009) Rosiglitazone and the risk of adverse cardiovascular outcomes. *Clin. Pharmacol. Ther.*, **89** (6), 776–778. doi: 10.1038/clpt.2011.43.

Woodcock, J. (2007) The prospects for "personalized medicine" in drug development and drug therapy. *Clin. Pharm. Ther.*, **81** (2), 164–169.

Wright, T.M. (2012) Focusing on Patients, Following the Science. Novartis R&D Investor Day Boston, November 8, 2012, https://www.yumpu.com/en/document/view/13994041/focusing-on-patients-following-the-science-25-mb-novartis.

Wright, R. (2014a) Merck Serono's CEO Belén Garijo – Enabling Risk and Refusing to Play It Safe. Life Science Leader Magazine.

Wright, R. (2014b) Reader Disputes Merck-Serono/Quintiles Collaboration As Being Game Changing. Life Science Leader Magazine.

Zerhouni, E.A. (2013) JP Morgan Healthcare Conference, San Francisco, CA, January 8, 2013, http://www.sanofi.com/Images/31656_2013-01-08_JPMorgan_Zerhouni.pdf (accessed 14 November 2015).

Zhou, Y.-T. *et al.* (2014) Pharmacokinetic drug-drug interactions between 1,4-dihydropyridine calcium channel blockers and statins: factors determining interaction strength and relevant clinical risk management. *Ther. Clin. Risk Manage.*, **10**, 17–26, http://www.pubmedcentral.nih.gov/articlerender.fcgi?artid=3873236&tool=pmcentrez&rendertype=abstract (accessed 10 August 2015).

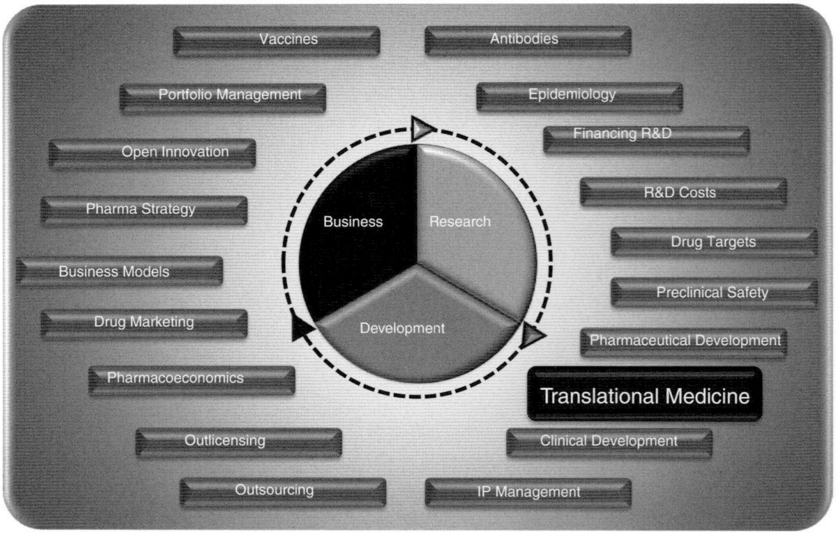
Vaccines
Antibodies
Portfolio Management
Epidemiology
Financing R&D
Open Innovation
R&D Costs
Pharma Strategy
Business
Research
Drug Targets
Business Models
Preclinical Safety
Drug Marketing
Development
Pharmaceutical Development
Pharmacoeconomics
Translational Medicine
Outlicensing
Clinical Development
Outsourcing
IP Management

# 11 Translational Development

*Nigel McCracken*

## 11.1 Introduction

The purpose of this book chapter is to look at the role Translational Development plays within the pharmaceutical industry to help translate basic research into clinical utility and the multidisciplinary tools and highly collaborative approaches that are employed to help deliver a tailored development solution designed to maximize benefit while minimizing risk.

Competition to find and develop drugs that are effective and differentiated is becoming harder. Currently there is the expectation from both regulatory bodies and payers to better understand the population that will receive benefit as well as provide evidence of additional clinical utility of the drug over either placebo or standard of care (SOC). Translational Development is one of the disciplines that has evolved over the last number of years to try and address this expectation, with a focus around dose optimization, differentiation, stratification, and repositioning.

The following questions are explored in this book chapter:

- Why are we less efficient in developing effective drugs when we have better disease understanding and technologies?
- What is Translational Development, and where does it fit into the drug development process?
- What are the main types of activities where Translational Development can provide value to a project?
- Combinations, segmentation, and differentiation – how can Translational Development contribute to the success of efficiently bringing to the market medicines that provide value to both the patient and the industry?

Within the pharmaceutical industry we are tasked to provide innovative, differentiated products that will have benefit for individuals, society, and the industry. In an attempt to develop more innovative and differentiated drugs, we have implemented new technologies to screen and produce more viable targets which are often not validated. Expectation to understand drivers of response, to stratify your population, as well as to identify and introduce outcome endpoints into our clinic studies introduces additional cost, complexity, and risk. Translational

*Value Creation in the Pharmaceutical Industry: The Critical Path to Innovation,* First Edition.
Edited by Alexander Schuhmacher, Markus Hinder, and Oliver Gassmann.

activities should be focused around mitigating risk (understanding MOA, establishing a clinical hypothesis, optimizing dose and study design) as well as providing added value to the project (through differentiation, stratification, and repositioning). The concept of smaller dedicated teams collaborating across multifunctions and bringing together basic science and clinical utility utilizing new technologies and increased disease understanding should be the foundations of all Translational groups.

As we look to the future where there will be the expectation that targeted therapies will be more widely available in multiple therapeutic areas, Translational Development will play a central role in building confidence and understanding around these molecular targets and their evoked clinical response as well as being instrumental in moving the drug development process away from trial-and-error drug discovery toward a more rational drug design.

### 11.1.1 Legacy

The goal of drug therapy is to provide effective treatment without toxicity. Ideally a drug is delivered so that it achieves a blood concentration that is within the therapeutic range to elicit a pharmacological response. The responses that we associate with a particular drug are due to physiological and biological properties of cells targeted by receptors of the recipient. They are determined by interactions between physiological, pathophysiological (liver function, kidney function, cardiac function, lung function, and other diseases), environmental (smoking, alcohol, nutrition, and drug therapy), and genetic factors. All of these different factors can cause variability in response, and it is therefore no great surprise that at best most nontargeted marketed drugs are effective in no more than 40% of the population.

It has been over 50 years since Sir James Black discovered the first clinically significant beta blockers, propranolol and pronethalol, which have had a lasting impact in the treatment and management of angina pectoris (Stapleton, 1997). Claude Bernard remarked, "The innovator's skill is in seeing what everyone has seen and thinking what nobody has thought" (Vlietstra, 1995). Sir James Black perceived links between clinical medicine and pharmacology, and he approached the clinical problem from the laboratory rather than from the bedside. His research in cardiovascular pharmacology contributed to the clinical understanding of the etiology of angina pectoris as well as provided a deeper understanding of the basic pharmacology of the autonomic nervous system. Over those 50 years we have seen 1091 new molecular entities (NMEs) approved by the FDA (1962–2011) with an average of around 22 NMEs being approved every year (FDA website, 2014). In this same time the costs of bringing drugs to the market have also dramatically increased. It is now estimated that for big pharma the cost to bring a new compound to market can be anything from USD 3.5 to 12 billion depending on the size of the company and the number of therapeutic areas they are working in. The reality is that the pharmaceutical business is in the grip of rising failure rates and rising costs, and it is believed that new technologies and a better understanding of biology will turn things around.

It may well be that with the implementation of new technologies to screen and produce more viable targets that we have moved too far away from what brought a lot of success in the past, namely, collaboration between different fields and groups of research and good science. It is also clear that in a number of indications we have very good drugs that are very hard to make better. As mentioned the expectation nowadays is to bring a compound onto the market that has differentiation from SOC as well as value for society, and more often than not we should also understand the population we are trying to treat. As a result the Me2 approach to drug development is no longer being pursued widely as a viable strategy within the industry.

In the last 20 years we have also seen some real innovation with respect to both targeted therapies and targeted delivery specifically within the field of biotechnology; however, we have also lived through the failure of "screen more and screen fast" and metrics related to number of hits (rather than focusing on quality of hits) as a means to increase productivity. The whole genome program and the sequencing of the first human genome back in the 2003 also brought an expectation of new targets and a more "Personalized Medicine" approach to treating individuals. Innovative technologies have been developed to sequence and analyze all this new data along with the ever-growing "omics" fields. If we are honest and look back on the Biomarker and Personalized Medicine drive employed within the industry over the last 12 years, the results have been disappointing in relation to the investment of time and money. It may well be that with the implementation of new technologies to screen and produce more viable targets that we have moved too far away from what brought a lot of success in the past, namely, collaboration between different fields and groups of research and good science.

## 11.2 Translational Development

Translational Development or Translational Medicine will mean different things to different people; however, one thing is clear – it seems to be important to almost everyone. Translational activities can cover everything from "bench to bedside research" all the way through clinical development to observational studies and capturing real-world data on treatment and practice.

Most companies have a Translational group somewhere within their R&D; however, the focus of the group as well as where it is placed in a company can be quite different. Experience has taught us that the most important factor is not where the Translational group is located but that the group is focused, empowered, and aligned throughout the whole organization. The value of Translational Development within the pharmaceutical industry is to integrate knowledge gained from pharmacology, toxicology, drug metabolism & pharmacokinetics (DMPK), and biomarkers with other line functions with the goal to select the best drug

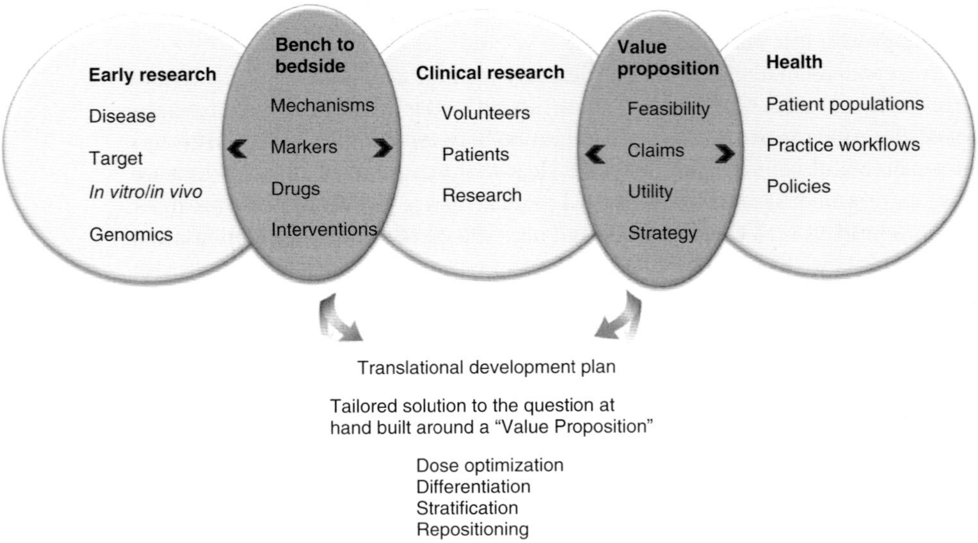


**Figure 11.1** Translating research into clinical utility and commercial success.

candidates and optimize study designs that help to mitigate risk as well as provide value to the portfolio through differentiation, stratification, and repositioning.

Interestingly, the foundation that Sir James Black founded grew out of his own experiences working in academia and industry and was based on the concept that "the majority of significant research discoveries result from the concentrated efforts of small, well resourced teams of talented and highly-focused scientists, who are unencumbered by the organisational distractions, decision time lags and lack of direction frequently encountered in large public and industrial institutions." The concept of smaller dedicated teams bringing together basic science and clinical utility utilizing new technologies and increased disease understanding should be the foundations of all Translational groups. Translational groups should look to translate the basic research into clinical utility, the so-called Bench to Bedside, as well as utilize clinical data Bedside to Bench and be focused around what the "Value Proposition" is for the project in question (see Figure 11.1). Activities should involve generating data with a goal to obtain regulatory approval, patient acceptance and pricing, and reimbursement and market access success and be tailored to the questions that need to be addressed. This tailored approach would be documented in a Translational Plan (TP) and be linked to a Target Product Profile (TPP).

### 11.2.1 TP

A TP should be generated alongside the development plan documenting the types of studies that will be conducted and the impact they will have in supporting the claims in the TPP. The TPP would be generated by the project team setting out the key success criteria that need to be achieved for regulatory approval as well as provide information on what the value proposition will be for the asset in the marketplace. The value proposition for the asset will be built around a unique selling point (USP) and can be based on question like how your compound will be differentiated in the marketplace from SOC as well as why there will be a positive benefit to society. The TP should therefore be tailored directly to the question that needs to be answered to help achieve the TPP as well as activities around derisking or adding value to the program and could be focused around:

- *Dose optimization* – Understand the intrinsic characteristics of your compound, pharmacokinetics (PK) and pharmacodynamics (PD) to optimize the delivery, frequency of dose, and dose of the compound in the clinic to maximize the chance of success.
- *Stratification* – Understand response/nonresponse within the population as well as disease segmentation.
- *Differentiation* – What data do I need to generate to differentiate my compound from the SOC?
- *Repositioning* – Based on the mode of action can we reposition our compound in another indication.
- *Biomarkers* – Is there a target engagement or efficacy biomarker I can develop to derisk the program or segment the population? Can we provide earlier indicators of efficacy that can facilitate earlier decision making? Is there a need for a diagnostic in the marketplace for the disease in question, and if there is, what type of technology am I going to use for first-line treatment of my diagnostic?

In many ways all of these activities are there to maximize benefit and minimize risk of failure for your compound in development, and it is therefore important to have a focused aligned approach as to how you will achieve this within the defined timelines. The TP should run alongside the development plan providing a framework that sets out *what* samples will be taken from the different types of ongoing activities (pharmacology, biomarker, PKPD, clinical, safety, etc.), *why* they are being taken, as well as *how* they will be used to generate value for the program. As mentioned before the link between science and the clinic is critical, so it is important that the TP is aligned with both discovery and clinical groups and in the case of diagnostic development should also include business and market access alignment. It would also be a natural progression that the TP would run through both Translational Development and Translational Medicine activities. Implementing the TP early in development is essential to avoid unwanted costs around having to retrospectively fit in activities in late development, particularly in relation to the need for a diagnostic within a program.

### 11.2.2 Translational Toolkit

We have and are living through a period where there have been a number of key contributions that brought us to the position where we have deeper disease understanding and innovative technologies that enable us to have a greater impact in clinical utility as well as how we go about developing drugs. Some of the scientific contributions, research infrastructures, and technologies that have had a major impact include the following:

- Human genome project
- Transgenic mouse models of disease
- Epidemiology and epigenetic and gene–environment studies
- Computational biology
- Medical imaging (structural and functional)
- High-throughput technologies (automation, NGS, nanotechnology, high-resolution optics, LCMSMS, etc.)
- Pharmacogenomics (PGx), transcriptomics, proteomics, etc.
- Pharmacometrics (PKPD modeling and modeling and simulation)
- Proteomics, metabolomics, and epigenetics
- Pathway analysis (systems biology and systems pharmacology)
- Genome-wide sequencing (genome-wide association studies (GWAS))
- Informatics and knowledge management (the ability to utilize open-access data sources)
- Precompetitive activities to further the field of science (ATLAS, IMI, FP7, Horizon 2020, etc.).

It is not within the scope of this chapter to go into any detail around these technologies as there are countless reviews and information available which are constantly being updated. These are the technologies that are available to be used by any Translational group to get to the heart of the question that needs to be answered, and it is therefore important to identify what methods are to be used and what impact that generated data will bring. Translational groups should try to be technology agnostic with the most appropriate approaches being applied for individual projects across the portfolio.

Translational groups can be presented with all types of projects from early-stage development (target engagement) all the way through to clinical candidates. Projects can come from internal development or through in-licensing efforts where you are presented with different types of information, timelines, and needs. This chapter will look at Translational activities encompassing the earlier phase of development from initial target engagement to early clinical development.

## 11.3 Dose Optimization

As mentioned in the introduction the goal of drug therapy is to provide effective treatment without toxicity, which involves delivering a drug that achieves a blood concentration that is within the therapeutic range to elicit a pharmacological response. When we think of what makes a good drug, we think about physical properties, target affinity and selectivity, and clearance which bring together a number of disciplines within early research. As a Translational scientist looking to optimize the dose and dose regimen in the clinic (with respect to maximizing the efficacy through exposure-response) there are some fundamental aspects that need to be understood and considered, namely, physicochemistry, target affinity, and clearance.

### 11.3.1 Physicochemical Properties

The physicochemical properties of the active pharmaceutical ingredient (API) (LogD, $pK_a$, molecular weight, chemical structure, and melting point) are important to understand as these affect the functional characteristics of the compound (dissolution rate, chemical stability, chemical reactivity, ionization, and diffusivity) which in turn will dictate the compounds' intrinsic behavior (solubility, permeability, and taste) including its PK properties (how the compound is absorbed, distributed, metabolized, and excreted) (Florence and Attwood, 2011). The efficacy of a drug at a target will be related to its exposure at the site of action, the affinity to the receptor, and how long the compound remains at the site of action to elicit a response (residence time). Physicochemical properties are critical to understand and optimize as these will affect the exposure (absorption and distribution) and residence time (as a result of clearance).

### 11.3.2 Target Affinity and Selectivity

If we are presented with a drug that engages a viable target, we would normally be provided with information on the affinity of the drug for the receptor, potency, and efficacy. Pharmacologists are interested in how the drug binds to the target as well as how this binding translates into effect, initially in an *in vitro* setting and then *in vivo*. Information on type of binding (reversible, irreversible), type of response (agonist, antagonist, partial, inverse, full, etc.), receptor numbers, receptor reserve, chirality, and drug residence time are important to understand when we start to take the steps from our target engagement to the bedside. Target affinity starts to provide information on receptor occupancy (70% occupancy occurs at 3X the *in vitro* affinity value – Hill–Langmuir) which in turn can provide initial information on clinical efficacy (see Table 11.1) depending on the target class (Cammack *et al.*, 2006).

### 11.3.3 Clearance

We know that *in vivo* efficacy is influenced by PK and PD and that to elicit an appropriate response the blood concentration needs to be above a certain "threshold" concentration for an appropriate period of time which will be dictated by how

**Table 11.1** Target occupancy requirements for clinical efficacy.

| Target class | Drug/ligand class | Target occupancy (%) |
|---|---|---|
| GPCRs | Antagonists | 60–80 |
| | Agonists (full efficacy) | 2–30 |
| | Agonists (low efficacy) | 60–95 |
| Ligand-gated ion channels | Antagonists | 65–90 |
| | Agonists | 5–80 |
| Transporters | All | 60–85 |

Adapted from Grimwood and Hartig (2009).

the drug in question is absorbed (if orally administered), distributed, and eliminated. The drug concentration at the site of action will elicit the pharmacological effect which in turn will provide a clinical response (toxicity and/or efficacy). The elimination or clearance of the drug from the body will be dictated by how the drug in question is metabolized and excreted from the body which in turn determines the duration of effect. Managing and understanding these three criteria is crucial to be able to initially predict the human dose, understand the "therapeutic window" of the compound in question, as well as predict variability in response.

### 11.3.4
### Prediction of Human Dose

As previously mentioned when we think about predicting an appropriate dose to elicit a response, we think about blood concentration at the site of action and its residence time. In its simplest term the dose prediction will be based on the dose that will provide a free plasma concentration of >3X affinity at the target (to achieve 70% occupancy) and the dose frequency which is dictated by the drug residence time (pharmacology) and drug clearance (PK). Over the last number of years the pharmaceutical industry has made great strides in predicting human dose based on animal *in vitro* and *in vivo* data through the use of more relevant *in vivo* animal models and of PKPD modeling. Models are only as good as the information you feed into them, and although it is clear that there are a number of good PKPD models that are already available in a number of disease areas, we are also being faced with new targets in new indications that do not have any proof of concept or validation in the clinic. In these instances it is important to build a hypothesis from initial target engagement through pathway modulation to the causal effect utilizing PD biomarkers in conjunction with PK data to help identify the biological active dose range to help in our "go" or "no-go" decisions (Sarker and Workman, 2007; Adjei, 2006; Goulart, Roberts, and Clark, 2004).

It is also important when determining the human dose to identify what factors may contribute to the variability in the response to the drug. There are many different factors that can affect the exposure–response relationship, which include differences in physiological, pathophysiological, environmental, and genetic factors. The next section will cover the important role that pharmacogenomics plays in understanding differences in exposure–response with respect to absorption, distribution, and metabolism and then separately its utility in disease segmentation.

## 11.4 Pharmacogenomics

Pharmacogenomics refers to the application of genome-wide approaches in order to understand genetic influences on drug response from absorption and metabolism to pharmacological action and therapeutic effects (Lesko and Woodcock, 2004). Pharmacogenomics has had its underpinnings within the field of pharmacogenetics, and historical studies were initially developed to understand individual differences in PK and metabolism, often focused on single nucleotide polymorphisms (SNPs), particularly in drug metabolism genes (Roses, 2008). There has been and still is a lot of ambiguity relating to the terms of "pharmacogenetics" and "pharmacogenomics" even though both link together genotype with drug response. Within this section the term "pharmacogenomics" will be used to include work conducted using DNA, RNA, and protein, which includes the identification of candidate genes and polymorphisms in relation to evoked clinical response, predicting drug response and clinical outcomes, reducing adverse events, and selecting dosing of therapeutic drugs on the basis of genotype.

### 11.4.1 Patient Segmentation

Individual variations in response to drugs and drug toxicity are a common occurrence in both the clinical setting and drug development research. When we think about drug response, we think about the PK of the drug (absorption, distribution, metabolism, and excretion) as well as its PD. Over the last two decades the amount of information contained in clinical pharmacology databases for drug metabolites and PK/PD have been greatly increased by genetic (SNP, copy number variation, NGS) and genomic (transcriptomic, proteomic, pathway analysis, etc.) information. In particular, sequencing of highly polymorphic loci such as the HLA region and genes involved in drug metabolism has increased the specificity of genetic associations with PK or PD phenotypes (Roses *et al.*, 2014). All this cumulative evidence strongly suggests that genetic polymorphisms in drug metabolism enzymes (CYP450, TPMT, UGT2B7, etc.) (Ma and Lu, 2011), transporters (PgP, OATP, NTCP, SLC01B1, etc.) (Hillgren *et al.*, 2013), receptors (G protein-coupled receptors (GPCRs), nuclear, enzymes, etc.), and other drug targets are contributing to interindividual differences in the efficacy and toxicity of drugs and as such need to be understood for any given compound in development.

A very good example of how technology in pharmacogenomics has progressed particularly in the field of drug metabolism in the last few years alongside the ability to analyze large data sets is the example of the Affymetrix DMET chip. The Affymetrix DMET chip is now being used routinely in early clinical development to screen for polymorphisms across all of the relevant drug metabolism and transporter genes to help explain variability in PK (absorption, metabolism, and elimination). The Affymetrix DMET+ chip is capable of simultaneously analyzing a panel of 1936 pharmacogenetically relevant mutations from 225 different genes consisting of 50 CYP450 genes, 45 phase II enzymes, 64 drug transporters, and 66 transcriptional regulators and enzymes. A summary of examples of known alleles which have been shown to influence the outcomes of specific drug therapy can be seen in Table 11.2.

About 70% of drugs licensed for human therapeutic use mediate their action through their interaction with one or more cellular receptor proteins. Sixty percent of these drugs are dependent on one or more members of the GPCR superfamily, whereas the rest act via nuclear or kinase receptors. Drug receptors, therefore, often constitute the first interface between a therapeutically administered drug and cellular physiology. Genetic variability in the receptors and proteins involved in receptor signaling pathways has been used to great effect in oncology, particularly relating to the role of tyrosine kinase.

Tyrosine kinases itself constitute a family of enzymes that owe their importance to their tight involvement in proliferative cell signaling and cancer management. The genotyping of tyrosine kinase variants was found to have applications in predicting therapy efficacy as well as drug resistance. This was highlighted by the characterization of the BRC/ABL translocation in chronic myeloid leukemia (CML) and the development of imatinib, a small-molecule tyrosine kinase inhibitor which binds specifically to the ATP pocket of the BCR/ABL kinase domain, intended for use in BCR/ABL-positive leukemia (O'Hare *et al.*, 2012). This characterization of the molecular mechanism of disease highlighted the potential for the development of patient-targeted therapy. Pharmacogenomic techniques were further used to determine that in addition imatinib inhibits the receptor tyrosine kinase cKIT and hence targets activation of cKIT mutants in gastrointestinal stromal tumors (GISTs) (McLean *et al.*, 2005).

### 11.4.2 Disease Segmentation

In oncology pharmacogenomics has also been used to great effect to segment disease and promote targeted therapy. Herceptin (trastuzumab) is a successful example of a targeted therapy approach within the treatment of breast cancer where pharmacogenomics was used to segment the population. In this instance trastuzumab – a monoclonal antibody targeting CD20 – was found to be particularly effective in a subgroup of patients whose tumors express abnormally high amounts of the HER2/neu protein (Lindpainter, 2002). It was found that up to 30% of breast cancer patients fall into this category and have a higher incidence of

**Table 11.2** Examples of known alleles influence the outcomes of specific drug therapy mentioned within product labels.

| Gene | Drug(s) effected | Outcome |
|---|---|---|
| ABCB1 | Various | Variants causing ABCB1 overexpression increase cellular efflux and reduce activity of affected intracellular acting drugs. Variants causing low ABCB1 expression tend to increase adverse effects due to low efflux and intracellular drug accumulation |
| OATP1B1 | Rosuvastatin | OATP1B1*5 homozygotes exhibit higher plasma levels of rosuvastatin |
| SLC01B1 | Simvastatin | Increased risk of myopathy in patients with the rs4149056 C-variant |
| SLC01B1 | Simvastatin | Variant rs4149056 associated with myopathy in patients on 80 mg daily simvastatin |
| UGT1A1 | Irinotecan | UGT1A1*28 allele causes low drug metabolism and potential life-threatening effects including myelosuppression, arrhythmia, neutropenia, thrombocytopenia, and diarrhea |
| G6PD | Dapsone | G6PD-deficient individuals may be at an increased risk of hemolytic adverse reactions |
| G6PD | Nitrofurantoin | G6PD-deficient individuals may be at an increased risk of hemolytic adverse reactions |
| VKORC1 | Warfarin | G-1639A allele causes increased bleeding risk especially when present with CYP2C9*2 or CYP2C9*3 alleles |
| POLG | Valproic acid | Increased risk of potentially fatal valproate-induced acute liver failure, especially in patients with the Q1236H substitution |
| NAT2 | Isoniazid, rifampicin | NAT2 slow acetylator genotypes *5, *6, and *7 are associated with higher risk of antituberculosis drug-induced liver injury |
| CYP2C9 | Warfarin | CYP2C9*3 homozygotes exhibit an extended warfarin half-life with possible hemorrhagic events |
| CYP2D6 | Codeine | CYP2D6 ultrarapid metabolizers may show opioid toxicity due to extensive metabolism of codeine to morphine |
| CYP2D6 | Amitriptyline, nortriptyline | CYP2D6 ultrarapid metabolizers experience no response, while poor metabolizers experience drug toxicity at conventional dosages |
| CYP2D6 | Tamoxifen | CYP2D6*10 homozygotes experience poor response to tamoxifen |
| DDC | L-Dopa | DDC rs921451 and rs3837091 variants are associated with poor L-dopa-treated Parkinson's disease management outcomes |
| TPMT | Azathioprine | TPMT*2, TPMT*3, or TPMT*4 alleles have a low TPMT activity, and patients show azathioprine toxicity |
| UGT1A1 | Irinotecan | Variant *28 associated with high risk of neutropenia |
| ALOX5 | ABT-761, zileuton | Patients carrying non-penta repeat Sp1 promoter variants express low ALOX5 and show reduced response to 5-LOX inhibitors |
| ADRB2 | Carvedilol | CHF patients who are Gln27 homozygotes show lower response to treatment than Glu27 homozygotes or heterozygotes |
| OPRM1 | Morphine and other opioids | A118G allele associated with reduced surface receptor expression and low response to exogenous opioids |
| NR3C1 | Glucocorticoids | Variants Ala229Thr and Ile292Val associated with decreased receptor ligand binding affinity; variant T746C and haplotype 237delC/C238T/G240C associated with lower glucocorticoid receptor expression |
| NR3C1 | Glucocorticoids | Glucocorticoid receptor promoter BclI G allele associated with glucocorticoid resistance |

metastasis and drug resistance and a shorter survival time than patients lacking HER2/neu overexpression (National Institute for Clinical Excellence, 2002). Women with this phenotype who are prescribed Herceptin stay cancer-free 65% longer than those patients on standard chemotherapy, although there are reports of adverse events including cardiac dysfunction, anaphylaxis, and an increased risk of infection. The approach to screen patients for the overexpression of HER2/neu was something that was employed early in clinical development and was responsible for both eliminating nonresponders and reducing the size of the clinical phase III study considerably.

Investigation of kinase deregulation within particular patient groups has led to further identification of mutant tyrosine kinases associated with disease progression and therapy modulation. Of interest is the identification of mutations in the receptor tyrosine kinase cKIT, associated with high relapse risk in core binding factor leukemias (Paschka *et al.*, 2006). Mutations in exons 8 and 17 of the KIT gene significantly decrease the overall survival of acute myeloid leukemia (AML) patients with the translocations inv(16) and t(8;21). The potential specific targeting of cKit using small-molecule tyrosine kinase inhibitors necessitates the promotion of screening patients with core binding factor AML for KIT mutations. This genetic profile of specific groups of patients gives prognostic and predictive information of clinical relevance (Cammenga *et al.*, 2005).

In addition to high relapse rate, therapy outcome markers include mutations that predict resistance to therapy. In colorectal cancer (CRC) treatment, K-Ras mutations significantly predict resistance to anti-EGFR therapy. Several clinical studies have shown that the presence of a K-Ras mutation is a significant predictor of resistance to anti-EGFR therapy in CRC patients. The identification of K-Ras mutants in CRC became part of the clinical practice protocols and clinical trials. Table 11.3 presents a list of mutations that are increasingly being used to predict therapy outcome within oncology.

It is clear that predicting response to treatment is one of the most important scientific advances in oncology and the role in which pharmacogenomics has helped understand the disease mechanisms and stratify the cancer population into specific mutations has been pivotal in the advancement of treatment. Efforts

**Table 11.3** Examples of genetic variants that are used to enable optimum selection of specific oncology therapy.

| Type of cancer | Biomarker | Treatment |
|---|---|---|
| Breast cancer | HER-2 overexpression | Trastuzumab, lapatinib |
| CML | BCR/ABL fusion gene | Imatinib |
| GIST | C-Kit mutation | Imatinib |
| NSCLC | EGFR mutation | Gefitinib, erlotinib |
| mCRC | KRAS mutation | Panitumumab, cetuximab |
| NSCLC | EML/ALK | Crizotinib |
| Melanoma | BRAF V600 | Vemurafenib |

are ongoing to understand molecular basis and redefine taxonomy of other diseases so that impact will be felt beyond oncology, for example, IMI Taxonomy Call. The IMI initiative is to develop an etiological-based taxonomy of human disease based on etiological mechanisms defined by molecular evidence rather than anatomical foci of the disease, symptomatology, and epidemic patterns of the disease. In this instance it is hoped that this approach would help reduce the complexity and cost of drug development and so increase success rates of drug development (make the drug discovery and development process more efficient to bring better medicines faster to patients).

### 11.4.3 Utility

Today there are around 159 references to biomarkers in FDA drug labels; however, the majority are "for information only" or "recommended" as opposed to mandatory (FDA website). The reason for this is that the vast majority of references are based on retrospective analysis of samples and data collected during clinical studies and lack prospective data generated on stratified populations in randomized controlled trials as required by regulatory authorities. Oncology is the exception where there are a large number of pharmacogenetic markers (HER-2, BCR/ABL, EGFR, KRAS, EML4/ALK, and BAF V600) being used in clinical practice (see Table 11.3). One of the main reasons why there are so few mandatory biomarkers mentioned in drug labels relates to the hurdles for regulatory approval of genetic biomarkers. The pathway for biomarker validation is complex as diagnostic are class 3 devices and require both an Investigational Device Exemption and Premarket Approval.

Despite potential challenges in genotyping, there are real advantages in conducting pharmacogenomics, with the impact tending to be higher the earlier the biomarker program is started in the development process (implementing PGx in phase I rather than in phase III). Studies with a PGx component have been used to support the SmPC in a number of ways:

- Elucidating the molecular or mechanistic basis for lack of drug efficacy or occurrence of adverse drug reactions (ADRs)
- Understanding dose
- Clarifying variability in clinical response to drugs by ruling out the role of well-known polymorphic genes as clinically significant contributors to variable drug PK and/or PD parameters
- Estimating the magnitude of potential drug–drug interactions (DDIs)
- Designing clinical trials to test for greater treatment effect in genetic subpopulations or subsequent differentiation from competitors with a specific genetic liability, for example, clopidogrel's CYP2C19 liability (Ford, 2014)

GWAS have produced many gene lists with subsequent rationales provided for selected genes. There are many reasons why specific genes "discovered" by GWAS lists have translated into clinically relevant programs, outside of oncology. These include lack of available affected tissue, study design, subject selection, and application to complex diseases where phenotypes have not been sufficiently narrowed down. It is therefore important that when we talk about the role of pharmacogenomics and its future impact, we realize that it will be a wealth of evidence that is used to help build up a clinical hypothesis around a specific new target and not one single technology. If we move toward clinical studies with fewer patients, we will certainly not have the numbers of patients to make any specific correlations using NGS alone. As a result we will need to know where to look which will mean that we will need to have a better understanding of both the disease we are trying to treat and how we go about identifying the appropriate patients for our studies. As a result there will need to be a paradigm shift of how we think about doing drug development which moves away from empirical-based clinical studies that are based on effect sizes and power calculations based on false-positive and false-negative results to the identification of patients who truly have the disease that we are trying to treat.

The ultimate promise of pharmacogenomics is that the knowledge of a patients DNA sequence will be used to enhance drug therapy to maximize efficacy to target patients who will respond to drugs and avoid ADRs. It is hoped that when the different technologies are fully mature and pharmacogenomics is integrated into the drug development process, we will start to deliver new medicines through research. Drugs will be created based on proteins, enzymes, and RNA molecules associated with genes and disease and will facilitate the development of therapy more targeted to specific diseases (McCarthy, 2005). This will lead to a reduction to the cost and risk of clinical trials as a result of targeting only those individuals capable of responding to a drug, which will in turn reduce the number of ADRs, failed trials, and time to approval. Understanding the molecular basis of disease may also lead to a "pathway"-based treatment approach if multiple affected genes in one disease all sit in a common pathway.

## 11.5 Biomarker Development

Biomarkers in one form or another have been around for hundreds of years. Biomarkers have been described as "a characteristic that is measured and evaluated objectively as an indicator of normal biological processes, pathological processes or pharmacological response to a therapeutic intervention." When we talk about biomarkers within the pharmaceutical industry, we also tend to subdivide biomarkers into different category types:

- Biomarkers for disease characterization
- Prognostic biomarkers
- Predictive biomarkers (response biomarkers)
- Risk biomarkers
- Biomarkers for measurement of response to treatment
- Efficacy biomarkers

### 11.5.1
### Biomarker Activities

Biomarker activities can have different strategic aims depending on the phase of development, the availability of samples, and the specific question that needs to be answered. Biomarker activities can be focused on:

- Facilitating early decision making
- Safety, efficacy, and tolerability
- Response and nonresponse
- Disease segmentation
- Differentiation and placement

As mentioned we do not always have the luxury of time on our hands; therefore, it is important for Translational groups to establish what is achievable within the time period of development and what types of technology you will utilize. There are also too many examples in the industry of having large biomarker programs with no focus or relevance to the question at hand, and as such it is important that there are a common understanding and alignment within the project team as to what biomarker samples will be generated from preclinical (pharmacology, PK, and safety) and clinical studies and how this data will be used to generate value and impact within the project.

Currently most clinical studies we perform are evidence based, requiring large randomized multicenter studies where the aim is to provide superior efficacy of new therapies compared to SOC, generally without molecular stratification of patients (Taube *et al.*, 2009). In light of the fact that there is an expectation around showing greater differentiation of the asset and stratification of the population, it is more crucial than ever that we continue to allow trial populations to be selected based on genetic data. In doing so, initial diagnostic tests could help link polymorphisms or expression patterns with ADR at an early stage in clinical trials, allowing companies to abandon suspect compounds at a less costly point. Additionally, genotype–phenotype correlation studies could maximize efficacy by identifying patients for whom a drug will be of ideal benefit. This would reduce end-stage trial sizes (and overall drug development costs) by selecting individuals with optimal response genotypes. A smaller and more homogenous subject group would lead to clinical results that are clearer and more indicative of the endpoints being evaluated, making drugs more likely to progress through the regulatory process. To this end there would need to be an assessment on the availability of the tools to be able to identify your patient population as well as the flexibility within the business model of initially obtaining a more restrictive label with respect to population with the later intention of expanding the label claims.

### 11.5.2
### Assessing the Opportunity

Biomarkers have been responsible for providing patients with innovative and targeted therapy; however, it is also clear that a lot of money can be wasted looking for biomarkers where there is neither the disease understanding to identify nor the technology to assess its validity. When deciding whether to pursue a biomarker approach in a particular area with a specific drug, it is important to assess whether there is an appropriate opportunity, which will involve gathering relevant information from a number of different groups within an organization. Biomarkers can be used to derisk development programs (target engagement markers, predictive markers, enrichment markers, etc.), differentiate assets, as well as help with market penetration or drug uptake (companion diagnostic or stand-alone diagnostic). As a result it is extremely important to make a thorough assessment of the opportunity and utility for a biomarker approach within a project. In particular if there is an intention to develop a companion diagnostic or stand-alone diagnostic biomarker, it is important to involve colleague from business, regulatory, and market access to help make the assessment. The type of relevant information would include:

- *Clinical assessment* – Supportive evidence required that patient segmentation has clinical validity and utility.
- *Business assessment* – Business considerations including modeling of segmented population and cost versus benefit.
- *Regulatory assessment* – Regulatory status and experience of the solution, any critical issues foreseen, or body of evidence required by regulators to approve a biomarker for its intended use.
- *Operational assessment* – Operational implementation of the solution including special technical requirements, ability to support registration studies, what studies/data do we need, and any critical issues foreseen.
- *Technical assessment* – Technical implementation of the solution including any special requirements of laboratory capabilities, global reach, ability to support registration studies, and any critical issues foreseen.
- *Gaps* – What are they, and how do we fill them? Identify any external resources, solutions, or collaborations that may be required.
- Cost and resource assessments.

## 11.6 Systems Pharmacology

We have discussed the importance of the relationship between pharmacology (affinity and residence time at target) and PK (exposure and clearance) in respect to drug optimization and response. We have also mentioned the traditional PKPD principles relating to exposure, target engagement and response, and the role that aligned collaboration and PKPD modeling is having to build confidence around the compounds we are developing with respect to dose optimization. Due to the fact that we are now working with nonvalidated targets, it is vitally important to try and build this same confidence around the target. Systems pharmacology is such an approach that looks to provide confidence in a positive

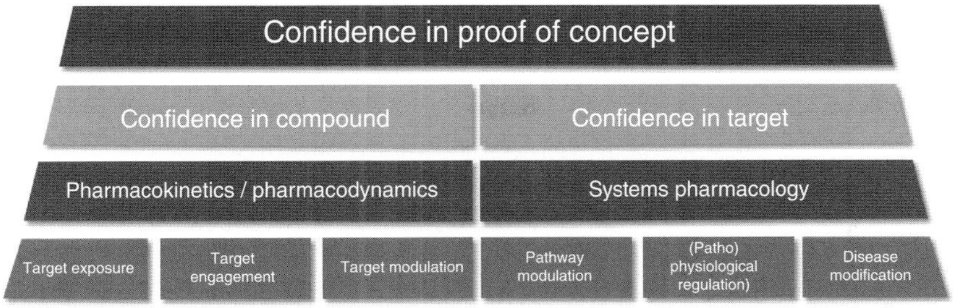


**Figure 11.2** Relationship PKPD and systems pharmacology as parallel approaches to build confidence around proof of concept.

proof of concept by trying to build understanding around the relationship between target engagement, pathway modulation, physiological regulation, and disease modulation (see Figure 11.2) (Vincini and van der Graaf, 2013).

## 11.7 Rational Drug Development

The process of drug discovery starts with the identification of a potential target at which the drug can elicit its response. This target can be an enzyme in a pathway, a receptor, a protein in signal transduction, or any protein produced in a pathological condition. Through better understanding of underlying molecular pathways of disease, it is hoped that translational approaches will move the development process away from trial-and-error drug discovery toward a rational drug design. We know that following the sequencing of the human genome, there is an estimate of around 8000 drug targets, of which around 5000 could be acted upon and only 450 are currently known. These new targets will undoubtedly exhibit variations owing to genetic polymorphisms, epigenetics, expression, posttranslational modifications, and so on, causing variations in effect. As a result it will be very important that pharmacogenomic studies comparing SNP maps and gene expression be compared between normal and affected individuals to identify genetic factors associated with the disease. Within the field of oncology we have examples of how effective this can be; however, this same disease understanding needs to be taken into other therapeutic areas. This will involve measuring the effects of drugs on pathways using transcriptomic and proteomic analysis in animals targeted directly at the tissue of interest and imaging techniques as well as the use of appropriate *in vivo* and mechanistic models to translate this information for human relevance.

We also know that there is a very big push these days to both genotype large populations either as a result of randomized clinical studies conducted by pharmaceutical or academic institutions or as a result of country-specific programs to genotype large parts of the population. Somehow we need to be able to bring this knowledge together across the diseases to help us better understand the molecular definition of "health" and "ill health." In doing so the hope is that we can identify new targets as well as develop early markers that identify the transition between the two states. This can only be achieved by studying cellular and molecular mechanisms of both states and by studying the interplay between environmental factors to which the population is exposed and their genetic makeup. Classifying diseases at the genetic, molecular, and cell functional level will help forge a stepwise change in the disease treatment and prevention that is already more advanced for some diseases like cancer. In a number of indications it is known that the genetic, molecular, and cellular mechanisms underlying common diseases are currently limited. As a result the rational discovery of omics-based markers and the development of more reliable preclinical disease models will be key to this success. This work will involve the development and combination of the analysis of both nucleic acids, proteomics, epigenomics, and cell functional analysis along with an integrative approach to drug development, which combines the identification and validation of new targets as well as a TP to test the hypothesis early in the clinical development program to help reduce risk. This would involve looking at the effect of drugs on whole pathways in the first instance, validating this in relevant *in vivo* models, and then developing appropriate phenotypic assays that can be effectively and objectively measured in an early clinical setting to provide evidence of potential efficacy in the clinic.

## 11.8 Concluding Remarks

As we look to the future where there will be the expectation that targeted therapies will be more widely available in multiple therapeutic areas, Translational Development will play a central role in building confidence and understanding around these molecular targets and their evoked clinical response (efficacy and safety signals). Understanding why people respond and do not respond to treatments will always be a challenge as there are many reasons to consider. The role of Translational Development will be to apply innovative technologies and leverage knowledge to answer this question and in doing so mitigate risk as well as provide value, and if I was to sum up how to achieve this, it would be to go back to what has brought us success in the past, namely, "Good Science and Collaboration!"

## References

Adjei, A.A. (2006) What is the right dose? The elusive optimal biologic dose in phase I clinical trials. *J. Clin. Oncol.*, **24**, 4054–4055.

Cammack, R., Atwood, T., Campbell, P., Parish, H., Smith, A., Vella, F., and Stirling, J. (2006) *Oxford Dictionary of Biochemistry and Molecular Biology*, 2nd edn., Oxford University Press.

Cammenga, J., Horn, S., Bergholz, U., Sommer, G., Besmer, P., Fiedler, W., and Stocking, C. (2005) Extracellular KIT receptor mutants, commonly found in core binding factor AML, are constitutively active and respond to imatinib mesylate. *Blood*, **106** (12), 3958–3961.

Florence, A. and Attwood, D. (2011) *Physicochemical Principles of Pharmacy*, 5th edn, Pharmaceutical Press.

Ford, N. (2014) Clopidogrel: what is a cardiologist to do? *J. Clin. Pharmacol.*, **54** (8), 881–883.

Goulart, B., Roberts, T., and Clark, J. (2004) Utility and costs of surrogate endpoints (SEs) and biomarkers in phase I oncology trials. *J. Clin. Oncol.*, **22** (Suppl. 14), 6012 (abstract).

Grimwood, S., Hartig, PR. (2009) Target site occupancy: emerging generalizations from clinical and preclinical studies. Pharmacol Ther, **122** (3), 281–301.

Hillgren, K., Keppler, D., Zur, A., Giacomini, K., Stieger, B., Cass, C., Zhang, L., and International Transporter Consortium (2013) Emerging transporters of clinical importance: an update from the International Transporter Consortium. *Clin. Pharmacol. Ther.*, **94** (1), 52–63.

Lesko, L.J. and Woodcock, J. (2004) Translation of pharmacogenomics and pharmacogenetics: a regulatory perspective. *Nat. Rev. Drug Discovery*, **3** (9), 763–769.

Lindpainter, K. (2002) The impact of pharmacogenetics and pharmacogenomics on drug discovery. *Nat. Rev. Drug Discovery*, **1**, 463–469.

Ma, Q. and Lu, A. (2011) Pharmacogenetics, pharmacogenomics, and individualized medicine. *Pharmacol. Rev.*, **63**, 437–459.

McCarthy, J. (2005) Life, diversity and the pursuit of haplotypes. *Nat Biotechnol.*, **23** (11), 1376–7.

McLean, S., Gana-Weisz, M., Hartzoulakis, B., Frow, R., Whelan, J., Selwood, D., and Boshoff, C. (2005) Imatinib binding and cKIT inhibition is abrogated by the cKIT kinase domain I missense mutation Val654Ala. *Mol. Cancer Ther.*, **4**, 2008–2015.

National Institute for Clinical Excellence (2002) Appraisal, Technology Appraisal Guidance No. 34–Guidance on the Use of Trastuzumab for the Treatment of Advanced Breast Cancer, 15 March, 2002.

O'Hare, T., Zabriskie, M., Eiring, A., and Deininger, M. (2012) Pushing the limits of targeted therapy in chronic myeloid leukaemia. *Nat. Rev. Cancer*, **12**, 513–526.

Paschka, P., Marcucci, G., Ruppert, S., Mrózek, K., Chen, H., Kittles, R., Vukosavljevic, T., Perrotti, D., Vardiman, J., Carroll, A., Kolitz, J., Larson, R., Bloomfield, C., and Cancer and Leukemia Group B (2006) Adverse prognostic significance of KIT mutations in adult acute myeloid leukemia with inv(16) and t(8;21): a Cancer and Leukemia Group B Study. *J. Clin. Oncol.*, **24** (24), 3904–3911.

Roses, A.D. (2008) Pharmacogenetics in drug discovery and development: a translational perspective. *Nat. Rev. Drug Discovery*, **7** (10), 807–817.

Roses, A., Saunders, A., Lutz, M., Zhang, N., Hariri, A., Asin, K., Crenshaw, D., Budur, K., Burns, D., and Brannan, S. (2014) New applications of disease genetics and pharmacogenetics to drug development. *Curr. Opin. Pharmacol.*, **14**, 81–89.

Sarker, D. and Workman, P. (2007) Pharmacodynamic biomarkers for molecular cancer therapeutics. *Adv. Cancer Res.*, **96**, 213–268.

Stapleton, M. (1997) Sir James Black and propranolol. The role of the basic sciences in the history of cardiovascular pharmacology. *Tex Heart Inst J.*, **24** (4), 336–42.

Taube, S., Clark, G., Dancey, J., McShane, L., Sigman, C., Gutman, S. (2009) A perspective on challenges and issues in biomarker development and drug and biomarker codevelopment. *J Natl Cancer Inst.*, **101** (21), 1453–63.

Vincini, P. and van der Graaf, P. (2013) Systems Biology for Drug Discovery and Development: Paradigm Shift or Flash in the Pan? *Clin. Pharm & Ther.*, **93** (5), 379–381.

Vlietstra, R. (1995) The "continuing" education of a physician. *Chest*, **108** (4), 1178.

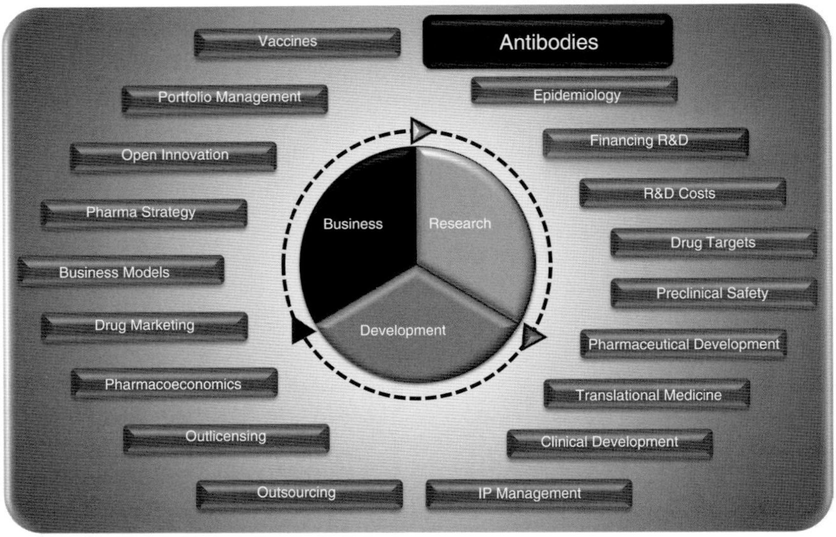
Vaccines
Antibodies
Portfolio Management
Epidemiology
Open Innovation
Financing R&D
R&D Costs
Pharma Strategy
Business
Research
Drug Targets
Business Models
Preclinical Safety
Drug Marketing
Development
Pharmaceutical Development
Pharmacoeconomics
Translational Medicine
Outlicensing
Clinical Development
Outsourcing
IP Management

# 12
# Forty Years of Innovation in Biopharmaceuticals – Will the Next 40 Years Be as Revolutionary?

*Mathias Schmidt, Sanjay Patel, Petter Veiby, Qiang Liu, and Michael Buckley*

## 12.1
## Introduction

What is a Biologic? In its broadest sense a Biologic is any therapeutic that is produced in a biotechnological process using elaborate systems with a lot of development effort. For the sake of scope, this chapter does not address nucleic acid-based medicines (e.g., gene therapy or RNA-based therapy), vaccines, or cell therapies. Most of the currently approved Biologics are protein therapeutics and the majority of them are produced in recombinant systems. In contrast to small-molecule medicines whose chemical structure can be precisely defined and analyzed using appropriate methods, a protein therapeutic is defined by its primary amino acid sequence, secondary and tertiary folding, appropriate formation of disulfide bonds, and multiple posttranslational modifications (including complex glycosylation). These properties are influenced by the fermentation process, media conditions, host cell organism, purification process, pH conditions, and formulation of the drug product. Hence, while there is a low technical hurdle to copy a small-molecule therapeutic, there is a high hurdle to reproduce a protein therapeutic since the exact upstream and downstream conditions of production, which strongly influence protein characteristics, are normally not publically disclosed. Exclusivity to the knowledge of the production process is a significant hurdle toward generic competition for Biologics and regulatory authorities speak of a biosimilar instead of a biogeneric, acknowledging that generation of a Biologic possessing the same features as the originator product is unlikely possible.

Other differentiating factors to small molecules are summarized in Figure 12.1. Apart from the elaborated production process, Biologics, with few exceptions such as replacement enzymes for the treatment of lysosomal storage diseases, only act on extracellular targets or mechanisms, generally have to be administered via invasive methods (intravenous or subcutaneous) and have cost of goods (COGs) that can be more than 100-fold higher than small molecules. So why do we need Biologics at all?

*Value Creation in the Pharmaceutical Industry: The Critical Path to Innovation*, First Edition.
Edited by Alexander Schuhmacher, Markus Hinder, and Oliver Gassmann.

(a)

| | Biologic | Small molecule |
|---|---|---|
| Size | 5000 –200 000 Da | 200–1000 Da |
| Site of action | Extracellular | Intra- and extracellular |
| Oral availability | – | + |
| COGS/g | >US$ 200 | <US$ 2 |

(b)

**Figure 12.1** Comparison of Biologics with small molecules. (a) Three-dimensional ribbon structure of an Immunoglobulin G IgG molecule. (b) Comparison of general properties of biopharmaceuticals compared with small-molecule therapeutics. COGs, cost of goods sold.

### 12.1.1 The Value Proposition of Biologics

#### 12.1.1.1 The Patient Perspective

Similar to antibiotics, Biologics have revolutionized the way we treat serious and life-threatening diseases. Biologics have some unique properties that differentiate them from small molecules and make them an integral part of evidence-based medicine. Due to their size and exquisite selectivity, for example, monoclonal antibodies (mAbs) can interfere with pathological mechanism not amenable to small molecules (such as the neutralization of inflammatory cytokines or blockade of essential ligand–receptor interactions). They can supplement or replace missing or malfunctioning proteins such as erythropoietin (certain forms of anemia or renal diseases), lysosomal enzymes (in lysosomal storage diseases such as Gaucher's disease or Pompe's disease), clotting factors (hemophilia), or insulin (Type 1 or Type 2 diabetes mellitus), just to name very few. Especially in genetic disorders, where a DNA alteration results in a lack or malfunctioning of a protein, therapeutic replacement proteins offer an intuitive option to replace the protein of pathological interest and influence disease severity. mAbs targeting the EGF receptor, Her-2, VEGF receptor, cluster of differentiation CD20, TNFα, IL-6, or integrins have revolutionized the way we nowadays treat devastating disorders such as autoimmune diseases or cancer. Not only have they added years to live (prolongation of survival) but also life to years (improvement of the quality of life if the disease is chronic with no realistic hope for a cure). Today, more than three dozen mAbs are approved by the EMA (European Medicines Agency) and/or FDA (US food and drug administration). Thirty plus years of clinical experience have demonstrated the safety of modern Biologics. Concerns about the safety of Biologics are usually associated with the liability of interfering with the molecular target (for mAbs) or immunogenicity (which is a potential risk for any Biologic, regardless of origin and size). Recombinant production technologies have significantly improved the safety of Biologics by reducing the risk of contamination with infectious agents from human or animal sources. A prominent example was the treatment of dwarfism due to insufficiency in human growth hormone (hGH). Before recombinant DNA technologies became widely available, hGH used to be isolated from pituitary glands of human cadavers. This has led to the transmission of Jacob–Creutzfeld disease in some individuals treated with hGH from a human source (Blizzard, 2012). Another example is the infection of a large number of hemophilic patients with HI virus after they were treated with clotting factors isolated from the blood of HIV-carrying donors. As a result of these experiences, strong precautions are now taken to avoid contamination of recombinant therapeutics due to a potential viral contamination of the production cell line employed in a biopharmaceutical manufacturing process.

Two case stories of revolutionary innovation should be exemplified to demonstrate the enormous benefit that Biologics have brought to patients: the development of insulin and the discovery of the technology of mAbs.

Early symptoms of diabetes were first described around 200 years ago, and around 300 years ago diabetes mellitus was diagnosed by the sweet taste of urine (caused by excessive glucose secretion via the kidney). It was not until 1921 that Frederick Banting and Charles Best saved the life of a diabetic dog by treating the animal with a pancreatic extract containing insulin. In 1922, the first human life was saved by treating a child suffering from Type 1 diabetes mellitus with a pancreatic extract. In 1923 Banting and Best were awarded the Nobel Prize for Medicine for their discoveries. Insulin production from porcine pancreas extracts became common at Novo Nordisk in 1930. Human insulin became the first recombinant protein that was approved by the FDA in 1982. Due to the worldwide explosion in diabetes prevalence, global supply of insulin would be highly challenging without recombinant DNA technologies and production of recombinant insulin in *Escherichia coli*.

The second example of game-changing innovation is the discovery of a technology that allows the generation of a self-renewable source for unlimited quantities of a single antibody against a specific target: the hybridoma technology for the generation of mAbs. In 1975 Georges Koehler and Cesar Milstein were awarded the Nobel Prize for the discovery of hybridoma technology to produce mAbs. By fusing the B-cells (that produce the antibody) of an immunized mouse with a myeloma cell line (i.e., immortal), Koehler and Milstein generated a hybridoma cell line that has unlimited replication potential and the ability to produce unlimited quantities of antibodies of a single source and selectivity. This technology laid the foundation for life-saving mAbs and an industry that generates the largest revenue among all therapeutic protein classes. At least as revolutionary was the generosity of Koehler and Milstein who did not patent their technology because they foresaw the potential of their invention for the way serious and life-threatening diseases could be treated in the future and feared that a patent would limit the penetration of mAbs within the pharmaceutical industry – what a gift to the world! It was less than 10 years after their discovery that the first mAb was approved (for the treatment of organ transplant rejection). Subsequent parts of this chapter highlight further innovation that today allow us to generate fully human mAbs from high-titer cell lines that display a much lower risk of immunogenicity than mAbs of murine origin.

#### 12.1.1.2
#### The Pharmaceutical Industry's Perspective

Despite their complex route of production, biopharmaceuticals have become an indispensible part of the portfolio of virtually every mid- to large-size pharmaceutical company. In an industry that is subject to extraordinarily high project attrition rates, billion dollar development costs, 10–15 year development cycle times, often single digit years of patent exclusivity for marketed products, and high levels of government regulation (including pricing), there remains a need for products that secure a sustainable revenue stream from a portfolio that is less subject to harsh competitive conditions when compared to small molecules. While manufacturing of Biologics is certainly more complex than for small molecules, there are certain key performance indicators that favor therapeutic proteins:

*Success rates*: Compared to small molecules, Biologics generally have a higher probability of success (PoS; Figure 12.2). A more comprehensive analysis of clinical development success rates can be found in Hay *et al.* (2014). Higher PoS is driven primarily by two factors: the safety profile of a Biological protein is generally well known and there are only few "class liabilities," while unforeseen safety concerns are poorly predictable for small molecules. In addition, the often more precise understanding of the mode of action in a particular disease pathway of Biologics results in less attrition due to lack of efficacy in Phase 2 trials.

*Development costs*: Surprisingly, despite higher manufacturing costs and front-loading of process development for Biologics manufacturing, the development costs of Biologics are slightly lower (or at least not more expensive) when compared to small molecules. Lower attrition rates are a significant factor and the lack of necessity for extensive drug–drug interaction studies as is common for small molecules is another contributor (DiMasi *et al.*, 2010; Figure 12.2).

*Pricing*: The high manufacturing costs of Biologics are compensated by premium pricing due to their often game-changing efficacy in serious and life-threatening diseases. While insulin analogs are at the lower end of the price range, annual treatment costs can be over US$ 300 000 for mAbs in life-threatening rare disease conditions (Eculizumab (Soliris); Soliris is indicated for the treatment of patients with paroxysmal nocturnal hemoglobinuria (PNH), a very rare disease condition, to reduce hemolysis). Typical annual treatment costs of most mAbs are in a range between US$ 10 000 and 100 000.

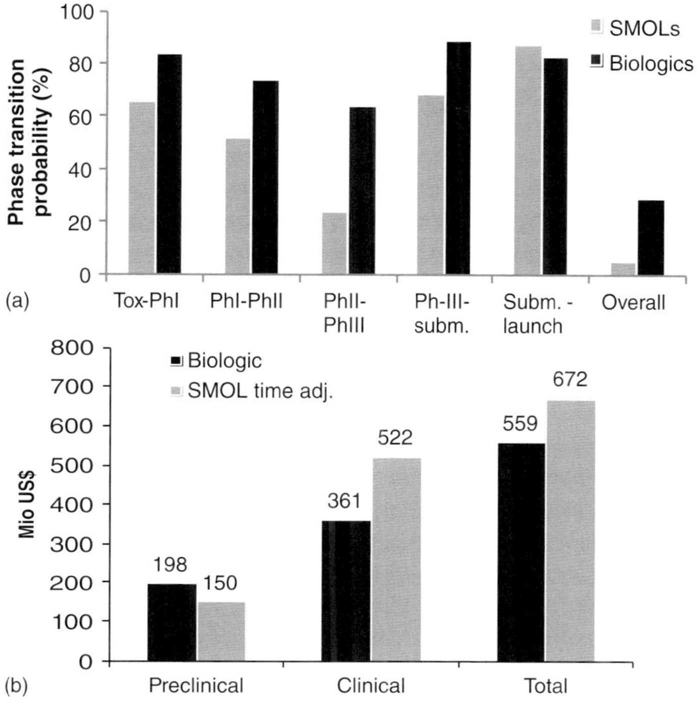


**Figure 12.2** (a) Probability of success rates of Biologics (black bars) and small molecules (gray bars) and average development costs (b) in various phases of development. SMOL, small-molecule therapeutic.

*Generic competition*: While both sales and margins usually plummet due to generic competition after loss of exclusivity for a small-molecule drug, biopharmaceuticals at least until now have been less susceptible to the generic threat (Grabowski, 2008, Figure 12.3). This is largely due to the nature of their complex manufacturing with small changes in upstream or downstream processing resulting in significant changes in protein characteristics ("the process defines the product"). A decline in sales most frequently is not due to generic competition but due to the availability of additional treatment options in the addressed indication.

In a pharmaceutical market that is characterized by stagnation, all the key performance indicators have positioned biopharmaceuticals to constitute the main growth driver (Rickwood and Di Biase, 2013; Figure 12.4). Seven out of the 10 best-selling drugs in 2013 were Biologics (Palmer, 2014; Table 12.1), indicating that any mid- to large-size Pharma cannot miss being in the biopharmaceuticals sector. Figure 12.5 displays the number of Biologics versus small molecules approved by the US FDA in recent years, demonstrating that Biologics have become an integral part of the pharmaceutical industry.

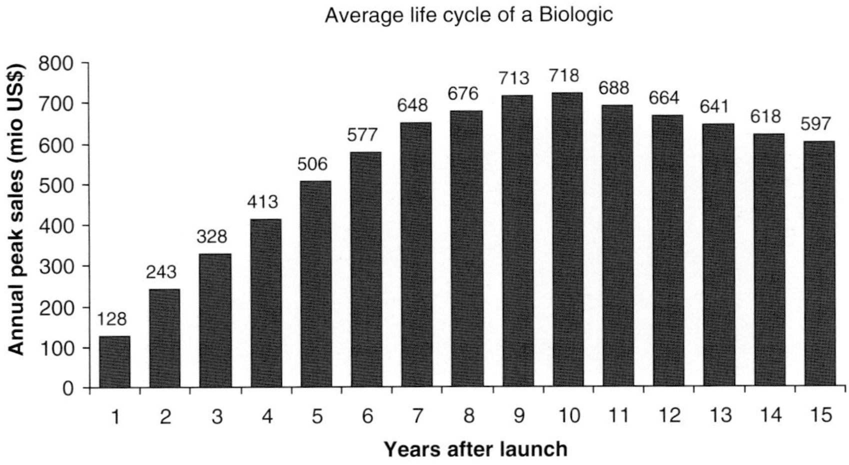


**Figure 12.3** Typical annual sales rate progression in the life cycle of a Biologic.

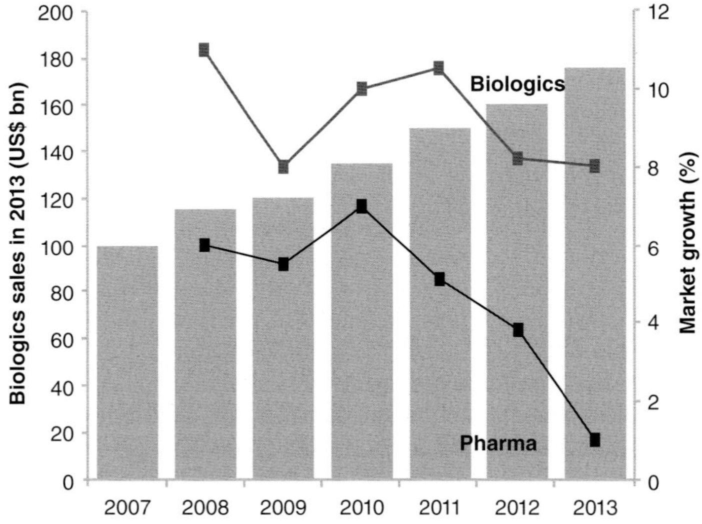


**Figure 12.4** Biologics sales from 2007 to 2013 and market growth rate for Biologics versus overall pharma market growth (gray bars).

### 12.1.2
### Biosimilars: A Blessing or a Threat to Innovation?

Generic erosion of branded medicines has traditionally been one of the most significant drivers for innovation in the pharmaceutical industry in order to maintain a sustainable business. Payor's funds that were saved by the availability of generic alternatives were mostly absorbed by premium prices for newly developed innovative medications. The increasing reluctance of payors to reward

**Table 12.1** Annual sales in US$ of the 10 best-selling drugs in 2013.

| Rank | Brand name | Class | 2013 sales ($bn) |
|---|---|---|---|
| 1 | Humira | mAb anti TNF | 11.02 |
| 2 | Enbrel | Receptor Fc anti TNF | 8.78 |
| 3 | Remicade | mAb anti TNF | 8.37 |
| 4 | Advair | SMOL beta2 agonist | 8.25 |
| 5 | Lantus | Long-acting insulin | 7.59 |
| 6 | Rituxan | mAb anti CD20 | 7.5 |
| 7 | Avastin | mAb anti VEGF | 6.75 |
| 8 | Herceptin | mAb anti Her-2 | 6.56 |
| 9 | Crestor | SMOL statin | 6.04 |
| 10 | Abilify | SMOL dopamine agonist | 5.5 |

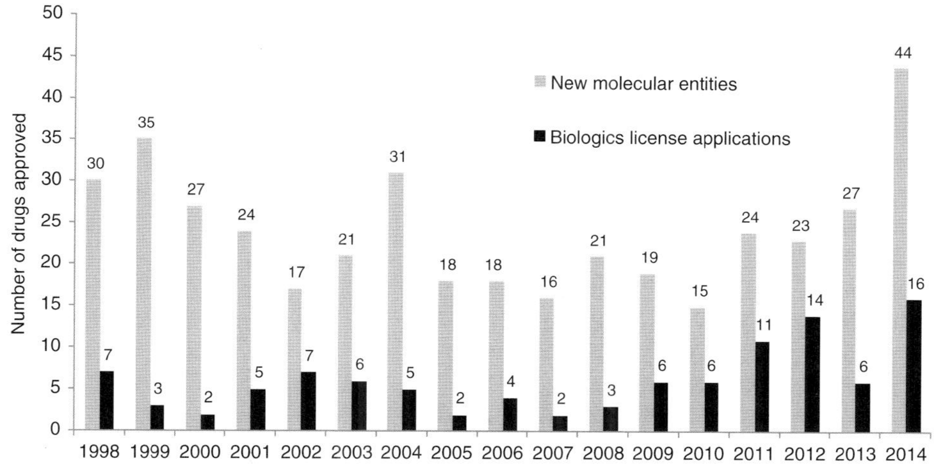


**Figure 12.5** Number of FDA drug approvals 1998–2014 split by drug modality.

innovation resulting from new medicines that address unmet medical needs, offer improved efficacy, or higher safety have recently threatened this model of continuous innovation.

Biosimilars have penetrated the market to a much lesser extent than initially anticipated. With loud fanfares the first biosimilar approval in the United States, a filgrastim biosimilar from Novartis and its Sandoz unit was welcomed in a historic session of the FDA's Oncologic Drugs Advisory Committee (ODAC) on 7 January 2015. While the limited biosimilar threat certainly is a short-term blessing for the pharmaceutical industry, it remains to be seen whether this lack of competition will turn into a situation where the sheer mass of expensive Biologics will no longer become affordable for payors. The implications of the low biosimilar threat need to be balanced with the risk of the regulatory authorities becoming more liberal in their approval standards for biosimilars: A biosimilar version of a life-saving Biotherapeutic that is indeed less efficacious or less safe would impose an unacceptable risk to patients who would be de facto withheld from their life-saving or prolonging treatment.

### 12.1.3
### Further Innovation in Biologics – Incremental or Revolutionary?

While recombinant DNA technologies and the ability to generate mAbs are generally considered as breakthrough innovations, many incremental innovations needed to occur to make Biologics manufacturable, safe (from a manufacturing standpoint), less immunogenic, applicable for a multitude of indications, and more efficacious especially in cancer indications. The following sections highlight those incremental (or revolutionary) innovations from a standpoint of development and manufacturing, formats, especially modifications of mAbs, and enhancement of the effector function of mAbs via the development of antibody-drug conjugates (ADCs) – a concept developed more than a century ago by Paul Ehrlich to exploit antibodies that bring a therapeutic payload specifically to diseased cells.

## 12.2
## The Evolution of Biologics Manufacturing

### 12.2.1
### Introduction

While breakthroughs in recombinant DNA technologies, hybridoma development, and target biology each contributed to the commercial success of biotherapeutics, they were not sufficient enablers in and of themselves. Instead, they required a parallel set of advancements in manufacturing sciences that have permitted sufficient quantities and quality of Biologics to be produced in a cost-effective manner. As an example of one such improvement, cell culture titers in the 1980s were typically well below 1 g/l, necessitating multiple costly 10 000 l production runs to satisfy annual dosing demands. Today, however, it is common for mAb titers to exceed 5 g/l, with some groups even reporting over 10 g/l (Huang *et al.*, 2010). This rapid increase in titers over the past 30 years has had a direct impact on manufacturing COGs, with one analysis suggesting that an increase in titer from 0.5 to 5 g/l roughly translates to a fivefold reduction in COGs (Kelly, 2009). As cost pressures are only expected to increase moving forward, this reduction in COGs can often be the difference between a product successfully being brought to market versus not being developed at all due to financial unattractiveness. In this next section, we will highlight some of the incremental innovations that have enabled low-cost manufacturing of Biologics and conclude with some thoughts on what the next 40 years of manufacturing innovation might bring.

### 12.2.2
### CHO Cells: The Industry Workhorse

As most biotherapeutics have requirements for proper folding and posttranslation modifications, they are typically synthesized in mammalian cells as opposed to microbial, yeast, or plant hosts. And while industry has historically employed a diversity of mammalian hosts in this manner (BHK, 3T3, NS0, HEK293), CHO cells are by far the dominant preference. From the first approved recombinant protein produced in mammalian cells in 1987 (tissue plasminogen activator – Activase) to more recent entrants such as Vectibix or Adcetris, CHO cell hosts account for nearly 70% of all recombinant therapeutics today (Jayapal *et al.*, 2007). Their dominance in large part reflects the long-standing safety record CHO cells have demonstrated in this field. For example, CHO cells have proven less prone to viral infection than other hosts such as BHK cells (Berting *et al.*, 2010). Additionally, analytical methods and downstream processes have evolved to better characterize and remove CHO host cell contaminants from the drug product. Another reason CHO cells enjoy a preferred status is the emergence of entire industries devoted to providing infrastructure supporting their use. This includes companies offering optimized media and feed formulations, genetic characterization and other testing services, and even consortia comprised of biopharma and academic representatives to facilitate and standardize the adoption of CHO cells across the industry. Indeed, the inertia CHO cells currently possess is so strong that any new host systems will need to demonstrate a truly revolutionary advantage (i.e., greater than an order of magnitude difference in expression capacity, significantly compressed timelines, superior product quality attributes, etc.) to garner significant attention from the manufacturing sector.

### 12.2.3
### Protein Production Strategies

As host like the CHO cells are used in a manufacturing setting, the total amount of protein they produce can be roughly approximated as a function of two parameters: the specific expression rate per cell ($q_p$) and the cumulative amount of biomass and time that the cells are alive and actively making product (IVCD – integral viable cell density). Each of these parameters has been systematically optimized over the past three decades, with the former being addressed through cellular engineering strategies and the latter through refinement of the culture conditions.

In terms of cellular engineering, the most common method employed to generate a highly productive cell line relies upon stable integration into the host cell genome. Using this approach, an expression plasmid containing DNA sequences encoding (i) the protein of interest and (ii) a selection-resistance gene is transfected into the host cell. For the protein of interest, even though the amino acid sequence is fixed, there is still flexibility in how best to design the precursor DNA. For example, considerable effort has been devoted to understanding how to choose codons triplets that are aligned with any inherent usage biases the host organisms possess (Angov, 2011). For the noncoding regions, strong promoters such as the cytomegalovirus (CMV) or elongation factor alpha (EF1α) promoter are typically used to drive protein expression (Boshart *et al.*, 1985; Deer *et al.*, 2004), while polyadenylation signal sequences such as the SV40 early or late sequence are used to confer mRNA stability. The choice of signal peptide has been also found to dramatically influence the expression rates of proteins, with some of the most effective sequences being nonnative (Kober *et al.*, 2013). Finally, genetic elements thought to prevent gene silencing (i.e., STARs) or promote integration into transcriptionally active regions of the genome (i.e., MARs) have been more recently utilized to confer genetic stability to the protein of interest (Kwaks *et al.*, 2003; Harraghy *et al.*, 2008). Given that up to 2 months may be required from thaw of a production cell bank to harvest of a 10 000 l bioreactor, ensuring stable expression of protein during this window is critical to demonstrating a reproducible and robust manufacturing process. All of the preceding considerations have also been extended to the design of the selection-resistance portion of the expression plasmid. The choice of selection, however, has generally been restricted to two main systems: the dihydrofolate reductase (DHFR) system and the glutamine synthetase (GS) system (Urlaub *et al.*, 1980; Wilson *et al.*, 1992). DHFR is an enzyme that reduces dihydrofolic acid to tetrahydrofolic acid, a nucleotide precursor. By transfecting DHFR (along with the therapeutic of interest) into a host cell that does not express DHFR (or expresses DHFR at low levels), highly expressing cell clones can be selectively isolated when cultured in media without exogenous nucleotides and with the competitive inhibitor of DHFR, methotrexate. In a similar manner, GS, an enzyme catalyzing the formation of glutamine from glutamate and ammonia, can be used to select high-producing clones when cultured in media without glutamine and with methionine sulfoximine. Taken together, these strategies, along with others not discussed here, have been instrumental in engineering the host cells to routinely achieve specific productivities exceeding 30–40 pg/cell day.

With regards to increasing the IVCD portion of the production equation, this is where perhaps the greatest advancements in manufacturing have been made over the past 30 years. One initial achievement was the development of serum-free media to replace the requirement for including fetal bovine serum, thereby addressing a critical safety and process consistency issue. Afterward, media development further evolved to become completely animal-derived component free and in some instances, chemically defined (i.e., devoid of any undefined hydrolysates from plant, yeast, or soy-based sources). Concurrent with this evolution was the recognition that media formulations should be modified based on the stage of manufacturing. For example, one formulation could be used to optimize the rapid accumulation of biomass, which is needed during manufacturing scale-up or the early phases of production runs. During the later phases, however, a more nutrient-dense formulation along with specialized feeds could be used to shift the culture from cell growth-phase to protein-production phase. Most of these media and feeds are available commercially from multiple vendors specializing in this field; but given that media optimization is often cell line dependent, many biopharmaceutical manufacturers have invested internally in creating entire teams of scientists devoted to this activity. The result of all this effort has been an exponential increase in IVCD over the years, a trend excellently illustrated by Wurm (2004). His example compared a representative mAb manufacturing process from 1986 to a more recent version in 2004. In the case of the 1986 process, cells reached a maximum density of $2 \times 10^6$ cells/ml after 7 days of batch culture (i.e., no feeding), after which viability rapidly declined as critical nutrients became depleted. In contrast for the 2004 process, the culture was able to attain a maximum density of $10 \times 10^6$ cells/ml, and more importantly, maintained this high density for almost 3 weeks. It is this latter, high-density period of culture that is critical for accumulating high quantities of protein product; and in the case outlined, the final accumulated yield was 4.7 g/l of mAb. As media development work has matured in more recent years, attention has shifted to investigating other physicochemical culture parameters, such as pH, dissolved oxygen tension, temperature, and small-molecule enhancers to further improve protein titers.

The incremental improvements in $q_p$ and IVCD outlined here have resulted in corresponding increases in titer over the past 30 years. When plotted over time, one analysis was able to model the increase reasonably well with a "Moore's law" approach, whereby reported titers double at the rate of approximately every 5 years (Svay *et al.*, 2010). It will be interesting to see if further innovation can maintain this exponential growth, resulting in titers approaching 20 and even 40 g/l in the coming decade.

### 12.2.4 The Impact of Increasing Titers on Manufacturing Facilities

The overall trend of increasing cell line titers in the last 30 years has resulted in a decreased demand for very large fermentation operations in the 10 000–15 000 l range. As a matter of fact, increasing titers have driven the shrinking of the overall footprint and physical infrastructure required to produce a biopharmaceutical product. Simply put, current expression technology allows for the production of a larger amount of antibody using a smaller biomass and lower amount of cell culture media. One outgrowth of this innovation has been the increased utilization of single-use disposable manufacturing process equipment for upstream cell culture operations. The use of disposables has had a huge impact on minimizing facility design, construction, and start-up costs and has obviated the need for cleaning validations which are expensive and time consuming. On the flip side, their use has increased the validation burden for leachables and extractables. Like most new technologies, the initial cost and perceived risks of disposable systems was prohibitive to many drug developers, but over time, the cost of disposable systems has continued to come down and quality has continued to improve. A considerable amount of development and innovation has led to the present state where a number of commercial companies offer a range of disposable manufacturing systems that utilize high-quality plastic resins with minimal leachable/extractable components. In the 1990s and early 2000s, there was a shortage of manufacturing capacity in the BioPharma industry, but this has been eliminated as a result of significant investment in infrastructure and the vast improvement in expression titers. A stable cell line producing 2–5 g/l is easily achieved and generally sufficient to supply all clinical trial and commercial supply requirements (Kelly, 2009). For most established Biologics companies, the focus is now on deriving maximal value from infrastructure investments by developing lean and highly efficient manufacturing processes. Hybrid facilities that contain a mix of both disposable and fixed stainless steel equipment have become common and offer cost and efficient facility utilization benefits to leverage existing infrastructure (Kelly, 2009).

### 12.2.5
### Protein Purification Platforms

In the manufacture of Biologics, the product is the process. Minor changes in either the upstream or downstream portions of the manufacturing process can have a significant impact on the quality attributes of the final product. While this has served as an effective deterrent to the threat of generics/biosimilars, it also poses some unique challenges in developing and validating a manufacturing process along with an accompanying suite of analytical methods.

Manufacturing of a biopharmaceutical consists of an upstream and downstream process. As mentioned in the previous section, most Biologics are produced using recombinant DNA technology in a mammalian cell line. For mAbs, the predominant cell lines are CHO lines, and the upstream portion of the process consists of the suspension culturing of these cells in a stirred-tank style bioreactor with a series of well-defined media. Innovations in driving production cell line titers into the 5–7 g/l range have put considerable pressure on the downstream portion of the process to deliver a highly pure mAb molecule in an efficient, cost-effective manner to achieve COGs targets. Table 12.2 outlines the steps in a typical manufacturing process for a mAb along with their intended purpose.

The evolution of Biologics manufacturing processes over the last 30 years has been a story of continuous incremental innovation in manufacturing technology, analytical methods development, and automation. Along the way, there have been a few fundamental developments that have revolutionized the ability to deliver high-quality products that are both safe and effective.

The first of these developments is the use of recombinant Protein A in an affinity chromatography capture step. As outlined previously, Protein A affinity chromatography is often the first chromatography step in the manufacturing process. This step results in a highly pure product (90–95%) with excellent yields in a single step and also serves the dual purpose of viral inactivation through the low pH hold. The innovation in this step started with the evolution from purified Protein A derived from *Staphylococcus* bacteria to recombinant Protein A derived from *E. coli*. A second component of innovation has been the continuous development of more robust chromatography resins that can accommodate high process

**Table 12.2** A typical downstream recovery process for a monoclonal antibody.

| Process step | Methodology | Purpose |
|---|---|---|
| Harvest | Centrifugation/filtration | Removal of cells and cell debris prior to chromatography |
| Protein A chromatography | Protein A affinity capture | Initial purification of IgG |
| Low pH hold | Hold of protein eluate at low pH | Inactivation of endogenous/adventitious viruses |
| Polishing chromatography – 1 | Cation or anion exchange; CHT or HIC | Removal of product/process impurities and viruses |
| Polishing chromatography – 2 | Cation or anion exchange; CHT or HIC | Removal of product/process impurities and viruses |
| Viral filtration | Membrane nanofiltration | Removal of endogenous/adventitious viruses |
| Bulk drug substance (BDS) formulation | Ultrafiltration/diafiltration | Formulation of BDS |

flow rates, low pH, and guanidine hydrochloride elution conditions, as well as the use of aggressive cleaning/sanitization solutions that use urea and sodium hydroxide. Currently there are three major types of Protein A resins that differ in their respective resin backbone compositions: glass or silica based, such as ProsepvA and ProsepvA Ultra (Millipore); agarose based such as Protein A Sepharose Fast Flow and MabSelect (GE Healthcare); and organic polymer based such as polystyrene-divinylbenzene Poros A and MabCapture (Applied Biosystems) (Liu *et al.*, 2010). All are widely used in manufacturing processes to produce clinical trial material as well as commercial products and offer a dynamic binding range of 15–50 g of antibody per l of resin depending upon the flow rate, the antibody to be purified, and the matrix used (Liu *et al.*, 2010). Resins and purification processes utilized in the 1980s and 1990s were not capable of purifying cell culture feed streams containing 2–5 g/l of antibody, but this is now possible using these types of Protein A matrices in combination with additional downstream polishing resins (Kelly, 2009).

A second area of innovation has been the development of platform manufacturing processes to apply to a drug developer's portfolio of antibody products. One of the strengths of mAbs is that they share conserved structural elements that allow for the utilization of common upstream and downstream processes. A base process that is similar to the one outlined in Table 12.1 can be established and then refined based upon the specific attributes of a given mAb such as glycosylation patterns, charge distribution, tendency to aggregate, process contaminants, and so on. The use of a platform process can have significant operational benefits by reducing the overall investment per mAb product candidate. Potential areas of savings include streamlining development efforts, simplifying raw material and warehousing costs, and simplifying scale-up and technology transfer (Kelly, 2009). A platform approach facilitates the early identification of critical process parameters to allow for a more focused validation approach and potentially set the table for a Quality by Design (QbD) regulatory submission strategy. The end results are product quality, production capacity, and costs that are predictable (Kelly, 2009). This more standardized approach, when executed effectively, can save time in process development and validation as well as accelerate timelines from discovery to First in Human (FIH) trials substantially.

In the area of automation, the use of high-throughput platforms for purification development has had a significant impact. This approach utilizes either mini-columns or microtiter plates to perform chromatography in batch mode. One advantage of this approach is that it requires 10- to 50-fold less sample for development and, by employing parallel processing of many different unit operations, can dramatically reduce the time required for process development (Lacki *et al.*, 2007). This approach would also extend to the refining of a platform process to meet the needs of any given individual antibody. High-throughput techniques can be used for primary screening of chromatography resins, determination of dynamic binding capacity, optimization of buffer compositions for binding, and elution as well as other key process parameters (Lacki *et al.*, 2007).

### 12.2.6 Conclusion: What Will the Next 40 Years of Innovation Bring?

All of the advancements described in the areas of cell engineering, cell culture, purification, and analytical characterization have combined to enable highly productive and cost-efficient manufacturing processes. It is no coincidence that the maturation of these processes has correlated with the rapid growth in the number of approved commercial Biologics. And as the number of potential biotherapeutics grows and evolve into even more complicated modalities, such as ADCs, bispecific mAbs, and alternative scaffolds, the importance of maintaining innovation in manufacturing will continue to be a priority in enabling the delivery of these therapies to the market.

Along those lines, we conclude by highlighting two potential innovative strategies that have shown recent promise in shaping the future of biomanufacturing. The first of these involves the use of transient protein expression as an alternative to traditional stable cell lines. Creating a high-expressing, stable production cell line is an extremely tedious and time-consuming process, where identifying an optimal clone among the millions of transfectants can take up to 10 months. When combined with the time needed for subsequent process development around this line, the overall timelines just for getting a therapeutic mAb into the clinic can be quite significant. Transient protein expression, by contrast, focuses on protein produced during the short window of time immediately after transfection before any DNA is stably integrated into the genome. There is no need to identify a single, clonal production cell under this approach because the entire transfection population can be considered as a single production pool. In this manner, the manufacturing process becomes greatly simplified, where the two main components are the DNA encoding the protein of interest and cells to transfect. One challenge with this approach is that protein titers during transient expression are significantly lower than what is observed in stable expression system. Recent published data, however, show that through a combination of cell engineering and culture optimization, mAb titers of greater than 2 g/l can be achieved (Kiseljak *et al.*, 2012). There will likely be regulatory hurdles as well, as the pool of transfected cells are significantly more heterogeneous than a population derived from a single, stable production clone. Another perhaps greater challenge is preparing sufficient DNA needed to transfect the hundreds or thousands of liters of biomass required for clinical and commercial production. If industry deems this strategy to be of sufficient interest though, we may see significantly more resources invested into making this vision more attainable.

A second area of innovation that is rapidly garnering attention involves the use of continuous bioprocessing as a manufacturing strategy. Traditional manufacturing of a therapeutic can be thought of as a series of discrete production runs, where the initial thaw and scale-up of a production cell bank vial ends with the harvest and purification of a terminal, nonviable cell culture reactor. Continuous bioprocessing, by comparison, aims to keep the conditions in the reactor unchanged, both from a nutritional and cell density perspective. In this manner, the culture can theoretically be prolonged indefinitely. And when combined with a similar approach for a purification platform, the generation of purified therapeutic no longer occurs in distinct batches but rather as a continuous stream of drug substance. Indeed, there are already reports of proof-of-concept being demonstrated and patents being filed as the manufacturing community races to establish a basic platform (Konstantinov *et al.*, 2014). Similar to transient expression, however, there is still considerable ambiguity regarding the regulatory approval landscape for such a platform. Additionally, it will be interesting to see if the improvement in costs and timelines will be substantial enough to warrant the significant infrastructure investment needed to either redesign existing manufacturing facilities or build new ones.

## 12.3 The Evolution of Alternative Scaffolds

While mAbs have demonstrated to be a very successful therapeutic modality, some factors like their large size (about 150 kD), binding to a single antigen, and complex manufacturing process can limit their usage in some applications. To overcome these limitations and introduce new functionality into antibodies, miniaturized antibodies including small antibody fragments as well as nonantibody novel scaffolds have emerged in recent years as the next generation of antibodies. Compared to mAbs, these binding scaffolds are much smaller, easier to manufacture, modular in nature, and especially can be conveniently linked together to create bi- or multispecific binding moieties (Gill and Damle, 2006). Hopefully these next wave innovations will ensure biopharmaceutical industries to continue their growth trend for the next 40 years. In the following we will discuss some of these novel small protein scaffolds, clinical applications of some of the scaffolds, and especially the promising future of bispecific antibodies (bsAbs).

### 12.3.1 Novel Small Protein Scaffolds

The need for smaller, cost-effective, and multifunctional protein therapies has driven many innovative approaches toward creating novel protein scaffolds which can be engineered to target any antigens via genetic mutagenesis coupled with *in vitro* selection systems, like phage display or yeast display. The availability of genome-wide protein sequences and understanding of protein structure and function relationship have all made it possible to identify a diverse panel of scaffolds to be evolved as potential next-generation therapeutics and includes small antibody fragments, like the single-chain fragment variables (scFvs) and diabodies (Dbs).

### 12.3.2 Single-Chain Fragment Variables and Diabodies

scFvs are composed of variable region of heavy (VH) and variable regions of light chains (VL) of an antibody and contain one of the antigen-binding arms of an antibody. The VH and VL chains are connected together from either C-terminus of the VH to the N-terminus of the VL, or vice versa with a long enough flexible linker, normally a 15 amino acids of three repeats of the Gly-Gly-Gly-Gly-Ser ((G4S)3) (Figure 12.6). scFvs maintain the original antibody's binding specificities, but usually have lower antigen-binding affinities and decreased stabilities than antibodies, for removing the fragment crystallizable (Fc) constant regions, decreasing the overall stability.

If the linker length is reduced to approximately five amino acids, just one repeat of the (G4S)1, it will prevent the pairing of the VH and VL domain within one chain, forces the VH to flip over and pair with VL in a different molecule, and create a dimeric bivalent molecule with monospecificity. These short-chain linked

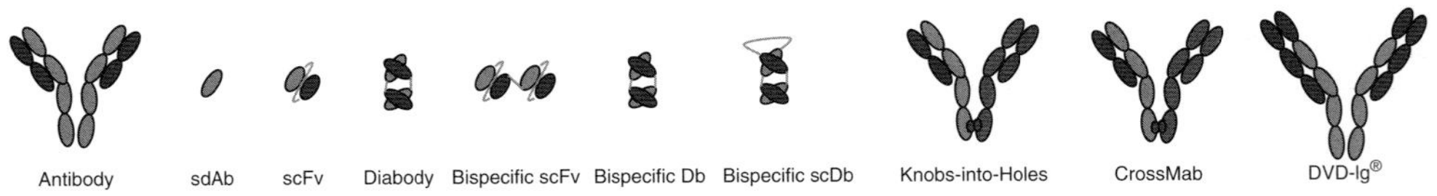


**Figure 12.6** Antibody scaffolds and bispecific antibodies. While antibodies, single-domain antibodies and single-chain Fv fragments are monospecific, all other derivatives can be engineered to recognize more than one antigen.

bivalent cross-over scFvs are called *diabodies* (Muller and Kontermann, 2011), roughly double the size of a scFv (~50 kD) (Figure 12.6). Each of the Db has two identical antigen-binding sites at opposite directions facing away from each other. Unlike full antibodies, which are often produced in mammalian cell cultures, scFvs and Dbs can be produced in *E. coli.*

### 12.3.3 Single-Domain Antibodies

Single-domain antibodies (sdAbs) are the smallest antigen-binding components of the antibodies and mostly consist of the antibody VH domain (11–15 kD) (Figure 12.6). These VH-only domains contain all three complementarity-determining regions (CDRs) of the antibody heavy chain that allow specific binding to target antigens, although with reduced affinities than antibodies. Without the normal pairing light chains to cover the hydrophobic patch at the surface of the HV domains, human sdAbs tend to aggregate and need engineering to improve their water solubility. sdAbs can be isolated from phage display libraries cloned from either immunized mouse spleen or CDR mutagenized human antibody repertoires. Although humans do not express antibodies as independent VH-only domains, Camelids such as camel and llamas naturally produce heavy-chain-only sdAbs that lack the heavy-chain CH1 domain and all light chain. These heavy-chain-only antibodies contain the so-called VHH binding domains or nanobodies. VHH domains carry signature hallmark residues that make them stable and soluble in the absence of pairing light chains. High affinity and target-specific VHHs have been successfully identified through panning of phage display libraries generated from immunized camelid serum (Saerens *et al.*, 2008).

### 12.3.4 Nonantibody Scaffolds

Not limited by using small antibody fragments containing the CDR loops to engage target antigen binding, many other natural structured protein scaffolds have been identified to substitute antibodies as next-generation biotherapeutics. In general, these novel protein scaffolds are small, stable by themselves, compatible with microbial expression systems, and can be genetically manipulated to bind diverse targets. Examples include AdNectins, A-domain proteins, Lipocalins, or Ankyrin repeat proteins. AdNectins are a family of nonantibody protein scaffolds based on the 10th fibronectin type III domain, a single-domain structure without any disulfide bonds. There are three distinct loops in this six antiparallel beta-sheet fold that can introduce structure diversity for high-affinity specific binding to therapeutic targets (Xu *et al.*, 2002). A-domain proteins are a family of naturally occurring cysteine-rich repeats that reside in many cell surface receptors. With each domain being 25–35 amino acids in size and containing three disulfide bonds, the A-domain was first identified in the human low-density lipoprotein receptor. Since then, it has been found in a large family of proteins that bind over 100 different targets (Silverman *et al.*, 2005). A-domain-containing proteins bind their targets often using multiple independent domains (typically two to three); each domain binds with low affinity, but the avidity of engaging multiple domains creates strong overall high-affinity binding. Lipocalins are a different family of highly conserved secreted proteins, in nature used for steroid and metabolite transport (Schlehuber and Skerra, 2005). Consisting of 160–180 amino acids with eight antiparallel beta strands as a funneled barrel, lipocalins use the four loops between the beta-sheets at one end of the barrel for target binding. These highly flexible loops can be engineered to bind diverse targets by introducing amino acid mutagenesis and *in vitro* phage display selection. The specific binders isolated from the libraries are also called *Anticalins* and are being developed as oncology therapeutics. Ankyrin repeat proteins were discovered in the cytoskeletal protein ankyrin and are present in over 400 ubiquitous proteins. Multiple ankyrin repeats typically fold together to form a single, linear solenoid structure called ankyrin repeat domains. These domains are one of the most common protein–protein interaction platforms in nature (Sedgwick and Smerdon, 1999). Each ankyrin repeat is composed of 33 amino acids structured into an alpha–hairpin–helix–loop–helix motif. The nonconserved amino acid residues in the beta turn and loop region can be randomized to create structural diversity for specific protein targets binding. These genetically engineered antibody mimetic proteins are also called DARPins (designed ankyrin repeat proteins); each DARPing utilizes of at least three, usually four or five repeats of the ankyrin motif for a specific target binding (Binz *et al.*, 2004).

Compared to antibodies, small protein scaffolds, including antibody origins, like scFvs and Dbs, and the nonantibody scaffolds have a smaller size and thus, at least in theory, better tissue penetration, simple structure, and thermal stability, and are easy to manufacture. Small scaffold proteins can be manufactured from bacteria cells, but the final production cost advantages are limited, since many of the bacteria expressed scaffolds will need extra refolding and endotoxin removal steps; while antibody expression levels are constantly improving to a now average >2 g/l industry standard. Another area of concern with novel scaffolds is the potential immunogenicity of engineered protein scaffolds. Although the small scaffolds are of human origin, the randomized residues introduced to allow specific target binding harbor the risk of potential immunogenicity (Schellekens, 2010). As an example, Amgen's anti-IL-6 A-domain protein program had been stopped due to the immunogenicity problem in Phase I human clinical trial. It is not known whether this is a molecule-specific incidence or a general concern of the nonantibody scaffolds. Even though nonantibody scaffolds have emerged from over 20 years ago, their clinical progress is still slow and approval as therapeutics by regulatory authorities is still only rarely seen. Their true potential may not be in replacing mAbs as therapeutics but rather than specific applications for which mAbs are less suitable such as imaging, oral delivery of biotherapeutics, cytoplasmic targeting, or as protein-drug conjugates, taking advantage of the scaffolds' small sizes for better cell and tissue penetrations. One of the attractive features of the small domain scaffolds is their ability to create bispecific or multifunctional molecules: applying each domain to bind to a different target or to a different epitope of the same target. Because of their symmetric, bivalent but monospecific nature, conventional antibodies are difficult to be engineered to be bi- or multispecific.

### 12.3.5
### Bispecific Single-Chain Fragment Variables and Diabodies

To overcome the limitation of monospecific property of antibodies, and to address many diseases' polypharmacology, bsAbs have been created to target two antigens simultaneously. Dual targeting can be used, for example, to address two antigens on the same cancer cell, improve the selectivity of cancer cells over normal cells, or create synergistic therapeutic by addressing two antigens. Theoretically, dual targeting will result in higher specificity, for bsAbs are more likely to bind to cells that express both targets than to cells that express only one, whereas conventional mAbs bind indiscriminately, and it is speculated this will translate into better safety and efficacy.

Two different target-binding scFvs can be linked together in tandem by a short middle linker, five amino acid residues, (G4S)1 to creating a VL1-(G4S)3-VH1-(G4S)1-VH2-(G4S)3-VL2-type bispecific scFvs (Figure 12.6). The CDR loops of VL1 and VH1 will bind target 1, while the CDR loops of VH2 and VL2 will bind target 2. Blinatumomab, a bispecific scFvs with above design and targeting CD19 and CD3, has been in development as a cancer immunotherapy since 2000 (Löffler *et al.*, 2000). The drug is developed as bispecific T-cell engager (BiTE) system by Micromet and designed to direct the body's disease-fighting immune cells, T-cells, to attack CD19-bearing cancer cells. Amgen acquired Micromet in 2012, later submitted Blinatumomab to US FDA for approval to treat an aggressive form of adult acute lymphoblastic leukemia based on promising Phase II data. On 3 December 2014, in less than 3 months after accepting the filing, FDA approved Blinatumomab under the brand name Blincyto, for the treatment of Philadelphia chromosome-negative relapsed or refractory B-cell precursor acute lymphoblastic leukemia (R/R ALL), setting an approval record time. The FDA had given Blinatumomab breakthrough designation and a priority review, which typically cuts the approval decision time to 6 months from the normal 10 to 12 months. Blinatumomab is a first approval in the United States for a bispecific antibody, except there was a previous approval of a rat-mouse hybrid bispecific antibody, Catumaxomab, only in Europe in 2009. Clearly, this is just the beginning of bispecific antibody development; there are at least 18 bispecific antibody programs in clinical development deploying various bispecific formats (Garber, 2014).

The bispecific Dbs are another bispecific antibody format formed by the coexpression of two separate and different chains with the structures: VH1-(G4S)1-VL2 and VH2-(G4S)1-VL1, the short linker between VH and VL will force them to cross-pairing with their respective variable chains, creating a monovalent bispecific molecule (Figure 12.6). The two VH1-(G4S)1-VL2 and VH2-(G4S)1-VL1 chains can be linked together by a long linker (G4S)3 in between to create a single bispecific molecule, single-chain Db (Figure 12.6). These bispecific fragment antibodies can be produced in mammalian cell culture, but they generally have lower expression titers than antibodies. MacroGenics, a small biotech company, is developing several oncology and autoimmune therapies using the bispecific Db design called Dual-Affinity Re-Targeting (DART). MGD 006 is a DART molecule that recognizes the CD123 overexpressed on the malignant hematological cells in AML and myelodysplastic syndrome (MDS) and CD3 on the cytotoxic T-cells to bring the T-cells close to the CD123-expressing cells for cell killing. This molecule is now in Phase 1 clinical study and is in codevelopment with Servier. Another bispecific DART, MGD 010 (CD32B × CD79B), a nondepleting B-cell downregulator, is being developed to treat autoimmune diseases, such as Systemic Lupus Erythematosus (SLE) and Rheumatoid Arthritis (RA), and this program is ready to go into Phase 1 clinical trial (as of 2014) in partnership with Takeda Pharmaceutical Company Limited.

### 12.3.6 Other Bispecific Antibody Formats

The desire to expand the functionalities of traditional antibody and test combination therapies drive the bispecific antibody field to flourishing with many innovative designs. Besides the bispecific scFvs and Dbs, there are many other bsAb formats in development. Blinatumomab is difficult to manufacture and has a very short half-life in serum, thus requires continuously intravenous administration. This is because Blinatumomab's relative small size (55 kD) and lack of the Fc domain means that it will be filtered out quickly by kidney and cannot engage the neonatal Fc receptor on epithelial cells to recycle back the drugs. Over the past decade, a lot has been learned from earlier efforts, including the tandem scFvs, Blinatumomab development. Newer formats are more stable, easier to manufacture, and have a longer plasma half-life via the inclusion of antibody Fc domains in the designs.

MacroGenics now fused an Fc domain onto its DART molecules, creating Fc-bearing DARTs for its bispecific programs. Fc domains can not only be used to increase the serum half-life, as such they can also be used to create bsAbs. The earliest bsAb design utilized a knobs-into-holes concept in the CH3 region of the Fc domain to bring two different antibody heavy chains together via heterodimerization (Carter, 2001). Since then this heavy-chain heterodimerizing design has been further improved by using opposite charged or other hydrophobic residues to bring the two Fc heavy chains together more efficiently, for example, by companies like Amgen and Zymeworks (Gunasekaran *et al.*, 2010; Lewis *et al.*, 2014). These Fc heterodimerizing designs only solved the heavy-chain pairing issue and still need to find ways to ensure proper pairing of two light chains with their respective cognitive heavy chains. One way to minimize the mispairing of cognitive heavy and light chains is to swap the heavy- and light-chain domains in the Fab regions like Roche's CrossMab platform (Figure 12.6 and Schaefera *et al.*, 2010). Dual variable domain immunoglobulin (DVD-Ig™) protein is another bispecific format developed by AbbVie. The structure has an Fc region and constant region 1 (CH1) like in a wild-type antibody configuration but contains two different variable domains linked in tandem in both heavy and light chains, so the DVD-Ig protein is a dual-specific, tetravalent IgG-like molecule (Figure 12.6). The advantages of this platform are good biophysical and pharmacokinetic IgG-like properties and the ability to produce them by standard antibody manufacturing processes with antibody-like high expression titers (Wu *et al.*, 2007). AbbVie has two bispecific DVD-Ig programs in clinical Phase II trial: ABT-122 targeting TNF and IL-7 for RA and ABT-981 targeting IL-1alpha and IL-1beta for osteoarthritis.

The approval of the first bispecific antibody Blinatumomab by FDA has shown the bispecifics have finally come of age. There is already a significant pipeline of bsAbs in clinical development to seek enhanced therapeutic efficacy via dual targeting (Garber, 2014). The recent breakthrough of targeting immune checkpoints in cancer immunotherapy will open the door to use bsAb targeting any two immune checkpoints combination, like Programmed Death 1 protein (PD-1) or Cytotoxic T Lymphocyte Antigen 4 (CTLA-4) (Currana *et al.*, 2010). PD-1 and CTLA-4 receptor proteins are both members of the CD28/B7 super family and T-cell coinhibitory receptors, which act like strong suppressors of anticancer immune response. Cancer cells have an enhanced expression of various immune inhibitory molecules apart from PD-1 and CTLA-4, and these brakes keep our immune system from recognizing the cancer. Thus, blocking these brakes on the cancers results in rerecognition and subsequent killing by cytotoxic T-cells. Anti-CTLA-4 antibody, Ipilimumab, was the first approved cancer therapy by targeting immune checkpoint. For many patients who do not respond to Ipilimumab, they have overexpression of PD-1, which is one of the reasons why there is so much interest in the combined application of anti-PD-1 and anti-CTLA-4 antibodies. Recently reported at 2014 ASCO meeting, a combination of the BMS anti-PD-1 antibody, nivolumab (Opdivo) in combination with anti-CTLA-4 antibody ipilimumab (Yervoy) produced a 1-year survival rate of 85% and 2-year survival rate of 79% for advanced melanoma patients, while the ipilimumab alone generated 2-year survival rate of 18%. Like anti-PD-1 antibodies, anti-PD-L1 (programmed death-ligand 1) antibodies have also been proven to be quite successful in combating melanoma. PD-L1 is the natural ligand of the PD-1 receptor and upregulated on macrophages and dendritic cells and activated T-cells and B-cells. There are many more immune checkpoint proteins like CTLA-4, PD-1, and PDL1 which can be targeted in combination by bsAbs to maximize the therapeutic efficacy in combating tumors. BsAbs can be the next wave of transformative medicines that can bring differential benefits to patients for many unmet medical needs.

## 12.4 Antibody-Drug Conjugates

Clinical reality taught us that the effector domain of antibodies, conveying complement dependent cytotoxicity (CDC) or antibody-dependent cellular-mediated

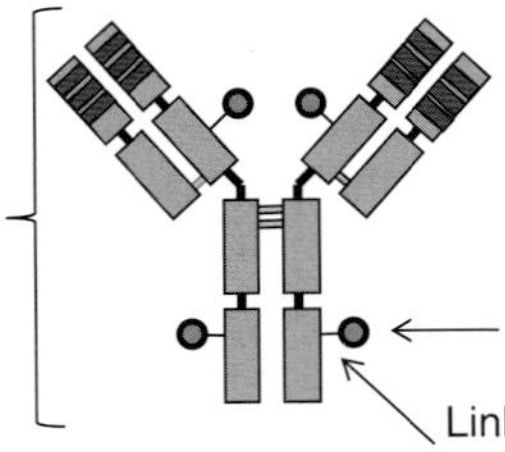


**Figure 12.7** Essential elements of an antibody-drug conjugate (ADC).

cytotoxicity (ADCC) often is not powerful enough for a maximally desired anti-tumor efficacy. The concept of using mAbs to deliver cytotoxic small molecules specifically to tumors goes back several decades to the development of mAb hybridoma technology and arguably back a century or so to Ehrlich's hypothesis of "magic bullets," molecules that would specifically be targeted to malignant cells and spare cells in the normal tissues (Lewis, 1908). Small-molecule cytotoxic agents (or traditional chemotherapy) have proven to prolong the life of cancer patients and are still mainstays of cancer patients' standard of care both in the adjuvant and advanced metastatic disease setting. Small-molecule cytotoxics do not discriminate between tumor tissue and normal tissue and thus are associated with significant dose-limiting toxicities often requiring cessation of treatment or dose reductions in order to manage the patients' toxicities. To enhance the therapeutic index, it is an attractive strategy to link cytotoxic molecules to tumor-specific mAbs for the targeted delivery of cytotoxic molecules to cancer cells.

The first ADCs (an overview graph of components of an ADC is shown in Figure 12.7) were developed by attaching chemotherapeutic drugs already in clinical use, primarily because these drugs had a well-defined clinical efficacy and toxicity profile. All ADCs are built using the same principal components; a mAb to a tumor-specific target, a linker that bridges the mAb and the cytotoxic drug which is the third component. One of the earliest ADCs was a doxorubicin conjugate to an anti-Lewis-Y antigen chimeric mAb (Trail *et al.*, 1993). It was tested in patients with metastatic colon and breast as these tumor types were known to express high levels of the antigen. This ADC did not move forward due to the broad expression of target in noncancer tissue as well as the relative low potency of the cytotoxic payload. The antigen is found in the gastrointestinal tract which was postulated to contribute to increased GI toxicity as compared to the doxorubicin as well as potentially limiting the delivery of ADC to target tissues (Tolcher *et al.*, 1999).

This first clinical experience with an ADC led to a focus on the identification of highly potent cytotoxic molecules and the hunt for tumor-specific cell surface antigen suitable for ADC development. Targets with a selective tumor expression became the primary focus for new mAb and ADC discovery. Besides being targeting tumor selective antigens, successful ADC development requires internalization of the antigen and ADC complex at sufficient levels to deliver cytotoxic

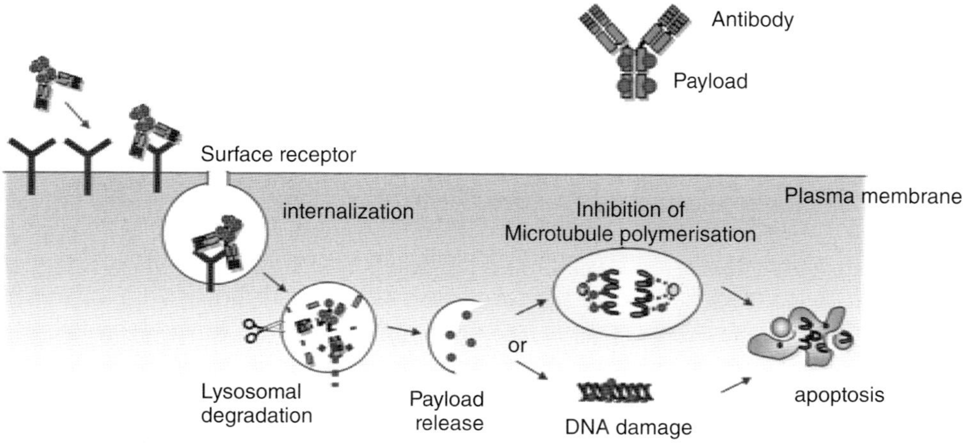


**Figure 12.8** Mechanism of action of an ADC: Upon binding of a tumor antigen, the ADC can be internalized via various mechanisms. Trafficking to the lysosome results in mAb degradation and payload release into the cytoplasma. Depending on the mechanism of action of the payload, the target cell undergoes programmed cell death in a cell cycle dependent or independent manner.

concentrations of the attached payload (see Figure 12.8 for a schematic view of the primary mechanism of ADCs). There is considerable debate around the required level of antigen expression for successful utilization of an ADC, and the level of expression is going to vary depending on the biology of the target itself. How efficient it is internalized, whether engagement with a mAb, would stimulate internalization as well as the trafficking inside the cells. It is also possible that the engagement with an ADC could lead to downregulation of the protein at the cell surface leading to reduced levels of available antigen for an ADC to bind to. All of these properties need to be understood at the molecular and cellular level in order to understand the suitability of a target for ADC development.

A significant step in ADC development was the approval of Mylotarg for the treatment of AML (Bross *et al.*, 2001). The drug consists of an anti-CD33 mAb linked to calicheamicin payload via an acid labile hydrazone linker. This was viewed as a major step forward for the ADC field as it contained a cytotoxic payload with several orders of magnitude greater cellular cytotoxic potency than doxorubicin. Development of calicheamicin as free drug is prevented by dose-limiting toxicities to normal tissues, but within the context of an ADC, it can be dosed to AML patients. Though the drug was approved in 2000, it was withdrawn from the market by Pfizer to the lack of efficacy and unacceptable toxicities found in postmarketing studies (FDA, 2010, CenterWatch, 2000).

The importance of plasma stable linkers and spacers between the mAb and the payload is highlighted from the Mylotarg experience as well as the clinical experience with disulfide-bridged maytansine payloads. C242-DM1(CanAg antigen-targeted humanized antibody C242) and MLN2704 (anti-PSMA humanized J591 mAb) were both conjugated using ImmunoGen's tumor-activated prodrug technology; however the payloads were falling off in plasma as the disulfide-containing linker was reduced prior to the ADC reaching the tumor (Smith, 2004). This resulted in an unfavorable pharmacokinetic (PK) profile of the ADC compared to the total mAb and was proposed to significantly contribute to toxicity. Since then, improvement in the stability of the disulfide bridge that connects the linker to the payload have been developed resulting in a number promising clinically active ADCs. An alternative linker system that has proven to be significantly more stable in plasma contains a protease cleavable trigger in between the mAb attachment and the payload. This linker–payload system is the basis for a number of ADCs in the clinic today and importantly is validated by brentuximab vedotin (Adcetris; see Table 12.3). Further stability of

**Table 12.3** Select ADCs in clinical development.

| ADC | Target | Linker–payload | Lead sponsor | Phase |
|---|---|---|---|---|
| Adcetris | CD30 | vcMMAE | Seattle Genetics | Approved |
| Kadcyla | HER2 | SMCC-DM1 | Roche | Approved |
| Inotuzumab ozogamicin | CD22 | Hydrazone-calicheamicin | Pfizer | Phase 3 |
| Milatuzumab doxorubicin | CD74 | Hydrazone-doxorubicin | Immunomedics | Phases 1–2 |
| Glembatumumab vedotin | GPNMB | vcMMAE | Celldex | Phase 2b |
| Pinatuzumab vedotin | CD22 | vcMMAE | Roche | Phase 2 |
| Polatuzumab vedotin | CD79b | vcMMAE | Roche | Phase 2 |
| PSMA-ADC | PSMA | vcMMAE | Progenics | Phase 2 |
| RG-7599 | NaPi2 | vcMMAE | Roche | Phase 2 |
| MLN0264 | GCC | vcMMAE | Takeda | Phase 2 |
| Indatuximab ravtansine | CD138 | SPDB-DM4 | BioTest | Phase 2 |
| Coltuximab ravtansine | CD19 | SPDB-D.M4 | Sanofi | Phase 2 |
| RG-7450 | STEAP1 | — | — | — |
| RG-7636 | ETBR | vcMMAE | Roche/Astellas | Phase 1 |
| RG-7458 | MUC16 | vcMMAE | Roche | Phase 1 |
| ASG-15ME | SLITRK6 | vcMMAE | Roche | Phase 1 |
| ASG-22ME | Nectin 4 | vcMMAE | Astellas/SGEN | Phase 1 |
| SGN-LIV1A | LIV-1 | vcMMAE | SGEN | Phase 1 |
| AGS-16M8F | ENPP3 | mcMMAF | Agensys | Phase 1 |
| PF-0626350 | 5T4 | mcMMAF | Pfizer | Phase 1 |
| SGN-CD33A | CD33 | vaPBD | SGEN | Phase 1 |
| SGN-CD70A | CD70 | vaPBD | SGEN | Phase 1 |
| IMMU-132 | Trop2 | Carbonate-SN38 | Immunomedics | Phase 1 |

vc = valine-citrulline; MMAE = monomethyl auristatin E; mc = maleimidocaproyl; MMAF = monomethyl auristatin F; va = valine-alanine; and PBD = pyrrolobenzodiazepine.

the linker–payload attachment has been obtained by introducing the so-called noncleavable linkers; essentially linkers that are resistant to proteases, reduced pH, and are not reducible. These linker systems require break down of the ADC to an amino acid-payload adduct in the lysosome that exerts the desired cytotoxic effect. One disadvantage to the noncleavable linker systems is that the cytotoxic end product is not cell permeable requiring a uniform level of target expression on all tumor cells whereas the cleavable linker systems results in payloads that once released from the ADC are free to attack neighboring cells whether they express the target or not (also called *bystander effect*). Depending on the target and tumor type, one linker system may be preferred over another.

Suitable cytotoxic payloads for ADC development have so far been found in sources from nature and belong to a class of cytotoxic compounds that target the cells microtubule infrastructure. These have superior potency as unconjugated free small-molecule cytotoxics. The Maytansines and auristatins kill cells in the 1 nM range of different cell types representing various tumor types of hematological and solid origins which is about 10–100 fold more potent than microtubule inhibitors (MTIs) that are in clinical use today such as the taxanes and vinca alkaloids (Senter, 2009). On rapidly proliferating cells *in vitro* the most apparent MOA is inhibition of the mitotic spindle formation leading to cell cycle arrest in M-phase and apoptosis. However, MTIs can act on a number of other microtubule-dependent processes such as ensuring proper cellular localization of key oncogenic proteins such as p53, Rb, and HIF1a (Roth *et al.*, 2007). Androgen signaling in prostate cancer cells is another pathway that is affected by MTIs, possibly explaining the success of taxane-based therapies in patient with advanced prostate cancer (Thadani-Mulero *et al.*, 2012). There are >30 ADCs in early and late phases of clinical development following the experience of brentuximab vedotin (Adcetris) in CD30 positive malignancies such as anaplastic large-cell lymphoma (ALCL) and Hodgkin's lymphoma (HL) and trastuzumab emtansine (Kadcyla) for HER2 overexpressing breast tumors resistant to trastuzumab (Herceptin), the unconjugated mAb. Most ADCs targeting solid tumor antigens such as MLN0264 show some level of single-agent activity, however, in solid tumors the degree of tumor control that Adcetris can induce in CD30 expressing hematological tumors have not been observed (Zambrano *et al.*, 2014). It is not clear that this is due to inherent resistance mechanisms to the microtubule-targeting payloads or access to the target in solid tumors. As more clinical data is collected and a better understanding of the distribution of the ADCs to the target in the tumors as well as the introduction of payloads with differing mechanisms, we will be able to dissect and make necessary improvements translate the promise of ADC to many more cancer patients.

It is important to note that the most advanced ADC platforms today rely on single payload per linker attachment site, and it has been observed that attempts to attach six or more linkers have met with the somewhat surprising finding that they are unstable pharmacological molecules. They are more rapidly cleared from the plasma as compared to ADCs with three to four linker attachments per mAb which tend to be cleared at a rate very similar to an unconjugated IgG1 molecule (Hamblett *et al.*, 2004). Hence most ADCs in clinical testing today carry a ratio of 3–4 payloads per mAb. Hypothetically, a larger number of payload molecules per mAb would deliver higher concentrations of cytotoxic payload to target expressing tissues, such as tumor, and thus provide better antitumor activity. By developing linkers that contain multiple payload attachment sites, one can now test the hypothesis. Using a polyacetal polymeric backbone, it is now possible to generate ADCs with as many as 20 payloads per mAb without increasing the number of linker attachments to the mAb (http://mersana.com/product-pipeline/antibody-drug-conjugates.php). Though not yet in the clinic, such ADCs require a lower threshold of antigen expression than the typical 3–4 payload per mAb ADCs and thus provide a potential advantage clinically.

In addition to the ability to test ADCs with larger numbers of payloads per mAb, next-generation ADCs are incorporating novel payloads which target DNA and other essential molecules/structures. This being another clinically validated small-molecule cytotoxic drug mechanism, significant progress has been made in the synthesis of DNA alkylating agent with pM cytotoxic potencies. Together with advances in mAb engineering and the introduction of specific conjugation sites on an IgG molecule have led to the generation of uniform ADCs where each ADC is conjugated to two payloads, one on each mutant site per chain. These uniform ADCs have shown to be much better behaved *in vivo* and may provide safety benefits as well as antitumor benefits.

The approval of the first ADCs in relapsed/refractory HL/ALCL and breast cancer has provided for much increased enthusiasm in ADCs as a therapeutic modality in cancer patients, and building on some key learnings from these first approved drugs is leading to a wave of next-generation ADC platforms promising to transform the way cancer patients are being treated with chemotherapy. Developments in antibody engineering, linker chemistries, and novel payloads are all contributing to the possibility to continue to deliver transforming therapies to patients in need. Table 12.3 provides an overview of selected ADCs in various stages of clinical development.

## 12.5
## The Next Wave of Biologics

While Biologics have revolutionized the treatment of many diseases that would have otherwise been without realistic alternative treatment options, there are still limitations to Biologics that are intrinsic to their format: Biologics in general are not amenable to administration via the oral route, they do not penetrate into the cytoplasm of cells, and they usually do not pass the blood–brain barrier (BBB). As a result of these limitations, they are not applicable for addressing certain diseases or disease pathophysiologies. What if the next wave of Biologics could address these limitations?

### 12.5.1 Orally Available Biologics

Initially, all mAbs had to be administered intravenously, sometimes with predosing using steroids to prevent potential anaphylactoid reactions. Antibodies have a long serum/plasma half-life (15–20 days), hence administrations only had to be repeated every 3 or 4 weeks for chronic treatment. Nonetheless, the administration in a hospital or clinic setting can be a significant time and travel burden for the patient. Development of high dosage formulations and companion injection devices has allowed for the subcutaneous administration of antibodies by the patients themselves, which is a significant improvement. Any Biologics format that comes in the form of a pill would constitute an innovation that would likely be groundbreaking for the entire biopharmaceuticals market. The human body's digestive system is made for preventing the passive uptake of large biomolecules via the gut as this would impose a significant immunologic threat. It therefore is not surprising that evolution has worked against oral availability of Biologics. While any technology attempting to make Biologics orally available is in its early infancy, we need to acknowledge that two peptide drugs (cyclosporine and desmopressin) are approved for oral administration. In fact, some formulation technologies have already accomplished oral bioavailability to possibly therapeutically relevant levels. However, fast clearance of peptides can be a hurdle.

### 12.5.2 Biologics That Enter the Cytoplasm

The biggest revolution in the Biologics arena would likely occur if therapeutic proteins, antibodies, or antibody fragments thereof were able to enter the cytoplasm of diseased cells. We know a plethora of essential disease pathways whose interference could result in a cancer cure. However, most of those targets are cytoplasmic, and their mechanisms of action are not amenable to small molecules. Having a Biologics format that can enter the cytoplasm, replace malfunctioning proteins, or inhibit key drivers of oncogenic signaling would revolutionize cancer treatment in an unprecedented fashion.

Similarly to the oral bioavailability of Biologics, evolution has selected against the ability of Biologics to pass cell membranes in a passive fashion, and gene therapy is more advanced in the delivery of proteins to the cytoplasm than Biologics themselves.

Two approaches exist that allow a mechanism of action for Biologics that is not confined to the extracellular space: therapeutic replacement enzymes for lysosomal storage diseases are given intravenously. The mannose-6-phosphate receptor on target cells recognizes glycoside residues containing mannose-6-phosphate on the therapeutic protein. The mannose-6-phosphate receptor then acts almost like a zip code to bring the replacement enzyme to the lysosome where it degrades pathologically deposited substrates. Antibody-drug conjugates deliver their payload into the lysosome as well where they ultimately find their way into the cytoplasm likely via passive permeation. It is certainly conceivable that future ADCs not only deliver cytotoxic payloads to tumor cells but also other therapeutic molecules that on their own (without targeting moiety) would have too small a therapeutic index.

### 12.5.3 Biologics That Pass the Blood–Brain Barrier

The brain and therefore many central nervous system (CNS) diseases have traditionally been excluded from any considerations for treatment with Biologics, and clinical trial attempts have been paved with failures (unless a peripherally acting factor can be addressed). The BBB is perceived to be an insurmountable barrier to the delivery of therapeutic proteins to the brain, and only negligible amounts of antibodies can be detected in brain tissue or fluids following systemic administration. Approaches that have exploited the transferrin receptor or insulin receptor as transcytosis vehicles have yielded an antibody deposition of 1% of systemic exposure in the brain. One true potential of bsAbs may lie in their ability to exploit one binding domain for shuttling via the transferrin receptor and another arm to convey therapeutic efficacy against a disease target of interest or to carry other therapeutic modalities into the brain.

### 12.5.4 Translational Medicine as Driver of Innovation

Translational medicine will likely revolutionize the way we define diseases and the avenues by which we address them. Currently many diseases are still characterized by very gross anatomical, pathological, or other symptomatic features and summarized as one disease (like RA); although we know that such broad definitions do not reflect the heterogeneity of the disease pathology. Consequently, approved therapies are administered to patients and only a small fraction of them will benefit while the majority experiences no or adverse effects. It is the hope and expectation that translational medicine will identify biomarkers of disease that lead to a much better categorization/stratification of patient populations. The implications imply both risks and opportunities for the pharmaceutical industry. Improved disease categorization will result in patient segmentation and definition of "boutique indications" requiring personalized medicine. The threat is that such a development will make it even harder for the pharmaceutical industry to pursue the megablockbuster model that addresses large patient populations. The opportunity is that redefinition of diseases and development of tailored medicines will result in a significant increase in uncovered indications and molecular targets to address these niches. To address such biomarker-defined patient populations, we will need to develop therapeutics with the highest selectivity and ability to interfere with unconventional disease mechanisms. The benefit for the patient will be a high degree of efficacy in this segmented population and the avoidance of drug exposure for patients who are unlikely to benefit from the given treatment. The reward for the drug developer will be a premium price for transformative efficacy and avoidance of exposure of patients who would not be able to benefit from these therapies. No current drug format other than Biologics is better suited to deliver such needed precision medicines.

## Disclaimer

The views and opinions expressed in this chapter are solely those of the authors. These views and opinions do not necessarily represent the views of their employer, Takeda Pharmaceutical Company Limited.

## References


Angov, E. (2011) *Biotechnol. J.*, **6** (6), 650–659.

Berting, A. *et al.* (2010) *Biotechnol. Bioeng.*, **106**, 598–607.

Binz, H.K., Amstutz, P., Kohl, A., Stumpp, M.T., Briand, C., Forrer, P., Grutter, M.G., and Pluckthun, A. (2004) *Nat. Biotechnol.*, **22**, 575–582.

Blizzard, R.M. (2012) *Indian J. Pediatr.*, **79** (1), 87–91.

Boshart, M. *et al.* (1985) *Cell*, **41** (2), 521–530.

Bross, P.F. *et al.* (2001) *Clin. Cancer Res.*, **7** (6), 1490–1496.

Carter, P. (2001) *J. Immunol. Methods*, **248** (1–2), 7–15.

CenterWatch (2000) Mylotarg (gemtuzumab ozogamicin), Medical Journals, May 2000, https://www.centerwatch.com/drug-information/fda-approved-drugs/drug/842/mylotarg-gemtuzumab-ozogamicin (accessed 10 August 2015).

Currana, M.A., Montalvoa, W., Yagitab, H., and Allisona, J.P. (2010) *Proc. Natl. Acad. Sci. U.S.A.*, **107** (9), 4275–4280.

Deer, J.R. *et al.* (2004) *Biotechnol. Progr.*, **20** (3), 880–889.

DiMasi, J.A. *et al.* (2010) *Clin. Pharmacol. Ther.*, **87** (3), 272–277.

FDA (0000) FDA·Pfizer Voluntarily Withdraws Cancer Treatment Mylotarg from U.S. Market, http://www.fda.gov/NewsEvents/Newsroom/PressAnnouncements/ucm216448.htm (accessed 21 June 2010).

Garber, K. (2014) *Nat. Rev. Drug Discovery*, **13**, 799–801.

Gill, D.S. and Damle, N.K. (2006) *Curr. Opin. Biotechnol.*, **17**, 653–658.

Grabowski, H. (2008) *Nat. Rev. Drug Discovery*, **7**, 479–488.

Gunasekaran, K., Pentony, M., Shen, M., Garrett, L., Forte, C., Woodward, A., Ng, S.B., Born, T., Retter, M., Manchulenko, K., Sweet, H., Foltz, I.N., Wittekind, M., and Yan, W. (2010) *J. Biol. Chem.*, **285** (25), 19637–19646.

Hamblett, K.J. *et al.* (2004) *Clin. Cancer Res.*, **10**, 7063–7070.

Harraghy, N. *et al.* (2008) *Curr. Gene Ther.*, **8** (5), 353–366.

Hay, M. *et al.* (2014) *Nat. Biotechnol.*, **32**, 40–51.

Huang, Y. *et al.* (2010) *Biotechnol. Progr.*, **26**, 1400–1410.

Jayapal, K. *et al.* (2007) *Chem. Eng. Prog.*, **103**, 40–47.

Kelly, B. (2009) *mAbs*, **1**, 443–452.

Kiseljak, D. *et al.* (2012) Proceedings of the 21st Annual Meeting of the European Society for Animal Cell Technology (ESACT), Vol. 5, pp. 497–500.

Kober, L. *et al.* (2013) Integrated Continuous Manufacturing of Therapeutic Protein Drug Substances *Biotechnol. Bioeng.*, **110** (4), 1164–1173.

Konstantinov, K. *et al.* (2014) US Patent 20140255994.

Kwaks, T. *et al.* (2003) *Nat. Biotechnol.*, **21** (7), 553–558.

Lacki, K.M. *et al.* (2007) High throughput process development: determination of dynamic binding capacity using microtiter plates filled with chromatography resins.

Presented at the American Chemical Society's National Meeting, Boston, MA, August 19–23.

Lewis (1908) *From Harben Lectures for 1907*, Royal Institutes of Health, Londan, p. 181; Bibl 200.

Lewis, S.M., Wu, X.F., Pustilnik, A., Sereno, A., Huang, F., Rick, H.L., Guntas, G., Leaver-Fay, A., Smith, E.M., Ho, C., Hansen-Estruch, C., Chamberlain, A.K., Truhlar, S.M., Conner, E.M., Atwell, S., Kuhlman, B., and Demarest, S.J. (2014) *Nat. Biotechnol.*, **32**, 191–198.

Liu, H.F. *et al.* (2010) *mAbs*, **2**, 480–499.

Löffler, A., Kufer, P., Lutterbüse, R., Zettl, F., Daniel, P.T., Schwenkenbecher, J.M., Riethmüller, G., Dörken, B., and Bargou, R.C. (2000) *Blood*, **95**, 2098–2103.

Muller, D. and Kontermann, R.E. (2011) *Bispecific Antibodies*, Springer-Verlag, Heidelberg, pp. 83–100.

Palmer, E. (2014) The 10 Best-Selling Drugs of 2013, http://www.fiercepharma.com/special-reports/10-best-selling-drugs-2013 (accessed 20 March 2015).

Rickwood, S. and Di Biase, S. (2013) White Paper IMS Health Biosimilars and Non-Original Biologics, http://www.imshealth.com/cds/imshealth/Global/Content/Healthcare/Life%20Sciences%20Solutions/Generics/IMSH_Biosimilars_WP.pdf (accessed 20 March 2015).

Roth, D.M. *et al.* (2007) *Traffic*, **8** (6), 673–686.

Saerens, D., Ghassabeh, G.H., and Muyldermans, S. (2008) *Curr. Opin Pharm.*, **8**, 600–608.

Schaefera, W., Regulaa, J.T., Bähnera, M., Schanzerb, J., Croasdaleb, R., Dürra, H., Gassnera, C., Georgesa, G., Kettenbergera, H., Imhof-Junga, S., Schwaigerc, M., Stubenraucha, K.G., Sustmannc, C., Thomasc, M., Scheuerc, W., and Kleinc, C. (2010) *Proc. Natl. Acad. Sci. U.S.A.*, **108** (27), 11187–11192.

Schellekens, H. (2010) *Discovery Med.*, **9**, 560–564.

Schlehuber, S. and Skerra, A. (2005) *Drug Discovery Today*, **10**, 23–33.

Sedgwick, S.G. and Smerdon, S.J. (1999) *Trends Biochem. Sci.*, **24**, 311–316.

Senter, P.D. (2009) *Curr. Opin. Chem. Biol.*, **13**, 235–244.

Silverman, J., Liu, Q., Bakker, A., To, W., Duguay, A., Alba, B.M., Smith, R., Rivas, A., Li, P., Le, H. *et al.* (2005) *Nat. Biotechnol.*, **23**, 1556–1561.

Smith, S.V. (2004) *Curr. Opin. Mol. Ther.*, **6** (6), 666–674.

Svay, K. *et al.* (2010) Optimizing overall manufacturing system performance through the use of animal-free cell culture supplements. Oral Presentation at The Biomanufacturing Summit, San Diego, CA.

Thadani-Mulero, M. *et al.* (2012) *Cancer Res.*, **72** (18), 4611–4615.

Tolcher, A.W. *et al.* (1999) *J. Clin. Oncol.*, **17**, 478–484.

Trail, P.A. *et al.* (1993) *Science*, **261**, 212–215.

Urlaub, G. *et al.* (1980) *Proc. Natl. Acad. Sci. U.S.A.*, **77** (7), 4216–4220.

Wilson, R. *et al.* (1992) Method for Dominant Selection in Eucaryotic Cells. US Patent 5122464A.

Wu, C., Ying, H., Grinnell, C., Bryant, S., Miller, R., Clabbers, A., Bose, S., McCarthy, D., Zhu, R.R., Santora, L., Davis-Taber, R., Kunes, Y., Fung, E., Schwartz, A., Sakorafas, P., Gu, J., Tarcsa, E., Murtaza, A., and Ghayur, T. (2007) *Nat. Biotechnol.*, **25**, 1290–1297.

Wurm, F. (2004) *Nat. Biotechnol.*, **22**, 1393–1398.

Xu, L., Aha, P., Gu, K., Kuimelis, R.G., Kurz, M., Lam, T., Lim, A.C., Liu, H., Lohse, P.A., Sun, L. *et al.* (2002) *Chem. Biol.*, **9**, 933–942.

Zambrano, C. *et al.* (2014) *J. Clin. Oncol.*, **32**, 5s (suppl; abstr 3546).

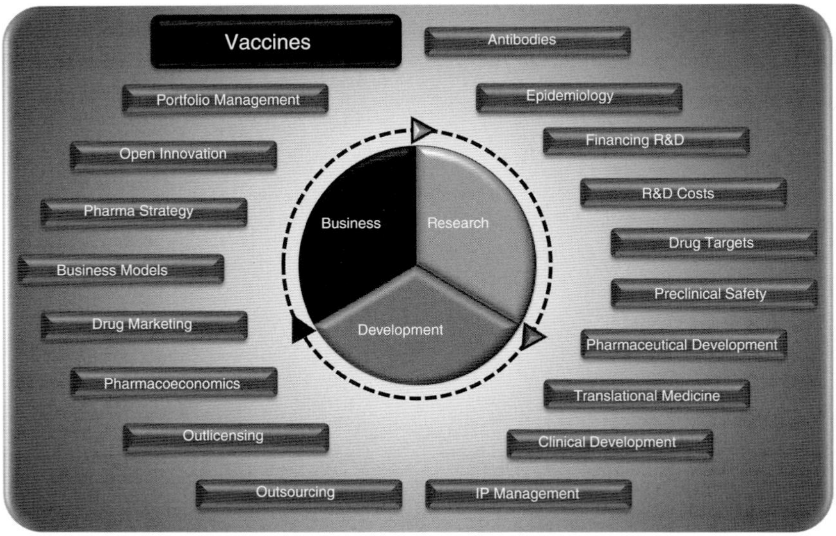
Vaccines
Antibodies
Portfolio Management
Epidemiology
Financing R&D
Open Innovation
R&D Costs
Pharma Strategy
Business
Research
Drug Targets
Business Models
Preclinical Safety
Development
Drug Marketing
Pharmaceutical Development
Pharmacoeconomics
Translational Medicine
Outlicensing
Clinical Development
Outsourcing
IP Management

# 13
# Vaccines: Where Inertia, Innovation, and Revolution Create Value, Simultaneously and Quietly

*Pierre A. Morgon and Hannah Nawi*

## 13.1
## Introduction

This chapter presents the space and role that vaccines occupy within the healthcare landscape, the emergence of novel immunization approaches, and the drivers of immunization. The chapter will further highlight the drivers of the fast growth observed in this particular segment of healthcare products and the evolution of a competitive landscape, including the emergence of smaller, nimbler companies. It will also demonstrate through a strategic analysis that innovation is needed all along each of the building blocks of the vaccine value chain and will underscore the specific drivers and evolutions to be expected for each of them. The conclusion of the chapter presents an overview of the players that will drive this innovation and their strategies.

Vaccines are peculiar healthcare products insofar as they require a medical intervention prior to the advent of a medical problem. Vaccination has been demonstrated to be the most cost-effective disease prevention and controlling measure of medical nature and is second only to access to clean water among all types of human interventions to combat diseases. Having said this, the effectiveness of vaccines is largely contingent upon the successful implementation of immunization recommendations. The vaccine market is expected to enjoy healthy growth in the coming years, outpacing several main pharmaceutical markets. This growth is expected to be driven by innovation from both the established international players and smaller vaccine players, some of them new to the field. The drivers of the vaccine market growth, linked to the decision to immunize, are very complex and driven by four key dimensions: acceptability for all stakeholders, accessibility, availability, and affordability of the vaccines. The trends impacting the development and usage of new vaccines are fueling innovation all along the value chain. Innovation in the field of vaccines will span the whole spectrum from unaddressed infectious diseases to nontransmissible diseases. Novel manufacturing technologies will contribute to ease the acceptability, availability, and affordability tensions. The evolution of the perception of the value attached to the vaccine proposition will ultimately ensure the

*Value Creation in the Pharmaceutical Industry: The Critical Path to Innovation,* First Edition.
Edited by Alexander Schuhmacher, Markus Hinder, and Oliver Gassmann.

sustainable growth of this healthcare market segment. The contribution to these evolutions of the vaccine market will come from the entire spectrum of players, from established players backed by life science conglomerates to the smaller, nimbler players exploring new approaches within each block of the value chain.

## 13.2 The World of Vaccines

### 13.2.1 What Are Vaccines?

Vaccines occupy a peculiar space in the landscape of healthcare products. They are biological drugs, administered to healthy people to stimulate or restore their own immune systems and generate a specific response, which is the ability to fight an infection or a disease. Vaccines are typically categorized as primary care products and are most often administered to large subsets of a given population in order to block the transmission of infectious diseases. As a consequence, immunization schedules are fully loaded for infants, which are particularly vulnerable once the maternal antibodies are eliminated from their bloodstreams.

Available evidence shows that vaccination brings important economic benefits. Although there is little data in low- or middle-income countries on long-term and societal economic benefits such as morbidity-related productivity gains, averting catastrophic health expenditures, growth in gross domestic product (GDP), and economic implications of demographic changes resulting from vaccination, cost-effectiveness studies suggest to policy makers that vaccines are an efficient investment (Ozawa *et al.*, 2013). Another review (Hyde *et al.*, 2012) concluded that the introduction of a new vaccine was most efficient when the vaccine was introduced into an existing delivery platform and when introduced in combination with a vaccine already in the routine childhood immunization schedule, in which it did not impact coverage of vaccines already included in said routine schedule. Other findings include the need for increased cold chain capacity, the positive impact on introduction and widespread use of auto-disable syringes into the immunization and the broader health systems, as well as the importance of training and education for healthcare workers. Most importantly, there was evidence in high-income countries that new vaccine introduction was associated with reduced healthcare costs.

### 13.2.2 Current Vaccines Are Mainly Prophylactic: Curative Vaccines Are Emerging

Until recently, all vaccines were preventive (prophylactic) and targeting infectious, transmissible diseases (although some, such as rabies, can be used after the exposure to the pathogen to elicit an immune response prior to the onset of the disease). Currently available vaccines cover a large number of diseases often categorized according to the nature of the transmission mechanisms. In some instances, these do not include human contact, and the pathogens are airborne (e.g., measles, tuberculosis, and pertussis), food- or waterborne (e.g., cholera, typhoid, and hepatitis A), or vector borne (e.g., yellow fever, Japanese encephalitis, and rabies). In other instances, human contact is involved, either directly (e.g., influenza and hepatitis B) or indirectly (via an object, e.g., *respiratory syncytial virus* and *rhinovirus*) or via droplets (aerosolization via coughing or sneezing, e.g., *meningococcus*).

Vaccines stimulate the immune system, hence allowing their efficacy to be measured in terms of immune response, a.k.a. immunogenicity. But immunogenicity doesn't equate clinical efficacy, which has also to be verified in clinical trials, except when "correlates of protection" – linking a certain level of immune response with one of clinical efficacy – are available.

The idea of immune stimulation has been gaining attention as well as research and development (R&D) funding and is starting to prove effective in oncology, beyond the long-pursued major infectious diseases such as human immunodeficiency virus (HIV) and hepatitis. One of the major challenges in developing therapeutic vaccines has been training the immune system how to recognize and destroy targeted "bad" cells. The current advances in research might make it possible to help the body fight HIV (Fischer *et al.*, 2014) as well as other types of diseases along the model developed to tame one type of tumor in a very specific way, possibly up to a fully personalized solution.

The Food and Drug Administration (FDA) approved the first such therapeutic vaccine in April 2010. Dendreon's Provenge (sipuleucel-T) was shown to extend life about 4 months in men with a certain type of metastatic prostate cancer.[1)] The vaccine provokes an immune response against a particular antigen, or identifying molecule, found on most prostate cancer cells. Provenge is stimulating more than patients' immune systems: investors, clinicians, and companies of all sizes have been encouraged by the proof that immune-enhancing treatments can benefit patients with relatively few side effects and command premium pricing in the marketplace. A course of Provenge treatment costs $93 000. Contrary to prophylactic vaccines, each dose of Provenge is custom made. White blood cells are withdrawn from a patient's bloodstream and shipped to a Dendreon manufacturing facility, where for 2 days they are incubated with a "fusion protein" that stimulates the cells' growth and maturation and another called prostatic acid phosphatase (PAP) (the antigen that studs prostate cancer cells). The patients' white blood cells take up the antigen, and within hours their surfaces bristle with fragments of the telltale molecule. The cells are then shipped back to the physician and infused into the patient. A full treatment includes three such procedures, 2 weeks apart.[2)] In addition to the doubts expressed by some clinicians about the actual efficacy of Provenge, the procedure is complex to execute, and competition from alternative

1) http://www.dendreon.com/(accessed 2014/12/07).
2) http://www.reuters.com/article/2012/03/30/us-provenge-idUSBRE82T07420120330 (accessed 2014/12/07).

therapies is very intense, all of them resulting in sending the Dendreon stock into a sharp decline[3] after the original hype fueled by expectations of a custom-tailored cancer therapy.

Therapeutic vaccines are still in focus in many R&D pipelines, yet the lessons learned from the Provenge failure will likely orient research toward solutions that will be easier to manage and implement in daily care procurement.

### 13.2.3 Drivers to Immunize: Individual and Collective

In the world of therapeutics, patients are motivated to seek treatment when they experience a health problem, but vaccination requires that consumers take action prior to experiencing the problem. Awareness of the risks of infectious diseases and of the existence of a preventive solution is therefore of paramount importance. The perception of the benefits of prevention requires counterfactual thinking (Roese, 1997), and compared to therapeutics, vaccines are typically raising more concerns about their safety than receiving kudos for their efficacy.

Most vaccines require a limited number of interactions – and often very spaced out in time, except for the annual influenza vaccine – between the subject and the healthcare system. Consequently and as studied by many authors (Kennedy *et al.*, 2011), personal experience plays a very limited role in the decision to immunize, relative to other sources of information. The rise of antivaccination movements, spreading their messages via digital channels, is making the task of the healthcare community even harder.[4],[5]

In a world witnessing societal trends in favor of individualized medicine, the management of the perception of the benefits and the risks associated with vaccines is a challenge for the industry, the healthcare providers, and the authorities alike.

### 13.2.4 The Pivotal Role of Recommendations

Vaccines are expected to have a direct beneficial effect on the health of the immunized individuals. In addition, preventive vaccines targeting transmissible diseases are expected to yield a "herd immunity" by limiting interindividual transmission, hence reducing the risk of infection in nonimmunized subjects.

Therefore, when vaccines fail to deliver their expected benefits, the root cause can reside in a failure to protect (hence related to the efficacy of the vaccine at the level of the immunized person) and also in the failure to vaccinate (hence related to

3) http://www.bizjournals.com/seattle/blog/health-care-inc/2014/08/dendreon-stock-sinks-as-analysts-downgrade-on-debt.html?page=all (accessed 2014/12/07).
4) http://www.washingtonpost.com/blogs/wonkblog/wp/2014/05/05/how-the-anti-vaccine-movement-is-endangering-lives/(accessed 2014/12/07).
5) http://www.forbes.com/sites/robertpearl/2014/03/20/a-doctors-take-on-the-anti-vaccine-movement/(accessed 2014/12/07).

reaching the desirable coverage rate in the target population). This dual risk drives authorities' interventions, ranging from subsidies to mandatory vaccination, so as to maximize the odds of reaching targeted immunization coverage. In some countries, authorities also take responsibility for the distribution and administration of preventive vaccines, either through public-operated primary care centers or public health campaigns. Such campaigns – and immunization recommendations alike – are easier to implement successfully in certain age groups that have frequent interactions with healthcare systems, such as infants or elderly subjects. By contrast, adolescents and adults are notoriously difficult to immunize, and vaccination coverage in these age groups tends to be low, even in developed countries.

Most communicable diseases cannot be eradicated, due, for example, to the existence of animal reservoirs of the pathogen. Therefore, these diseases resurface when immunization pressure declines (e.g., measles in the United Kingdom, pertussis in France, etc.), and the authorities have to maintain their commitment once a public recommendation is issued and funded. The resulting long-term commitment (almost open ended) for payers drives downward price pressures on vaccines.

## 13.3 The Vaccine Market: Substantial, Fast Growing, with Intense and Concentrated Competition

For many years, the vaccine market has been attracting less attention than that of other primary care products, whether antibiotics, antihypertensives, or antidepressants. With the advent of newer technologies, combination products, and novel vaccines, the vaccine market has been steadily expanding and is now one of the top growing segments (alongside antivirals) outpacing the majority of the traditional therapeutic segments (including antihypertensives and the once fast-growing antihyperlipidemics), as illustrated in Figure 13.1 (Evaluate Pharma, 2014).

The vaccine market has reached a size which makes it attractive for the larger healthcare companies, as we will be discussing further in this chapter. The growth of this market segment is driven by a combination of factors. Firstly, the product mix is evolving, with novel vaccines launched against previously unaddressed medical needs. Secondly, the product mix is also evolving with the improvement on existing products, either by expanding the scope of the germs coverage (see, for instance, the evolution of the pneumo conjugate vaccine from 7, then 10, and currently 13 serotypes of *pneumococcus*) or the targeted age groups (here again, the pneumo conjugate vaccine is a good example, with the demonstration of the clinical efficacy across all age groups, from infants to elderly). Also, some improvement can be brought through combination vaccines, which contribute to reducing the number of injections – hence simplifying the immunization schedule – by offering coverage against several diseases in one shot. Thirdly, the expansion of the number of countries issuing recommendations for universal immunization

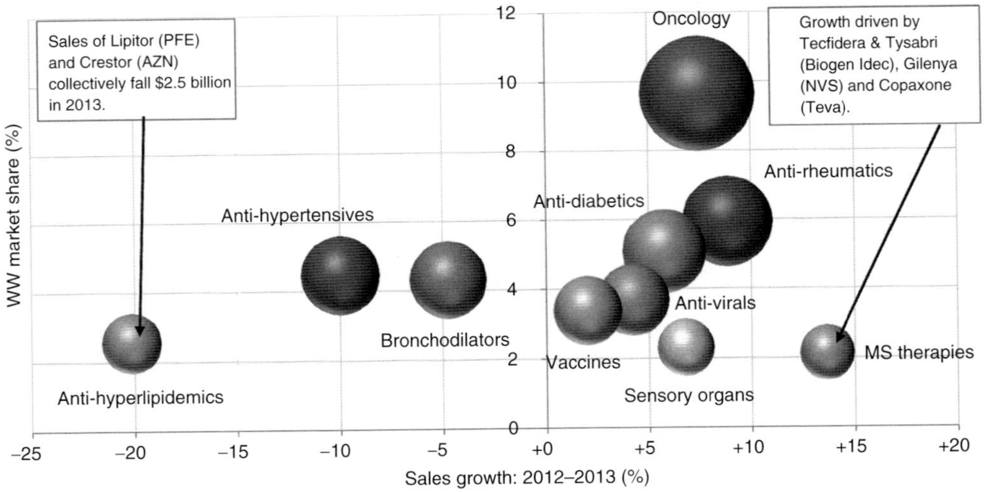


**Figure 13.1** Top 10 therapy sales in 2013, market share and sales growth (2012–2013). Source: Evaluate Pharma® (1 June 2014).

coverage against certain diseases is also contributing to volume growth. As the targeted diseases are transmissible and cannot be eradicated, the commitment taken by public health authorities is a long-term and open-ended one, hence contributing to the sustainability of the volume opportunity. In turn, this drives the public payers to look for cheaper alternatives as time is elapsing, hence opening an opportunity for biosimilar vaccines.

For the time being, the vaccine sales are generated predominantly in high-income countries. For the year 2008, published data (Morgon, 2011) indicated that as much as 77% of worldwide vaccine revenues were coming from the mature economies, which accounted for only 14% of the world population, but most importantly only a meager 8% of the world's birth cohort. The United States account for the largest share, followed by the European Union, Japan, and then by the two population heavyweights, namely, China and India.

Going forward, analysts anticipate that the vaccine market should be growing at +7% Compound Annual Growth Rate CAGR until 2020, as illustrated in Figure 13.2. We anticipate that the attraction of the vaccine segment will remain for the large healthcare multinational companies which have a competitive vaccine division, leading them to reinforce their presence. Also, the magnitude of the opportunity will entice smaller players to flex their muscles and compete in selected subsegments of the vaccine market.

## 13.4 The Vaccine Industry: Domination of the Heavyweights, for Now ...

The industry structure is evolving, driven by the attractiveness of the vaccine segment.

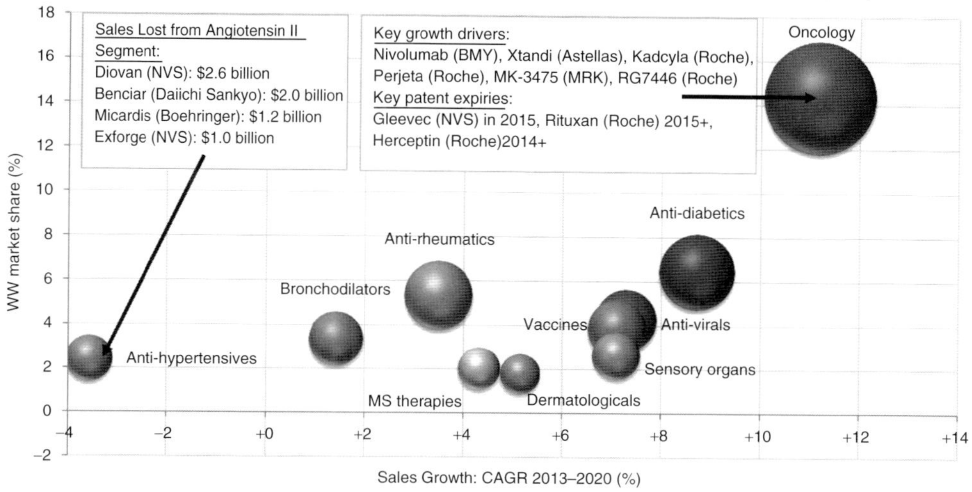


**Figure 13.2** Top 10 therapy areas in 2020, market share and sales growth. Source: Evaluate Pharma® (1 June 2014).

It should be noted that out of the top 20 healthcare companies, 8 have a vaccine division (see Table 13.1) (Evaluate Pharma, 2014). This pillar of the portfolio of the large players has been evolving substantially in the recent past.

Starting from the top of the chart, many strategic moves related to the vaccine franchises can be observed. The companies listed in Table 13.1 and not described in the following text do not have a vaccine division.

First, in spite of recent massive investments[6] and of the launch of its novel meningococcal B vaccine Bexsero,[7] the Novartis vaccine franchise wasn't performing up to the desired level. One of the root causes was the poor market performance of Bexsero[8] which failed to bring the vaccine division back to profitability. In line with previous comments from the CEO of Novartis, the company eventually sold most of its portfolio to GlaxoSmithKline (GSK[9]) in a rare product portfolio swap. It should be noted that Novartis retains its influenza franchise, in which it has been investing massively in recent years, with a flagship manufacturing unit dedicated to cell-cultured flu vaccines and jointly developed with the U.S. Department of Health and Human Services (HHS), Biomedical Advanced Research and Development Authority (BARDA).[10] (Note: Since this chapter was written, Novartis has sold its influenza vaccine business to CSL of Australia).

6) http://www.businessweek.com/globalbiz/content/nov2009/gb2009113_520982.htm (accessed 23/12/2014).
7) http://www.fiercevaccines.com/special-reports/novartis (accessed 23/12/2014).
8) http://www.reuters.com/article/2013/07/24/us-novartis-vaccine-bexsero-idUSBRE96N0JF20130724 (accessed 23/12/2014).
9) http://www.novartis.com/newsroom/media-releases/en/2014/1778515.shtml (accessed 23/12/2014).
10) http://www.novartis.com/newsroom/media-releases/en/2014/1793172.shtml (accessed 23/12/2014).

**Table 13.1** Worldwide ranking of healthcare companies according to 2020 projected prescription sales

| | | WW Rx sales ($billion) | | | WW market share | | | Rank |
|---|---|---|---|---|---|---|---|---|
| | **Company** | **2013** | **2020** | **CAGR (%)** | **2013 (%)** | **2020 (%)** | **Change (+/−)** | **Change (+/−)** |
| 1 | Novartis | **46.0** | **54.4** | +2 | 6.4 | 5.3 | −1.1pp | +0 |
| 2 | Roche | **39.1** | **52.4** | +4 | 5.5 | 5.1 | −0.3pp | +1 |
| 3 | Sanofi | **37.7** | **50.0** | +4 | 5.3 | 4.9 | −0.3pp | +1 |
| 4 | Pfizer | **45.0** | **47.8** | +1 | 6.3 | 4.7 | −1.6pp | −2 |
| 5 | GlaxoSmithKline | **33.1** | **41.2** | +3 | 4.6 | 4.1 | −0.6pp | +1 |
| 6 | Merck & Co. | **37.5** | **39.4** | +1 | 5.2 | 3.9 | −1.4pp | −1 |
| 7 | Johnson & Johnson | **26.5** | **35.9** | +4 | 3.7 | 3.5 | −0.2pp | +0 |
| 8 | AstraZeneca | **24.5** | **26.0** | +1 | 3.4 | 2.6 | −0.9pp | +0 |
| 9 | Novo Nordisk | **14.9** | **25.8** | +8 | 2.1 | 2.5 | +0.5pp | +5 |
| 10 | Gilead Sciences | **10.8** | **23.7** | +12 | 1.5 | 2.3 | +0.8pp | +8 |
| 11 | AbbVie | **18.8** | **23.2** | +3 | 2.6 | 2.3 | −0.3pp | −1 |
| 12 | Bayer | **15.6** | **22.4** | +5 | 2.2 | 2.2 | +0.0pp | +1 |
| 13 | Bristol-Myers Squibb | **12.3** | **21.6** | +8 | 1.7 | 2.1 | +0.4pp | +4 |
| 14 | Amgen | **18.2** | **20.9** | +2 | 2.5 | 2.1 | −0.5pp | −3 |
| 15 | Takeda | **13.4** | **18.4** | +5 | 1.9 | 1.8 | −0.1pp | +1 |
| 16 | Eli Lilly | **20.1** | **18.1** | −2 | 2.8 | 1.8 | −1.0pp | −7 |
| 17 | Boehringer Ingelheim | **14.5** | **15.6** | +1 | 2.0 | 1.5 | −0.5pp | −2 |
| 18 | Teva Pharmaceutical Industries | **17.6** | **15.2** | −2 | 2.4 | 1.5 | −1.0pp | −6 |
| 19 | Astellas Pharma | **10.3** | **14.5** | +5 | 1.4 | 1.4 | −0.0pp | +0 |
| 20 | Biogen Idec | **5.4** | **13.0** | +13 | 0.8 | 1.3 | +0.5pp | +8 |
| | Total top 20 | 461.3 | 579.5 | +3 | 64.3 | 57.0 | −7.3pp | |
| | Other | **256.2** | **437.9** | +8 | 35.7 | 43.0 | | |
| | Total | 717.5 | 1017.4 | +5 | 100.0 | 100.0 | | |

Source: Evaluate Pharma® (1 June 2014).

Second, Sanofi Pasteur is the fully owned subsidiary of Sanofi. It expanded its capacity and presence in the low- to middle-income segment with the acquisition of Shantha Biotechnics in 2009[11] and invested massively to bring to market the first ever dengue fever vaccine[12] while maintaining its extended portfolio of vaccines. As illustrated by recent corporate communications from Sanofi,[13] vaccines remain a strategic pillar in the sustainable development of the group, and the removal of the CEO in October 2014 is unlikely to substantially alter this strategic orientation.

11) http://www.fiercebiotech.com/story/sanofi-snaps-indias-shantha-784m/2009-07-27 (accessed 23/12/2014).
12) http://www.sanofipasteur.com/en/media/press_tool_kits/focus_on_dengue/ (accessed 23/12/2014).
13) http://en.sanofi.com/press/press_releases/2014/news_list_2014.aspx (accessed 23/12/2014).

Fourth, Pfizer enjoys a unique position among the vaccine companies. It has managed to bring the pneumo conjugate – which was part of the portfolio of Wyeth which Pfizer acquired in 2009 – to super-blockbuster status. It should be reminded that Wyeth once was a vaccine player with a diversified portfolio and yet it made the strategic decision to focus almost exclusively on pneumo conjugate.[14] Today, Prevnar 13 is the largest vaccine in terms of revenues (by a wide margin), and it is still growing due to the expansion of coverage across age groups and countries.

Moreover, Pfizer acquired the vaccine activity of Baxter (except influenza, in an interesting similarity with the GSK–Novartis deal) in 2014[15] in a move which broadens the scope of its portfolio toward viral vaccines and strengthens its manufacturing footprint in Europe.

Fifth, GSK is now the undisputed leader of the vaccine market. Prior to the acquisition of most of the Novartis portfolio in 2014 (see preceding text), it already enjoyed a dominant share of the vaccine market. The acquisition boosted the share of vaccines in GSK total revenues[16] from 9 to 14% – hence reinforcing the strategic importance of this franchise – and the GSK share of the vaccine market by value from 24 to 29%.

Going forward, it will be very interesting to observe how GSK will manage its much diversified portfolio of vaccines and the choices that it will be making to preserve, divest, or discontinue certain products.

Sixth, Merck remains a strong contender in the U.S. vaccine market and is still linked to Sanofi Pasteur through a joint venture established in 1994 in 15 European countries, which is a unique example of a situation actually impacting the competitive intensity in that geography.

Merck remains focused on vaccines as a strategic part of its portfolio[17] with a particular emphasis on viral vaccines.

Seventh, J&J made a strategic move in the vaccine market when it acquired Crucell in 2011[18] and has reconfirmed its intention to play a pivotal role in this market, as illustrated by its commitment to the research of an Ebola vaccine.[19]

Eighth, AstraZeneca owns only one vaccine, the intranasal influenza vaccine FluMist – which it acquired as part of the MedImmune assets in 2007 – without any apparent willingness to divest this product, when the strategic orientation of MedImmune is to become the R&D arm of AZ, exploring biologics in the fields

14) http://download.bioon.com.cn/upload/month_0810/20081024_92e7416b7af8d956eb3ezk2JE5VS4FhU.attach.pdf (accessed 23/12/2014).

15) http://www.pfizer.com/news/press-release/press-release-detail/pfizer_completes_acquisition_of_baxter_s_marketed_vaccines (accessed 23/12/2014).

16) http://www.gsk.com/media/280221/novartis-transaction-investor-slides.pdf (accessed 23/12/2014).

17) http://www.businesswire.com/news/home/20131001005957/en/Merck-Announces-Global-Initiative-Sharpen-Commercial-Focus#.VJrIkv9Ak (accessed 23/12/2014).

18) http://www.biopharminternational.com/johnson-johnson-acquires-vaccine-developer-crucell (accessed 22/12/2014).

19) http://www.jnj.com/news/all/Johnson-Johnson-Announces-Major-Commitment-to-Speed-Ebola-Vaccine-Development-and-Significantly-Expand-Production (accessed 22/12/2014).

of respiratory, inflammation, and autoimmunity.[20] Going forward, AZ may opt to divest FluMist if it remains the sole vaccine asset of the whole portfolio.

Further down in the chart of global healthcare companies, Takeda occupies the 15th position. It launched its vaccine franchise in 2012,[21] and since then, it has essentially been working on its organization and on focused R&D programs[22] in an attempt to expand in Europe and other markets.

Lastly, Astellas is also showing some interest in this market as it has been inking deals to build a vaccine franchise.[23]

Overall, the market remains extremely concentrated, with a domination of the top historical players, namely, GSK, Sanofi Pasteur, Merck, and Pfizer. This level of concentration is much higher than the share of the global pharmaceutical share of the top healthcare companies, but the comparison would be more appropriate with single therapy areas, where concentration is typically very high, as it is in specific vaccine franchises.

From the profitability standpoint, the incumbents leverage their established manufacturing capacity to maintain operating profit margins about 30%, which is comparable to the average margin of pharmaceutical companies. The higher cost of goods sold (COGS) of vaccines – which are linked to the capital intensiveness, the complexity of the manufacturing process requiring skilled labor, and the stringent manufacturing quality requirements – are offset by substantially lower commercialization costs resulting from a concentration of buyers: as much as 50% of the worldwide vaccine sales result from public bodies' purchases.

For the aforementioned reasons, incumbents with a satisfactory level of utilization of their industrial capacity, yielding a positive return on invested capital, and with a robust commercial network enjoy a strong competitive advantage.

As illustrated in Figure 13.3 and Table 13.2, analysts are forecasting a persistence of the status quo with the sustained preeminence of the existing larger players, capitalizing on their broad portfolios and on their extended industrial and commercial footprints.

Yet, and notwithstanding the inevitable forecasting errors linked to the time span of the projection and the diversity of the underlying hypotheses on the vaccine candidates of each of the portfolios (usually, analysts vastly underestimate the inertia in vaccine R&D), we'd like to argue that the forecast is not accounting for a number of existing and brewing changes as of 2014.

As a matter of fact, some 50 other vaccine manufacturers are located mostly in low- to middle-income countries. Although they generate a relatively small portion of the worldwide vaccine sales, they account for the largest share in volume

20) https://www.cambridgenetwork.co.uk/directories/companies/1322/products/1584/ (accessed 22/12/2014).
21) http://www.takeda.com/company/reports-publications/2013_ebook/2013ar/index.html#page=29 (accessed 22/12/2014).
22) http://www.fiercevaccines.com/story/takeda-reorg-preps-vaccines-unit-emerging-markets-test/2014-09-18 (accessed 22/12/2014).
23) http://www.fiercevaccines.com/story/astellas-inks-deal-create-portfolio-vaccine-focused-biotechs/2014-01-06 (accessed 2014 12 06).

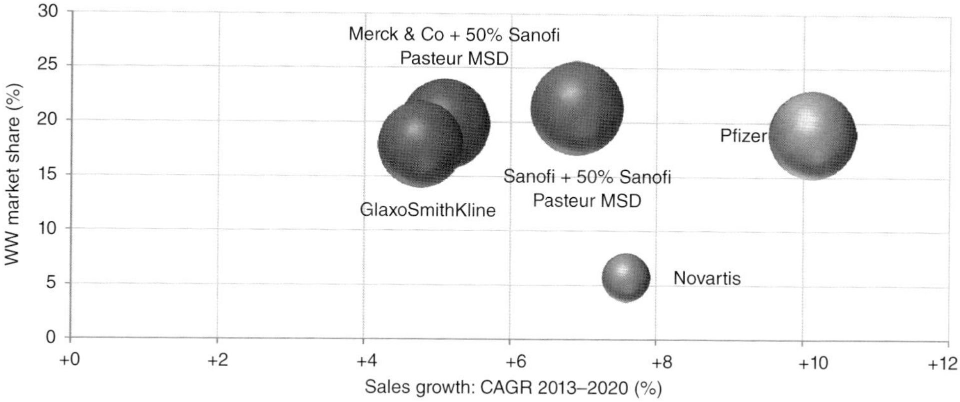


**Figure 13.3** Worldwide sales, market share, and sales growth (2013–2020). Note: Bubble = WW sales in 2020. Source: Evaluate Pharma® (1 June 2014).

**Table 13.2** Top 10 companies and total worldwide vaccine sales 2013–2020.

| Rank | Company | WW sales ($ million) 2013 | WW sales ($ million) 2020 | CAGR 2013–2020 (%) | WW market share 2013 (%) | WW market share 2020 (%) | Rank Change 2013–2020 |
|---|---|---|---|---|---|---|---|
| 1 | Sanofi + 50% Sanofi Pasteur MSD | 5 518 | **8 805** | +7 | 21.6 | 21.3 | +1 |
| 2 | Merck & Co + 50% Sanofi Pasteur MSD | 5 766 | **8 167** | +5 | 22.6 | 19.8 | −1 |
| 3 | Pfizer | 3 974 | **7 808** | +10 | 15.5 | 18.9 | +1 |
| 4 | GlaxoSmithKline | 5 351 | **7 412** | +5 | 20.9 | 17.9 | −1 |
| 5 | Novartis | 1 416 | **2 363** | +8 | 5.5 | 5.7 | +0 |
| 6 | Novavax | — | **946** | n/m | n/m | 2.3 | New |
| 7 | Johnson & Johnson | 346 | **913** | +15 | 1.4 | 2.2 | +1 |
| 8 | Emergent BioSolutions | 247 | **639** | +15 | 1.0 | 1.5 | +4 |
| 9 | CSL | 434 | **464** | +1 | 1.7 | 1.1 | −3 |
| 10 | Astellas Pharma | 350 | **414** | +2 | 1.4 | 1.0 | −3 |
| | Top 10 | 23 401 | **37 929** | +7 | 91.5 | 91.8 | |
| | Other | **2 166** | **3 391** | +7 | 8.5 | 8.2 | |
| | Total industry | 25 567 | **41 320** | +7 | 100.0 | 100.0 | |

Source: Evaluate Pharma® (1 June 2014).

(as much as 86% according to 2009 data from the World Health Organization (WHO), the United Nations Children's Fund (UNICEF), and the World Bank). As previously mentioned, most of these players cater only to their domestic markets and are often fully owned by the local government. Yet, some have taken steps to upgrade their quality and obtain the WHO Prequalification. For instance, the strengthening of the manufacturing quality and capacity of the Indian and Chinese manufacturers doesn't seem fully taken into consideration by the analysts. The likes of Serum Institute of India, Biological Evans, Bharat Biotech, China National Biotech Group (CNBG), and Sinovac Biotech are flexing their muscles, improving on breadth of portfolio and manufacturing quality, and squarely aiming at expanding beyond their national borders and low-income market segments (Francis *et al.*, 2014).[24),25)]

In addition, technology transfers from multinational players to emerging country manufacturers have been shown to be effective and efficient tools to reduce product development cost and shorten the time-to-market. A good example is in the case of *Haemophilus influenzae B* (Hib) technology, which had a major global impact and could be applicable to other vaccines (Hamidi *et al.*, 2014) since the access to Hib conjugate technology is the first step toward other conjugate vaccines.

Therefore, we anticipate that the competitive intensiveness of the vaccine market in low- to middle-income segments will only intensify both in terms of pricing and of innovativeness of the vaccine offering.

Moreover, analysts' perception of the products that will dominate the market in 2020, as indicated in Table 13.3, underscores the deeply rooted belief that recommendations will drive volume and therefore commercial success, as has been the case in the past. Yet, we posit that this is valid for transmissible diseases and the dogma should start to show cracks on the surface when the market migrates progressively to therapeutic vaccines or to vaccines addressing infectious diseases such as community-acquired infections.

Other than the specific case of influenza vaccines where supply vastly outstrips demand (as a consequence of the massive buildup of capacity in the wake of the influenza pandemics), the vaccine market segments are demand driven. Vaccines used to be perceived as high volume/often low price – and therefore limited global revenues – products. Recent evolutions have demonstrated the ability of some vaccines to obtain the membership in the coveted club of blockbuster products, as epitomized by Pfizer's Prevnar (although the merit goes to Wyeth, as indicated earlier) and followed by Merck's Gardasil.

In the 2013 global ranking of worldwide product sales (Evaluate Pharma, 2014), Pfizer's Prevnar is ranked 21st (immediately below Roche & Novartis' Lucentis and well above some well-known brands such as Gilead's Truvada or J&J's Velcade), and Merck's Gardasil ranked 43rd, a tad below Eli Lilly's Cialis and Allergan's Botox. By 2020, analysts are forecasting a further expansion of Prevnar sales to over USD 6 billion (representing a CAGR of +7%), taking it to the 12th position in the global ranking of worldwide product sales.

Meanwhile, the same analysts are forecasting that Merck's Gardasil will exit the ranking of the top 50 worldwide product sale (Evaluate Pharma, 2014).

24) http://www.nbr.org/publications/CHA/PHS11_China_Report.pdf (accessed 22/12/2014).
25) http://www.fiercevaccines.com/story/indias-biological-e-pentavalent-vaccine-takes-gsk-jj/2013-04-18 (accessed 22/12/2014).

**Table 13.3** Top 5 vaccine products worldwide in 2020.

| Rank | Product | Generic name | Company | WW sales ($million) | | CAGR | WW market share | | Current status |
|---|---|---|---|---|---|---|---|---|---|
| | | | | 2013 | 2020 | 2013–2020 (%) | 2013 (%) | 2020 (%) | |
| 1 | Prevnar 13 | Pneumococcal vaccine | Pfizer + Daewoong | 3763 | **6078** | +7 | 14.7 | 14.7 | Marketed |
| 2 | Pentacel | DTaP, Hib, and polio vaccine | Sanofi | 1525 | **2343** | +6 | 6.0 | 5.7 | Marketed |
| 3 | Gardasil | Human papillomavirus (HPV) vaccine | Merck & Co. | 1831 | **2055** | +2 | 7.2 | 5.0 | Marketed |
| 4 | Fluzone/ Vaxigrip | Influenza vaccine | Sanofi | 1234 | **1822** | +6 | 4.8 | 4.4 | Marketed |
| 5 | Pediarix | DTaP, hepatitis B, and polio vaccine | GlaxoSmithKline | 1349 | **1726** | +4 | 5.3 | 4.2 | Marketed |

We would suggest a more nuanced analysis, factoring in the recent expansion of the serotype coverage as well as the scope of indications of the nine-valent Gardasil to include additional cancers linked to the human papillomavirus (HPV).[26)]

Also, we hypothesize that vaccines targeting nosocomial infections will generate blockbuster-level revenues, in spite of relatively limited volumes, due to the health economic rationale supporting their use that will support high prices, as currently witnessed in therapeutic fields such as oncology or antivirals.

Other vaccines against transmissible diseases such as Sanofi Pasteur's dengue vaccine should also hover around the $1 billion mark due to a large target population in countries with a lasting commitment to public immunization. Evidently, the propensity to sell into public markets at discounted prices, combined with mounting competitive pressures enticed by the suboptimal efficacy profile of the Sanofi Pasteur dengue vaccine, can conjure up to limit its revenue opportunity.

### 13.4.1 Barriers to Entry: From R&D Risk to Capital Intensiveness

The vaccine market is one of high entry barriers, especially in the mature economies, both for novel and me-too vaccines.

The intellectual property (IP) rights of specific antigens, adjuvants, and technologies (e.g., conjugation) do play a role, yet IP is not as pivotal as in the field of small molecules. The most important barrier to entry is rather the manufacturing know-how, due to the complexity of the production processes and to a lesser extent to the clinical and regulatory prowess.

For the time being, vaccine manufacturing facilities have been built to yield high volumes at competitive costs to offset the impact of the massive capital expenditures, leading to a situation where a minimum efficient scale acts as a barrier to entry. Vaccines being biological products, me-too followers cannot resort to the cheaper abbreviated new drug applications, resulting in limited regulatory and clinical savings. Also, in the case of vaccines for which a correlate of protection hasn't been established between antibody titers (immunogenicity) and clinical protection (efficacy), the size of the phase III clinical trials is now matching that of small molecules in primary care therapeutic indications, hence resulting in massive clinical development costs. This is especially visible with pediatric vaccines, one example being the *rotavirus* vaccine Rotateq for which the combined pre- and postlicensure studies enrolled a whopping 70 000 children (although arguably subsequent to the market withdrawal of a previous rotavirus vaccine over safety concerns).

The combination of the above factors results in long life cycles for vaccines, which is unique in the world of medical solutions.

26) http://www.fda.gov/NewsEvents/Newsroom/PressAnnouncements/ucm426485.htm (accessed 22/12/2014).

### 13.4.2
### Five Forces Analysis: Competitive Intensiveness and Downstream Hurdles

The vaccine market is not subject to strong upstream or lateral forces (Angelmar and Morgon, 2014). The only suppliers that wield some bargaining power are those contributing to the manufacturing prowess, for example, by controlling a critical part of the process thanks to dedicated equipment. There are few substitutes to vaccines: some are of a preventative nature and aim at reducing the exposure to the infectious agent, and others are essentially therapeutic approaches to deal with the infection, the latter being not always available (e.g., in the case of rabies).

Therefore, the vaccine market is characterized by an intense competitive situation linked to the conditions described earlier and by downstream hurdles to reach the end users. Owing to the fragmentation of the purchasing process, several customer types have to be considered.

The vaccine markets range from private (prescribing pediatricians or other healthcare providers and end users paying out of pocket) to public (governments or international nongovernmental organizations (NGOs) purchasing globally to execute an immunization program). The proportion varies across countries and antigen types and is essentially linked to the scope of the public immunization recommendations. In other words, once a public health authority recommends the use of a given vaccine and funds the purchase, the corresponding market transitions from a lower volume, higher price private usage to a larger volume, lower price public usage (sometimes with a residual private market, typically catered to with a different vaccine presentation).

As the purchasing process is extremely fragmented between prescribers, end users, payers, and influencers, the drivers of the decision to immunize are complex and driven by the acceptability, accessibility, availability, and affordability of the vaccines.

#### 13.4.2.1
#### Acceptability

The registration of vaccines by regulatory authorities is only the first acceptability hurdle that vaccine manufacturers have to clear (arguably the most important and costly one). Subsequently, vaccine companies need to convince prescribing physicians to engage in a vaccination discussion with the end users and also to national immunization advisory bodies to recommend the said vaccine. Prescribers range from vaccine advocates to vaccine cautious, and in general terms, they recommend more actively well-established vaccines that have demonstrated benefits (Angelmar and Morgon, 2014), and their prescription behavior is also driven by sociodemographic and social psychological variables, ranging from intrinsic religiosity, self-description as being conservative in addition and perceived self-efficacy in engaging in vaccination discussions, to the understanding of the disease and the intent to immunize oneself of family members. Lastly, end users may have objections to vaccination, linked to philosophical, religious (Grabenstein, 2013), or personal beliefs. Since these have been increasing, fueled by antivaccine movements issuing generally unsubstantiated writings, innovation resulting in lowering such mental barriers would vastly increase the public health impact of immunization solutions, and healthcare providers have been recognized as having a pivotal role in imparting information and advice to the public on the vaccines promotion and are models for patients and parents in particular, even though vaccination coverages of Health Care Provider HCPs remain relatively low (Tafuri *et al.*, 2014). Also, immunization happens in discrete and often spaced-out events which are not conducive to high levels of treatment adherence. Several ways to address this issue and the resulting failure to immunize have been tested to boost awareness and commitment to action, both at the end user and at the provider level (Wallace *et al.*, 2014), ranging from simple vaccination cards to more digital solutions, including texting and other web-based approaches.[27),28),29)]

The accelerators of the acceptability to immunize include the perception of a threat linked to a disease, which is positively correlated with the prevalence (Philipson, 2000) and the perceived severity of the disease, as well as to an individual perception of one's susceptibility to that disease. The perceived benefits and risks associated with the vaccine also play a pivotal role. For instance, one possible obstacle to greater immunization rates is the false belief that it is possible to contract the disease from the vaccine, as recently demonstrated by the representative survey experiment that was conducted in the United States to assess the extent of the flu vaccine misperception (Nihan *et al.*, 2015). These are typically assessed at a different level between the end users – focusing on the individual benefits, that is, achieving protection against the disease while avoiding adverse effects – and the providers and the authorities, often placing more emphasis on the collective benefit linked to blocking the transmission of the disease in the target population. Lastly, various norms ranging from subjective ones such as the influence of the end users' referents to objective ones such as the existence of mandatory immunization guidelines contribute to influencing the acceptability of immunization.

The various organizations issuing vaccination recommendations, including national or supranational government authorities, national immunization technical advisory bodies, medical and scientific societies, and even private insurers, evidently contribute to accelerating immunization uptake and coverage.

#### 13.4.2.2
#### Accessibility

In countries where the primary care procurement system is well developed, the opportunities to immunize are theoretically numerous, provided the end users undertake wellness visits rather than acute care ones. And even in such instances, the interaction with the healthcare provider doesn't always result in a discussion

27) http://www.ncbi.nlm.nih.gov/pubmed/21159080 (accessed 26/12/2014).
28) http://www.cdc.gov/vaccines/pubs/pinkbook/downloads/strat.pdf (accessed 26/12/2014).
29) http://www.clinicaladvisor.com/text-message-reminders-may-improve-vaccine-adherence/article/338060/ (accessed 26/12/2014).

on immunization for the previously described reasons. In less developed countries, the patchy and often remote primary care infrastructure (Usman *et al.*, 2010) and erratic supply of electricity, impacting negatively on the ability to maintain a cold chain that is reliably robust and routinely monitored (Smith *et al.*, 2011), often result in poor accessibility to immunization.

#### 13.4.2.3 Availability

Results from the combined action of purchasing and selling decisions, the former being controlled by public and private purchasers and the latter resulting from procurement policies of vaccine manufacturers.

Once a vaccine has been recommended by a national or supranational influencing body, the odds of public purchases increase commensurately and are further increased if public or political pressure adds momentum. Moreover, such recommendations vastly reduce the financial risk derived from a gap between demand and supply, since public, recommendation-backed demand is much easier to gauge. Therefore, the availability of vaccines in the public market segments is essentially driven by demand backed by public authorities' decisions and commitment to fund the corresponding healthcare expenditure. Yet, it should be reminded that both the recommendation by National Immunization Technical Advisory Group (NITAG) and the funding policy decision are country-specific market access processes. A recent study compared NITAGs of 13 developed countries by describing vaccination committees' bodies and working processes (Ricciardi *et al.*, 2015). This study supports previous findings about the disparities in NITAG processes which could potentially explain the disparity in access to vaccinations and immunization programs across countries. The NITAG recommendations are used by policy decision makers for implementation and funding of vaccine programs, and vaccine manufacturers are expected to submit data-backed dossiers for NITAGs to issue recommendations. The heterogeneity of the decision analysis frameworks and processes could be contributing to the credibility concern among the public and healthcare professionals.

At the other end of the market spectrum, the availability of vaccines in the private market segments depends on the commercial strategies of vaccine manufacturers. In fact, beyond the target end users that are defined by the medical need and the clinical evidence supporting a vaccines' usage, vaccine companies can decide in which country they want to operate as well as which subsegment in one country and even on the product characteristics, ranging from the formulation to the presentation. Therefore, the most sophisticated combination products and the most expensive presentations (such as the prefilled syringes) are typically sold in affluent market segments, while less affluent ones are addressed with less sophisticated formulations in simple vials.

#### 13.4.2.4 Affordability

For prescribers and end users alike, perceived affordability results from a combination of financial and nonfinancial costs, including the time needed to consult and be vaccinated. Healthcare providers are concerned with the financial outcomes of their vaccination decisions, whether they prescribe or dispense vaccines. In the latter case, the costs associated with managing an inventory of refrigerated – or even frozen – products with an expiry date can play a significant role, and physicians often prefer to concentrate their purchasing on one of the vaccines per disease or medical service rendered. In addition to the cost to their practice, one additional barrier for prescribers to advocate vaccination is the existence of a significant out-of-pocket payment for the end user.

When vaccines are purchased by public bodies, the commitment to endless funding (as most infectious diseases cannot be eradicated) drives public buyers to put pressure on the price of the novel vaccine as well on those of the older ones, so as to reallocate budgetary resources. In low-income countries, donor (e.g., the Global Alliance for Vaccines and Immunization (GAVI)) financing is pivotal, and these bodies have developed procurement processes including increased visibility and reliability of the demand forecasts, so as to increase the attractiveness of this market segment and augment the availability of vaccines (reliability of supply).

Overall, prices of vaccines vary greatly across countries and also between subsegments in the same country, and tiered pricing is commonly practiced as well as encouraged to address affordability (i.e., ability to pay) and access issues[30] as well as willingness to pay (correlated with cost-effectiveness criteria – used in most European countries – which are often incomplete or unreliable at the time of licensure of a new vaccine due to the lack of available reliable information (Black, 2013) and are also incompletely capturing the unique benefits of vaccines, such as the herd immunity). The latter point is a key issue for vaccine manufacturers, as described in a recent study (Laxminarayan *et al.*, 2014) which concluded that vaccinations have important nonhealth consequences for productivity and economic well-being and that the value of statistical life (VSL) is a relevant measure of willingness to pay for small reductions in the probability of death through interventions such as vaccines while typical health measures like disability-adjusted life years (DALYs) and quality-adjusted life years (QALYs), widely used to gauge the value of therapeutic solutions, do not capture the full long-term benefits of vaccines; the recommendation is that VSL can be used as complement to DALYs and QALYs to better inform priority setting across vaccines. Other authors also recommend the use of vaccine preventable disease incidence (VPDI) as a better measure of a vaccine's public health value, since vaccine efficacy does not reflect underlying disease incidence and vaccines with modest efficacy can have high VPDI and thus public health value (e.g., rotavirus vaccine). It is suggested that VPDI be considered as a key measure by vaccine decision makers and regulatory agencies (Gessner *et al.*, 2014).

Lastly, some price differences may reflect different costs related to supply chain, product liability, contractual terms, and so on.

30) http://www.ifpma.org/fileadmin/content/Global%20Health/Vaccines/Vac123-F_20130904_IFPMA_Position_on_tiered_pricing_for_vaccines.pdf (accessed 26/12/2014).

## 13.5
## New Vaccine Developments: Strategic Trends and Why Innovation Is Needed All along the Value Chain

As in any market and despite the value residing in commercializing me-too products, vaccine manufacturers are pursuing differentiation strategies. Some are product based and include novel vaccines addressing unmet medical needs, as well as "me-better" vaccines, aiming at improving efficacy, safety, convenience of use, or supply chain requirements. Besides, manufacturers strive to ensure reliability of supply, since meeting this expectation – especially from the perspective of public purchasers – provides a clear competitive advantage. Evidently, these differentiation strategies are proving effective when the willingness to pay remains unaffected.

Moreover, manufacturers producing in countries with lower cost of labor tend to compete on the basis of price, provided their level of quality and the reliability of their supply are deemed on par with those of international players. The latter have been trying to offset this threat through localization of production units in lower-income countries, as well as through partnerships with or acquisition of lower-cost manufacturers.

Several strategic trends are likely to coexist and to shape the global vaccine market.

First and impacting the existing products, there is already a visible trend toward commoditization, characterized by a strong pressure on the cost of vaccines, driven by international aid donors and public purchasers, as well as by local players extending their reach beyond their markets of origin and becoming global challengers. All vaccine players remaining in the existing vaccine market segments will need to execute successfully cost-minimization strategies (such as localizing production in low-cost countries or cheaper delivery devices) and to shape future demand while designing improved products to cost less, both at manufacturing and commercialization levels.

Second and in sync with major societal trends, the vaccine market is evolving toward consumerism. The driving forces include affordability and willingness to pay, with an increase in individual decision power as out-of-pocket costs are on the rise and easier access to vaccine is made available. Also, more effective diagnostic tools and better information systems are acting as accelerators of "self-vaccination," or rather "self-vaccine prescription" – until actual self-vaccination becomes a reality. Vaccine companies will need to think in terms of new value propositions, appealing to individual end users while balancing the collective benefit and simultaneously to boost their abilities to interact directly with end users, ranging from customer intimacy to branding and advertising.

Third, there will be an inexorable expansion of vaccine usage in new diseases and disease areas, fueled by the novel findings in immunology and the shift in demographics and epidemiology. The new applications will emerge for both infectious diseases and nontransmissible ones, driving the vaccine manufacturers to improve on existing products while outsourcing R&D to explore adjacent fields such as monoclonal antibodies and vector-based approaches, as well as grooming their ability to shape the regulatory and public health environments.

Fourth and characteristically for heavily controlled markets, the apparently antithetic and yet simultaneous forces of globalization and deglobalization will be felt. While there will be a convergence of regulatory standards (even if some countries may be tempted to use their local pharmacopeias as barriers to entry of foreign players), there will be an increasing differentiation of needs between regions and countries. The market actors will be driven to explore novel solutions allowing versatile and localized manufacturing, as well as enhanced market intelligence capabilities and know-how to shape product acceptability at regulatory and end-user levels.

Vaccine innovation will be needed all along the value chain and should trigger such substantial changes that some of the existing players will require a complete reboot, if not a reprogramming of their corporate DNA.

### 13.5.1
### Where Is Innovation Needed? R&D

Many transmissible diseases are satisfactorily addressed and have been brought to the verge of eradication, or at least to oblivion, by existing vaccines. Yet, new targets abound, as defined by unmet medical needs where stimulating an immune response from a person would bring a valuable medical response. Is there a consensus on the remaining "big opportunities" for preventative vaccines? Probably not.... As the new targets are more elusive and development timelines and costs tend to become similar to those of small molecules (without the benefit of the regulatory accelerated pathways that have been put in place for orphan drugs or oncology therapies), the attractiveness and market access feasibility of novel vaccine solutions are not ranking high in R&D portfolio discussions. And yet the authorities remain more involved in the clinical development of vaccines than of therapeutics because of the expected impact in public health. Other than the ongoing efforts to target AIDS, malaria, and resistant tuberculosis, the disappearance of large-scale scourges is likely to limit the number of NGO-coordinated public–private partnerships in R&D, while the private sector development should dominate.

As indicated, new targets could be mapped according to attractiveness and feasibility (the latter may differ according to the capabilities of each vaccine player), and low attractiveness may be offset by public incentives, while feasibility is contingent upon assembling the relevant scientific capabilities.

Some examples of high attractiveness and high feasibility novel targets include nosocomial infections (with existing projects on *Clostridium difficile* and *Pseudomonas aeruginosa*, yet much less progress on *methicillin-resistant Staphylococcus aureus* (MRSA) and *Escherichia coli* suggesting that feasibility may vary within this targeted segment). The high feasibility but low attractiveness targets include diseases that could be addressed with existing technologies but with a low disease burden or unattractive epidemiology, such as hepatitis E (such a vaccine has been developed by a Chinese manufacturer) or *Salmonella paratyphi*.

R&D efforts are more likely to be directed toward the "high risk, high reward" bets characterized by a low feasibility but a high attractiveness, such as HIV, but also metabolic disorders (e.g., diabetes type 1, obesity) or various types of cancers.

Lastly, targets that will be categorized as "low attractiveness and low feasibility" are unlikely to receive much funding, except if incentives and funding mechanisms can be organized. For instance, these would include *Helicobacter pylori* or *E. coli* causing urinary tract infections (as gastric ulcer and urinary tract infections are well managed with therapeutic solutions available as affordable generics) or *Hantavirus* (low prevalence).

New vaccine constructs, such as viruslike particles (VLPs), are also investigated. VLPs are designed to mimic the actual virus in form except that they do not contain any virus genome. This is a way of making virus particles which are both effective – since they present the epitopes on their surface – and safe as there is no possibility of infection from the vaccine. Furthermore, multiple copies of the same epitopes are found in repeating structures on the virus surface, thus making virus and VLPs more "immunogenic" or more effective.

Another current focus of R&D programs is on improving existing vaccines, for example, due to incomplete serotype coverage such as in the case of dengue, or expanding the duration of immunity (almost all current vaccines require booster vaccination in order to achieve durable protective humoral immune responses, regardless of the vaccine type). Recent findings (Slifka *et al.*, 2014) indicate that the form of the vaccine antigen (e.g., soluble or particulate/aggregate) appears to play an important role in determining immunogenicity and the interactions between dendritic cells, B cells, and T cells in the germinal center are likely to dictate the magnitude and duration of protective immunity. Understanding these mechanisms could lead to the development of vaccines offering a more durable immunity.

Improving on existing vaccines also encompasses the reduction of the number of injections to improve convenience and reduce the vaccine burden (especially for infants) through the demonstration of more effective immunization regimens requiring less injections. Conjugation of existing antigens is also being explored to improve efficacy – especially in immature immunological subjects such as infants – and to optimize COGS. And formulations are explored to enhance heat stability in order to alleviate supply chain requirements.

Other drivers of vaccine R&D activity include the need for expansion of coverage in the case of an epidemiology evolution (e.g., meningococcal meningitis) or when serotype replacements occur due to vaccine pressure (as it has been demonstrated further to pneumococcal conjugate immunization) or when additional epidemiology or clinical data demonstrate the role of additional serotypes in the disease (e.g., HPV). Also, addressing ailments of other age groups, especially adolescents and adults/seniors, can drive incremental clinical development, as epitomized by the progressive development of the scope of indications of Prevnar.

Alternative immunization routes are also investigated, notably to address needle phobia. One such route is the intranasal administration, especially for influenza (Maltais *et al.*, 2014); the sublingual route had also been recently investigated with promising results (Gallorini *et al.*, 2014). These alternative administrations would help to expand volume and allocate immunization workload to other HCPs such as pharmacists and possibly to the end users themselves.

Meanwhile, therapeutic vaccines are still in the making, especially in oncology, despite the massive failure of Dendreon's Provenge.[31] Eventually, therapeutic vaccines will prove sustainable when the logistics and processes involved in their manufacturing prove to represent a scalable business model.[32]

Recent research findings (Strioga *et al.*, 2014) indicate that there is a close and bidirectional interaction between cancer and the immune system: the immune system recognizes tumor-associated antigens expressed by malignant cells. Tumor antigens are self-protein derived and possess the potential of immune tolerance, and homologous xenoantigens have signs of "nonself" enabling them to break tolerance. The authors conclude that it is reasonable to include xenovaccination into tumor immunotherapy regimens.

Another recent example of personalized cancer vaccine has reportedly moved to phase IIb trials at leading U.S. cancer hospitals.[33] The Elios melanoma vaccine to be assessed in this pivotal trial is expected to deliver personalized immunotherapy developed from the patient's cancer cells and dendritic cells to create a specific immune response in hopes of destroying any remaining cancer cells after surgery and thereby prevent recurrence of melanoma and improve overall survival rates. This vaccine candidate utilizes a particular patient tumor's unique antigenic and molecular profile and a novel delivery mechanism to set the immune system to defeat that patient's disease. This therapy, if proven effective, should be applicable to any patient, with any tumor.

Investors' expectations focus heavily on therapeutic vaccines, evidently starting with oncology. As an example, the "most valuable" list of R&D projects described in Table 13.4 includes Northwest Biotherapeutics' cancer vaccine (reportedly in phase III) with over $2 billion in potential 2020 revenues, as well as programmed death-1 (PD-1) and programmed death-1 ligand-1 (PD-L1) ligands which are akin to immunotherapeutic compounds (Evaluate Pharma, 2014).

The advent of therapeutic vaccines addressing nontransmissible diseases will drastically change the market access approach to support the demonstration of the value of said vaccines. As demonstrated earlier, the value proposition for vaccines remains difficult to establish and document due to the existence of both individual and collective benefit for preventative vaccines, whereas therapeutic vaccines should claim individual benefit only.

31) http://www.fiercevaccines.com/story/dendreon-bankruptcy-weighs-cancer-vaccine-field/2014-11-10 (accessed 03/01/2015).

32) http://www.fiercevaccines.com/story/are-personalized-approaches-solution-troubled-cancer-vaccine-field/2014-10-23 (accessed 03/01/2015).

33) http://www.firstwordpharma.com/node/1253209#axzz3Mo7zfO27 (accessed 24/12/2014).

**Table 13.4** Top 20 most valuable R&D projects (ranked by net present value).

| Rank | Product | Company | Phase (current) | Pharmacological class | WW product sales ($million) 2020 | | Today's NPV ($million) |
|---|---|---|---|---|---|---|---|
| 1 | Nivolumab | Bristol-Myers Squibb | Phase III | Antiprogrammed death-1 (PD-1) mAb | 6 012 | | **23 150** |
| 2 | MK-3475 | Merck & Co | Filed | Antiprogrammed death-1 (PD-1) mAb | 4 063 | New entry | **16 747** |
| 3 | RG7446 | Roche | Phase III | Antiprogrammed death-1 ligand-1 (PD-L1) mAb | 2 937 | New entry | **15 639** |
| 4 | Obeticholic acid | Intercept Pharmaceuticals | Phase III | Farnesoid X receptor (FXR) agonist | 2 992 | New entry | **11 426** |
| 5 | Ledipasvir/ Sofosbuvir | Gilead Sciences | Filed | Hepatitis C nucleoside NS5A and NS5B polymerase inhibitor | 2 818 | | **9 876** |
| 6 | Palbociclib | Pfizer | Phase III | Cyclin-dependent kinase (CDK) 4 and 6 inhibitor | 2 950 | | **7 925** |
| 7 | DCVax-L | Northwest Biotherapeutics | Phase III | Cancer vaccine | 2 046 | New entry | **5 502** |
| 8 | VX-809 + ivacaftor | Vertex Pharmaceuticals | Phase III | Cystic fibrosis transmembrane conductance regulator (CFTR) corrector | 1 900 | | **5 011** |
| 9 | MEDI4736 | AstraZeneca | Phase III | Antiprogrammed death-1 ligand-1 (PD-L1) mAb | 967 | New entry | **4 711** |

(*continued overleaf*)

**Table 13.4** *(continued)*

| Rank | Product | Company | Phase (current) | Pharmacological class | WW product sales ($million) 2020 | | Today's NPV ($million) | |
|---|---|---|---|---|---|---|---|---|
| 10 | Lampalizumab | Roche | Phase II | Anticomplement factor D mAb | 1 122 | New entry | **4 520** | |
| 11 | Revascor | Mesoblast | Phase III | Mesenchymal stem cell | — | New entry | **4 332** | |
| 12 | Idelalisib | Gilead Sciences | Filed | Phosphatidylinositol 3-kinase (PI3K) inhibitor | 1 273 | New entry | **3 615** | |
| 13 | Evolocumab | Amgen | Phase III | Antiproprotein convertase subtilisin-like kexin type 9 (PCSK9) mAb | 1 093 | New entry | **3 563** | |
| 14 | LCZ696 | Novartis | Phase III | AT1 receptor-neprilysin (ARNI) inhibitor | 1 329 | New entry | **3 005** | |
| 15 | Nivolumab | Ono Pharmaceutical | Filed | Antiprogrammed death-1 (PD-1) mAb | 348 | | **2 996** | |
| 16 | Alirocumab | Sanofi | Phase III | Antiproprotein convertase subtilisin-like kexin type 9 (PCSK9) mAb | 1 048 | | **2 950** | |
| 17 | Plegridy | Biogen Idec | Filed | Interferon beta | 1 047 | New entry | **2 931** | |
| 18 | Abemaciclib | Eli Lilly | Phase III | CDK 4 and 6 inhibitor | 651 | New entry | **2 922** | |
| 19 | Ocrelizumab | Roche | Phase III | Anti-CD20 mAb | 894 | New entry | **2 777** | |
| 20 | Secukinumab | Novartis | Filed | Anti-interleukin-17 (IL-17) mAb | 1 030 | New entry | **2 735** | |
| | Top 20 | | | | 36 520 | | **136 332** | |
| | Other | | | | 111 649 | | 282 192 | |
| | Total | | | | 148 169 | | **418 525** | **46%** |
| | | | | | **NPV of R&D pipeline June 2013** | | 286 367 | |

Source: Evaluate Pharma (1 June 2014).

Thinkers and authorities are getting concerned that vaccine innovation should be better organized and corralled (Leroy *et al.*, 2014). To consolidate the integration of the fragmented European vaccine development landscape, TRANSVAC – the European Network of Vaccine Research and Development, funded by the European Commission (EC) – has initiated the development of a roadmap through a process of stakeholder consultation. The outcome of this consultation highlighted the need for transnational cooperation and the opportunities that could be generated by such efforts. The authors suggest that this cooperation can be achieved through the establishment of a European Vaccine Research and Development Infrastructure (EVRI) to support cooperation between existing vaccine R&D organizations from the public and private sector and other networks throughout Europe. It is expected that it could enter into operational phase from 2017 onward and become sustainable over time by receiving support from multiple public and private sources – assuming sufficient political and financial commitment can be secured – and by income generated. The concept is gaining support from different stakeholders who agree that such an infrastructure can function as leverage institution between public and private institutions and contribute to the sustainability of public healthcare systems.

### 13.5.2
### Where Is Innovation Needed? Manufacturing and Product Improvement

We have already mentioned the development of combination vaccines to improve convenience. The corresponding manufacturing challenges include the development of specific technologies for formulation and stabilization of the antigens. As the number of antigens and carrier proteins is increasing, the hurdles may prove insurmountable, and novel technologies such as micropellet manufacturing, which corresponds to freeze-drying on a per-antigen basis, could allow for easier formulation of ad hoc combination vaccines. The pending questions include registration and regulatory acceptability of multiple combos derived from the multiple combinations of the freeze-dried microbeads.

Vaccine manufacturing facilities require large investments and are often uniquely dedicated. New flexible manufacturing technologies should add versatility while preserving the ability to respond swiftly to higher than expected demand, which is particularly relevant to offset uncertainty of forecasting for novel vaccines as well as for vaccines addressing diseases with fluctuating epidemiology.

### 13.5.3
### Where Is Innovation Needed? Acceptability

Improving vaccine acceptability is a daunting and swelling challenge, as it needs to be successfully achieved for all stakeholders, making it difficult due to their diverging nature and standpoints, from the public health authorities to the antivaccine movements, and the individual-, religion-, or belief-driven expectations for "pure" vaccines.

Some authors have recommended making the perceived risk of not vaccinating higher than that of vaccination and adding the indirect risks to the community to the personal risks (Reyna, 2012) in plain language (since herd immunity is a poorly understood notion in the lay public) and communicated through the healthcare providers which are the most important source of information for vaccination decisions. Therefore, vaccine manufacturers and public health authorities could have a joint field of innovation to develop educational communications strategies to drive the beliefs and attitudes of healthcare providers and combat "vaccine fatigue" linked to the crowding of the immunization schedule for infants.

Maintenance of the willingness to vaccinate despite the quasi disappearance of targeted diseases is indeed critical. This could be achieved through leveraging data on disease resurgence in situations of failures – in order to maintain immunization coverage – and working on individual benefits (or fears, whether real or not), as these are increasingly trumping collective benefits in the minds of individuals, as demonstrated by current societal trends.

It is also a must for all responsible stakeholders to address the antivaccination movements that claim that vaccines are not safe and effective and that vaccination recommendations infringe civil liberties or religious beliefs. Such claims strike emotional chords (Betsch *et al.*, 2011) and cannot be antagonized solely with education and evidence-based communication, but rather by engaging end users in discussion forums and mobilizing social motivations.

At the level of the authorities and advisory bodies, regulatory changes facilitating the market access for me-too (read "biosimilars," which is what vaccines have been for decades before the term was coined for me-too monoclonal antibodies) may positively increase competitive intensiveness and thus shorten the currently extended life cycle of branded incumbent vaccines, bringing cheaper alternatives to cash-strapped public health systems.

### 13.5.4
### Where Is Innovation Needed? Accessibility, Both as a Function of Supply (Availability) and Logistics

Public health authorities strive to strengthen and optimize immunization supply chains so that vaccines are delivered to the end users effectively, efficiently, and sustainably. With the expansion of investments in global health and a broader portfolio of vaccines, the need for vaccine supply chains to operate at their most optimal levels is increasingly stressed. Integration with other supply chains is often presented as a strategy to improve efficiency. However, recent studies indicate that the proposed benefits from integration of vaccine supply chains with other supply chains may not outweigh the costs (Yadav *et al.*, 2014). Further analysis of this segment of the value chain is warranted.

Other analyses reviewed the cost and cost-effectiveness of vaccination projects performed outside of medical practices as a means to expand coverage and facilitate access. For instance, the review of a school-based influenza program (Yoo *et al.*, 2013) demonstrated that the direct per-dose cost of this project exceeded that of office-based vaccination. Yet, when broader averted costs were included, the program was less costly than office-based vaccination, and the findings support expanded program implementation and suggest efficiencies.

Lastly, research results (Ambrose *et al.*, 2013) have demonstrated that self-administered intranasal live attenuated influenza vaccine appeared similarly effective against influenza-like illness and had comparable safety profiles as the same vaccines administered by healthcare providers, indicating another route to enhance accessibility and affordability through reduced administration costs and possibly improve end-user acceptability.

### 13.5.5
### Affordability and Sustainability

By nature, transmissible diseases cannot be eradicated and therefore are there to stay, hopefully controlled by a combination of prevention and cure. Once vaccine pressure diminishes, the diseases resurface, as exemplified by measles, mumps, or pertussis in various European countries or in the United States. The public health authorities that decide to recommend a vaccine and to fund the corresponding immunization program make a decision which is tantamount to a long-term and open-ended commitment, the execution of which requires the maintenance of the quality and reliability of supply of the corresponding vaccine.

As novel vaccines become available and are recommended by the countries' advisory bodies' for routine immunization, the public payers are often looking for savings on existing public expenditure. From the perspective of the vaccine manufacturers, addressing the impact of tight public health budgets (not necessarily linked to fiscal or economic crises) leading payers to make trade-offs between vaccines and other healthcare investments, and possibly among vaccines to create budget room for the newer ones, is a mounting and complex challenge. As indicated earlier, the demonstration of the value proposition of vaccines is complex to establish, and vaccine manufacturers are often caught in a value squeeze, the price being challenged downward by the public payers and the COGS being subject to inflationary pressures linked to the cost of labor and quality. When the tension becomes too high, the manufacturers can be driven to portfolio decisions involving the pruning of certain products or ranges. In turn, it diminishes the number of providers of the corresponding vaccines, oftentimes leaving the market to the emerging market players which operate at lower costs. This situation has been described for various vaccines,[34),35)] such as measles, yellow fever, whole-cell pertussis-based pentavalent combinations, rotavirus, and so on, where the competitive field is evolving, raising a number of concerns related to quality and supply, leading the international purchasers to consider guaranteed supply

34) http://www.unicef.org/immunization/23244_shortage.html (accessed 03/01/2015).
35) http://www.gavi.org/about/gavis-business-model/vaccine-supply-and-procurement/ (accessed 03/01/2015).

agreements and to commit to transparency on forecasted requirements to maintain the incentive to produce in sufficient quantities.

As the need to maintain quality and reliability of supply is well recognized and the consequences for vaccine pricing and market access are better known, more research on the health economics of long-term vaccine usage is clearly warranted. The demonstration of the sustained value of immunization could be better documented by a connection with the long-term end-user monitoring by creating time-bound link with the care procurement monitoring systems.

## 13.6 Where Will Innovation Come from? Strategy and Players

Owing to their long-term presence and massive capital investment, the existing players are extremely busy maintaining their current portfolio, and this is consuming a large fraction of their annual capital and operating expenditures. But since redeveloping an existing vaccine is an even more costly proposition, the long-term maintenance of existing portfolios and technologies can foster the perception that precious few innovative technological options are available to improve existing vaccines. The search for additional mass and portfolio expansion is driving large companies in the space of mergers and acquisitions (M&A) and alliances, as indicated earlier. Recent examples include GSK, which executed a complex product swap with Novartis and acquired a large fraction of their vaccine franchise, and Pfizer, which acquired most of the former vaccine portfolio of Baxter. As the pool of acquisition targets in developed countries is dwindling – with the exception of few riskier bets among biotechnology start-ups – we expect a growing trend for "north–south" M&A deals.

For these existing players, manufacturing remains a strategic pillar accounting for up to 30% of their P&L, hence a limiting trade-off factor due to the required capital expenditure, especially when the choices are made in favor of large, dedicated, and inflexible facilities. The impact of such manufacturing units is massive on fixed costs and COGS, as well as on the cost of quality assurance and quality control. The matter is even worse when the capacity is underutilized or when yield loss is substantially higher than projections, for example, in the case of suboptimally defined processes.

Going forward, we expect that these existing multinational players will be refocusing their portfolios where their expertise is residing (maybe except Pfizer which is venturing into the field of viral vaccines with its recent acquisition), delivering innovation through novel vaccines or more complex combinations and pruning the nonprofitable ranges of the ancient and less sophisticated vaccines. Yet, as payers and donors expect continuous supply and maintenance of existing sources, we anticipate some complex decisions, possibly at the expense of profitability. Lastly, the multinational players will be expected to contribute intellectual, scientific, and financial resources to the various government-driven initiatives related to bioterrorism and pandemic preparedness.

We anticipate a sharp increase of competitive intensiveness from "emerging" market players which display a clear strategic intent to go beyond their domestic borders. Originally, these companies were aiming at the local market and were benefiting from the support and protection of the local authorities. Over time, these companies have mastered modern technologies – sometimes thanks to technology transfers and partnerships with multinational companies – and increased their quality standards and production output to cater to other markets. Examples are numerous: in India and China (whose regulatory authority, the Chinese FDA, is already WHO approved for WHO Prequalification) but also in Korea, Brazil, and so on. These companies will eventually challenge the established multinational players all the way to their home markets, further intensifying the search for innovation to sustain profitable growth and impacting the life cycle of established vaccine brands in developed markets.

The last question for strategists could be, "is there room for new players in such a market?" We'd like to posit that there is indeed room for more players with new characteristics, both in terms of capabilities and organizational design.

First, we expect intellectual property to play an increasing role in the field of vaccines as it does in the field of small molecules and well-defined biologicals (such as monoclonal antibodies). The search for better defined processes and synthetic material and growth media is already a fertile ground for several companies which are offering cost-effective alternatives to existing R&D (e.g., for antigen characterization) and manufacturing (e.g., conjugation or purification) processes. Going forward, we expect IP to become a better protection – albeit one with a known expiry date – than industrial know-how and prowess.

Second, as the vaccine market is still largely demand driven (with the major exception of the influenza market, which is supply driven due to a worldwide overcapacity, albeit geographically uneven), there is room for capacity expansion, either from existing players or from new players adopting flexible manufacturing equipment and design.

Third, the boundaries of the vaccine market are increasingly pushed beyond the limits of infectious, transmissible diseases. We anticipate that the delimitation of the traditional vaccine market will be increasingly blurred with the launch of novel immunotherapy solutions in a number of fields ranging from oncology to rheumatology to allergy, to name a few. Many start-up companies will be entering this fray, alongside some of the current major healthcare companies, some of them venturing into vaccines for the first time.

The march of innovation is already felt, the vaccine market is evolving, and it will keep adding value to the people and to the nations through better health and wealth.

### 13.6.1 Take-Home Messages

Vaccines occupy a peculiar and growing place in the landscape of healthcare solutions, and the technology evolutions and societal trends in favor of prevention will further fuel the expansion of that market, within the space of infectious diseases but also that of noncommunicable ones, resulting in an evolution of the recommendations supporting immunization.

As the vaccine market will keep expanding, the large healthcare companies will seek to maintain their historical dominance resting on their industrial and commercial strengths, while smaller and nimbler players will enter the market and expand their presence with differentiated offerings and market access strategies.

These innovative approaches will become visible in each of the building blocks of the vaccine value chain and should be further amplified by supportive measures from the regulatory authorities and favorable societal trends. The vaccine industry will need to overcome mounting obstacles pertaining notably to the acceptability and affordability of its solutions.

The sustainable growth of the vaccine market will stem from the repositioning of the portfolio of the historical players concomitantly with the advent of vaccines for noncommunicable diseases. Growth will be further enhanced by the emergence of additional supply of vaccines against transmissible diseases from newer players mastering either large-scale production and quality standards or novel manufacturing technologies backed by IP rights.

## References


Ambrose, C.S. *et al.* (2013) The safety and effectiveness of self-administration of intranasal live attenuated influenza vaccine in adults. *Vaccine*, **31** (6), 857–860.

Angelmar, R. and Morgon, P.A. (2014) in *Innovation and Marketing in the Pharmaceutical Industry* (eds M. Ding, J. Eliashberg, and S. Stremersch), Springer, pp. 365–423.

Betsch, C. *et al.* (2011) The influence of narrative v. statistical information on perceiving vaccination risks. *Med. Decis. Making*, **31** (5), 742–753.

Black, S. (2013) The role of health economic analyses in vaccine decision making. *Vaccine*, **31** (51), 6046–6049.

Evaluate Pharma (2014) World Preview 2014, Outlook to 2020, June 2014.

Fischer, K. *et al.* (2014) Therapeutic HIV vaccines: prior setbacks, current advances, and future prospects. *Vaccine*, **32** (43), 5540–5545.

Francis, D.P. *et al.* (2014) Global vaccine supply. The increasing role of manufacturers from middle income countries. *Vaccine*, **32** (41), 5259–5265.

Gallorini, S. *et al.* (2014) Sublingual immunization with a subunit influenza vaccine elicits comparable systemic immune response as intramuscular immunization, but also induces local IgA and TH17 responses. *Vaccine*, **32** (20), 2382–2388.

Gessner, B.D. *et al.* (2014) Vaccine preventable disease incidence as a complement to vaccine efficacy for setting vaccine policy. *Vaccine*, **32** (26), 3133–3138.

Grabenstein, J.D. (2013) What the World's religions teach, applied to vaccines and immune globulins. *Vaccine*, **31** (16), 2011–2023.

Hamidi, A. *et al.* (2014) Lessons learned during the development and transfer of technology related to a new Hib conjugate vaccine to emerging vaccine manufacturers. *Vaccine*, **32** (33), 4124–4130.

Hyde, T.B. *et al.* (2012) The impact of new vaccine introduction on immunization and health systems: A review of the published literature. *Vaccine*, **30** (45), 6347–6358.

Kennedy, A. *et al.* (2011) Confidence about Vaccines in the Unites States: understanding parents' perceptions. *Health Aff.*, **30** (6), 1151–1159.

Laxminarayan, R. *et al.* (2014) Valuing vaccines using value of statistical life measures. *Vaccine*, **32** (39), 5065–5070.

Leroy, O. *et al.* (2014) Roadmap for the establishment of a European vaccine R&D infrastructure. *Vaccine*, **32** (51), 7021–7024.

Maltais, A.-K. *et al.* (2014) Intranasally administered Endocine™ formulated 2009 pandemic influenza H1N1 vaccine induces broad specific antibody responses and confers protection in ferrets. *Vaccine*, **32** (26), 3307–3315.

Morgon, P.A. (2011) Opening Doors to the Developing World Through Superior Market Access Strategies. Investor Day, March 2011.

Nihan, B. *et al.* (2015) Does correcting myths about the flu vaccine work? An experimental evaluation of the effects of corrective information. *Vaccine*, **33** (3), 459–464.

Ozawa, S. *et al.* (2013) Cost-effectiveness and economic benefits of vaccines in low- and middle-income countries: a systematic review. *Vaccine*, **31** (1), 96–108.

Philipson, T. (2000) in *Handbook of Health Economics*, vol. **1B** (eds A.J. Culyer and J.P. Newhouse), Elsevier, Amsterdam, pp. 1761–1799.

Reyna, V.F. (2012) Risk perception and communication in vaccination decisions: a fuzzy-trace theory approach. *Vaccine*, **30** (25), 3790–3797.

Ricciardi, G.W. *et al.* (2015) Comparison of NITAG policies and working processes in selected developed countries. *Vaccine*, **33** (1), 3–11.

Roese, N.J. (1997) Counterfactual thinking. *Psychol. Bull.*, **121** (1), 133–148.

Slifka, M.K. *et al.* (2014) How advances in immunology provide insight into improving vaccine efficacy. *Vaccine*, **32** (25), 2948–2957.

Smith, J. *et al.* (2011) Vaccine production, distribution, access and uptake. *Lancet*, **378** (9789), 428–438.

Strioga, M.M. *et al.* (2014) Xenogeneic therapeutic cancer vaccines as breakers of immune tolerance for clinical application: to use or not to use? *Vaccine*, **32** (32), 4015–4024.

Tafuri, S. *et al.* (2014) Addressing the anti-vaccination movement and the role of HCWs. *Vaccine*, **32** (38), 4860–4865.

Usman, H.R. *et al.* (2010) Determinants of third dose of diphtheria-tetanus-pertussis (DTP) completion among children who received DTP1 at rural immunization centers in Pakistan: a cohort study. *Trop. Med. Int. Health*, **15** (1), 140–147.

Wallace, A.S. *et al.* (2014) Experiences with provider and parental attitudes and practices regarding the administration of multiple injections during infant vaccination visits: lessons for vaccine introduction. *Vaccine*, **32** (41), 5301–5310.

Yadav, P. *et al.* (2014) Integration of vaccine supply chains with other health commodity supply chains: A framework for decision making. *Vaccine*, **32** (50), 6725–6732.

Yoo, B.-K. *et al.* (2013) Cost effectiveness analysis of elementary school-located vaccination against influenza—results from a randomized controlled trial. *Vaccine*, **31** (17), 2156–2164.

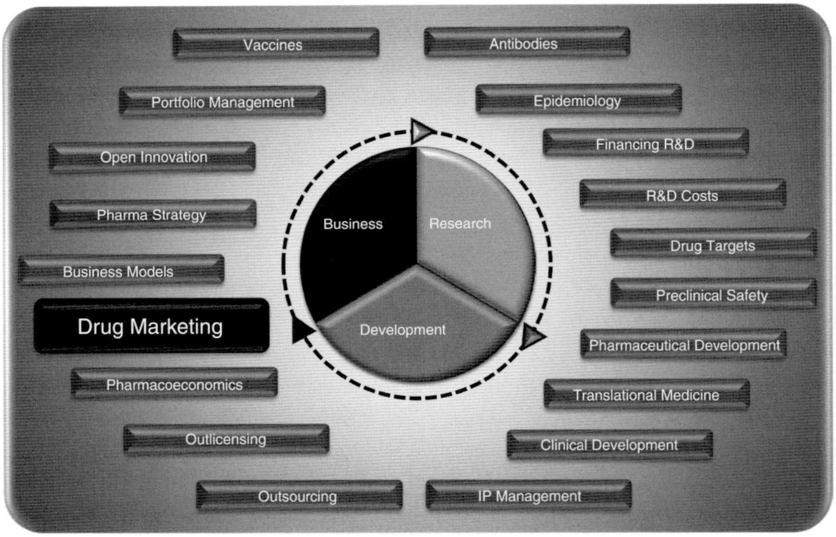
Vaccines
Antibodies
Portfolio Management
Epidemiology
Financing R&D
Open Innovation
R&D Costs
Pharma Strategy
Business
Research
Drug Targets
Business Models
Preclinical Safety
Drug Marketing
Development
Pharmaceutical Development
Pharmacoeconomics
Translational Medicine
Outlicensing
Clinical Development
Outsourcing
IP Management

# 14
# The Patient-Centric Pharma Company: Evolution, Reboot, or Revolution?

*Pierre A. Morgon*

## 14.1
## Introduction

The purpose of this chapter is to present the trends affecting the players in the healthcare universe and driving the increasing focus on real-life patient data during the clinical development of a new therapy or healthcare solution, up to the consequences in terms of the market access strategy design and execution. This chapter will underscore the impact of players not originally positioned in the healthcare space but leveraging their know-how in terms of consumer data management, thereby creating not only more innovation opportunities but also more economic value linked to healthcare and substantial change for the existing life science companies.

We will explore why the satisfaction of the patient should be the ultimate performance indicator of healthcare procurement. Driven by the megatrends which impact the healthcare ecosystem, ongoing regulatory and lobbying activities increasingly aim at including real-life patient data during clinical development to ensure a better assessment of the clinical, economic, and social utility of a new therapy or healthcare solution. As a consequence, market access is now a lasting activity, starting as early as phase II in clinical development and stretching throughout the life of the product and aiming at building the body of evidence which supports the value proposition over time. This will be facilitated by the progressive penetration of companies not originally positioned in the healthcare space yet leveraging their know-how in terms of consumer data management to contribute to the digitalization of the life of the people. They will play an essential role in creating not only more innovation opportunities but also more economic value linked to healthcare, which could help to offset the ever-growing burden of healthcare costs in the real economy, in addition to contributing to a better utilization of healthcare resources. The impact for the life science companies is broad and far-reaching in terms of governance from R&D, to intellectual property management, to corporate social responsibility.

*Value Creation in the Pharmaceutical Industry: The Critical Path to Innovation*, First Edition.
Edited by Alexander Schuhmacher, Markus Hinder, and Oliver Gassmann.

## 14.2
## Health, Always ...

Ever since human beings existed, health has been a topic of paramount importance and an integral part of fighting for one's subsistence along with an overlapping with the search for food and shelter.[1)]

Not only has health always been a subject of extreme importance, but also it has superseded wealth as a topic of interest. Taking a simple look at the Google Ngram Viewer chart of the number of books addressing the topic of health versus those focused on wealth, one can only notice that since the 1800s, health-related books have always outnumbered those related to wealth by anywhere between 10 and 20% up until the 1920s when the number of books focused on health skyrocketed (with an acceleration since the 1970s) and the lead of health over wealth as a literature topic is now close to fivefold as expressed in total number of books.

The progressive structuring of human society has led to an organization of more effective care procurement so as to contribute to a healthier society, along with the organization of education and other public services. In this context the search for medical interventions has progressively evolved from empirical to systematic and scientific, and the life science industry has progressively emerged from the capitalistic efforts to streamline the search for such interventions addressing increasingly complex, unmet medical needs. This evolution has led to a fully structured healthcare industry, entirely organized around its capabilities to generate innovation, to protect it with the relevant intellectual property rights, to manufacture its assets in a reproducible and high-quality manner, and to commercialize them while complying with a complex set of regulations and guidelines. The industry is supply-chain savvy (especially in very specific market segments, such as vaccines and some biological products) and is nowadays much less driven by its manufacturing capabilities but rather by its R&D prowess and its superior capabilities in terms of dealing with its stakeholders during the latter part of the life cycle of its products, from late stage development, to regulatory approach, to market access to commercialization.

## 14.3
## The Mission of the Healthcare Industry

The healthcare industry's mission has always been to focus on generating innovative products and solutions, both therapeutic and preventive, for the benefit of the populations around the world, hence addressing varying types and magnitudes of unmet medical needs which vary considerably across countries and at times within countries. These discrepancies are not only a source of complexity in

1) https://books.google.com/ngrams/graph?content=health%2C+wealth&year_start=1800&year_end=2000&corpus=15&smoothing=3&share=&direct_url=t1%3B%2Chealth%3B%2Cc0%3B.t1%3B%2Cwealth%3B%2Cc0 (accessed 06 July 2014).

the management of the R&D portfolio of the healthcare industry and its commercialization policies but also an opportunity to differentiate from competition and to create a lasting competitive advantage.

Contrary to the situation in a number of industry segments, the entire value chain of the healthcare industry is subject to a large and ever-increasing number of regulations. As an example, as indicated in the guidelines of the IFPMA, the industry should enforce strict principles of ethical conduct, ensure execution of high standards of manufacturing practices and quality assurance, provide contributions to the overall expertise, and foster collaborative relationships and partnerships with the various stakeholders dedicated to the improvement of public health.[2)]

The discoveries of the healthcare industry have changed the face of our world, and the impact has been extremely visible especially from the demographic perspective, along with the access to clean water and better food: life expectancy in the developed countries has more than doubled in the past century. This substantial impact on demographics has happened simultaneously to a number of other changes contributing to increasing life expectancy, ranging from better sanitation to access to safer food without supply limitations and to better living conditions as a whole. As a consequence, the population has been aging and the healthcare issues that were linked to the previous societal conditions have progressively given way to more chronic conditions related to aging and to a more sedentary lifestyle. The search for solutions to the ailments linked with modern life in mature economies has evolved accordingly and the R&D efforts focusing on conditions such as hypertension, diabetes, depression, and cancer, to cite only a few, have progressively superseded the search for solutions against infectious diseases in the development programs of the healthcare industry.

In parallel with these dominant trends, a number of initiatives driven either by public authorities or by NGOs such as patient associations or international organizations have ensured that ailments afflicting small patient populations or lower-income regions or countries still receive sufficient attention and R&D funding. Nowadays, the majority of leading healthcare companies have some form of R&D program dedicated to neglected conditions and/or "diseases of the South."

It is therefore obvious that the healthcare industry's mission is aligned with societal ambitions for a healthier and more sustainable world (Mistra Pharma, 2009).

Yet, the healthcare industry is working with constrained resources and therefore is prioritizing and making R&D choices which are driving the focus of its portfolio. Such arbitrages are not always well understood by the lay public which has sometimes diverging aspirations which it conveys through various media, ranging from the classical tools in democracies such as voting all the way to the virtual world of social networks. Thus, it is the balance of the perception between these sometimes opposite goals and more specifically the emerging gap between public health objectives and individual expectations which is increasingly shaping the agenda of the various stakeholders with which the life science industry is interfacing.

2) http://www.ifpma.org/about-ifpma/welcome.html (accessed 06 July 2014).

## 14.4
## Megatrends Affecting the Strategic Scorecard of the Healthcare Industry

During the past decades, healthcare expenses have increased steadily despite numerous measures attempting to curb their growth. The healthcare spend is currently representing a very large share of the real economy expressed in terms of GDP. In most countries of the OECD, this share is growing at such a pace and up to such a level that it is challenging the sustainability of the healthcare systems (World Economic Forum, 2013). With the exception of the United States where healthcare expenses represent a staggering 18% of GDP, most developed countries are investing 10% or more of their real economy in healthcare.[3)] And even in emerging economies, the cost of healthcare is leading to difficult trade-offs: when a country is facing simultaneously the emergence of diseases which are frequent in mature economies (such as hypertension, diabetes, dyslipidemia, depression, and cancer) and has still to deal with the remaining diseases typical of the emerging world (diarrheal or respiratory infectious diseases such as cholera, typhoid fever, rotavirus diarrhea, and tuberculosis), the public health decision makers are driven to tough choices to allocate their healthcare priorities and resources. In addition, these emerging countries are often facing infrastructure challenges and need to invest not only in the management of a wide scope of medical conditions but also in the construction of robust, efficient, and accessible primary care infrastructure, including the brick-and-mortar components of hospitals and dispensaries and the human component of properly trained healthcare providers.

While the emerging countries are facing these difficult investment choices, the mature economies which are crippled by their economic debt also face tough trade-offs to drive the mutation of the healthcare infrastructure and deal with the public expectation for efficient and personalized care procurement.

In this context, the healthcare industry is evidently considered as a driver of the progression of the healthcare costs when in actuality the share of the total healthcare costs represented by medications and devices in the total amount of healthcare expenditures usually comprises between 15 and 20% depending on the markets. The question therefore becomes one of understanding where the actual drivers of the bulk of the healthcare expenditures come from and what could be done to manage those effectively.

The healthcare spending keeps increasing as it is driven by the "cost disease" as described by William J. Baumol and William G. Bowen in the 1960s: as a matter of fact, healthcare procurement is a very labor-intensive activity. Every patient has to be examined individually; he or she is subject to specific examinations, possibly by different healthcare providers of different specialties until the diagnosis is firmly established, the course of treatment is decided, and a successful outcome is reached. Therefore, at every step of the patient journey, one can observe the intervention of one or more healthcare providers whose role, due to the evolution of science and the better understanding of diseases and their treatments, is

3) http://stats.oecd.org/Index.aspx?DataSetCode=SHA# (accessed 06 July 2014).

ever more specialized. These healthcare providers are extremely well trained and represent high-quality resources and are therefore remunerated accordingly. The fees for honorarium of healthcare staff are subject to increases like salaries without much productivity gains since, by definition, a diagnosis is personalized to each patient case. One could argue that automation – and subsequently productivity gains – is happening in other quadrants of healthcare procurement chain such as biological or radiological examinations, yet the benefits of such productivity gains are more than offset by the increasing sophistication of the said examinations which are growing more numerous, more complex, and therefore more expensive.

The trend of the share that healthcare costs represent in the real economy is worrying for the public authorities because the growth in health spending has always outpaced that of the GDP, especially during economy downturns. What is even more worrying is the absence of correlation between healthcare expenses and quality of care (Health Consumer Powerhouse, 2009), and that has drawn the authorities' attention on the efficiency of care procurement. It has also been attracting the lay public's attention on the broad – and not always well-defined – topic of the value for the individual. This lack of correlation between the magnitude of investments and the quality of care procurement is also stretching to the relationship between the health expenditure and the degree of satisfaction that consumers have with their healthcare systems.

Consequently, the absence of a solid correlation between investments in healthcare and quality of care is triggering a host of activities driven by the authorities across all countries to search for efficiencies, looking for benchmarks from abroad to identify more effective ways to utilize healthcare resources. The authorities are also concerned by the poor correlation between healthcare spending and patient satisfaction, as the topic is politically loaded and oftentimes central to electoral programs. From the patient perspective, the increasingly ubiquitous availability of information and the behavioral evolution of challenges of both medical and public authorities are fueling mounting patient-driven challenges.

In addition to the earlier mentioned, the life science industry has to cope with a number of other global macrotrends that affect its relationship with its ecosystem and stakeholders.

First, there is the well-known shift of economic power to emerging markets, which drive global economy growth and have their own healthcare expectations while the developed countries are saddled with debt. Therefore and in both cases, social, political, and economic pressures all drive investments in healthcare infrastructure and increase price pressure on medical interventions. The mature economies are looking for more cost-effective ways to utilize their healthcare resources without triggering backfire from the lay public over care rationing concerns. And the emerging economies have to arbitrate between healthcare priorities and invest in infrastructure to ensure the proper level of care coverage, access, and quality. In the first instance, the healthcare industry has to find the right balance between innovation and the price it charges for the said innovation so as to generate an economic and social surplus for the healthcare system. In the second instance, the industry has to balance its portfolio to ensure its adequacy with the local epidemiology and public health priorities.

Second, the emerging markets will represent the largest share of the growing cohort of healthcare consumers thanks to the swelling of the ranks of the middle class, driven by the economic growth. Simultaneously, these numerous additional and eager healthcare consumers will increasingly live in urban areas; as a consequence, they will have easier access to care procurement – which is positive to deal with acute diseases yet will result in a commensurate increase in the proportion of chronic diseases – and to health information, leading to an increase in qualitative and quantitative demand and in the challenges of the medical and public authorities. The flipside of the increase in the urban population will also take the form of a less active lifestyle and evolution of dietary regimens, resulting in an increase in incidence of chronic diseases linked to sedentary behaviors and to a greater exposure to poorer air quality, which will trigger a higher incidence of respiratory conditions.

Lastly, whether in developed countries or increasingly in emerging ones, we are looking at shifting demographics with aging populations (also a consequence of the previous trends and better economic status contributing to aging) which will increase healthcare demands in relation to the growing incidence of chronic diseases and organ failures linked to aging.

The healthcare industry has always been paying close attention to those trends and they contributed largely to the R&D orientations and to the definition of business development and geographical expansion priorities.

## 14.5 Focus on the Societal Trends and Their Consequences for the Management of Healthcare Innovation

Besides these economic and demographic trends, one has also to consider the impact of societal evolutions.

In recent years, the combination of education, access to information, and mobility has contributed to the growing sense of self over that of the group. While such trends are obviously welcome from the perspective of consumer goods, as it creates greater opportunities for marketers to segment the markets and create a value proposition resting on providing products that match more closely the expectations of the customers, it came with a number of hard-to-manage consequences in the field of healthcare.

For instance, when it comes to immunization, we have been witnessing the progressive evolution of resistance against immunization recommendations. Case in point is the growing number of parents who refuse to immunize their children against infectious diseases because they do not perceive those as a threat and are worried about potential adverse events – some immediate, but most importantly the deferred, longer-term ones – linked to vaccine administration. Since vaccines are meant to protect against infectious and therefore transmissible diseases, it is of utmost importance for the public health authority to ensure an adequate coverage in order to block transmission of the said diseases. Typically, a coverage of 70% or more of any population is considered as an adequate threshold to block the transmission of the disease such as influenza. Because it is so important to reach such high coverage rates of the target population, all the stakeholders involved in executing immunization campaigns are striving to ensure adherence to the immunization calendars or guidelines. Thus, the goal is to mobilize a large enough share of the population to undertake a medical intervention on an otherwise healthy group of subjects using a standardized medical intervention. The resistance to the recommendations is often stemming from the fear of safety issues as much as from a less precisely defined concern over such "one-size-fits-all" medical interventions. In fact, in an era when patients are expecting customized solutions, the very standardized nature of the immunization recommendations is negatively striking the chord of individualism, resulting in skepticism and sometimes resistance.

The objective of the above paragraph is not to question the validity of the behaviors responsible for such a trend, but rather to signal its growing magnitude and its consequences for the stakeholders involved in the design and in the execution of healthcare policies. For all concerned, the ever more assertive attitude of patients has been considered very seriously and patient centricity has become a daily staple, up to the buzzword of the past decade. Almost every life science company is claiming in its mission statements that the patient is central to its strategy and driving the focus of its strategic activities, from the priorities of the R&D investment, to the quality of manufacturing, to the content and quality of the information flows in the context of the commercialization of its products. Such statements related to patient centricity are now so ubiquitous that they are hardly differentiating and it is important to look beneath the surface of the glossy corporate brochures and press releases to assess the reality of patient centricity and how it translates in the real life of these corporations.

Taking a step back, one has to remember that patients are first and foremost ... people, like you and this author. And these people live in an information age, accelerated by mobility, in which data is ubiquitous and where access to media is easy yet where understanding is leaving room for improvement and is extremely uneven across population subgroups depending on the level of education and command of healthcare matters.

It is commonplace to say that what happens in an Internet minute is staggering.[4] People access a deluge of data (including those related to healthcare) under various forms and from a multitude of sources, accelerated by the connectivity between devices. In this context, not one stakeholder can master the full scope of the information flows to build a comprehensive, reliable, and accurate representation of what is broadcasted on a given healthcare topic. Subsequently, taking into consideration the voice of the customer is commensurately difficult, in light of conflicting priorities as described below, and compounded by expanding expectations for

4) https://image-store.slidesharecdn.com/561a0856-b834-11e3-a614-22000a9780da-original.jpeg (accessed 01 June 2014).

more effective and better profiled care procurement. Overall, the difficulty is simply finding what the real voice of the customer is (The Economist Intelligence Unit Limited, 2012).

As new generations reach adulthood and working age, customers of the healthcare system (a.k.a. patients) and healthcare providers are increasingly "born digital"; Internet, e-books, and social networks are all part of the standard tools that they are familiar with and which are an integral part of their daily functioning.[5] As the generations that are defined as "digital natives" grow up and reach the level of influence, starting from voting age all the way to positions of authority, one is witnessing drastic changes in information management practices and we are likely to see a parallel evolution of the way public health priorities are determined and addressed.

When looking at the number of articles published on healthcare topics per year from 1970 to 2010, one notices a steady increase of the number of publications over 40 years, with the rate of increase becoming more pronounced from the beginning of the 2000s.[6] This surge in the number of publications is obviously the combined result of intensive research efforts and the sharing of scientific knowledge. The sheer volume of information which is created is triggering a growing challenge for the human cognitive capacity due to the exponential increase in both the number of sources and their complexity when these are needed for quality clinical decision-making.

As the diagnosis of patients and the decision regarding the best course of action to follow are both becoming more complex, so is the task of the various stakeholders aiming at providing reliable, trustworthy, and most importantly understandable information, both to the practicing medical community and to the lay public. Evidently, the same difficulty is amplified downstream at the patient level, since the healthcare providers who face patients often do not have enough time or adequate educational skills to communicate effectively with the patients, most of the latter due to a lack in the relevant educational background to approach data in a discerning manner and to form a relevant and scientifically sound opinion on their own.

## 14.6 The DNA of the Healthcare Industry: R&D and the Management of Innovation

Before exploring further the consequences of these societal trends, one has to put in perspective the evolution of the life science industry's R&D focus and output.

The R&D process of the life science industry has always been marked by high failure rates. From several thousand compounds analyzed during the early phase of exploratory research, a few hundreds reach preclinical testing phase, only a handful enter the clinical trial stage, and a small number are

5) http://mashable.com/2013/12/21/technology-age-comic/ (accessed 01 June 2014).
6) Online searches at PubMed http://www.ncbi.nlm.nih.gov/pubmed (accessed 01 June 2014).

successfully going through the entire clinical development and regulatory review process. The healthcare industry is the largest industry sector in terms of R&D spending,[7] yet its R&D productivity has been declining in recent years despite the so-called biotechnology bubbles.[8] The decline of R&D productivity, both qualitatively and quantitatively, has been extensively commented and attributed to a combination of factors ranging from more complex medical needs to address to harder-to-understand pathological mechanisms of the diseases, increasing difficulties to translate fundamental research into medical discoveries, rising hurdles to identify responder populations, tightened regulation to assess drug safety, and so on (Paul *et al.*, 2010). Over the years, the focus of R&D has evolved toward specialty care, capitalizing on the progress made by fundamental research in very selected disease areas with smaller but much better defined target patient populations. Nowadays the largest number of R&D projects is found in the field of oncology, while the cardiovascular segment is getting much less attention and central nervous system diseases even less so (Citeline, 2013).

The concern for the life science industry is that the return on R&D investment has eroded despite sustained spending in terms of percentage of total revenues (and the nominal amount of the said R&D spend had increased, commensurately to the turnover of the healthcare industry), reflecting the lower outputs in new molecular entities, as well as the more limited volume and value sales for each new entity which addresses more targeted patient populations. Yet the industry maintains its R&D investment as the analysis of the evolution of approvals by regulatory agencies (Food and Drug Administration (FDA) and European Medicines Agency (EMA)) indicates that innovation is driving sales, albeit at a lower rate of return. It is expected that this trend will continue, as indicated by a pool of evidence ranging from the declarations of the CEOs of the healthcare companies to the sustained flow of investment by venture capitalists and capital developers in innovation-focused start-ups.

The maintenance of the R&D investment should not hide the substantial shift in the mix of projects, in terms of disease areas, and precisely defined patient populations, as indicated above. Even the latest development candidates addressing widespread conditions such as dyslipidemia are subject to a clinical development resting on a larger number of clinical trials, each of them addressing smaller, well-targeted, and defined patient subpopulations, rather than larger scale clinical studies including less precisely profiled patients.[9] Evidently, the nature and structure of such clinical studies are driven by the evolution of science and of the request for specific types of supportive evidence from the authorities. As the drug candidates address chronic diseases, in the management of which, the patient attitude

7) http://www.ifpma.org/fileadmin/content/Publication/2012/IFPMA_CorpBrochureWEBVERSION.pdf (accessed 06 July 2014).
8) http://www.fda.gov/AboutFDA/WhatWeDo/History/ProductRegulation/SummaryofNDAApprovalsReceipts1938tothepresent/default.htm (accessed 06 July 2014).
9) http://www.scripintelligence.com/home/ACC-PREVIEW-What-drug-trials-will-be-hot-in-Washington-DC-350858 (accessed 19 July 2014).

is a contributing factor to the treatment outcome, the design of the clinical trials is starting to take this dimension into consideration.[10)]

The narrowing of the patient populations initiated by the more precise, science-based profiling is further enhanced by the behavioral dimensions. As the societal trends have evolved and care procurement has grown more customized, these emerging individualized approaches lead patients to expect a fully personalized approach, fueling further the search for information.

## 14.7
## Societal Expectations for Personalized Medicine

We now need to better define what personalized – or "customized" – medicine is, what the expectations of the patients are, and the consequences for the industry. This author has been already writing on this topic.[11)] Could medicine be customized, like the newer generations of cars, fully tuned to the customers' liking (color, decoration, and features), the women's bracelets with a unique selection of "charms," the decoration of our homes, the attention that we get when we meet our private banker or when we flash any frequent user membership card, and so on? In actuality, patients increasingly want medicine to be fully tailored to one individual, and in many instances, they are ready to foot the bill (at least some of it) to an extent commensurate with the level of personalized attention that they get. It will probably be analyzed over time if this is a consequence of the societal evolutions or of the technological leaps that are pushing ever further the boundaries of the understanding of the pathological mechanisms underlying the medical scourges that affect us, down to details specific to an individual.

Beyond the expectations of the individuals, what is the reality of the procurement of care? It is still a far cry from truly personalized medicine. In the earlier article referenced above, this author has argued that there is a substantial gap linked to the false belief prompted by the buzz on personalized medicine which is falsely leading people to believe that we are seeing already personalized medicine, whereas we are actually looking only at increasingly precisely profiled medicine, yet still far from a genuinely individualized medicine.

The latest FDA report on personalized medicine is a case in point.[12)] Taking a topline view at its content, the report is (and predictably so) very regulatory and science-driven and is about genomics, imaging, and devices, with a focus on regulating *in vitro* diagnostics (companion to the treatment) to ensure availability of safe and effective (and reliable?) diagnostic tools.

10) http://www.scripintelligence.com/home/Healthcare-2030-facing-up-to-a-pharma-future-346693 (accessed 17 July 2014).

11) http://brainfoodtv.com/personalized-medicine-who-needs-it-and-what-for/#.U8qQ6LHmcSQ (accessed 15 July 2014).

12) http://www.fda.gov/downloads/scienceresearch/specialtopics/personalizedmedicine/ucm372421.pdf (accessed 23 July 2014).

Yet, at no point is the report giving any perspective on the crude – and cruel – fact that the entirety of the evidence collected through all these newer diagnostic techniques and the foundational science available on genetic profile, protein markers, and so on cannot be converted into meaningful medical interventions. In other words, we have more data and information than what we know how to and we can use in practice to tackle human ailments.

Also, and despite the fact that the report describes personalized medicine as "tailoring medical treatment to the individual characteristics, needs, and preferences of each patient," it is barely scratching the surface of the psychological and behavioral aspects. Yet, it is increasingly documented that these play a role in the odds of developing and then presenting a given disease (for instance, linked to eating and exercising habits), as well as how much the patient is likely to consult (and the triggers to be activated before the patient-in-the-making takes any action), how he or she is likely to influence the prescriber's attitude toward a specific treatment option, and, subsequently, how treatment adherence behaviors drive the risk of relapse and thus a sizable fraction of the treatment outcomes.

Addressing the patient's expectations and his or her medical needs in an integrated manner is therefore more complex by the day, and examples abound to illustrate this difficulty, as quoted from an article from a fellow BRAINFOOD TV editor of this author.[13)]

> A cancer patient named Rose was able to manage her disease for seven years by finding the right drug for treatment of her tumor based on her unique reaction to different chemotherapies and specific testing of her tumor cells.

Even if current care procurement cannot be considered as fully personalized, a successful individual outcome and consequently the satisfaction of the patient should be the ultimate performance indicator of healthcare procurement (and it is increasingly regarded as such), as the demonstration of the value is both recognized by the healthcare system and perceived by the patient as a consumer of the output of the care procurement process.

## 14.8
## New Players Contributing to Information Management to Substantiate Value Propositions for Novel Therapies

The pivotal question therefore becomes one of accessing all the relevant information for the diagnosis, deciding on a tailored approach for care procurement, and measuring the outcomes so as to collect insights as to the effectiveness of the intervention.[14)]

13) http://brainfoodtv.com/promise-better-outcome-personalized-medicine/#.U8qQI7HmcSQ (accessed 18 July 2014).
14) http://brainfoodtv.com/thus-spoke-uber-data/#.U8qTErHmcSQ (accessed 15 July 2014).

The first approach to collect data on the medical effectiveness of any healthcare intervention has been built on large pools of patients using the products and monitored in real-life conditions. Such patient cohorts, or registries, are now commonly used by the life science industry and also by the healthcare authorities (especially the payers) to determine whether the assumptions established on the basis of the findings stemming from the randomized clinical trials are actually happening in real-life conditions. The value of the insights emerging from the analysis of the real-world data harvested from such registries is well recognized, as illustrated by the newer regulatory approaches, aiming at facilitating access to innovation, which often rests on studying the target patient population in naturalistic conditions ahead of the regulatory approval of the drug candidate.

Such real-world medical data are essential to the understanding of the value generated by a new product. Yet they need to be supplemented by behavioral data, also collected in real-life conditions so as to understand the impact of the patients' attitudes on the treatment outcomes. The standard tools used to monitor patients in clinical trials (case report forms) or in registries (electronic medical records) have not been designed to capture these behavioral dimensions, and new technologies supplemented by the relevant analytics will be needed to reach that goal.

Technology trends are clearly observable. Beyond the obvious and usual suspects such as the cloud, mobility, and big data, one witnesses the surge in sensors, both in terms of types and in the degree of quality and reliability of the measurement of biological parameters, which add to the individual-specific data points stemming from the identification of biomarkers and genetic testing. A large number of companies with a wealth of experience in managing large volumes of consumer behavior data are now approaching the field of healthcare. For instance, one can think of the new health suite launched by Apple, the connected solutions of Orange, the massive investments made by Google, and so on.[15] For the time being, these approaches are still at the experimental stage and their purpose is reportedly to gather large volumes of data to test the feasibility of such approaches on a larger scale and to initiate preliminary correlation analyses. The next step will require the connection between the insights emerging from the analysis of these real-world behavioral data and the findings stemming from the real-world clinical data. Once this connection is made, the platform to analyze the contribution – positive or negative – of the behavior to the treatment outcomes will be ready, to the delight of all the stakeholders, provided that the algorithms leading to the conclusions are designed with public health in mind.

Such technological trends will be supplemented, on the therapeutic intervention front, by the current emerging "second wave" of biotechnology products which brings about very effective and finely targeted solutions; these will see their efficacy/safety ratio further enhanced by the expanding use of medical informatics processing patient data (McKinsey and Company, 2013) and of the insights of cognitive science regarding patients' behaviors.

15) http://online.wsj.com/articles/google-to-collect-data-to-define-healthy-human-1406246214 (accessed 28 July 2014).

In the search for ways to address inefficiencies of our current healthcare systems, good care will remain a function of efficient diagnosis, good testing, adequate treatment, and properly trained healthcare providers, delivering organized care in affordable packages.

The technology and metrics frame crystallizing around the patient casts an increasingly comprehensive web, but it is not just these solutions around the patient which will be connected. It is not even only the patient journey which will be engulfed in this web but also the entire life of the people, granting us the visa to cross the ultimate frontier of personalized medicine, that is, knowing ourselves to maximize the effectiveness of medical interventions, and consequently our degree of satisfaction with our healthcare procurement system.

The role of the companies that are digitalizing the life of the patients – and not only the patient but also the entire population – is becoming more obvious, when key questions are arising as to what to do with the data and how to do it. The debate is happening between private companies, lawmakers, regulators, influencers of all stripes, and so on. And this may not be to the liking of the patients, as a more data-driven approach will leave less room for decisions based on unsubstantiated beliefs, especially in a tightly regulated framework, designed to ensure effectiveness and respect of data privacy.

As computing power is becoming pivotal to harnessing the value of the data and as a result will further fuel the exponential evolution of technological changes in healthcare, the power – and the value – may ultimately reside with the owners of the codes, the algorithms, the expert systems, and ultimately the insights stemming from the analysis of the data. Here again, the data will be understandable by those who can put it in context and operationalize it toward delivering more efficient care. As a consequence, the power of the patient should be declining. And many a physician should be relieved to see the progressive extinction of the population of "clipboard patients" (those who come into the physician's office with a stack of printouts from the Internet and who believe that they know their condition, are asking for a specific treatment, and need to be "undiagnosed" when their perceptions are erroneous to the point that they may conflict with the execution of the therapy that they require).

But will that actually be the case? Will the patient's power fade until it falls in oblivion and becomes a thing of the past, or is there an alternative situation? Another scenario may be that the nature of the demand from the patient will evolve from a specific care intervention based on a self-diagnosis, as it is increasingly the case today, to more transparency and accountability about who controls the expert systems, how they have been designed, and most importantly for which purposes so that these are known, understood, and endorsed by the patients themselves. Whether such a transparent governance can be put in place in full or not, the question of the level of trust in the healthcare system will remain. And as trust in the healthcare system is volatile and elusive, the "trust infarction" may be just around the corner if the patients grow doubts as to the motives of the bodies governing the healthcare priorities and the allocation of care procurement resources.

The patients may also legitimately start to inquire about the fate of the other stakeholders along the healthcare journey. Over the past decades, this journey has become increasingly complex. A few decades ago, the bulk of the interaction between the people and the health procurement system could have been described as linear. A person feeling ill would ask for an appointment with a doctor, would be diagnosed, and would receive a prescription in a large proportion of instances. The patient would have the prescription filled in his or her usual pharmacy and would take his or her treatment until the condition is successfully managed, or otherwise it would bring the person back to the physician. Nowadays, the journey is more complex and could be described as multilayered and "matrixed" due to the number and nature of interactions happening simultaneously, which are not always in a synchronized manner: patients are seeing several doctors in the same period of time, and these practitioners may prescribe medications in parallel with each other and in many countries, these prescriptions are not always reconciled for consistency; the patient may be sent to hospitals or other facilities to undergo specific examination procedures or may be referred to paramedics to accompany the procurement of some parts of the care package. Meanwhile, that same person is collecting information from his or her private spaces, ranging from the inner circles of family and friends to the outer and virtual circle of social networks and the Internet.

In this complex volatile world of multiple interactions which are too often poorly coordinated and almost never gauged for true effectiveness, key questions are bound to arise and they are clear and present in the minds of some. These questions span the entire spectrum, from the actual execution of care procurement up to the broader question of the governance of the healthcare systems. But what will ultimately matter for the patient, as the consumer of healthcare, is what will happen in the circles close to him or her. Ultimately, the patient will be questioning whether the physicians might become mere integrators of insights provided in consolidated patient-specific dashboards, addressing all dimensions, from biological to psychological, and possibly up to care procurement recommendations entirely driven by guidelines, therefore being tasked with balancing the latest from evidence-based medicine and financial priorities and subsequently deciding on the course of intervention for the patient.

Beyond the circle that is close to the patient, the questions will also focus on the other stakeholders. Will the authorities and payers abandon totally the educational programs targeting healthcare practitioners and patients to focus solely on fine-tuning the parameters of the frame of reference, starting with the algorithms that will be the backbones of the expert systems supporting – if not driving entirely – care procurement decisions? And then again, on the basis of which evidence and with what set of goals?

At the heart of the system, a number of questions of ethical and moral nature will also be raised more frequently and acutely than they are already surfacing today, mostly under the form of interrogations pertaining to the protection of data privacy and the management of data integrity. Will the sets of data be truly comprehensive and made available in an accurate, transparent, intelligible, and trustworthy way? And what type of "ethics board" will supervise the entire process?

All these questions will polarize the debate around healthcare. And in countries where this topic is political ammunition serving electoral purposes, the debate is likely to be more intense, if not acrimonious. The role of the public authorities, starting with the regulatory one, will therefore remain of critical importance as it has been in the past decades to shape the value chain of the other stakeholders of the healthcare system, starting with the life science industry through evolving pieces of regulation.

## 14.9 The Role of the Key Stakeholders in Shaping a New Regulatory Framework

As we have already indicated and in order to place such regulatory trends in perspective, the R&D in the life science industry is seeing a constant decline of its productivity due to a combination of factors, ranging from having less "low-hanging fruits," as the simplest cures have long been discovered, to facing more complex and chronic diseases to understand, diagnose, and address, resulting in longer and more expensive clinical development, to dealing with more complex regulations reflecting the increase of the burden of the proof, consequently leading to overall erosion of R&D profitability.

Evidently, the regulatory authorities have played a pivotal role in this increased complexity by raising the standards of evidence required to document the dossier toward marketing authorization. The clinical data stemming from the modern clinical development programs are therefore valued among the corporate assets because they not only document the clinical efficacy and safety of a product, itself a rare asset, but also provide the foundational layer of the demonstration of the economic and societal value of this product, when combined with epidemiology data and health economic and outcomes research insights. The regulatory bodies (and the pricing and reimbursement authorities, to some extent) have taken a more active role in defining the nature not only of the data but also of the tools mandated to manage such data, and a superior agility at understanding the implications and the future evolutions of the regulations, mastering the data management systems, and answering data-based questions in a timely manner is becoming a competitive advantage, all along the value chain, from clinical development to commercialization (Lu and Su, 2010).

As these evolutions of the industry's functioning, core capabilities, and performance indicators have not escaped the attention of the authorities (cynics would say that it would be impossible not to hear the voice of the industry, clamoring that it needs high prices for all newer products to fund the ever more expensive research efforts), they have grown concerned about maintaining a proper access to innovation (Barker, 2010) and more specifically ensuring that the patients suffering from rare and poorly addressed medical conditions are assisted in accessing the care that they need. This line of reasoning has spawned a flurry of work, leading to the development of innovative regulatory pathways, including the possibility for the healthcare industry to interface early in the clinical development to discuss its orientations with the authorities[16] and the creation of specific statuses corresponding to expedited review times and commensurately adjusted data requirements.[17] Such regulatory dispositions are in place both at the US FDA (there are four of them: Fast Track, Breakthrough Therapy, Accelerated Approval, and Priority Review) and at the European authority level, the EMA,[18] which has three such processes.

It is now demonstrated that the evolution of the standard regulatory approaches has ensured a faster access to innovative care (Aggarwal, 2014) and it also had the collateral benefit of a better definition of the population of good responders since the aim of these newer regulatory dispositions is precisely to focus on the most suitable patients to maximize efficacy and tolerability of a drug development candidate. From the industry perspective, the resulting potential market is often smaller, but the better outcomes result in a lower incidence of adverse reactions, and the payers are more willing to pay a premium for products addressing the medical needs more precisely and effectively.

In parallel, additional exploratory work on the regulatory front is made to design a novel type of regulatory process, referred to as "*adaptive licensing* " (Eichler *et al.*, 2012), although some authors and experts such as Richard Barker are arguing that the term is not appropriate as it does not require any modification of the existing regulatory processes, but rather a better definition of an earlier and progressive patient access, under the proper level of pharmacovigilance control,[19] and subsequent expansion to broader patient populations. In such a new context, real-world medical data will become more essential than they currently are, and the ongoing analysis is aiming at exploring how nontraditional studies (i.e., not randomized, controlled, with strict inclusion and exclusion criteria) could provide the relevant, pragmatic, and observational data, relative to the existing standard of care, to fulfill the needs of the regulatory authorities (for the medical approval) as well as the pricing and reimbursement authorities (which are performing the health technology assessment analyses). The existing EMA legislation on conditional authorization provides a frame with the required flexibility to execute such innovative development approaches.

Evidently, beyond the needs of the regulatory stakeholders, such an approach bears the promise of generating evidence as to how to address the broader changes

16) http://www.fda.gov/AboutFDA/CentersOffices/OfficeofMedicalProductsandTobacco/CDRH/CDRHInnovation/InnovationPathway/default.htm (accessed 18 July 2014).
17) http://www.fda.gov/downloads/Drugs/GuidanceComplianceRegulatoryInformation/Guidances/UCM358301.pdf (accessed 18 July 2014).
18) http://www.mhra.gov.uk/PrintPreview/DefaultSplashPP/CON143792?ResultCount=10&DynamicListQuery=&DynamicListSortBy=xCreationDate&DynamicListSortOrder=Desc&DynamicListTitle=&PageNumber=1&Title=EU%20licensing%20flexibilities (accessed 18 July 2014).
19) http://www.scripintelligence.com/home/First-drugs-chosen-in-European-early-licensing-pilot-352203 (accessed 18 July 2014).

of the healthcare system which will be required to ensure the sustained access to patients and, more broadly, to an adequately flexible access to affordable care.

The modifications stemming from the generation of such real-life evidence along the backbone of clinical development have far-reaching implications, not only from the perspective of the regulatory and clinical development processes, as described above, but also from that of the engagement of the contributors to the building blocks of these data management processes, ranging from healthcare providers, to payers, to third parties, to patient themselves (whether as individuals or through the vehicle of patient associations).

For instance, healthcare providers will need to embrace the routine usage of more digital tools and interfaces in their daily practice, from the already existing electronic medical records to novel solutions such as expert systems allowing tailoring of the medical intervention to a clinical case while interfacing with the patient. From the vantage point of the public payer, the authorities will be driven to provide not only the legal frame within which the adoption of such digital solutions will progress smoothly (addressing, for instance, the questions of privacy management, interoperability of data systems, etc.) but also the funding (directly or indirectly through fiscal incentives) and the stirring of a robust debate by spearheading forums and other collaborative works. Lastly, though this is not the least significant variable in the equation, the patients will have to be willing to share their data. Although the younger generations seem to be more open to sharing their personal and medical data, this aspect of the debate is unlikely to be resolved in the near term.

Going forward, the stakeholders tasked with the management of the real-life healthcare data will be facing formidable challenges, starting with the volume (the already massive amounts of data available in the myriad of databases, from national health system bases to specific patient registries, all of them with different data nature and structure since they have originally been created for other purposes). The second difficulty is the "knowledge gap" which we have already described, as it is harder to make sense with the data and it is driving the industry to invest more resources and management attention to early discovery research and translational medicine (Paul *et al.*, 2010). And the third one is the conflict between the sense of urgency (necessity to create the framework before the data backlog is too enormous) and the budgetary constraints.

## 14.10
## The Consequences for the Healthcare Industry in Terms of Governance and Capabilities

The earlier described trends have been generating profound changes on the roles and governance of all the stakeholders of the healthcare ecosystem. As regards the healthcare industry, the companies have been factoring them into the design of their strategic roadmaps and this is visible through a lasting impact on the management of the life cycle and of the value chain.

The life cycle of the healthcare novel products has been, for many decades, very product-centric and linear, starting from early discovery, evolving into preclinical and then clinical development, then regulatory, pricing, and reimbursement, and finally commercialization. The process was aiming at amassing lots of evidence, from the basic chemical and biological science needed to characterize the molecule, to the clinical data required to demonstrate its efficacy and safety under controlled conditions, to the economic evaluation requested by the payers. And when the products are commercialized, the accumulation of information continues through pharmacovigilance and additional clinical work. Overall, this large and growing body of evidence provides reassurance as to the clinical and economic value of a product and backs the intellectual property rights that help to keep me-too and generic competition at bay for the duration of the protection.

The emerging reality is that this entire process inherited from the heydays of chemistry is increasingly patient-centric. The evolution can be noticed from the design of the clinical development, in which the type of data collected and requested by authorities and the recruitment of the patients have evolved. We have already described the changes driving the authorities to request more "real-life" data in the course of the clinical development, so as to better anticipate the level of disruption brought by a novel therapy in the existing standard of care. The recruitment of the patients is therefore happening along different criteria and this serves well the purpose of all the patient associations and NGOs which are working at creating more opportunities for patients to access innovation earlier, even before it is fully approved by the regulatory authorities.

Thus, the development of novel therapies is progressively escaping the sole control of the industry and its regulators and is becoming multidirectional, involving more partners and stakeholders from the design to the execution. As the industry does not control all the required capabilities and know-how (Accenture, 2013) – both of which have evolved faster and further than what the industry was willing to internalize, either by choice or by necessity – the clear consequence is that the industry has to walk the path of collaboration and partnering on the phase of its value chain that it used to control exclusively.

The healthcare industry, owing to its mission, has an organizational structure of its own, quite different from that of most other industries.

First, looking at the similarities, it has functions dedicated to performing operations, which remain in existence and deliver their contribution along the life cycle axis. These functions include R&D, manufacturing, and commercialization. But these functions have been operating in an evolving matrix.

Such matrix presents functions that could be clustered into two distinct groups. There is obviously the presence of the all control functions that can be found in other industries, such as finance and controlling, legal, and HR. And there is the other group of controlling functions that are specific to the healthcare industry and whose remit has been expanding and strengthening over the past years, driven by a combination of internal forces (product portfolio, processes) and external forces (regulatory, financial, and societal). These include medical affairs, regulatory affairs, market access, global marketing, and compliance.

Starting with medical affairs, one can only expect this function to play a key role in an industry focused on the health of human beings. Yet, this role has been subject to evolutions in the past decades. From their former role as the partner of the commercial functions from the launch of the product, medical affairs have been increasingly involved earlier in the clinical development phase and have established their pivotal role in the management of scientific and medical information, as well as in the content-based interfacing with a wide scope of stakeholders, from the medical community, to the patients and their associations, to the authorities.[20] As the portfolio of the industry is transitioning to specialty care products, supported by more complex fundamental science and medical evidence, and the corresponding patients are harder to diagnose and to manage, the role of medical affairs is bound to strengthen further.

Regulatory affairs are a function which has been in existence for many decades and are primarily tasked with ensuring that the companies are functioning in line with the existing regulations. Regulatory affairs ensure the appropriate licensing, marketing, and legal compliance of pharmaceutical and medical products in order to control the safety and efficacy of products. This function combines knowledge of scientific, legal, and business issues to ensure that products, which are developed, manufactured, or distributed by the companies, meet the required legislation. Regulatory affairs also advise on and coordinate the approval and registration of products. As the regulations are evolving and becoming more complex, stringent, and multilayered (with the national and supranational levels), it goes without saying that the remit of regulatory affairs has moved in sync.

Another function tasked with ensuring that the industry adheres to rules is compliance. This function has emerged progressively as a standalone one, bordering legal and regulatory, now functioning as an independent body reporting to the highest echelon of the corporate hierarchy and monitoring and reporting results of the compliance and ethics initiatives of the company. It is focusing specifically on the numerous interactions that the stakeholder-facing resources of the companies are having with medical professionals and government officials (the latter being broadly defined as anyone who is paid for by a public body, which includes, for instance, the prescribing physicians in public hospitals).

The fourth function to bear in mind in the healthcare industry matrix of control-focused functions is global marketing. It may sound surprising to cluster a marketing function in the "controlling group," but one has to remember the reality of the healthcare industry: it has long innovation cycles, and it brings very few new products to the market. As a consequence, it has few assets to maintain its profitability and to fuel its future growth. Moreover, these assets are backed by standardized data sets which back the product claims and have been expensive to generate. As a consequence, bar specific regulatory situations in some countries which may slightly modify the scope of the registration of the product (e.g., indications or dose regimen), the products are launched on the basis of a standard platform, which has to be respected so as to minimize the risk of product

20) www.cegedimrm.com/medical-affairs-insights (accessed 11 July 2014).

usage which has not been fully documented. Global marketing is therefore ensuring that the commercial teams are staying within the approved boundaries, often described in a reference document.

Lastly, let us focus our attention on market access. For the purpose of this chapter, this function deserves a specific attention.

To begin with, it is important to bear in mind that "market access" is a widely misused expression and oftentimes poorly understood or at least subject to so many interpretations that a definition is warranted. Market access literally means the activities required in order to "access" a market; but in the jargon of the healthcare industry, it encompasses the activities which are not related to R&D, nor to medical affairs, nor to regulatory, or nor to commercial. As a matter of fact, it encompasses the preparation of the value proposition and subsequently the execution of the negotiations with the payers and the maintenance over time of the terms of the said negotiations.

In countries in which the governance of the healthcare system is calling for a centralized pricing or reimbursement approval as a mandatory preamble to commercialization (i.e., every country except for the United States and the United Kingdom, although the latter is considering following the path of Germany which has started to implement a centralized pricing procedure, emulating all the other EU countries), the intervention of the market access function used to be a "one-off" step of the life cycle. It was essentially taking place after the regulatory approval (i.e., when the novel product was considered as acceptable from the medical perspective) and as a prelude to commercialization, providing as an output the price and, where applicable, the level of reimbursement by public payers.

The preparation of the negotiation with the authorities is resting on converting clinical data and epidemiology findings into equations and models presenting the value of a given product or intervention for the healthcare system considered –hence a task to be replicated across markets, since the treatment guidelines, patient flows, and costs of interventions, let alone the valuation of one's life, vary from one country to another.

In light of the evolution of the clinical development, with the progressive inclusion of real-life data, market access has mutated in a lasting activity, fully embedded in other corporate processes (McClearn and Croisier, 2013), starting as early as phase II in clinical development and stretching throughout the life of the product and aiming at building the body of evidence which supports the value proposition over time.

Authorities are increasingly eager to understand how a novel therapy will impact the status quo in a therapeutic segment and create a new balance in the said segment of the healthcare ecosystem. Evidently, the authorities and the payers – and sometimes the patients – are demanding an early and tentative assessment of the impact on the standard of care, including the identification of the economic and social benefit linked to using a product (i.e., in addition to its medical benefit assessed by the regulatory authorities), hence the tentative collateral impact on care procurement for the targeted ailment. As the full impact is becoming visible over time, the industry is requested to make commitments in terms of value delivered to the system on the basis of models and assumptions.

The nature of the data that is requested has evolved, to include real-world evidence as mentioned earlier in this chapter. Such data can be obtained in an experimental design (hence controlled) or be purely observational (such as retrospective cohort studies, often quite complex due to the heterogeneity of the structure of medical records and the hurdles to access relevant data). Behind the request to better understand the effectiveness and safety of the products in daily clinical practice, the authorities are searching for evidence strengthening payment or reimbursement decisions and allowing to optimize the allocation of healthcare resources and to minimize the uncertainty pertaining to the impact of the novel therapy.

And as the "real" patients are increasingly central to the process, the appetite for real-world behavioral data in addition to real-world clinical data is emerging. The industry is evidently struggling with the requirements stemming from these evolving demands. How could the industry address this world of unstructured behavioral data when its DNA is based on a strong foundation of randomized, controlled clinical trials and the adventure of real-world clinical data is still young? Beyond the technical issues pertaining to the nature and structure of the data, the industry is acknowledging a substantial gap in competencies and skills. But this is dwarfed in comparison to the cultural gap: owing to the focus on maximizing while preserving rare assets by complying with a flurry of regulations spanning the entire value chain, the culture of risk management seems to be the primary obstacle for the industry to close the gaps rapidly.

The space dedicated to analyzing the behaviors of patients and their impact on care procurement outcomes will not remain empty for long. One is already witnessing the progressive penetration of companies which were not originally positioned in the healthcare space and which are leveraging their know-how in terms of consumer data management, whether from the perspective of hardware and software of data mining. These companies will contribute to digitalize the life of the people, including the patients and also expanding into the world of prevention, in other words the "stay well" space.

## 14.11
## The Sustainable Path Forward for the Healthcare Industry

The novel approaches undertaken by the companies entering the healthcare arena are driving an evolution of the acquisition of patient-related data by supplementing the existing approaches with a data pool which is hoped to provide critical information about the behavior of patients in the near future. Also, it is expected that these data sets will be used along with the other data repositories to allow a genuinely comprehensive view of the patients and of their relationship with care procurement. Therefore, these companies play an essential role in creating innovation opportunities in relation to the data acquisition and management, the generation of insights, and the consumption of healthcare resources. These companies also create more economic value linked to healthcare, and although the magnitude of that economic potential remains to be quantified, one may expect that it could offset – at least partially – the ever-growing burden of the healthcare costs in the real economy by creating an economic surplus in addition to contributing to a better utilization of healthcare resources.

The impact for the "traditional" life science companies is broad and far-reaching. As Brian Smith has eloquently described (Smith, 2012), the pharmaceutical industry has always been evolving, or rather coevolving. The forces at work involve physical technology (the changes in the environment of basic and applied sciences relevant to healthcare), social technology (the environment of regulations, economics, and politics as well as the more specific evolutions of the healthcare systems, such as patient flow, care procurement, or pricing and reimbursement), and the mutations of the business models of the companies (resulting in evolving strategies, themselves driving changes in organization and processes). More specifically, life science companies are seeing mutations of their business model, from the basic science driving a portfolio shift to specialty care products requiring a different approach to clinical development; they are facing evolving regulatory standards and expectations for types of data which transform the nature of the burden of the proof and hence of the substantiation of the value proposition of the assets and of the approach to strengthening the intellectual property position; and they are confronted with financial and societal pressures that are not easy to reconcile. The healthcare industry has navigated these turbulent times by adjusting its resource allocation and governance, notably by showing a greater openness to outsourcing and partnering. This approach has offered a greater flexibility to access the relevant skills and capabilities and to adjust to the fluctuations in demand for such assets. Considering that the new entrants in the field of healthcare data collection and management are large corporations with extensive financial resources and that this move is part of their strategic agenda, the healthcare companies have little choice other than exploring ways to partner with the data management giants.

The practices of the industry keep evolving and the strategies of the firms show some remarkable similarities, resting on four pillars: the search for innovation to fuel growth, a balanced portfolio (ranging from the very disease-area focused ones of specialty care companies to the broad and balanced ones of the large pharma conglomerates), a diversification across market segments and geographies, and the search for external growth (with business development activities resulting in more focused, bolt-on deals than in megamergers). Only the fittest resist in the environment and even some of the healthier and most promising companies fall prey of the acquisition appetite or larger groups. As the most successful strategies will be understood and emulated by the other players, the evolution will spread and take root.

One should expect a major evolution of the way future medicines are developed (and this is already happening, as exemplified by the newest oncology treatments), regulated, priced or reimbursed, marketed, and ultimately … consumed! Care procurement resting on custom-tailored therapeutic options will mandate multidisciplinary collaborations to a greater extent, as well as the coordination and engagement of multiple stakeholders and evidently a massive surge of data requiring adequate management capabilities.

The crux of the matter is the very nature and the quantification of the value added by the healthcare industry. The first building block of the value proposition is of technological nature: as described in this chapter, new technologies, from R&D to commercialization, a novel approach to information management and insight generation, and more targeted – if not yet personalized – care procurement solutions will determine part of the content of the value proposition. The second building block is societal: a better informed, more assertive population with a greater interest in prevention, exerting pressure on authorities, will enhance the demands for accurate and trustworthy comparative analyses and further fuel risk aversion. This second pillar completes the scope of forces at work to define the value proposition, which is evidently growing more complex and fragmented.

In order to thrive in its ecosystem, the industry needs to maintain a flow of innovation, to navigate more complex hurdles (quality, regulatory, market access), to engage successfully with a wider range of networked stakeholders through multiple channels (including translating market data into actionable insights), to adjust its strategic roadmap and organization while preserving its profitability, and to polish its corporate social responsibility stance so as to improve its image and perceived trustworthiness.

Remaining in sync with the pace of change of its ecosystem and retaining a voice at the table of the essential contributors to the bettering of human life and society should therefore drive the healthcare industry to engage in a complete revolution of its governance and processes, much more than a simple evolution or even a reboot.

### 14.11.1
### Take-Home Messages

Among the building blocks of the healthcare industry value chain, market access has evolved into a lasting activity, starting as early as phase II in clinical development and stretching throughout the life of the product, therefore bridging R&D and real-life consumption of the product. The aim is to build a credible and sustainable body of evidence which supports the product's value proposition over time, factoring in all real-life data and especially the patient experience as a consumer of care.

New players not originally positioned in the healthcare space are adding unique know-how in terms of consumer data management and play an essential role in bringing innovation and economic value linked to healthcare, offsetting the ever-growing burden of the healthcare costs in the real economy and contributing to a better utilization of healthcare resources.

The life science companies must evolve their governance from R&D to intellectual property management and to corporate social responsibility, reinforcing their collaboration networks, strengthening specific capabilities (medical affairs, health economics, and healthcare data management at end-user level), and engaging a broader spectrum of stakeholders.

## References


Accenture (2013) Pharma's Sharp Right Turn on R&D.

Aggarwal, S.R. (2014) A survey of breakthrough therapy designations. *Nat. Biotechnol.*, **32**, 323–330. doi: 10.1038/nbt.2864, Published online, 08 April 2014.

Barker, R. (2010) A flexible blueprint for the future of drug development. *Lancet*, **375**, 357–359.

Citeline (2013) Scrip Research and Pharmaprojects.

Eichler, H.G., Oye, K., Baird, L.G., Abadie, E., Brown, J., Drum, C.L., Ferguson, J., Garner, S., Honig, P., Hukkelhoven, M., Lim, J.C., Lim, R., Lumpkin, M.M., Neil, G., O'Rourke, B., Pezalla, E., Shoda, D., Seyfert-Margolis, V., Sigal, E.V., Sobotka, J., Tan, D., Unger, T.F., and Hirsch, G. (2012) Adaptive Licensing: taking the next step in the evolution of drug approval. *Clin. Pharmacol. Ther.*, **91** (3), 426–437. doi: 10.1038/clpt.2011.345, Epub 2012 Feb 15.

Health Consumer Powerhouse (2009) Euro Health Consumer Index Report 2009.

Lu, Z. and Su, J. (2010) Clinical data management: current status, challenges, and future directions from industry perspectives. *Open Access J. Clin. Trials*, **2**, 93–105.

McClearn, C. and Croisier, T. (2013) *Big Pharma's Market Access Mission*, Deloitte University Press.

McKinsey & Company (2013) The Big Data Revolution in Healthcare, January 2013.

Mistra Pharma (2009) *A Healthy Future – Pharmaceuticals in a Sustainable Society* Published in collaboration between, Apoteket AB, MistraPharma and Stockholm County Council.

Paul, S.M. *et al.* (2010) How to improve R&D productivity: the pharmaceutical industry's grand challenge. *Nat. Rev. Drug Discovery*, **9**, 203–214.

Smith, B. (2012) *The Future of Pharma: Evolutionary Threats and Opportunities*, Gower.

The Economist Intelligence Unit Limited (2012) Economist Intelligence Unit Survey, July 2012.

World Economic Forum (2013) Sustainable Health Systems: Visions, Strategies, Critical Uncertainties and Scenarios, January 2013.

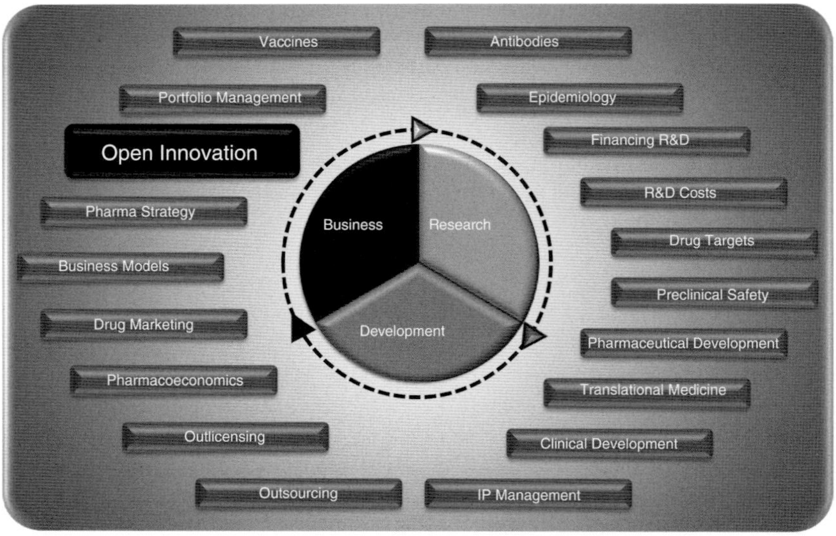
Vaccines
Antibodies
Portfolio Management
Epidemiology
Open Innovation
Financing R&D
R&D Costs
Pharma Strategy
Business
Research
Drug Targets
Business Models
Preclinical Safety
Drug Marketing
Development
Pharmaceutical Development
Pharmacoeconomics
Translational Medicine
Outlicensing
Clinical Development
Outsourcing
IP Management

# 15
# The Pharmaceutical Industry is Opening Its R&D Boundaries

*Alexander Schuhmacher and Ulrich A. K. Betz*

## 15.1
## Introduction

The purpose of this chapter is to provide a sound basis to understand the complex of open innovation, to offer an overview of today's open innovation initiatives in the pharmaceutical industry, and to highlight new open innovation paradigms that may play a role in the pharmaceutical business in the future.

In 2003, Henry W. Chesbrough described in *Open Innovation – The New Imperative for Creating and Profiting from Technology* for the first time a new innovation paradigm advancing from primarily inside-driven and vertically integrated innovation management ("*closed innovation*") to a more holistic and open understanding of innovation ("*open innovation*").

Cost and growth pressures can result in a rethinking and a change from the more traditional innovation model of closed innovation toward the new concept of open innovation. According to Chesbrough, many industries are in a transition phase between both paradigms today (Chesbrough, 2003). With respect to the pharmaceutical industry, a reduced efficiency of the research and development (R&D) engines resulting in a stagnant or even reduced number of new drugs approved by the regulatory agencies, strong competition from generics, increased cost pressure from payers, and an increased biological complexity of new target indications (keyword: chronic diseases) may have advanced the shift toward open innovation. Thus, in the past years pharmaceutical companies have broadened their external networks toward research collaborations with academic research institutes, technology providers, or codevelopment partners. To fulfill the demand to increase productivity and reduce R&D timelines and costs, companies started to outsource R&D activities that were not defined as core competence. "*Core competences*" may be defined as factors that an organization sees as being central to the way it, or its employees, works, and that result from the collective learning of the organization (Müller and Lechner, 2011). In addition, internal R&D processes were adjusted to the more open R&D model, and new processes such as alliance management and knowledge management were established. The corporate frontier of pharmaceutical companies became permeable and more open. As a result, the focus of

*Value Creation in the Pharmaceutical Industry: The Critical Path to Innovation,* First Edition.
Edited by Alexander Schuhmacher, Markus Hinder, and Oliver Gassmann.

pharmaceutical R&D expanded from a purely internal toward a mixed internal and external model.

With this book chapter, we intend to answer the following questions:

- What is open innovation and what are the differentiators to closed innovation?
- What is the status of open innovation in the pharmaceutical industry today?
- What are the successful examples of open innovation initiatives that have been installed in the pharmaceutical industry?

In summary, this book chapter describes the paradigm of open innovation in context of the pharmaceutical industry. Today, the U.S. pharmaceutical company Eli Lilly may have established the most open model toward external innovation, as it has integrated its innovation processes with its business model. Other companies are following this more open R&D model. We describe open innovation processes and the capabilities that are required in context of open innovation, namely, the absorptive, relational, and multiplicative capabilities. And we illustrate that the not-invented-here (NIH) and not-sold-here (NSH) syndromes are attitudes that disable open innovation. In addition, we portray the more traditional elements of open innovation that have been part of the R&D business model for a while: target scouting, research collaborations, drug licensing, outsourcing, and joint ventures. And we disclose the newer concepts of open innovation such as new frontier sciences, drug discovery alliances, private–public partnerships, innovation incubators, virtual R&D, crowdsourcing, open source innovation, and innovation camps.

## 15.2
## Open Innovation versus Closed Innovation

In the closed innovation paradigm, R&D is primarily understood as a company-internal fully end-to-end controlled process without external contributions. According to Chesbrough, most of the international companies in the past had built up large internal R&D organizations. These were fueling sales and company growth by developing new products for the company to launch to the market.

Closed innovation is a model whereby "successful innovation requires control" (Chesbrough, 2003). Companies must generate their own ideas, discover, develop, and produce their own products, market, and distribute them, service and finance them, all by themselves, "because one cannot be sure of the quality, availability, and capacities of others' ideas: If you want something done right, you've got to do it yourself" (Chesbrough, 2003).

Chesbrough defined some basic thinking prominent in the closed innovation paradigm:

- Hire the best and brightest people to work for your corporation.
- New products and services are developed by the company itself.
- Products and services that are discovered in-house need to be launched to the market by the company itself.

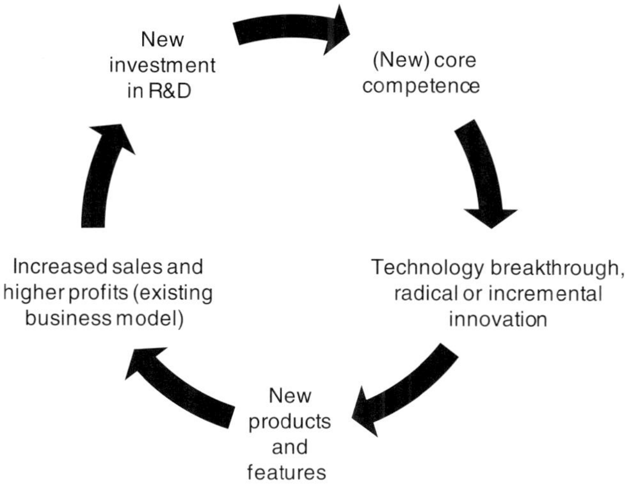


**Figure 15.1** The logic of closed innovation. Adapted from Chesbrough (2003) Open Innovation. The New Imperative for Creating and Profiting from Technology. Harvard Business School Press, p. xxi.

- Being first-to-market successfully is possible with discoveries originating in the boundaries of the firm.
- High internal R&D investments result in the discovery of best products which enable the company to lead the market.
- Stringent management of intellectual property (IP) rights is key to control products, markets, and competitors.

The cycle as illustrated in Figure 15.1 demonstrates these interdependencies of the model of closed innovation.

In R&D management, the closed innovation logic results in closed boundaries of the company. Ideas are generated by in-house scientists, screened and filtered, and product development projects advanced through an internal R&D process. Resulting products are launched to the market by the company itself. Typically, companies have developed stage gate models and R&D funnels, with strict control point (gates) where promising and less promising projects are separated. Some projects are killed, while others survive. Resources are allocated to surviving projects till the next stage gate review. The aim of a gated process is to increase the probability of success (PoS) of remaining projects, while less promising projects are eliminated on their way through the R&D funnel.

As outlined in Figure 15.2, new ideas enter the pipeline of an organization based on internal capabilities, know-how, and knowledge of experts employed by the company. The number of research projects per phase is decreasing with advancing project maturity, as failing projects are eliminated on their way through applied R&D phases. Finally some of the internal projects result in new products launched to the market. The scope of all R&D activities is related to the target market(s) of the company.

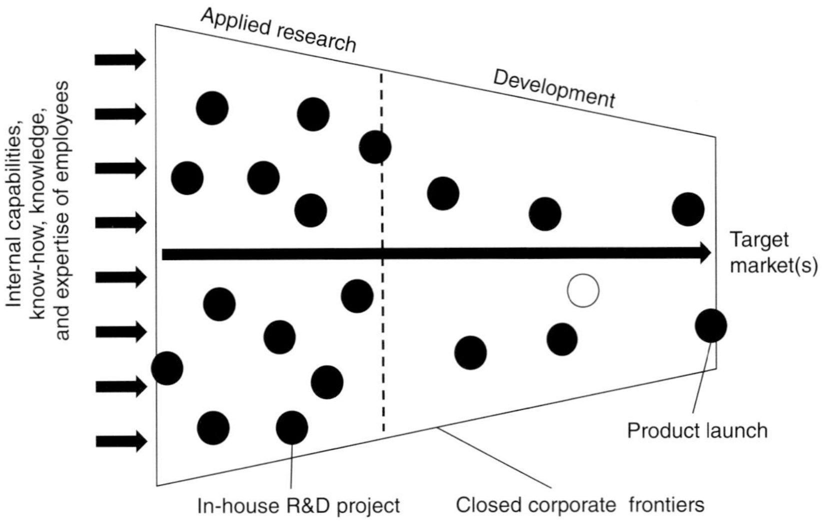


**Figure 15.2** R&D funnel in the logic of closed innovation. Adapted from Chesbrough (2003) Open Innovation. The New Imperative for Creating and Profiting from Technology. Harvard Business School Press, p. xxii.

IP is primarily used to secure a competitive edge, fight off competitors, and protect the company's R&D results.

The primary source of new ideas is the company itself, mainly the corporate R&D functions. Companies try to be highly attractive employers and offer experts in R&D lifetime employment. Internal structures and processes are built to provide an optimal environment for innovation to flourish and to stimulate the company's own scientists to be creative and innovative. "Picking a man of genius, giving him money, and leaving him alone" may describe the situation best (Chesbrough *et al.*, 2006).

Chesbrough describes the logic of closed innovation as "one of closed, centralized, internal R&D. At its roots, the logic implies a need for deep vertical integration. ... One must do everything internally .... Outside the fortified central R&D castles, the knowledge landscape was assumed to be rather barren. Consequently, the firm should rely on itself" (Chesbrough, 2003).

Two mindsets are associated with the closed innovation paradigm. In the "*NIH syndrome*," new ideas, research projects, or products originated from outside the company's boundaries are negatively viewed, and internalization of these externally originated assets is difficult. In the "*NSH*," mindset products generated by inhouse R&D need to be marketed by the own company. If a company is not successful in launching a product, why should others achieve this. Both syndromes are affecting the productivity of an organization negatively. External assets that might be valuable additions to the company's pipeline are excluded, and internal assets that might be marketed more successfully by other companies are not licensed out. Both mindsets potentially reduce the value generation potential of R&D.

Pharmaceutical companies use their internally developed know-how and knowledge to identify, evaluate, and access external ideas and technologies – an ability that is defined as absorptive capacities (Chesbrough *et al.*, 2006). Companies that fail to develop absorptive capabilities have a severe competitive disadvantage but can still get access to complementary resources via alliances or partnerships. Contract research organizations (CROs) are supporting the development of new drugs to reduce time-to-market and to reduce development expenses.

Chesbrough describes "open innovation" as "a paradigm that assumes that firms should use external ideas and internal ideas, as well as internal and external paths to market, as firms look to advance their technology. Open innovation combines internal and external ideas into architectures and systems whose requirements are defined by a business model. The business model utilizes both external and internal ideas to create value" (Chesbrough, 2003).

As a consequence, internally developed products can also generate value when being launched to the market via external channels that are not within the current business of the company.

As illustrated in Figure 15.3, new R&D ideas, projects, capabilities, knowledge, know-how, and expertise from both inside the company and outside the corporate frontiers are used to fill the R&D pipeline. New research concepts and development projects can be accessed via research collaborations, licensing, and acquisitions and can enter the R&D funnel at each phase of the R&D process. Most importantly, the boundaries of the R&D organization are permeable in both directions – inside out and outside in. In the same way, projects can leave the organization to find another (probably better) way of value creation.

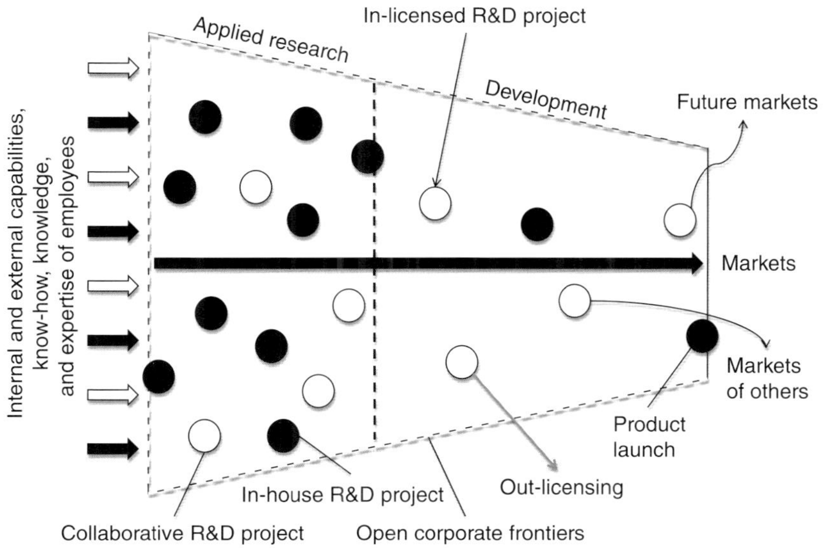


**Figure 15.3** Open innovation paradigm for managing R&D. Adapted from Chesbrough (2003) Open Innovation. The New Imperative for Creating and Profiting from Technology. Harvard Business School Press, p. xxii.

In the open innovation paradigm, external collaboration partners, service providers, and suppliers can provide technologies and other services at least to the same quality of what the company could have delivered internally. In consequence, companies that use the potential of external providers can allocate their resources more efficiently to the R&D projects that have been prioritized, and projects can advance faster on their way to higher market potential. In addition, projects that were terminated or put on the shelf under the paradigm of closed innovation can be offered to external partners and used for their (new) markets.

Chesbrough contrasts the principles of closed and open innovation as illustrated in Table 15.1.

Chesbrough also published the following points of differentiation for Open Innovation, relative to prior theories of innovation (Chesbrough *et al.*, 2006):

- In Open Innovation, external knowledge is of equal importance as internal knowledge.
- In Open Innovation, the company does not only strive to hire the best and brightest people to work for the company, but does also seek the genius outside the company.
- In Open Innovation, type I (false positive) and type II (false negative) measurement errors are considered in evaluating R&D projects.
- In Open Innovation, companies go outside to access external innovation and enable outward flow of internally generated innovation.

**Table 15.1** Contrasting principles of closed and open innovation.

| Closed innovation principles | Open innovation principles |
|---|---|
| The smart people in the field of technology work for the company | Not all smart people work for the own company. Companies need to work with smart people inside and outside the corporate boundaries |
| To profit from R&D, research, development, and product launch are done by the company | External R&D can create significant value, and internal R&D is needed to claim some portion of that value |
| Internal R&D provides full control over projects and the opportunity to be first to the market | Companies don't need to originate the research to profit from it |
| The company that launches an innovative product to the market first has the highest payoff | Building a better business model is better for companies than being the first to the market |
| The company that creates most and best ideas in the industry has a competitive edge | The companies that make the best use of internal and external ideas have a competitive edge |
| The company needs to protect their products by stringent IP management | Companies can profit from others' use of their IP and from buying others' IP whenever it advances their business model |

Adapted from Chesbrough (2003) Open Innovation. The New Imperative for Creating and Profiting from Technology. Harvard Business School Press, p. xxvi, lines 18–34.

- In Open Innovation, it is believed that the accessible external knowledge can be of high quality.
- In Open Innovation, IP management is done proactively.
- In Open Innovation, innovation intermediaries play an important role.
- In Open Innovation, new performance metrics are required to assess the performance of a company's innovation process.

## 15.3 Business Models in an Open Innovation Framework

Companies must reorganize themselves to be able to use the full potential of open innovation. In that context, Chesbrough defines the change from closed to open innovation more as the establishment of a new business model rather than changes in innovation processes only.

A "*business model*" is defined as a design of cross-linked activities which realize a certain value proposition to achieve value creation (Müller and Lechner, 2011). In other words, it is the way how companies create, deliver, and capture value. Business model innovation refers to the creation of a new business itself and, thus, to a new way of how a company creates, delivers, and captures value.

In *Open Business Models – How to Thrive in the New Innovation Landscape*, Chesbrough details six different types of business models (types 1–6) in accordance to their openness. "*Type 4 business models*" may be most prominent among the multinational pharmaceutical companies. In this business model, the companies have started to open their R&D toward external innovation. The use of external ideas and technologies expand the companies' access to a greater variety of resources, reduces costs of R&D, and time-to-market, and risks are shared among partners. Companies forecast their needs, for example, when and how to fill gaps in the R&D pipeline, and set up shopping lists for external ideas and technologies. Networks with partners help to identify potential in-licensing targets. Opening toward external innovation is a source of new growth in addition to the internally discovered and developed drug candidates that are launched to existing target markets. Suppliers and customers are systematically integrated in the innovation process, and their ideas may also appear in the portfolio of pharmaceutical innovation. Founded to support the review of R&D project portfolios or business cases for single R&D projects, scientific advisory boards are established as another source of external knowledge in the innovation management process. Collaborations and partnerships with universities and scientific institutes are utilized to get access to external know-how and ideas. Strategically, the focus of R&D is still within the areas of the current business of the company, such as a certain therapeutic area. Innovation management is a cross-functional activity with market experts being involved in the process. Market potential and market access are evaluated in detail and used for innovation management, mostly as integral part in the target product profiles (TPPs) of drug candidates.

According to Chesbrough, the U.S. pharmaceutical company Eli Lilly may be an example within the pharmaceutical industry that can be considered to belong to the group of "*type 5 business model*," which is even more open toward external innovation than type 4, as companies integrate their innovation processes with their business models. The type 5 business model in context of open innovation connects the different functions within a company and enables the company to react on complex challenges effectively. The business model is known to partners and outsiders, and they understand what kind of ideas and technologies the company is looking for. Partners have access to the company's innovation process and vice versa. As an example, employees of partner companies may be members of the scientific advisory board of the company. The company gets access to its customers' needs and future requirements, and customers are part of the innovation process of the company. Opening new opportunities in the market, type 5 companies try to deeply understand the unmet needs of patients, physicians, payers, and other stakeholders. Companies of type 5 build strong alliances with complementing companies, such as preferred partnerships with CROs in clinical development. Innovation is a core element in the company's business model and part of the DNA of the company. All employees feel that they can add value to the company and its future success. Internal and external ideas and technologies are integrated comparably into the R&D process of the company. Both sources of innovation are coordinated with the help of a joint platform that manages all R&D activities. IP is managed strategically as a financial asset. Potential collaboration and licensing targets are identified by patent mapping; external technologies are actively scouted to strengthen the companies' IP position. Licensing of external projects or technologies is organized as a profit center. Innovation management is still a cross-functional activity with experts from marketing, sales, and finance supporting R&D and being members of cross-functional project teams.

## 15.4 Open Innovation Processes

When the business model changes from closed to open innovation, the respective processes must be adapted as well. Open innovation processes should be tailored to resonate with the paradigm that innovation can come from internal and external sources and that internal ideas can also find an external path to the market (Figure 15.4).

Reviewing 124 companies, Gassmann and Enkel proposed three core process archetypes for open innovation: (i) the outside-in process, (ii) the inside-out process, and (iii) the coupled process (Gassmann and Enkel, 2004). In the "*outside-in process*," companies invest in cooperations with suppliers, customers, and other sources of external knowledge to enrich their knowledge base and to increase their innovation potential. Chesbrough defined outside-in open innovation as "outsiders contributions that enable an enterprise to create offerings whose scale

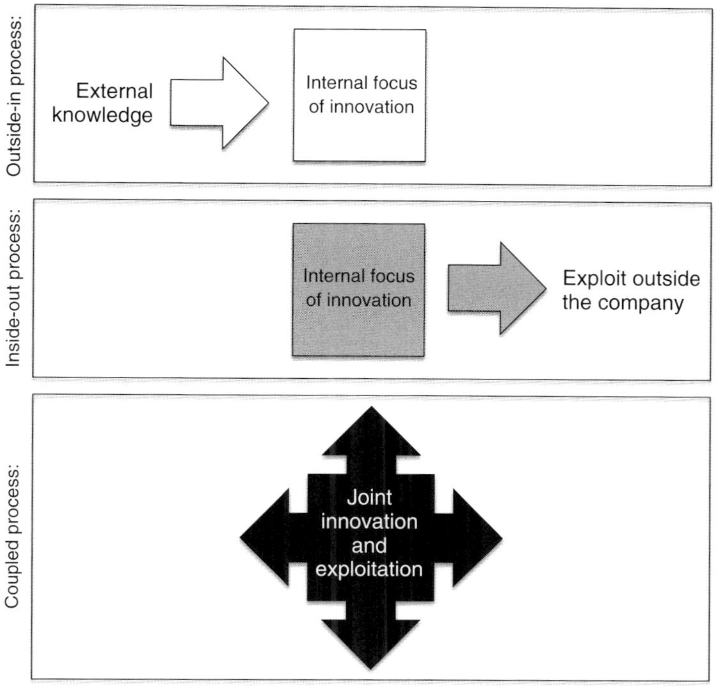


**Figure 15.4** The three open innovation processes. Adapted from Gassmann and Enkel (2004).

belies its internal capabilities" (Chesbrough and Garman, 2009). Pharmaceutical companies intensively use the outside-in process by collaborating with universities and other academic research institutes or by in-licensing drug candidates. Universities and scientific institutes can enhance the innovation potential by providing research knowledge and technology capabilities. Licensed IP may give freedom to operate, and licensed drug candidates can fill project portfolio gaps.

The benefits resulting from outside-in-driven innovation can be operational or strategic, such as a better utilization of internal resources, the access to new products and technologies, the reduction in technical or financial risk, or a faster time-to-market. The fundamental advantage of accessing external innovation is the access to new knowledge, new technologies, and flexible resources.

Companies that prefer to use the "*inside-out process*" in open innovation focus on externalizing their knowledge, innovation, and IP "in order to bring their ideas to market faster than they can through internal development" (Gassmann and Enkel, 2004). This means that companies place some of their assets or projects outside their walls, for example, out-license their IP or transfer their technologies to other companies. In this context, the locus of invention does not necessarily need to be the locus of exploitation. Examples of inside-out processes in the pharmaceutical industry are (i) technology licensing, such as Roche's offer of PCR licenses to other biotech and pharmaceutical companies or (ii) pharmaceutical companies that out-license drug candidates so that they can be developed in new therapeutic indications, which are not within the strategic focus of the licensors. (iii) CROs contribute to the innovation processes of pharmaceutical companies by performing clinical trials and using drug candidates. Pharmaceutical companies benefit from this type of outsourcing by getting access to the CROs' expertise in managing clinical trials effectively or its expertise in a new disease area that is out of the company's current scope. Pharmaceutical companies can also benefit from managing internal and external resources more flexibly and from increasing the speed of development. Improved cost effectiveness, cost control, and reduced fixed costs and the access to expert knowledge may also be arguments in favor of outsourcing.

In the "*coupled process*," companies combine access to external knowledge (outside in) with their ambition to bring new products to other markets outside their current scope (inside out). Long-term, bidirectional interactions between two or more partners such as consortia of competitors, joint ventures, or alliances as well as cooperations with universities or research institutes are examples of this type of process. The benefits gained from coupled processes are an improvement of the competitive position or a mitigation of risks (Gassmann and Enkel, 2004). In the pharmaceutical industry, for example, cooperations between pharmaceutical and biotech companies are common to apply new technologies or to codevelop new drug candidates.

According to Gassmann and Enkel, each company principally prefers one model, although elements of other models are also used. As an example, pharmaceutical companies access external innovation by collaborating with universities or by in-licensing of new drug candidates while they outsource clinical trials to CROs, participate in public–private partnerships, or collaborate with other pharmaceutical companies and universities in the frame of consortia such as joint drug developments for the treatment of neglected diseases (Kar, 2010).

## 15.5
## Capabilities and Attitudes Enabling Open Innovation

Open innovation can result in increased performance, but first companies need to overcome negative attitudes toward externally generated innovation and strengthen their capabilities in acquiring external innovation.

As reviewed by Lichtenthaler *et al.* (2011), companies are characterized by two types of corporate cultures that have a negative impact on enabling open innovation (Lichtenthaler *et al.*, 2011). The "*NIH*" syndrome is a negative attitude toward externally sourced innovation, while the "*NSH*" syndrome describes a negative attitude to transfer corporate technologies or product development projects for a product launch by another company. NSH may be based on the fear to sell the crown jewels to competitors, while NIH may result either from limited or from negative experience with outside-in transfer of innovation.

In contrast to these disabling syndromes, companies need to develop specific capabilities to enable open innovation. Companies need to have or need to develop the capacities and expertise to identify and assimilate externally available valuable ideas, projects, information, or technologies ("*absorptive capabilities*"). "*Multiplicative capabilities*" are required in respect to the inside-out process, as the exploitation of internal knowledge, ideas, projects, or technologies outside the company is related to the company's capability to select partners, as well as to multiply and transfer knowledge to the outside environment. And "*relational capabilities*" are needed when using the coupled process in long-term, trustful relationships with partners.

## 15.6 Open Innovation in the Pharmaceutical Industry

### 15.6.1 The More Traditional Elements of Open Innovation

Looking at what has already been done in the past, most, if not all, pharmaceutical companies' R&D models rely on internal R&D activities built on internal know-how and internal expertise. Research was and is still done in-house to a greater or lesser extent. Clinical development is planned and managed internally, conducted with in-house experts, potentially case by case complemented by external service providers. Pharmaceutical companies collaborate with academic research institutions to get access to know-how and technologies. External drug candidates are licensed in – preferably in late-stage clinical development – to fill pipeline gaps and mitigate risks. Nonstrategic activities are sourced out to service providers and CROs. The following paragraphs detail the more traditional elements of open innovation that pharmaceutical companies exploit usually before the newer concepts of open innovation are described.

#### 15.6.1.1 Target Scouting

Today, pharmaceutical companies mainly fill their R&D pipelines utilizing three different approaches to discover new therapeutic entities, namely, the (i) phenotypic screening, the (ii) target-based screening, and (iii) the analoging approach which is based on the further optimization of known therapeutic entities.

As drug targets are part of complex networks, the target-based screening approach is associated with an advanced complexity of target selection and validation (Agarwal, 2013; Sams-Dodd, 2005). In consequence, pharmaceutical companies add to their in-house activities external research concepts and scout for new targets to feed their screening pipelines.

The phenotypic screening represents the more traditional way of drug discovery and has recently been entering a renaissance with modern high content screening approaches. Because of the complexity of its assays and the lack of clear structure–activity relationships, it is often hampered by the increased difficulty of therapeutic entity optimization. The need for a thorough target deconvolution to identify the molecular target underlying a positive effect adds further aspects to its complexity. Also, pharmaceutical companies are relying on external know-how, collaborations with technology providers, and alliances with biotech companies or academic research institutes to reduce the complexity related with phenotypic screening.

#### 15.6.1.2
#### Research Collaborations

Research collaborations are sought to get access to world-class science and new technologies. In most cases, pharmaceutical companies collaborate with internationally acknowledged scientists and their labs to use their creativity and scientific excellence. In addition, academic institutes are a primary source of new drug targets and innovative drug discovery projects (Frye *et al.*, 2011; Schuhmacher *et al.*, 2013). Likewise, academic collaboration partners can provide know-how in therapeutic areas and new technologies. For example, the Division of Signal Transduction Therapy (DSTT) is a consortium of the University of Dundee, the Medical Research Council (MRC), and six pharmaceutical companies, namely, AstraZeneca, Boehringer Ingelheim, GlaxoSmithKline (GSK), Janssen Pharmaceutica NV, Merck Serono, and Pfizer, to perform research on the development of new drug treatments for major global diseases. Specifically the consortium aims to target protein kinases and the ubiquitylation system in the development of these therapies (http://app.dundee.ac.uk/pressreleases/2012/may12/drugdiscovery.htm, last access: 30 November 2014).

#### 15.6.1.3
#### Drug Licensing

In addition to in-house drug development, licensing has been the second path to get access to innovation in the past years. In most cases, pharmaceutical companies, in this case the licensee, get permission, either limited or exclusive, to a drug candidate or a group of drug candidates from the licensor, mostly biotechnology companies. Licensing deals are made preferably for late-stage drug candidates, as they provide less technical risk for the licensee and higher revenue for the licensor.

In general, according to an analysis performed by McKinsey, an increasing amount of drug candidates of pharmaceutical companies has been originated externally (see Figure 15.5).

In another analysis by Schuhmacher *et al.* (2010), it has been described that multinational pharmaceutical companies have acquired on average 50% of their pipeline projects from external sources, whereby 25% were in-licensed (Table 15.2) (Schuhmacher *et al.*, 2013). In 2014, J + J, Merck & Co., Pfizer, and Roche were among the most active license deal makers in the pharmaceutical industry (www.medicus-associates.com/pharmadeals.net/journal/pharmadealsreview/fulltext/pdf/1673; last access: 22 November 2014).

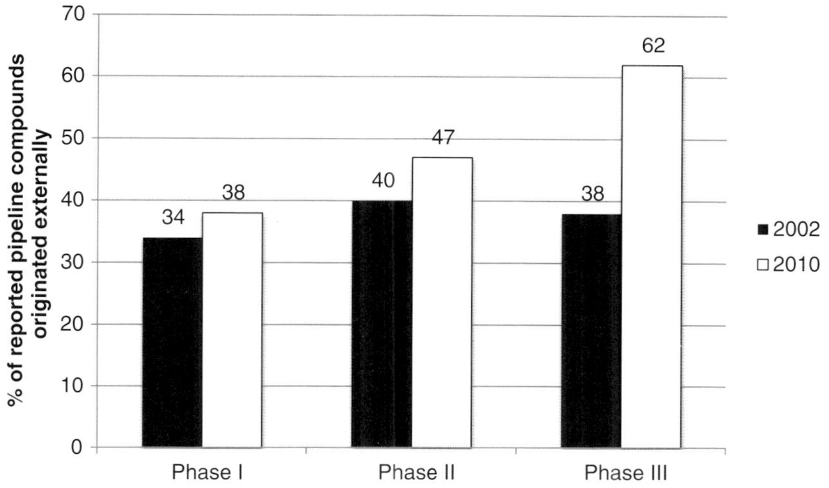


**Figure 15.5** Reported in-licensing in clinical development phases in 2002 and 2010. Adapted from: Dhankar *et al.* (2011).

**Table 15.2** Key R&D pipeline figures of multinational pharmaceutical companies.

| Total number of R&D projects | Number of organic R&D projects | Number of R&D projects accessed by company acquisition | Number of R&D projects licensed | Total externally sourced R&D pipeline (%) | Externally sourced R&D pipeline by licensing (%) |
|---|---|---|---|---|---|
| 62 | 30 | 24 | 8 | 52 | 13 |
| 102 | 44 | 30 | 28 | 57 | 27 |
| 56 | 46 | 0 | 10 | 18 | 18 |
| 103 | 42 | 42 | 19 | 59 | 18 |
| 111 | 79 | 15 | 17 | 29 | 15 |
| 241 | 136 | 22 | 83 | 44 | 34 |
| 113 | 50 | 34 | 27 | 56 | 24 |
| 176 | 104 | 28 | 44 | 41 | 25 |
| 143 | 75 | 53 | 15 | 48 | 10 |
| 143 | 74 | 47 | 33 | 48 | 23 |
| 116 | 33 | 39 | 54 | 72 | 47 |
| 65 | 40 | 19 | 32 | 38 | 49 |
| | | | | Average: 50 | Average: 25 |

Data derived from Evaluate Pharma®, annual company reports (2011), Datamonitor®.

Licensing drug candidates offer powerful advantages to the licensor, such as:

- Strategic pipeline filling
- Faster time-to-market
- Higher probability for licensed late-stage projects to reach the market
- An evaluated threefold increase in return, when reinvesting from internal research to licensing (Chew, 2010)

As the operational benefits of licensing drug candidates are obvious, pharmaceutical companies compete aggressively for all the promising drug candidates (Adams, 2010). As a result, this trend drives up the prices, reduces the success of licensing, and reduces earning potential for this "low-risk" R&D approach. Pharmaceutical companies are forced to look for less commercially attractive late-stage licensing options or for earlier-stage drug candidates. Due to the increased scarcity of such assets available for licensing and rising licensing costs, however, early-stage assets have received increased attention. Already between 2008 and 2009, 57% of in-licensing deals were signed for preclinical drug candidates. Recently also new players have entered the market with innovative biotech companies emerging in India and China. While these countries have initially mainly been viewed as a source of cost-efficient outsourcing, also the increasing innovation potential in these countries has been recognized early on (Betz and Neumann, 2009).

The real potential of drug candidates licensed or acquired from biotech companies has been questioned in an analysis by Bain & Co. Due to the high demand, large pharmaceutical companies find it increasingly difficult to get high-quality drug candidates in-licensed. From 6000 projects analyzed, only 200 were qualified as being of interest to big pharmaceutical companies, and 50% of this figure were regarded as having the potential to become a top seller drug (Behnke and Hueltenschmidt, 2010). On the same token, Professor Gary Pisano from Harvard University recently questioned the efficiency of the biotechnology industry and considered the pharmaceutical industry to be more efficient in providing drug candidates than the biotechnology industry. This view would challenge the business model of the pharmaceutical industry to access a main portion of their innovation by in-licensing drug candidates from biotechnology companies (Pissano, 2006).

#### 15.6.1.4
#### Outsourcing

Outsourcing in the pharmaceutical industry has already started in the 1950s by Hoffmann-La Roche (Quinn, 2000). Today, pharmaceutical companies use strategic outsourcing in basic research, discovery research, active product ingredient (API) development, and clinical development to integrate and use new technologies, to balance internal resource bottlenecks, to access high-quality services, or to reduce fixed costs and retain flexibility in their own R&D organizations (Godin, 2005). The increased trend toward outsourcing in the pharmaceutical industry is firstly fueled by an increased outsourcing of clinical drug development and secondly outsourcing in discovery research (Mehta and Peters, 2007; Clark, 2010). Table 15.3 illustrates statistics on outsourcing activities in the years between 1997 and 2009.

It is important to note that even R&D phases that have been thought to be of strategic importance for pharmaceutical companies are outsourced to service providers and CROs. Today, R&D outsourcing is even taking place in all preclinical research phases with outsourcing strategies that consider the risk-sharing

**Table 15.3** Outsourced R&D expenditures by type in the years 1997, 2001, 2005, and 2009 (Hu, 2007).

| Year | Spending in USD billion | | | |
|---|---|---|---|---|
| | **Discovery** | **Clinical development** | **Total** | **Percentage outsourced (%)** |
| 1997 | 0 | 2 | 2 | 10 |
| 2001 | 2 | 5 | 7 | 24 |
| 2005 | 4 | 9 | 13 | 33 |
| 2009 | 7 | 17 | 24 | 41 |

model of the later phases (http://www.lifescienceleader.com/doc/global-drug-discovery-outsourcing-market-gaining-momentum-0001, last access: 4 December 2014).

Outsourcing has become more attractive especially due to cost advantages realized with the emergence of capable CROs in China and India, although recent quality problems showed the risks of outsourcing to these countries, when regulators in France, Germany, Belgium, and Luxembourg were suspending the marketing approval of 25 generic drugs due to concerns over the quality of data from clinical trials conducted by India's GVK Biosciences (http://www.reuters.com/article/2014/12/06/india-pharmaceuticals-germany-idUSL6N0TP3HZ20141206).

#### 15.6.1.5
#### Joint Ventures

Joint ventures have already been used in the past to combine equities in a joint R&D activity. In the last years, some new joint ventures have been announced, where pharmaceutical companies seek the technical expertise or market access of specialty partners. For example, Amgen set an agreement with Watson in 2011 to develop, manufacture, and commercialize oncology biosimilars (http://investors.amgen.com/phoenix.zhtml?c=61656&p=irol-NewsArticle&ID=1641126; last access: November 6, 2015). Another example, Biogen Idec together with Samsung (Korea) or Merck & Co. with Hanwha Chemical (Korea) set up joint ventures to develop, manufacture, and commercialize biosimilars or together with Sun Pharmaceuticals to get access to emerging markets (http://www.thepharmaletter.com/file/109732/amgen-in-cancer-biosimilars-deal-with-watson.html, last access: 30 November 2014, online.wsj.com/articles/SB10001424052748704529204576256471908122588; last access: 22 November 2014).

### 15.6.2
### The Newer Concepts of Open Innovation

Newer concepts of further opening up R&D organizations of pharmaceutical companies are scouting for new frontier sciences, early alliances, crowdsourcing, private–public partnerships, innovation incubators, virtual R&D, and open source innovation. Pharmaceutical companies use a multitude of different approaches to generate innovation and also try to leverage as much as possible existing assets by out-licensing, codevelopment, or copromotion activities.

#### 15.6.2.1
#### New Frontier Science

"New frontier science" is a program that was initiated by Takeda in 2010. The program name "new frontier sciences" may have been founded in dependence on a series of space exploratory missions conducted by NASA (New Frontiers program) with the purpose of reaching distant planets of our solar system. The aim of Takeda's program was to explore new technologies that may impact the pharmaceutical business in the long run (20 years and later). The founding of this program is inspired by the case of Genentech, which, at a very early stage, focused its R&D activities on a new technology, namely, recombinant DNA technology, and consequently founded a new industry, the biotechnology industry. Today, Genentech belongs to the Roche Holding and is the provider of most of the new drug products that are launched by Roche. In general, new frontier science approaches rely on breakthrough innovation, can be considered as high-risk activities, have a low PoS, and provide a potentially very high return on investment (ROI).

#### 15.6.2.2
#### Drug Discovery Alliances

Early drug discovery and development alliances are becoming increasingly important for pharmaceutical companies to get access to external innovation and technologies. While in-licensing of late-stage drug development candidates and new frontier sciences meet short-term (1–2 years) and very long-term (more than 10–20 years) needs, drug discovery alliances with pharmaceutical or biotechnology companies, research institutes, or universities support the mid- to long-term strategies (3–10 years) of pharmaceutical companies. Alliances are not used only to get access to external capabilities, but also provide the advantage of expanding available resources, opportunities for licensing, access to external networks, and support in decision making.

"Early alliances" are a special form of drug discovery alliances that were established by Nycomed in 2008. "Early alliances" was a concept whereby long-term relationships in discovery research and preclinical development with start-up companies that seek funding were established. Nycomed, in the meantime acquired by the Japanese Takeda, offered their internal resources such as labs and personal and collaborated with the biotechnology companies to get knowledge and access to their drug programs. The programs were discovered and developed jointly. In case the partners failed to show preclinical proof of concept (PoC), Nycomed took the risk of failure. In the case of successful preclinical PoC, Nycomed asked for a first right of refusal. The key advantages of this type of alliance were a fast decision-making process without long-lasting licensee–licensor negotiations, which are standard in drug licensing and acquisition. In addition, the pharmaceutical partner could use available capacities to discover and develop jointly an early drug candidate with the purpose of having access to the drug candidate later on. Likewise, the pharmaceutical company could use early alliances to familiarize with a new technology or therapeutic indication without the need to invest too many resources. The biotechnology partner gets access to more technical know-how and resources without the necessity to seek a strategic investor or without the risk of losing control of the own drug candidate.

#### 15.6.2.3
#### Private – Public Partnerships

An important trend are public – private partnerships and business ventures, which are funded and operated by a partnership of governmental institutions plus one or more private sector companies. Examples include the Biomarkers Consortium (source: www.biomarkerconsortium.org), the Critical Path Institute Consortia (www.c-path.org/consortia.cfm), the Innovative Medicines Initiative (www.imi.europa.eu), or the Serious Adverse Events Consortium (www.saeconsortium.org). For example, in 2004, the FDA initiated the Critical Path Initiative, together with pharmaceutical companies, to improve the process and the effectiveness of the pharmaceutical drug development process of pharmaceutical R&D. The critical path program comprises the development and utilization of biomarkers, the modernization of the clinical trials methodologies, the use of bioinformatics, and the improvement of manufacturing technologies (Woodcock and Woosley, 2008). The aim of all these initiatives is to find and develop solutions for the main challenge of pharmaceutical R&D, the low productivity, and the related high R&D costs. Other interesting examples of private – public partnerships (Nwaka and Ridley, 2003) are the global health partnerships (GHPs) around drug R&D for neglected tropical diseases (NTDs) and other serious unmet health needs, such as malaria or tuberculosis (http://www.ifpma.org/fileadmin/content/Publication/2012/BSR_Review_of_Global_Health_Partnerships_final_report_Sept_3_2012.pdf, last access: 30 November 2014). Today, 20% of all partnerships concentrate on HIV/AIDS, 16% on NTDs, 16% on women and children's health, and 14% on malaria, and 14% of all partnerships currently focus on noncommunicable diseases (NCDs). Other sorts of public – private partnerships are biotech clusters, such as in Germany the publicly and privately supported Heidelberg BioRN cluster or the m4 Munich biotech cluster which bring together pharmaceutical companies, biotechnology companies, and academia to stimulate the generation and exchange of innovation (http://www.m4.de/en/index.html, last access: 22 November 2014).

#### 15.6.2.4
#### Innovation Incubator

The U.S. biotechnology company Biogen Idec has founded an innovation incubator named bi3 (http://bi3.biogenidec.com, last access: 22 November 2014). Bi3 is a platform to access external innovation. Bi3 offers the opportunity to external scientists to access resources, labs, facilities, and business support, which enables the external scientists a rapid start-up of a new business idea. In return, Biogen Idec expects to increase its opportunities in new drug candidates.

Innovation incubator models optimizing academia–industry interaction are, for example, the Pfizer Centers for Therapeutic Innovation (CTI) or the GSK–University Cambridge academic incubator. The Pfizer model invites academic researchers to submit project ideas that are selected for funding by a joint industry–academia selection committee. Projects are then implemented by mixed academia–industry teams working side by side sharing bench and clinic time. Thereby, Pfizer is providing financial funding, human resources, phage display, and peptide libraries as well as associated technologies, while academics bring in their hypotheses of new drug mechanisms. Decision making is done in joint steering committees. The inventions are filed in the name of both partners with Pfizer having a right of first refusal (Ratner, 2011). In the GSK model the collaborating university dedicates a team of academic experts to work on self-defined drug discovery and development projects, also bearing some of the financial project risk.

The Merck Serono bioincubator in Yavne, Israel, is helping biotech companies and start-ups to translate their basic research results into practical applications. The Merck Serono Ventures Fund now manages an initial investment of approximately €10 million, and Merck Serono's Inter-Lab Research and Development Center in Yavne is providing organizational and infrastructure support. The goal is to have at least six start-ups working to transform the ideas of Israeli scientists into new medications or technologies by 2018.

Recently a new model of academia–industry interaction called *outcubation* was described which is located between outsourcing and incubation (Betz and Tidona, 2015). The outcubation approach starts with the publication of an unresolved preclinical challenge published as a call at major universities. Young academic talent is encouraged to apply with an idea how to resolve that challenge. Participants are selected in the frame of a 1-week bootcamp and employed by and relocate to the Outcubator, which is incorporated as a private biotech company. The Outcubator is located on the campus of a global innovation hub. Each team is selected and sponsored by a pharma partner and is guided by an experienced industry mentor, for example, a senior researcher of the sponsor company, and an academic advisor, such as an university professor from the specific field of the team's research topic. The outcubation approach combines synergistically the best of both worlds, academia and industry, to generate a unique environment enabling breakthrough innovation.

#### 15.6.2.5
#### Virtual R&D

Virtual R&D may be described best as an R&D organization which operates with very few internal R&D staff and utilizes external resources, technology, and facilities to cover the whole R&D process. The basic principle of virtual R&D, essentially to keep the organization slim, has been successfully applied by some pharmaceutical companies such as Shire or Chorus. Another aspect of virtual R&D is the access to global resources on demand and the option to leverage the best value out of a given asset. The Chorus model of Eli Lilly has proven that virtual R&D can help reduce cost and time needed in pharmaceutical R&D. Since 2002, the R&D productivity of the Chorus model has apparently been 3–10 times higher than the traditional pharmaceutical R&D model (Longman, 2007). The model is described in further detail later.

Other examples of virtual R&D in the pharmaceutical industry are Endo, King Pharmaceuticals, and Ranbaxy (India). The latter received marketing approval for a new antimalaria drug in 2011 that has been developed by a virtual project team of scientists that worked for different companies and organizations in Switzerland, Australia, and India. Originally, the project was initiated by Medicines for Malaria, a not-for-profit organization (NPO), and drug development was steered by Roche (http://cdippel.wordpress.com/2011/11/17/a-long-strange-trip/, last access: 30 November 2014). Another example is Endo Pharmaceuticals, a company that uses virtual R&D to fill its short- and midterm project pipeline. In virtual discovery programs it combines in-house experts in diverse research fields with the know-how of external partners such as in India. Endo's research teams provide the concept, while several external research groups increase the probability to find new drug candidates in a shorter period of time, with less fixed costs and lower investment risk (http://www.healthtechzone.com/topics/healthcare-leaders-care-points/articles/105184-endo-pharmaceuticals-embraces-virtual-rd-expand-drug-pipeline.htm, last access: 30 November 2014). King Pharmaceuticals, which is now a subsidiary of Pfizer, ran a virtual R&D organization that had an R&D rate of 4% in 2006. The company did not maintain own labs, and all development activities were outsourced to academic institutions, to pharmaceutical and biotechnology companies, or to CROs (http://www.pharmexec.com/pharmexec/Strategy+Articles/King-Without-the-Crown/ArticleStandard/Article/detail/333311, last access: 30 November 2014).

#### 15.6.2.6 Crowdsourcing

Over the past few years, a series of crowdsourcing initiatives have been started in the pharmaceutical industry. "Crowdsourcing is a strategy that involves outsourcing of knowledge generation and problem solving to external actors with the help of a public request to a group of people" (Gassmann, 2010). Eli Lilly can be considered as the pioneer and most active pharmaceutical company in this context and has created four open innovation projects called *InnoCentive*, *YourEncore* (www.yourencore.com), the Lilly Phenotypic Drug Discovery Initiative, the Lilly Target Drug Discovery Initiative, or the Lilly TB Drug Discovery Initiative (https://openinnovation.lilly.com/dd). InnoCentive and YourEncore are now operating as independent companies.

The innovation intermediary InnoCentive (www.innocentive.com) is the most prominent initiative in the pharmaceutical industry. It has been started as an R&D platform that provides a potentially global knowledge pool for pharmaceutical companies. It is a problem–solution platform designed to connect pharmaceutical companies and their research challenges, with external solution providers receiving a monetary reward if they provide successful solutions.

Undoubtedly, crowdsourcing provides some advantages for pharmaceutical companies, such as the ability to access a global brain pool for problem – solution in a fast and nonbureaucratic way. The underlying processes and structures are flat and not hierarchical. Pharmaceutical companies can receive a large number of solutions to a posted problem from different solution providers, such as individuals, freelancers, or experts from other industries, that are usually outside the reach of the company. The broad access to a diverse human brain pool, the low fixed costs, and respective investments are speaking in favor of crowdsourcing.

Regardless of these advantages and widely cited pioneer initiatives, crowdsourcing is not really widely used in the industry yet. This is due either to some perceived difficulties, such as IP issues, or the role internal scientists play in the context of crowdsourcing. NIH attitudes are certainly important for all forms of outside-in innovation. Scientists may fear losing control when they use crowdsourcing. At the same time, they may also worry that they would become redundant in the event that crowdsourcing would be successfully applied on a consistent basis. In addition, there may also be the misconception that the companies already employ the best scientists in-house or collaborate with the best experts available externally. These open questions of culture, leadership, and motivation need to be addressed to allow companies to untap the huge potential of crowdsourcing and make it a great success.

Beyond working with crowdsourcing service providers such as InnoCentive, Innoget, or NineSigma, companies are now increasingly starting their standalone portals on company websites. In one example, crowdsourcing is used as an excellent tool to scout for new drug targets and to access target proposals. Bayer has realized this potential and has started the initiative "From Targets to Novel Drugs" in 2009, which is accessible via www.grants4targets.com (Lessi *et al.*, 2011). Until today, 1110 applications for reviewing a new target proposal were filed, 13% of which were accepted (http://www.grants4targets.com/scripts/pages/en/status_report.php, last access: 30 November 2014). A further example is the Merck Serono Open Compound Sourcing program that allows interested partners to submit small molecular compounds for inclusion in the Merck Serono High-Throughput Screening (HTS) library along with success participation and collaboration options. Likewise, the Merck Serono mini library provides a collection of former Merck Serono R&D compounds to interested partners that can be used to discover new therapeutic utilities, again combined with success participation opportunities (www.merckserono.com/open_innovation). At the same time, Merck Serono is also offering a series of grants to sponsor research, which can be freely applied for in a crowdsourcing-like approach (http://www.merckserono.com/en/about_us/responsibility/GlobalGrants/Global_Grants.html).

#### 15.6.2.7
#### Open Source Innovation

There is a growing trend in the pharmaceutical industry to support open source innovation. In open source innovation, information is shared in an incremental, cumulative fashion across companies, institutions, and research platforms. Individuals contribute for free, with the understanding that results will be owned by the public domain. The model aims at improving creativity by bringing together the best minds regardless of organizational affiliation. Business critical is the access to open innovation patent pools, avoiding limitations, and barriers in drug R&D. For example, GSK and Alnylam have pooled their IP in respect to the treatment of tropical diseases, drug candidates, and money to support an initiative to identify drugs for neglected diseases. The Massachusetts Institute of Technology (MIT) also joined the initiative, which in the meantime has open access to 2300 patents (Kar, 2010). The thinking behind is to provide new drugs targeting diseases of the developing world where there is no potential ROI and where IP does not play the same important role as in the major pharmaceutical markets. As, for example, GSK opened its Tres Cantos campus to allow their researchers to work more collaboratively with external scientists from universities, NPOs, and other research institutes (http://www.gsk.com/en-gb/research/sharing-our-research/open-innovation/, last access: 30 November 2014). In this context and in May 2010, it published over 13 500 promising potential "hits" in the journal *Nature* to stimulate drug discovery research for malaria (Gamo *et al.*, 2010).

The Structural Genomics Consortium (SGC) could be considered as another example of this philosophy. The SGC catalyzes research in new areas of human biology and drug discovery by focusing explicitly on less well-studied areas of the human genome. It accelerates research in these new areas by making all its research output available to the scientific community with no strings attached and by creating an open collaborative network of scientists in hundreds of universities around the world and in nine global pharmaceutical companies.

In summary, this network of academic and industry scientists is driving a new scientific and drug discovery ecosystem whose primary aim is to advance science and is less influenced by personal, institutional, or commercial gain.

#### 15.6.2.8
#### Innovation Camps

The innovation camps are another new and powerful approach of open innovation that brings external and company-internal experts together for joint ideation, generation of new project plans, or joint problem solving. Usually this comes in the form of a dedicated event such as a summer camp or a hackathon and has been successfully applied by Novartis (International BioCamp (http://www.novartis.com/careers/biocamp/)), the MIT (Hacking Medicine; http://hackingmedicine.mit.edu/), and Merck Serono with the Merck Serono Innovation Cup (innovationcup.merckserono.com). The Merck Serono Innovation Cup brings together the brightest life science and MBA postgraduate students from major innovation hubs from all over the world for 1 week at the summer camp near Merck's Darmstadt headquarters. The company-external young talents meet the generation of Merck's retirees for joint ideation and to develop new project ideas covering unmet medical need. The best project plan worked out during the week is awarded the Merck Serono Innovation Cup (€20 000), and suitable project plans are implemented at Merck Serono. Similar innovation camps are also used to boost ideas collected in a grassroots innovation process at Merck. In the frame of an innovation bootcamp, ideas are advanced to full business plans with the help of external advisors or collaboration partners (Betz *et al.*, 2014).

#### 15.6.2.9 Fluctuating Open Teams

Along these lines, another new open innovation approach, termed "fluctuating open teams (FLOTs)" has recently been described that combines synergistically internal and external talent for a drug discovery project. In this paradigm, rather than relying only on a fixed project team of internal employees, a core team of classic internal employees is harvesting new ideas and perspectives in regular intervals or at critical points in the project from a highly diverse group of temporary team members assembled according to predefined criteria from the global brain pool. These temporary team members are selected to exhibit high diversity in relation to experience (ranging from retirees to active professionals and PhD students), affiliation (ranging from academia to service provider, biotech industry, or free-lancers), as well as technical expertise (covering central and associated fields in relation to the project team's task) (Betz *et al.*, 2014).

## 15.7 New Business Models in View of the Potential of Open Innovation

### 15.7.1 General Trends in the Pharmaceutical Industry

The key problem of the pharmaceutical industry is that despite unprecedented investments in R&D, the number of new drugs reaching the market has remained flat leading to a constant rise in costs for each new drug brought to the market. Ways to respond to this challenge are as described earlier a shift toward open innovation and also restructurings, Mergers and Acquisitions M&As, and more stringent portfolio management even in early phases (Betz, 2011).

With regard to the challenges pharmaceutical companies face and the associated cost pressure on R&D, companies like Pfizer, AstraZeneca, or GSK have restructured their R&D organizations in the past years and have reduced traditional R&D capacities significantly.

GSK launched its Discovery Performance Units (DPUs) approach with 38 DPUs in 2008, promising to cut groups that underperform and support those making progress after winning 3-year budgets for their work (http://www.fiercebiotech.com/story/gsk-will-cut-3-research-units-add-4-rd-productivity-swells/2012-02-07). In 2010, GSK closed down several R&D centers in Europe and shifted R&D capacities to countries in emerging markets. In respect to the closure of the R&D site in Harlow, United Kingdom, GSK's chief executive officer (CEO) Andrew Witty said: "I don't like to see anybody leave the organisation or there to be any job losses. But the reality is that we do from time to time need to look at ways we can improve our probabilities of success" (http://www.pharmafile.com/news/gsk-downsize-harlow-rd-site, last access: 30 November 2014). R&D at GSK is getting cost efficient, open, more virtual, further partnership oriented, and global.

Similarly, the British–Swedish AstraZeneca announced to cut around 1600 jobs in R&D globally by 2016 after having already shrunk down the R&D organization by 2200 jobs in the previous years (http://www.pharmatimes.com/Article/13-03-19/AstraZeneca_reorganises_R_D_and_cuts_1_600_jobs.aspx, last access: 30 November 2014). Capacities are either concentrated to three major R&D sites in the United States and Europe or are outsourced to pharmerging markets like China and India to achieve cost effectiveness. In the same way, Merck & Co. recently announced a 17% reduction in R&D personnel resulting in a USD 1.25 billion saving in R&D costs (http://www.fiercepharma.com/story/skinny-earnings-cost-cuts-boost-merck-bristol-myers-forest-fx-hits-sanofi/2014-04-29, last access: 30 November 2014). The changes in R&D also include a refocus on therapeutic indications and the start of four innovation centers of external innovation. Likewise, in particular Pfizer has been active in reducing its R&D costs after running through two mergers since 2003 (http://www.fiercepharma.com/story/pfizers-post-megamerger-cost-cutting-record-51500-jobs-7-years/2014-04-29). After the acquisition of Wyeth in 2009, it cut its annual R&D expenditure from USD 7.5 billion in 2003 and an all-time high of USD 9.4 billion in 2010 to USD 6.7 billion in 2013, bringing its R&D rate down to an all-time low of 12.9% (Cressey, 2011). In parallel and in 2010, Pfizer established the Global CTI, an open innovation model that aims at founding global partnerships between Pfizer and academic medical centers (http://www.pfizer.com/research/rd_works/centers_for_therapeutic_in-novation.jsp, last access: 30 November 2014).

Another model, established by Eli Lilly, is the Fully Integrated Pharmaceutical Network (FIPnet). This model might be one of the most radical ways in today's pharmaceutical industry to open R&D toward external innovation. FIPnet combines traditional in-house R&D with outsourcing as well as new ways of accessing external innovation. Strategic acquisitions (such as the companies ICOS and ImClone) complement the new concept. Reduction in fixed costs is achieved by outsourcing preclinical and clinical activities to external partners. Functional outsourcing of toxicology and clinical trial management is done with the help of CROs. Discovery activities are partnered with Nicholas Piramal and Suven Life Sciences, both Indian companies, and with the Chinese Hutchison Medi Pharma. Strategic equity investments are done with the help of the Eli Lilly Venture Fund (www.lillyventures.com). As described earlier, also new is Eli Lilly's virtual drug development network, Chorus (www.choruspharma.com). Starting in 2002, Chorus is an autonomous division of Eli Lilly dedicated to cost-effective drug discovery and development up to human PoC. Chorus follows a lean and flexible drug discovery and development process with a small group of experienced scientists coordinating and organizing external R&D activities. Eli Lilly is out-licensing drug candidates from its internal organization to Chorus and has the option to get the drug candidate back after PoC. Since 2002, the Chorus model revealed that costs and time could be reduced significantly by external sourcing resulting in a 3–10-fold productivity increase versus the traditional pharmaceutical development model (http://www.choruspharma.com/about-us.html, last access: 30 November 2014). Today, Chorus is an integral part of Eli Lilly's FIPnet and an alternative way for drug discovery and early development.

## 15.8 Outlook

Regulations have been a phenomenon in the pharmaceutical R&D process since 1970s. It is axiomatic that close collaborations of pharmaceutical companies with regulatory authorities are a must in order to develop amiable partnerships and smooth drug approval processes. In addition, the regulatory authorities have also realized that the pharmaceutical innovation engine is stuttering and have initiated "public–private partnerships," such as the FDA Critical Path Initiative, to discuss new ways to increase R&D productivity. As seen in the past years, traditional innovation approaches proved unsatisfactory in convincing the financial markets of the capabilities of the pharmaceutical industry to renew its drug pipelines. Internal R&D complemented with research collaborations, drug licensing, and strategic outsourcing is no longer sufficient to guarantee efficiency in view of the enormous R&D investments. Evidently, the industry has developed new approaches of open innovation that will impact how R&D is conducted in the future. Pharmaceutical companies streamline their R&D organizations to be more cost effective. Often different R&D groups in one company compete for resources, and underperforming R&D groups are scaled down or closed. In addition, more than in the past, companies use outsourcing today to reduce their fixed costs in R&D, to gain more flexibility, and to get access to the best resources available that cannot be built up internally without vast R&D investments. Large CROs, such as Quintiles, Parexel, or Covance, are positioned as an integral part of the pharmaceutical R&D value chains, and they provide considerable know-how due to the high number of trials they perform as service providers. Smaller providers, such as biotechnology companies, offer special technologies or methods that are useful in the R&D process. Pharmaceutical R&D has become a global business, with Indian and Chinese service providers playing an increasingly important role in R&D outsourcing.

Pharmaceutical companies have built internal alliance management functions to liaise with collaboration partners, biotechnology companies, and universities.

Special teams scout for new targets, drug candidates, or new frontier sciences. In some companies, these activities are complemented by Internet platforms for crowdsourcing.

Next, innovation incubators are receiving increasing attention, and companies like Pfizer, GSK, or Merck have set up new constructs to optimize interaction at the academia–industry interface. Drug discovery itself has received increased importance in the academic research environment, as exemplified by the setup of HTS or drug discovery units at academic centers such as at the EMBL or the Max Planck Institute. New global innovation hubs are emerging and could also be of great interest to pharmaceutical companies in a way that they can provide easy access to a pool of know-how and expertise, as seen in the innovation hubs and life science centers in Shanghai or Singapore.

Virtual R&D is another highly interesting trend in the pharmaceutical industry, where small groups of scientists discover and develop a new drug candidate purely relying on external resources. A leaner and more flexible R&D organization could potentially be achieved by following the strategy of the "knowledge leverager" which is discussed in detail in the book chapter on R&D business models.

In sum, pharmaceutical companies realized that the challenging questions of the future will only be solved if working in networks and if installing a more open R&D organization.

## References


Adams, M. (2010) Datamonitor – Pharmaceutical Licensing Overview, http://about.datamonitor.com/media/archives/4097 (accessed 30 November 2014).

Agarwal, P. (2013) *Nat. Rev. Drug Discovery*, **12**, 575–576.

Behnke, N. and Hueltenschmidt, N. (2010) Bain & Co. Insight. Changing Pharma's Innovation DNA, www.bain.com/publications/articles/changing-pharmas-innovation-dna.aspx (accessed 30 November 2014).

Betz, U.A.K. (2011) *Drug Discovery Today*, **16**, 609–618.

Betz, U.A.K., Camacho, N., Gerards, M., and Stremersch, S. (2014) Grassroots innovation: a promising innovation paradigm for pharmaceutical companies, in *Innovation and Marketing in the Pharmaceutical Industry, Emerging Practices, Research, and Policies*, International Series in Quantitative Marketing, vol. **20**, Chapter 4 (eds M. Ding, J. Eliashberg, and S. Stremersch), Springer, New York. ISBN: 978-1-4614-7800-3.

Betz, U.A.K., Czodrowski, P., Esdar, C., Deutsch, C., Beier, N., Finsinger, D., Cezanne, B., Urbahns, K., and Wucherer-Plietker, M. (2014) Fluctuating open teams (FLOT) – a new approach to boost team creativity and innovation potential. *Technol. Transfer Entrep.*, **1**, 77–81.

Betz, U.A.K. and Neumann, S. (2009) The Marco Polo project, in *Fallstudien zum Innovationsmanagement* (eds J.H. Fisch and J.-M. Roß), Gabler, Wiesbaden, pp. 93–110, ISBN 978-3-8349-1047-9.

Betz, U. and Tidona, C. (2015) *Nat. Biotechnol.*, **33** (1), 20–21.

Chesbrough, H.W. (2003) *Open Innovation. The New Imperative for Creating and Profiting from Technology*, Harvard Business School Press, Boston, MA, p. xxviii.

Chesbrough, H.W. and Garman, A.R. (2009) *Harv. Bus. Rev.*, **87** (12), 68–76, 128.

Chesbroug, H. *et al.* (2006) *Open Innovation – Researching a New Paradigm*, Oxford University Press, Oxford.

Chew, J. (2010) Big Pharma: Why In-Licensing is Not the Cure, http://seekingalpha.com/article/222782-big-pharma-why-in-licensing-is-not-the-cure (accessed 11 August 2015).

Clark, D.E. (2010) *Drug Discovery Today*, **16**, 147–157.

Cressey, D. (2011) *Nature*, **470**, 154.

Dhankar, A. *et al.* (2011) *Evolution or Revolution. McKinsey Perspectives on Drug and Device R&D in 2012*, McKinsey & Company.

Frye, S. *et al.* (2011) *Nat. Rev. Drug Discovery*, **10**, 409–410.

Gamo, F.-J. *et al.* (2010) *Nature*, **465**, 305–310.

Gassmann, O. (2010) *Crowdsourcing – Innovationsmanagement mit Schwarmintelligenz*, Carl Hanser Verlag, München.

Gassmann, O. and Enkel, E. (2004) Towards a theory of open innovation: three core process archetypes. Proceedings of the R&D Management Conference (RADMA), Sesimbra, Portugal, July 8–9, 2004.

Godin, G. (2005) *Drug Discovery World*, Fall, 15–17.

Hu, M. (2007) Kellogg School of Management. The Innovation Gap in Pharmaceutical Drug Discovery & New Models for R&D Success, www.kellog.northwestern.edu (accessed 30 November 2014).

Kar, S. (2010) *Future Med. Chem.*, **2** (9), 1411–1415.

Lessi, M. *et al.* (2011) *Drug Discovery Today*, **16**, 288–292.

Lichtenthaler, U. *et al.* (2011) *MIT Sloan Manage. Rev.*, **53** (1), 45–48.

Longman, R. (2007) *In Vivo*, **25** (5), 1–5.

Mehta, S. and Peters, L.S. (2007) *Res. Technol. Manage.*, **5**, 28–33.

Müller Stewens, G. and Lechner, C. (2011) *Strategisches Management*, 4 Auflage, Schäffer-Poeschel Verlag, Stuttgart.

Nwaka, S. and Ridley, R.G. (2003) *Nat. Rev. Drug Discovery*, **2**, 919–928.

Pisano, G. (2006) *Harv. Bus. Rev.*, **84** (10), 114–124, 150.

Quinn, J.B. (2000) *MIT Sloan Manage. Rev.*, **41** (4), 13–28.

Ratner, M. (2011) *Nat. Biotechnol.*, **29**, 3–4.

Sams-Dodd, F. (2005) *Drug Discovery Today*, **10**, 139–147.

Schuhmacher, A. *et al.* (2013) *Drug Discovery Today*, **18**, 1133–1137.

Woodcock, J. and Woosley, R. (2008) *Annu. Rev. Med.*, **59**, 1–12.

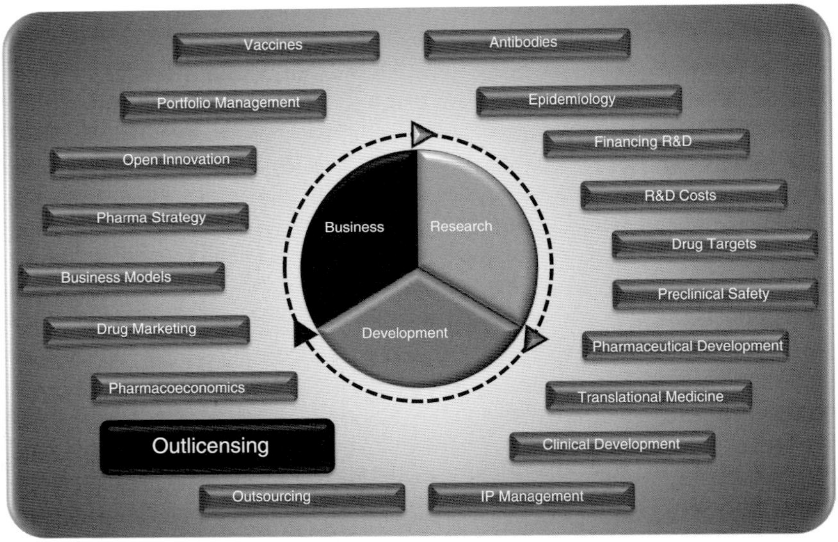
Vaccines
Antibodies
Portfolio Management
Epidemiology
Financing R&D
Open Innovation
R&D Costs
Pharma Strategy
Business
Research
Drug Targets
Business Models
Preclinical Safety
Drug Marketing
Development
Pharmaceutical Development
Pharmacoeconomics
Translational Medicine
Outlicensing
Clinical Development
Outsourcing
IP Management

# 16
# Out-Licensing in Pharmaceutical Research and Development[1]

*Oliver Gassmann, Carol A. Krech, Martin A. Bader, and Gerrit Reepmeyer*

## 16.1
## Introduction

Whereas traditional collaboration approaches such as alliances, in-licensing, and codevelopment are quite established, out-licensing has long been considered a difficult task by many established pharmaceutical companies.

The purpose of this chapter is to illustrate how out-licensing by pharmaceutical companies can contribute to an increase in research and development (R&D) efficiency.

This book chapter explores the following questions:

- What is the relevance of R&D collaborations in the pharmaceutical industry?
- What drives and hampers out-licensing in pharmaceutical R&D?
- How is out-licensing of R&D managed in pharmaceutical companies?

For many decades, pharmaceutical R&D was a fully integrated process where the pharmaceutical company owned and conducted nearly every step of the R&D value chain. However, the pharmaceutical sector is extraordinarily research intensive. Over the last three decades its innovation process has been transformed (Arora *et al.*, 2009). There has been continued growth in the number and value of alliances between firms over time (Grabowski and Kyle, 2012). The total number of alliances by pharmaceutical firms increased between 1995 and 2008 from 600 to 1600 deals (Knowles and Higgins, 2011). During the first decade of the 2000s, the number of alliances with market valuations of at least $100 million increased severalfold (Recombinant Capital, 2008; Grabowski and Kyle, 2012).

Successfully executed out-licensing programs provide pharmaceutical firms with several benefits such as additional revenue generation, cost and resource effectiveness, and mitigation of R&D-related risks.

1) This chapter is based on two previous publications: Gassmann, O. *et al.* (2008): *Leading Pharmaceutical Innovation – Trends and Drivers for Growth in the Pharmaceutical Industry.* Springer; Reepmeyer, G. (2005). *Risk-sharing in Pharmaceutical Industry – The Case of Out-licensing.* Physica-Verlag.

*Value Creation in the Pharmaceutical Industry: The Critical Path to Innovation,* First Edition.
Edited by Alexander Schuhmacher, Markus Hinder, and Oliver Gassmann.

At the end of this chapter, a case study of Novartis demonstrates how the out-licensing can be successfully adopted and managed.

## 16.2
## Performance-Based R&D Collaborations on the Rise

The increasing pace of innovation in the pharmaceutical industry calls for more flexible and looser forms of innovation agreements. As such, today's pharmaceutical R&D collaborations not only cover the typically technology-intensive research collaborations but also increasingly include various contractual partnership agreements such as codevelopment or copromotion agreements. In particular, most novel niche markets can only be entered successfully via partnership with specialized companies. Therefore, many pharmaceutical companies have begun to engage in collaboration agreements with various partners to gain access to different niche markets and to utilize an external partner's specialized sales force capabilities. However, megadeals that include codevelopment and comarketing components have the potential to become very complex, which increases the risk associated with the deals. Within the last decade the proportion of codevelopment and copromotion deals has decreased and have been replaced by simpler deals such as collaboration with shared responsibilities and licensing (Elvidge, 2013) (Figure 16.1).

Many companies decide to choose collaborations to not only access new markets but also to accelerate the market diffusion of their new products. This is particularly true for selling new medicines on a global scale, where a promising approach is to team up with a partner who already has broad expertise, deep

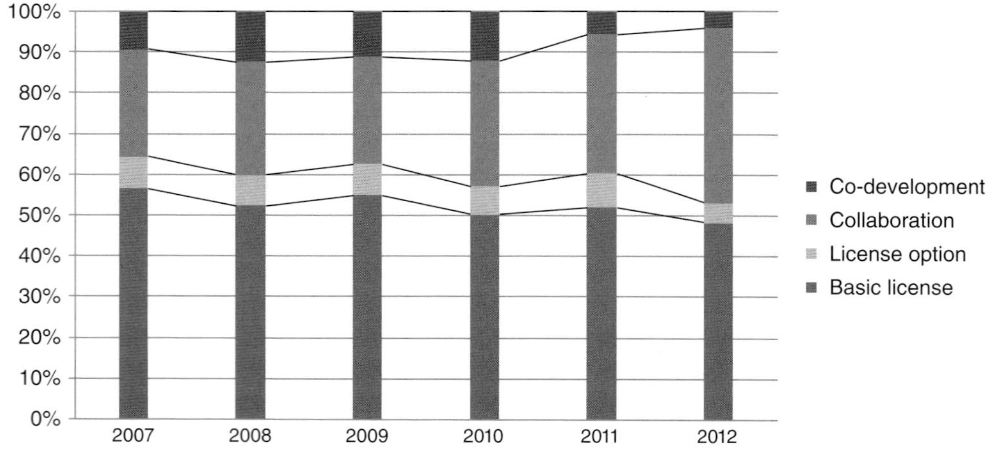


**Figure 16.1** Distribution of pharmaceutical deals. (The Deloitte Recap of 2012 Deals in Elvidge, 2013.)

knowledge, or special access to the targeted market, and who is most likely capable to improve the product launch in the desired market.

Pharmaceutical companies usually expect that the anticipated higher sales generated by the collaboration will more than offset the revenue lost to their collaboration partner. A successful collaboration requires that both partners' success is based on the success of the joint project. The fortunes of both parties in the R&D collaboration must be inextricably linked: both companies gain or lose together, and each partner regards the other partner's success as their own success. Both sides of the R&D collaboration should be highly incentivized to turn the mutually developed product into a success.

Linking the performance of both partners to the joint project's performance seems to represent a new paradigm in pharmaceutical R&D collaborations. This type of performance conjunction can be observed in the nature of many of today's pharmaceutical R&D collaborations. Performance-based structures are unambiguously on the rise.

## 16.3
## The Impact of Collaborations on the Value Chain

A network that includes outside innovation is highly important in pharmaceutical R&D because of the relatively low likelihood that all necessary substances are generated in-house. "By the mid-1990s, … as drug discovery became more science-intensive, with increased emphasis on "deep" understanding of physiology at the molecular level, it became not just more expensive but also more difficult to manage … was accompanied by increased willingness to exploit external sources of technology, through in-licensing or strategic partnerships" (Cockburn, 2004, pp. 15–16). Interactions with partners reduce not only the risks inherent to investing in the firm's own research infrastructure but also the risks of lack of access to desired substances.

Large pharmaceutical firms increasingly look to collaborate with development-stage firms to gain access to new technologies and buttress their R&D pipelines (Grabowski, 2011). Therefore they increasingly concentrate on their core advantages while leveraging outside innovators for their own innovation process (for open innovation, see Chesbrough, 2003, 2006; Gassmann, 2006), by:

- Outsourcing R&D
- Joint ventures in R&D
- Crowdsourcing R&D
- Strategic research alliances with complementary partners
- In-licensing intellectual property (IP) rights (mostly in later R&D stages and often from biotech firms)
- Out-licensing research in the early R&D stages
- Spin-offs and divestitures of R&D activities

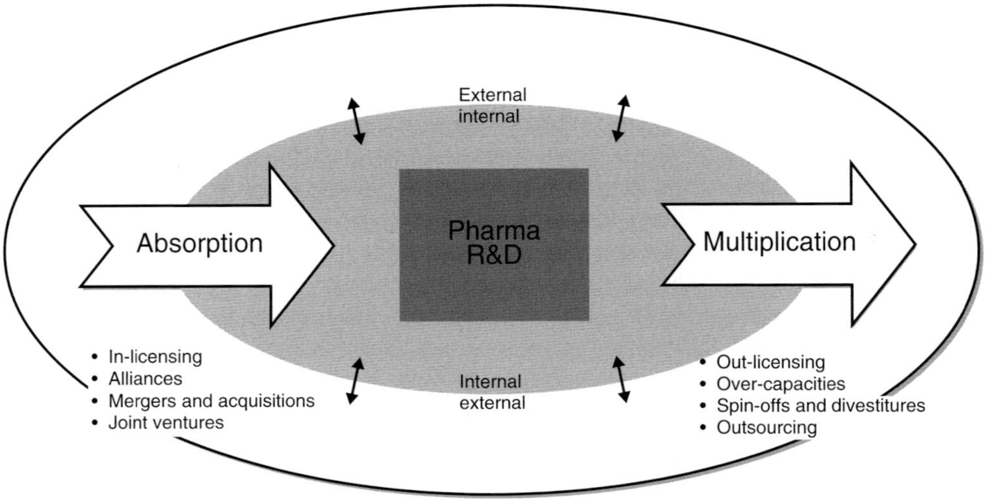


**Figure 16.2** Restructuring of pharmaceutical R&D departments and resulting interaction with external partners (Gassmann and Reepmeyer, 2005).

Concentrating on their core competencies and streamlining R&D activities involve deciding which tasks must remain inside their boundaries and which tasks should be absorbed from outside entities or could be multiplied by outsourcing them to external partners (Figure 16.2).

The externalization of all or part of a firm's R&D activities has led the creation of an evolving market for innovation and numerous licensing and partnership agreements (Grabowski and Kyle, 2012). Upfront payments in such deals rise directly proportional to the development stage. In 2013, the top 50 collaborations platform technologies were worth an average of $46 million in upfront payments, while marketed and phase III products cost an average of $355 million in upfront payments (Skripka-Serry, 2014).

These days, fundamental breakthroughs are increasingly likely to occur in an external organization's research lab and not within a pharmaceutical company's R&D department. The cascade of knowledge flowing from new sciences and technologies is simply far too complex for any company to handle individually.

Many different types of potential partnerships exist in the pharmaceutical industry. Due to the industry's highly integrated structure and value chain, the closeness of the relationship between the pharmaceutical company and the external partner can serve as a criterion for the classification of the interaction (Figure 16.3). The nature of the interaction can embrace different attributes, features, and forms depending on the commitment of resources.

In the case of outsourcing, the partner company usually conducts predefined tasks in return for a service fee, whereas collaboration refers to a more closely linked effort between two or more companies engaged in a joint project. This frequently involves risk, revenue, and profit-sharing agreements. In the case of

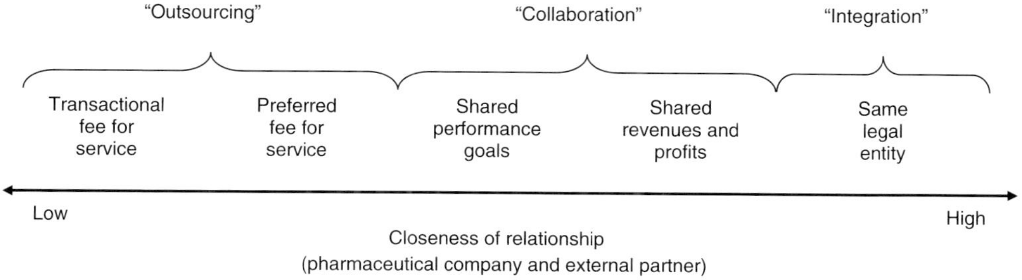


**Figure 16.3** Classification of partnerships in pharmaceutical R&D activities (Reepmeyer, 2005).

integration, both companies' activities usually melt together, and there is no visible separation between the two entities.

The decision to engage in collaboration is usually based on the trade-off between interfirm cooperation and vertical integration. This trade-off is seen as a comparison between transaction costs and development costs. Where organizational efficiency gains are greater than production efficiency losses, a firm will choose to cooperate (Greis, Dibner, and Bean, 1995).

## 16.4 Generating Value from Pipeline Assets by Out-Licensing

> As the hurdles for developing a successful new drug get higher, and as R&D budgets shrink, companies are looking for ways to continue to generate value from pipeline assets, while sharing the risks inherent in any R&D program. These types of deals are becoming a major strategy to meet these challenges.
>
> *(LaMattina, 2012)*

By nature, successful collaborations center on sharing certain IP or markets, and as such they require that both partners make contributions to the joint work. Therefore, the different types of R&D collaborations can be classified according to two criteria: the work that is contributed by the pharmaceutical company, and the work that is contributed by the external partner. From the perspective of the pharmaceutical company, this classification scheme results in four different types of R&D collaboration (see Figure 16.4):

- Research alliance
- In-licensing
- Codevelopment
- Out-licensing

Seventy percent of pharmaceutical companies expect their collaborations will primarily take the traditional forms of out-licensing products or technology

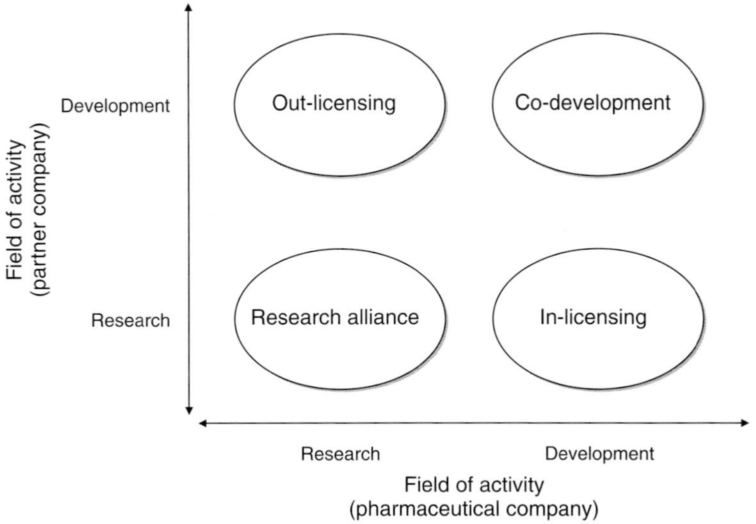


**Figure 16.4** Different types of collaboration in pharmaceutical R&D (perspective: pharmaceutical company) (Reepmeyer, 2005).

(Deloitt, 2011). Whereas research alliances, in-licensing, and codevelopment are quite established and traditional collaboration approaches, many established pharmaceutical companies have long considered out-licensing a difficult task.

Out-licensing, as a phenomenon, is as new as in-licensing; however its utilization by large and established pharmaceutical companies is fairly novel. Large pharmaceutical companies such as Eli Lilly, Bayer, Roche, or Novartis have only recently adopted this R&D collaboration strategy.

Out-licensing can play a critical role in accessing the diverse sources of innovation in the new pharmaceutical R&D landscape (Allarakhia and Walsh, 2011), and the phenomenon has experienced remarkable growth over the past few years. The number of out-licensing deals between pharmaceutical and biotech companies increased throughout the 2000s from 300 deals in 2000 to 400 deals in 2010, with a peak of 654 deals in 2006 (Deloitt, 2011; OECD, 2011).

Out-licensing strategies utilize external resources to further the development of internally developed substances. Thus out-licensing always promotes the dissemination of technologies and products by integrating a company's IP with complementary assets (Rothaermel, 2001). Therefore, out-licensing should be adopted by companies that possess a strong position in a specific technological area but lack the complementary assets necessary to exploit the technology (see Figure 16.5). Successfully executed out-licensing programs provide the pharmaceutical firm with several benefits, including additional revenue generation, cost and resource effectiveness, or mitigation of R&D-related risks.

In pharmaceutical R&D licensing generally improves the probability of success (Grabowski and Kyle, 2012; Arora *et al.*, 2009). Pharmaceutical companies aim to start deals early (i.e., a large portion of licensing deals are signed for products in

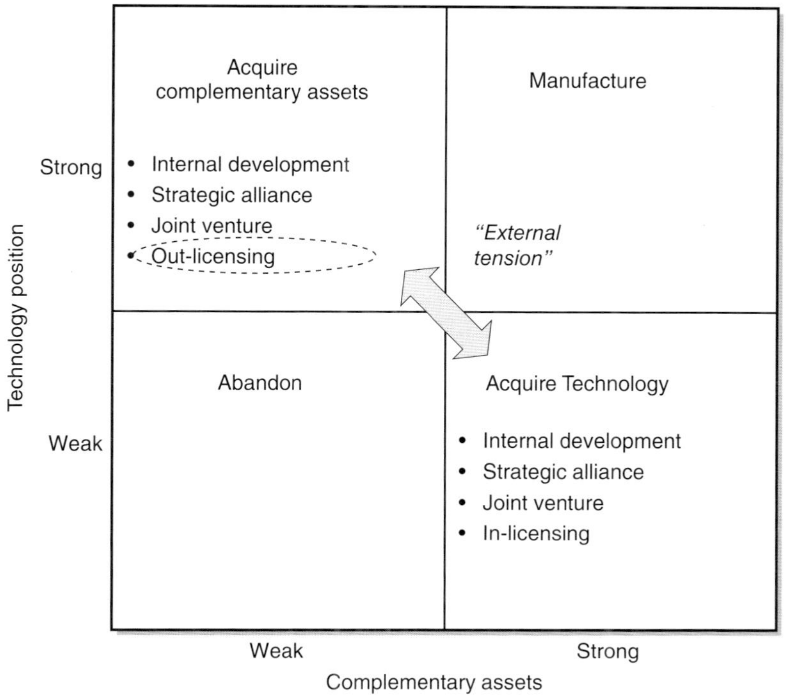


**Figure 16.5** Out-licensing: The neglected strategy to gain complementary assets for the utilization of a company's own technology (compare Megantz, 2002).

early development stage) with smaller upfront payments and prefer to sign many small deals as opposed to a few big ones (Skripka-Serry, 2014).

Out-licensing lowers risk because less investment and fewer resources are needed; much of the risk remains largely offloaded onto the partner company, who is responsible for the further development of the licensed product (Megantz, 2002). If a product fails or a strategy changes, the pharmaceutical company can usually opt out without any undesirable follow-up operations aside from the small termination fees that might occur.

There are many reasons why it may be advantageous or even necessary for a pharmaceutical company to out-license IP rights to a third party. These reasons include:

- No further intended internal use for the technology, compound, or IP rights
- Lack of resources and/or internal expertise for further exploitation
- The risk profile of the substance and/or compound does not match internal requirements
- Specialization in certain product areas or technologies, as in the case of a portfolio restructuring
- Exploitation of therapeutic areas that differ from initially intended therapeutic areas (e.g., a pharmaceutical company only plans to develop a product or

technology in one therapeutic area, but this product may have applications in other areas; this could even include areas that go beyond pharmaceuticals such as cosmetics or plant breeding)
- Lack of commercialization potential (e.g., the drug's target market is judged insufficient in size to justify further R&D investments)
- Low-risk opportunity to move into new, complementary, or unknown markets
- Improvement of the company's revenue stream and/or market penetration by focusing on short-term income
- Expansion of geographical reach
- Side benefits, for example, increased brand visibility due to licensee advertising or the use of improvements developed by the licensee
- Maximization of the firm's asset utilization and drug portfolio value by leveraging internal R&D capacities
- Gaining advantages in noncore markets by selling noncore technologies
- Testing a market that may later be exploited by direct investments

Probably the most important benefit of out-licensing is that the pharmaceutical company can increase the utilization rate of its internal research results without significant additional investment. As soon as the pharmaceutical company dices to discontinue internal development of a compound, the pharmaceutical company can then find new avenues to commercialize its IP at the respective stages of the R&D process. This requires abandoning the traditional path of commercialization and creating a new market for a compound (see Figure 16.6). An external partner's R&D resources can be utilized to bring the compound to the market instead of letting the compound decay inside the pharmaceutical company's R&D boundaries.

Selling the IP rights of a compound to an external partner for further development not only helps transfer R&D risk but also allows the generation of royalty revenues should the licensed compound be brought to market.

A critical aspect of any out-licensing agreement is the licensor's (i.e., the patent owner) ability to retain an interest in the licensed compound's future performance. As the estimation of a drug's potential is very complex and characterized by a high uncertainty, mistakes in evaluating the upside potential of a substance could have a high financial impact. Most pharmaceutical companies fear missing out on the opportunity of participating in a drug's upside potential and thus typically only participate in out-licensing deals if they can retain certain relicensing rights, such as a callback option. The example of Aventis demonstrates the importance of such a provision. In 1998, Hoechst Marion Roussel (in 1999 incorporated into Aventis. Today known as *Sanofi*; see Sanofi, 2012) failed to recognize the potential of one of its products when the firm sold it for a relatively low price to King Pharmaceuticals without retaining any stakes in the drug's future potential. The specialty pharmaceutical company King Pharmaceuticals (in 2011 incorporated into Pfizer; see Pfizer, 2011) acquired the drug Ramipril (Altace), an angiotensin-converting enzyme (ACE) inhibitor, from Hoechst Marion Roussel as part of a package of three drugs costing US$363 million (King Pharmaceuticals,

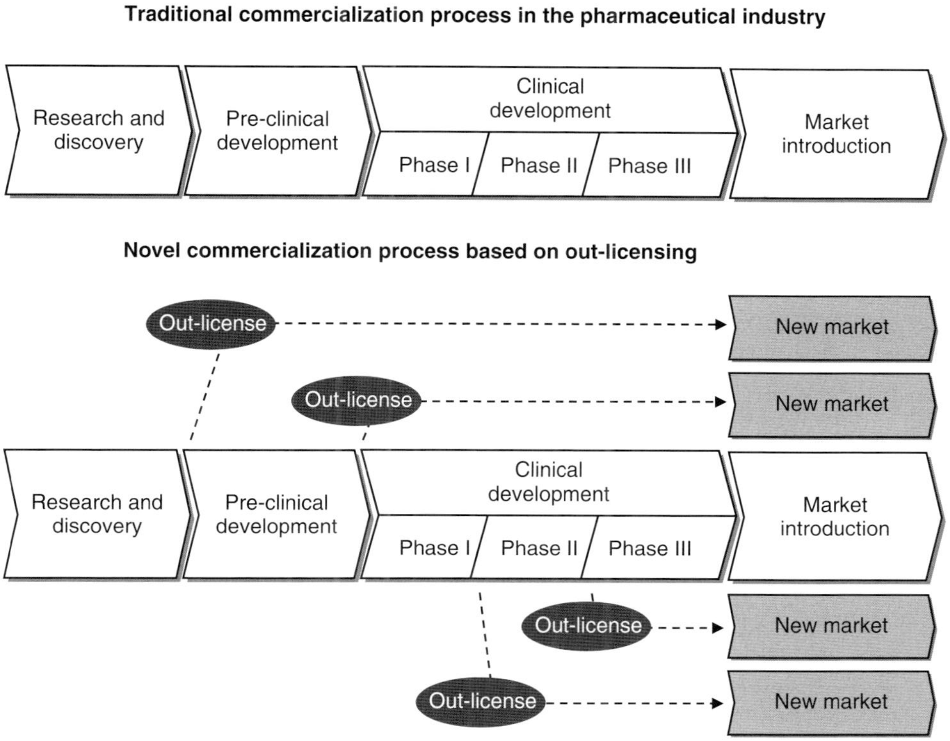


**Figure 16.6** Out-licensing as a way to open new markets (Reepmeyer, 2005).

1998). Hoechst Marion Roussel sold the drug because it was not clearly differentiated from other products in this crowded therapeutic field. After a subsequent study conducted by King Pharmaceuticals, which proved that the drug reduced risk of death from stroke and heart attack in certain populations, the drug got additional labeling that certified its uniqueness. King Pharmaceuticals generated US$527 million in sales with Altace in 2003 alone. Aventis missed out on this great opportunity because they had abandoned the stakes in the drug due to a retrospectively wrong judgment about the drug's potential.

Today, most established pharmaceutical companies still only use out-licensing as a tactical tool. They only out-license compounds that did not make it onto the company's top priority list. Potentially, any compound that can be developed more efficiently by an external partner could become subject to an out-licensing agreement, not only terminated compounds. A more strategic approach to out-licensing could help improve the effectiveness of internal R&D.

A more progressive approach to the commercialization of research results could be implemented if respective out-licensing activities received clear profit and loss responsibilities. Out-licensing could be organized as a profit center incorporating out-licensing as an independent item in the company's budget calculations with a clear goal to increase the commercialization of corporate research.

## 16.5
## Pharmaceutical Companies' Resistance toward Out-Licensing

For years, pharmaceutical companies have been reluctant to utilize out-licensing. A frequently mentioned reason for this is that, "no one will win any awards within a large drug company for a successful out-licensing deal that generates some upfront and modest expectations for royalties in the distant future" (Ratner, 2003, p. 12).

An even greater obstacle to out-licensing is the realization that selling the IP rights to an unrecognized blockbuster could be a career-ending move for the respective executive. Most large pharmaceutical companies are uncomfortable about relinquishing control of their drugs because they fear that selling products will leave gaps in their revenues (Windhover, 2000).

In addition, if a big pharmaceutical company that has the necessary infrastructure to bring a compound to the market decides to terminate an R&D project, other companies often believe that the compound has any value and thus refrains from utilizing the approach. Many companies fail to recognize that most R&D projects are terminated due to reasons unrelated to the compound itself. For example, when Bayer intends to out-license a substance, it highlights that its reasons for out-licensing are "most likely [due to] a Bayer-specific problem rather than a problem of the substance" (Reepmeyer, 2005, p. 76).

Furthermore, fully integrated firms only out-license their noncore products (Kollmer and Dowling, 2004). The reason is usually a misfit between the product and their overall strategy even though, given their size and resources, they could bring these products to the market independently.

## 16.6
## Managing Out-Licensing at Novartis: A Case Study

All out-licensing efforts at Novartis are organizationally embedded in the firm's Business Development and Licensing (BD&L) department. Novartis considers out-licensing an important activity that adds economic value to idle substances. Novartis usually decides to license out a substance for one of the following reasons:

- The substance is no longer strategically relevant.
- The substance does not fulfill the required performance potential.
- New substances arise which cannot be pursued within the scope of Novartis and without making significant upfront investments in infrastructure and know-how.

At the very least, Novartis seeks to recoup the costs that have been incurred by the respective substances so far. In addition, Novartis expects to achieve additional benefits and royalty revenues should the collaboration partner successfully introduce the new substance in the market. For this reason, Novartis always includes revenue participation clauses or relicensing options in any out-licensing contract.

In order to decide which substance should be out-licensed, Novartis conducts market dynamics analyses, including observation of competitors' activities. Substances that have already used significant resources are of particular interest for out-licensing, as this would be one method to generate some form of return on their initial R&D investment. Therefore, it is not unusual for the project team to suggest out-licensing a stopped substance.

Novartis sees particular benefits in licensing out to smaller companies. These firms are usually very interested in Novartis' substances because it significantly boosts their reputation to announce a licensing deal with Novartis, which in turn facilitates their ability to secure funding from other investors.

### 16.6.1 Out-Licensing as a 10-Step Process

Novartis uses a standardized 10-step out-licensing process (see Figure 16.7). First, a cross-functional committee decides which substance(s) might potentially be out-licensed. While the decision committee makes out-licensing decisions for all global development projects, local teams might make out-licensing decisions for substances that target a relatively small, local market. After the decision committee has reached a decision, a team supervised by the Head of Drug Delivery Licensing and Out-licensing (DDLO) creates a brief product profile, usually two to three pages in length, that aggregates critical nonconfidential product information. This product profile contains information about former development activities and data about its market potential.

Novartis next identifies possible licensing partners. The company reviews publicly and internally available information about other biopharmaceutical companies and their respective R&D activities. The possible licensing partners are evaluated and assessed according to their competencies and capabilities to pursue further development of the drug candidate.

After Novartis identifies possible licensing partners, Novartis sends them the brief product profile along with a nonbinding inquiry about the potential out-licensing opportunity. While some substances attract statements of interest within a relatively short period of time, other substances might be available for out-licensing for several years. In the case that no partner can be found, these substances are withdrawn and their projects terminated. If a possible licensing partner shows initial interest in a substance, a mutual nondisclosure agreement is signed that allows Novartis to send confidential and detailed documentation about the substance to the potential partner. Upon review of this additional documentation, if the potential partner is still interested in pursuing out-licensing, then the first personal interaction between the two parties takes place. Experts of both firms meet to discuss the substance, including chances and risks associated with the drug's development. In some cases, Novartis might send small probes of the substance to the partner to allow the partner to test it *in vitro* or *in vivo* trials.

After the first in-person meeting, there may still be several partners interested in the substance. In this case the most interesting partners are invited for due

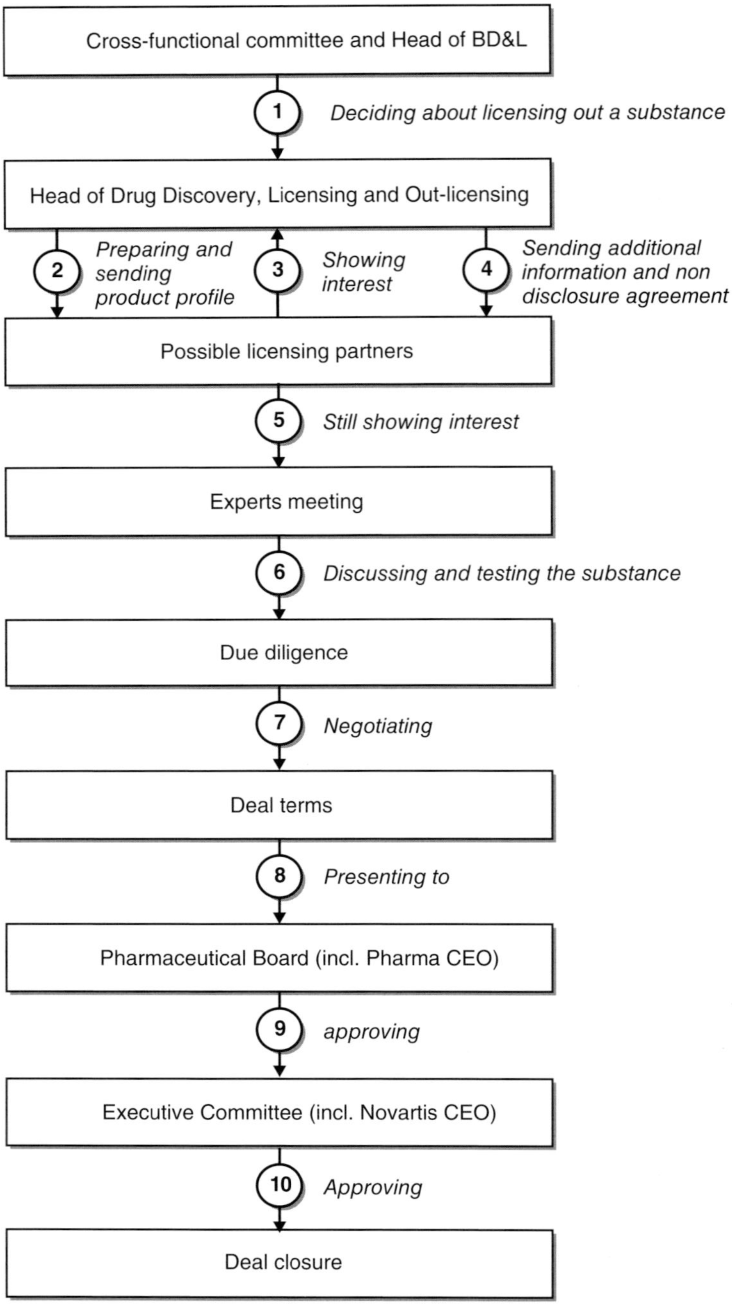


**Figure 16.7** Out-licensing process at Novartis (Reepmeyer, 2005).

diligence negotiations. Due diligence typically takes place at Novartis and lasts a day or two. After this step, Novartis and the potential partner/s discuss the key terms of the licensing. If the partner/s agree on the terms, the Head of DDLO gives an internal presentation to the Pharmaceutical Board. If the Pharmaceutical Board approves the deal, the CEO of the pharmaceutical division presents the deal to the Novartis Executive Committee. Its role is to ultimately approve, halt, or terminate any out-licensing project.

### 16.6.2 Out-Licensing Contract Design

All out-licensing efforts at Novartis are organizationally embedded in the firm's BD&L. Throughout the collaboration, the link between Novartis and the partner is managed by a designated Key Contact Person. For larger deals Novartis sometimes creates a Steering Committee comprised of BD&L employees whose charge is identical to the Key Contact Person. The Key Contact Person or Steering Committee conducts semiannual and annual analyses to ensure that contractual agreements are met. Typically, Novartis stipulates in its contract that it may withdraw from a licensing deal if key targets and milestones are not met.

Risk considerations play a crucial role for out-licensing. While out-licensing represents the risk that an internally developed substance might generate significant revenues for an external partner, Novartis is also exposed to the risk of not participating in this upside potential. To mitigate the risk of underestimating a drug's potential, Novartis typically retains a callback option in all out-licensing contracts. Typically, the contract states when Novartis would like to use its callback option. The callback option differs across substances and depends on product-specific development risks, the substance's market potential, and the performance potential of the licensee. The callback usually occurs if the risks associated with further internal development of the substance are low enough to justify the callback and would typically occur somewhere between clinical phase II and the end of the development process. If a callback is not part of the contract, Novartis reserves the right to start relicensing negotiations depending on the respective substance's development situation.

The entire out-licensing process, from initial negotiations through deal closure, usually takes 6–9 months, excluding the time needed to search for and screening potential partners. Throughout the out-licensing process, Novartis stakeholders affected by the out-licensing deal are encouraged to provide commentary or criticism (Gassmann *et al.* 2008).

### 16.6.3 Structure of the Out-Licensing Collaboration with Speedel

In 1998 researchers at Novartis developed a new substance, Aliskiren. Just as it was about to enter development stages, Novartis' R&D management decided to halt the project. However, a team of Novartis' employees strongly believed in the substance and were interested in further pursuing this opportunity. In 1999 they left Novartis and set up the company Speedel to in-license the substance from Novartis. Although Novartis had chosen to halt Aliskiren's development, they looked at this out-licensing partnership with Speedel as an opportunity to continue the substance's development with low-risk exposure, as the licensing agreement would transfer the development risks to Speedel.

Aliskiren is an oral renin inhibitor that has demonstrated tremendous potential to treat hypertension. Today's market for antihypertensives represents ~$40 billion (i.e., roughly 40% of the total cardiovascular market) (Speedel, 2004). Hypertension affects more than 150 million people in the developed world. It is a major risk factor for stroke, chronic renal disease, congestive heart failure, and myocardial infarction. Renin inhibitors such as Aliskiren work by regulating the kidney's production of renin. Renin is an enzyme that is associated with the release of a substance that narrows blood vessels and makes it more difficult for blood to flow through arteries, thereby raising blood pressure. Aliskiren suppresses the release of renin and thus keeps blood pressure in a normal range.

After Novartis out-licensed the substance Aliskiren, Speedel began development activities for its clinical phases I and II. In total, Speedel conducted 18 clinical studies comprised of roughly 500 patients and healthy volunteers. To pilot studies in chronic renal failure and heart failure, Speedel conducted a 4-week, 220-patient phase II. Speedel was the first company to establish clinical proof of concept in phase II and to develop and patent a commercially viable manufacturing process for a renin inhibitor, which had been an area of industry research for over 20 years.

Throughout the duration of the collaboration, Novartis retained a callback option to license the substance back at any development stage. As Novartis was pleased by the positive development results, in September 2002 the company exercised the callback option to conduct further development and commercialization (see Figure 16.8). In November 2003, Novartis announced that it was moving Aliskiren into clinical phase III trials and initiated them by March 2004. After the callback Speedel had no further cost of development and commercialization and received significant milestone payments due to the successful development in phases I and II. The size of the milestone payments were not disclosed, but the payments did include cash elements and equity participation. In January 2005, Novartis announced positive results from the phase III study of Aliskiren as a monotherapeutic for the treatment of hypertension and positive phase II data from a combination study of Aliskiren with Diovan, another antihypertensive compound and Novartis' leading respective drug. In 2007 Novartis received market approval for Aliskiren in the U.S. and made revenue of $40 million with the respective products Rasilez and Tekturna. With Diovan, Novartis own hypertension product, they made a revenue of $5 billion. Since the Diovan patent expired in 2012, Rasilez and Tekturna narrowed the gap for generics. In 2008 Novartis bought for more than CHF 900 million the majority of Speedel's shares leading to a share rise of 92% at Speedel.

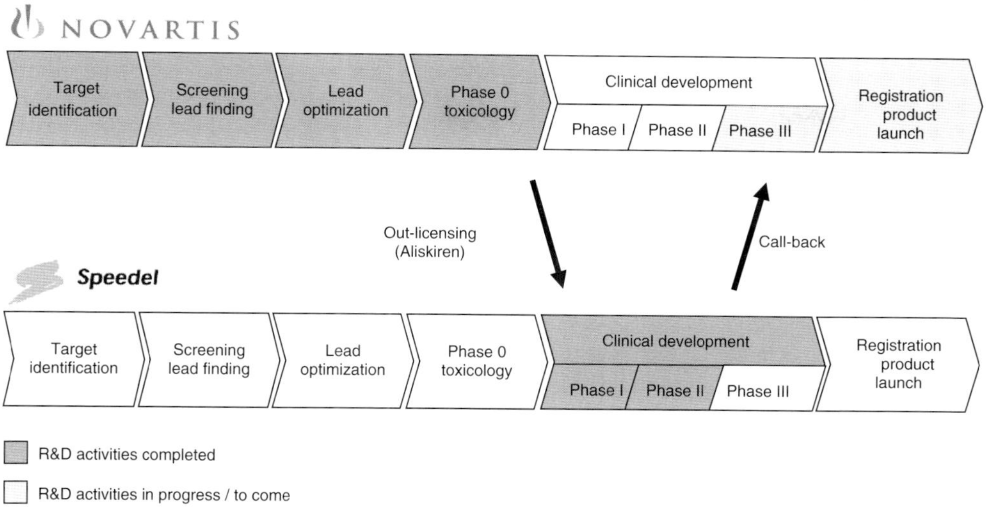


**Figure 16.8** Out-licensing collaboration between Novartis and Speedel (Reepmeyer, 2005).

In summary, Novartis' out-licensing approach enabled the company fill its pipeline without enlarging its already significant in-house R&D budget. The success of the collaboration with Speedel demonstrates the way that the continuing disintegration of the pharmaceutical value chain creates new partnership models that allow pharmaceutical companies to leverage not only their own core competencies but also the strengths of their partners.

One of the most notable features of this collaboration is the fact that Speedel had been a start-up company, and at the time the deal closed it did not have a track record of successful drug development project execution. Novartis sold Aliskiren to a company for further development despite the fact that it was not sure whether the partner firm would be able to successfully execute the compound's development. The fact that Speedel's management team had previously been employed with Novartis was one main reason why Novartis trusted Speedel to turn the substance into a success.

## 16.7 Future Directions and Trends

Improving the flow of innovation and closing the R&D productivity gap are foremost challenges in the pharmaceutical industry. Given major technological and economic changes in drug design and delivery, the pharmaceutical industry faces nothing short of a revolution. This chapter identifies several trends and suggests potential future directions for leading pharmaceutical innovation (Gassmann *et al.* 2008):

*Prepare for the accelerating importance of open innovation*: Today's companies already rely on vast networks of external partners to improve the performance of their R&D departments. This trend will be even more pronounced in the future. Partnerships will be created that go far beyond the "technology sourcing" strategies represented by early pharma–biotech alliances. Intensive competition, access to markets, scarce internal resources, lack of know-how, cost cutting or restructuring, growth aspirations, synergies and efficiencies, and risk reduction all represent reasons why pharmaceutical companies increasingly engage in external partnerships for R&D activities.

*Barriers to external partnerships are expected to diminish*: The traditional vertically integrated structure of pharmaceutical R&D departments is expected to further decrease, while interactions with external partners will dramatically increase. This will reduce the internal complexity of pharmaceutical R&D but increase the complexity of managing external relationships. Reliance on outside innovation requires pharmaceutical companies to think and act in more process-oriented ways. Barriers between intraorganizational units and external partners are expected to diminish. Only those pharmaceutical companies that are able to optimally manage their R&D collaborations will be able to benefit from novel developments and opportunities in the pharmaceutical industry.

*Commercialize internal research results more strategically*: A more strategic approach to improve R&D pipelines will likely result in a more aggressive commercialization of research results. Pharmaceutical companies should not only actively look for potential in-licensing candidates but also for out-licensing. They should actively create and customize IP rights bundles that can be marketed to potential licensees. Due to the increasing number of niche markets, specialty pharmaceutical companies are tempted to acquire late-stage projects in order to avoid the opportunity costs and risks at earlier stages. Thus these are several potential licensees that may need the pharmaceutical company's IP. Pharmaceutical companies should not only regard these companies as potential competitors who might "steal" blockbuster revenues but should instead treat these firms as potential customers and partners who help generate additional revenues and profits.

## References


Allarakhia, M. and Walsh, S. (2011) Managing knowledge assets under conditions of radical change: the case of the pharmaceutical industry. *Technovation*, **31**, 105–117.

Arora, A., Pammolli, F., Gambardella, A., and Magazzini, L. (2009) A breath of fresh air? Firm types, scale, scope and selection effects in drug development. *Manage. Sci.*, **55** (10), 1638–1653.

Chesbrough, H. (2003) *Open Innovation: The New Imperative for Creating and Profiting from Technology*, Harvard Business School Press, Boston, MA.

Chesbrough, H. (2006) *Open Business Models: How to Thrive in the New Innovation Landscape*, Harvard Business School Press, Boston, MA.

Cockburn, I.M. (2004) The changing structure of the pharmaceutical industry. *Health Aff.*, **23** (1), 10–22.

Deloitt (2011) A Review of the Recent Deal-Making and Licensing at Pharma and Biotech, http://www.biotechday.ch/media/biotech%20day%202011/pr%C3%A4sentationen/deloitte_recap_johnson.pdf (accessed 08 March 2015).

Elvidge, S. (2013) The changing shape of deal making, in *Biopharmaceutical-Deals Roundup: Will there be Growth in 2013?*, http://www.atebion-bds.com/pdfs/changing-shape-of dealmaking.pdf (accessed 20 February 2015). Nature Publishing Group, London.

Gassmann, O. (2006) Opening up the innovation process: towards an agenda. *R&D Manage.*, **36** (3), 223–226.

Gassmann, O. and Reepmeyer, G. (2005) Organizing pharmaceutical innovation: from science-based knowledge creators to drug-oriented knowledge brokers. *Creativity Innov. Manage.*, **14**, 233–245.

Gassmann, O., Reepmeyer, G., and Von Zedtwitz, M. (2008) *Leading Pharmaceutical Innovation: Trends and Drivers for Growth in the Pharmaceutical Industry*, Springer Science & Business Media.

Grabowski, H.G. (2011) The evolution of the pharmaceutical industry over the past 50 years: a personal reflection. *Int. J. Econ. Bus.*, **18** (2), 161–176.

Grabowski, H.G. and Kyle, M.K. (2012) Mergers, acquisitions, and alliances, in *Handbook of the Economics of the Biopharmaceutical Industry*, Oxford University Press.

Greis, N.P., Dibner, M.D., and Bean, A.S. (1995) External partnering as a response to innovation barriers and global competition in biotechnology. *Res. Policy*, **24**, 609–630.

King Pharmaceuticals (1998) King Pharmaceuticals and Hoechst Marion Roussel Sign an Agreement for One of the Largest Single Product Transactions in Emerging Pharmaceutical Sector, http://www.prnewswire.com/news-releases/king-pharmaceuticals-and-hoechst-marion-roussel-sign-an-agreement-for-one-of-the-largest-single-product-transactions-in-emerging-pharmaceutical-sector-77821482.html (accessed 20 March 2015).

Knowles, S. and Higgins, M.J. (2011) Vertical disintegration in the pharma industry and the role of IP. *Intellect. Asset Manage. Mag.*, **45**, 10–18.

Kollmer, H. and Dowling, M. (2004) Licensing as a commercialisation strategy for new technology-based firms. *Res. Policy*, **33**, 1141–1151.

LaMattina, J. (2012) Why is Pharma Out-Licensing its Compounds?, http://www.forbes.com/sites/johnlamattina/2012/10/29/why-is-pharma-out-licensing-its-compounds/ (accessed 20 February 2015).

Megantz, R.C. (2002) *Technology Management – Developing and Implementing Effective Licensing Programs*, John Wiley & Sons, Inc., Hoboken, NJ.

OECD (2011) Trade and Innovation: Pharmaceuticals. OECD Trade Policy Papers, No. 113, OECD Publishing.

Pfizer (2011) Pfizer Completes Acquisition of King Pharmaceuticals, Inc., http://www.businesswire.com/news/home/20110301006102/en/Pfizer-Completes-Acquisition-King-Pharmaceuticals#.VRvOc9g6m70 (accessed 31 March 2015).

Recombinant Capital (2008) Analysts Notebook: Trends, http://www.recap.com/consulting.nsf/ContentWebPublish/9A5A93B5D9B4080A882574440076AB44.001?OpenDocument (accessed 16 January 2012).

Reepmeyer, G. (2005) *Risk-Sharing in Pharmaceutical Industry – The Case of Out-Licensing*, Physica-Verlag.

Rothaermel, F.T. (2001) Complementary assets, strategic alliances, and the incumbent's advantage: an empirical study of industry and firm effects in the biopharmaceutical industry. *Res. Policy*, **30**, 1235–1251.

Sanofi (2012) Die Unternehmensgeschichte von Sanofi, http://www.sanofi.de/l/de/de/layout.jsp?scat=83168B09-5E3D-4A1D-9440-AB2C102863E5 (accessed 20 March 2015).

Skripka-Serry, J. (2014) Biotech & Pharma 2013 Licensing & Partnering Activity Review: Diminishing Upfronts, Increasing Platform Licenses Indicate Stronger Risk Aversion among Big Pharma, http://bio-associate.blogspot.ch/2014/01/biotech-pharma-2013-licensing.html (accessed 20 February 2015).

Speedel (2004) Introducing the Speedel Group. Company Presentation, http://www.speedelgroup.com (accessed 10 September 2004).

Windhover (2000) Opportunism knocks. *Windhover's Rev. Emerg. Med. Ventures*, **5** (4), 32, http://www.windhover.com/contents/monthly/exex/e_2000900064.htm (accessed 19 December 2004).

Ratner, M. (2003) In-licensing: still a difficult model. *Windhover's Rev. Emerg. Med. Ventures*, **8** (10), http://www.windhover.com/ contents/monthly/exex/e_2003900172.htm (accessed 19 December 2004).

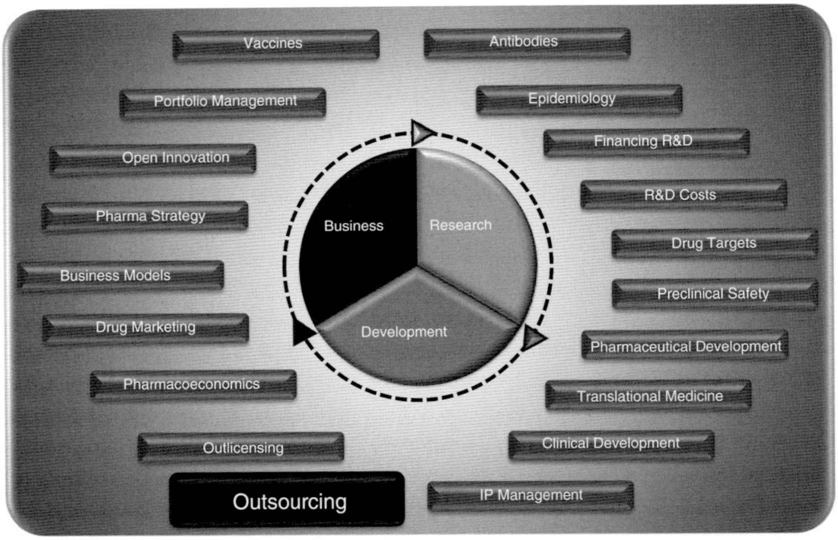
Vaccines
Antibodies
Portfolio Management
Epidemiology
Financing R&D
Open Innovation
R&D Costs
Pharma Strategy
Business
Research
Drug Targets
Business Models
Preclinical Safety
Drug Marketing
Development
Pharmaceutical Development
Pharmacoeconomics
Translational Medicine
Outlicensing
Clinical Development
Outsourcing
IP Management

# 17
# Trends and Innovations in Pharmaceutical R&D Outsourcing

*Antal K. Hajos*

## 17.1
## Introduction

The purpose of this chapter is to review the role outsourcing plays in drug development and how it contributes to innovation. Following a summary of the evolution of outsourcing, current and future trends are discussed. The key questions we will discuss herein are therefore: (i) How has outsourcing innovated itself? (ii) how does and will outsourcing contribute to innovation? and (iii) what are the future trends?

The outsourcing industry is still one of the younger industries in the area of pharmaceutical drug development when compared to the existence of about a century of pharmaceutical manufacturers. Outsourcing in the pharmaceutical industry has been there for several decades but only in about the past 20 years established itself as an integral part of today's drug development landscape. While initially the main activities of outsourcing firms were in the area of clinical development (Clinical CROs = contract research organizations) and certain areas of manufacturing (CMOs = contract manufacturing organizations), the industry has spread over time into virtually all aspects supporting the pharmaceutical industry along the drug development chain, from discovery to marketing support. The degree of outsourcing, the fraction of R&D spend outsourced versus spend in-house, has increased from some 10 to 40% in the period of 1997–2009 (Hu, 2007; Schuhmacher and Betz, 2015). At the same time, while initially purely in the area of clinical development, discovery proportionally increased from 0 to about 15% of total outsourcing, and overall outsourcing spend increased fourfold Schuhmacher and Betz (2015).

## 17.2
## Drivers to the Use of Outsourcing

### 17.2.1
### Overview on the CRO Market

It is probably fair to say that outsourcing in the 1980s and early 1990s was primarily a means of accessing either specialty knowledge, models, or scientific

*Value Creation in the Pharmaceutical Industry: The Critical Path to Innovation,* First Edition.
Edited by Alexander Schuhmacher, Markus Hinder, and Oliver Gassmann.

expertise for isolated, frequently preclinical studies, or more commonly complementation of Pharma's own staff resources in terms of capacity, geography, or both. The former tended to initially rely on collaboration primarily with academia and resulted in commercialization of academic research or academic spin-offs that became "CROs." Likewise, the foundation of many primarily clinical CROs was based on pharmaceutical personnel changing sides and establishing a CRO or "research institution" to serve in fact frequently their previous organization. In parallel, especially large Pharma increasingly had to utilize outsourcing in order to respond to the increasing needs of flexibility of resources to serve the fluctuating pipelines and serve the pressures of globalization by expanding the development activities to new emerging and global markets of both clinical trials and subsequently marketing of new drugs. This created the need for a greatly increased geographic reach of the development organization that was to be covered to various degrees by CROs. Moreover, with almost exploding costs to bring new drugs to the market, and increasing development cycle times that needed to be optimized to achieve best possible "time to market," as discussed in other chapters in this book in more detail, outsourcing was there to stay and evolve.

Over the years, the outsourcing market has substantially grown and greatly diversified. While outsourcing is used at virtually all stages of pharmaceutical R&D, discovery and preclinical outsourcing spend ("research outsourcing") is still considerably smaller (about 20–30% of total) than that of clinical (or "development") outsourcing (about 70–80%) which is certainly also due to the majority (about 60–70%) of all R&D costs flowing into phase 2–4/patient clinical trials in the clinical development versus the earlier research phase. While the traditional clinical phase definitions do not entirely match with the newer concepts of approaching development, the later trial stages, that is, phase 2b and beyond, are nowadays commonly referred to as *late phase* . "Early phase" in turn, being the first-in-man, proof-of-concept, and comparable clinical trials at the earlier stages (usually phase 1/2a), while strategically of paramount importance, are still a smaller portion of the total clinical and outsourcing spend. Consequently, most clinical CROs are concentrating on late phase that reflects the main volume of outsourcing both in total staffing and spend. Some of the mid and larger players are also branching into early development or preclinical. There are various combinations of services in the areas of preclinical, clinical, and ancillary vendor services (such as technology) that are also serviced by some CROs. These services tend to be commonly arranged either along the value chain or be selected based on specific expertise in certain areas such as analytical methods or technology. Also, CROs tend to also restrict services to one of the GxP areas (Good Manufacturing Practice (GMP), Good Laboratory Practice (GLP), Good Clinical Practice (GCP)) for the great complexity that results in maintaining such parallel regulatory frameworks that serve such areas, yet again there is exceptions to this. In any case, there is probably virtually no CRO that indeed serves all aspects of the pharmaceutical R&D value chain, be it that some claim to do so and may also be very close to it.

Additionally, there are a large number of specialty providers in the GMP (manufacturing, clinical trial supplies), GLP/GCP area (laboratory providers) as well as technology (such as electronic data capture (EDC), interactive voice response systems (IVRSs), diagnostic imaging, electronic patient-reported outcome (ePRO) systems, and patient recruitment), and logistics providers serving special needs.

As a result, the CRO industry has been described as highly fragmented, with effectively hundreds of CROs present in the United States as well as in Europe of which the majority serve consequently considerably less than 1% of the total market even in their region. This may be great for finding the right and specialized expertise for any single case but in turn is close to a nightmare in terms of maintaining efficient best practices both in vendor selection as well as efficiencies of scale. Having said this, there is good reason when such fragmentation persists over such a long period in an industry that can be considered reasonably mature. There has been considerable consolidation among the CROs especially in the past crisis years that resulted in a larger spread of the larger ones getting even larger, and the smaller ones getting more specialized or, sadly enough, vanishing due to an unsustainable business model in the "niche" that turned out to be none. We believe this consolidation reflects to a large extent the decision drivers for selection of CROs to be developed later herein and is proof that in this industry, neither "one size" nor "one model," fits all.

Within this chapter, while many considerations would also be applicable to all other providers, we will focus on the major portion of the industry in terms of number of staff and revenues, the clinical CROs.

The fluctuation in pipeline flow, as driven by various factors including availability of in-licensed development candidates; innovation of new chemical entities (NCEs) or new biological entities (NBEs) "at source"; and pipeline adjustments due to overall commercial development changes affecting R&D spend led to either overabundance or lack of internal resource. With increasing need for flexibility, the degree of outsourcing had to accordingly increase to attenuate the peaks and valleys in resource requirements. Figure 17.1 shows the principle factors that would drive outsourcing volume and therefore the degree of outsourcing for any given sponsor organization as a factor of pipeline volume (number and size of projects) and availability of internal resource in terms of capacity and capabilities (including locations of trial conduct). The principle of this mechanism would commonly be to have a certain degree of own resources, above which outsourcing would then be utilized.

The need for globalization of drug development could be handled by globalization of the own sponsor organization only by some of the largest Pharma companies. Most of mid and probably large Pharma, and of course an increasing number of biotech's, specialty Pharma companies, and emerging medtech companies needed to focus and therefore complement their geographic reach by either partnering with other Pharma via out-licensing and/or by outsourcing. This has certainly been a driver for the ongoing globalization of CROs.

Cost containment of ever increasing development costs and cycle time optimization are obvious requirements of control and optimization that are there to stay. In the early 1990s, and probably for many companies well up into the twentieth century, outsourcing decisions were done as what has been referred nicely as

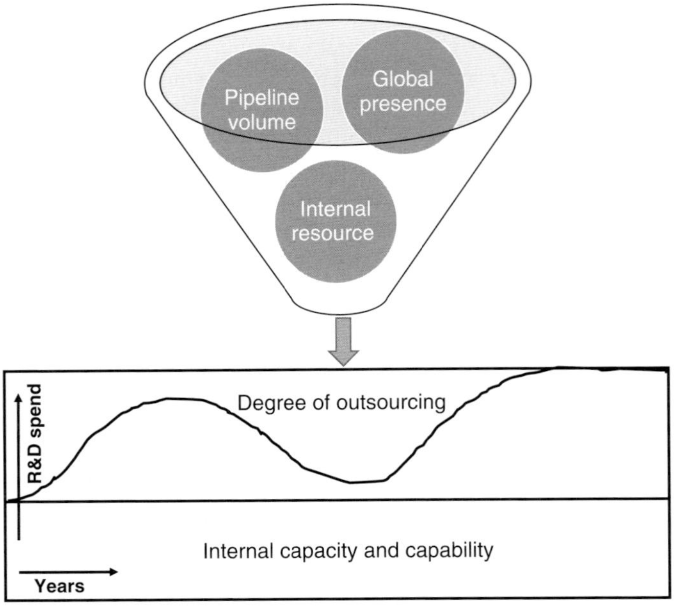


**Figure 17.1** Degree of outsourcing as a function of pipeline volume and internal resource and capabilities.

*tactical outsourcing* but may have been in fact not so much tactics but disparate decision makers in the same organization that made unaligned assignments and selections of CROs, frequently based on their personal network, to cover their needs. One may speculate that the drivers cost containment and cycle times were usually not the key decision factors. Due to the increasing price pressures on Pharma, in turn came the need for more systematic and objective provider selections. It is probably only fair that Pharma had to pass on the price pressures to their CROs as well, due to the dwindling efficiencies of the development process, with the result of a considerable decrease of CRO margins over the past two decades. The answer whether these margins are nowadays still fair or too low to maintain a compliant, experienced, or even "innovative" development organization (=CRO) will probably depend on whether we ask buyer or seller of services here. The overall industry picture and financial data exemplify these two ends to the spectrum also well: Record highs in the valuation of publicly traded CROs (that are reflective however of the large to very large segment only) and high interest in venture capital, while in parallel record highs in the number of smaller CROs filing for bankruptcy. Just as in real life, and in a Darwinian process, the fittest and most adaptive will win, and there clearly has been and is also going to be some further serious fatalities and wreckage along the way.

With the increasing need to improve the outsourcing process, not the least driven by the fact of the continuously increasing spend of Pharma to CROs, "strategic outsourcing" in form of dedicated outsourcing functions arrived, again more so or earlier in the larger Pharma companies than in small organizations.

**Table 17.1** Key objectives to utilize outsourcing in the early days.

| Trigger | Need | CRO capability | Comment |
|---|---|---|---|
| Fluctuating pipeline | Resource flexibility | Provide capacity | Capacity, staff resources were usually needed for noncore activities such as monitoring and data management |
| Globalization | Geographic reach and regulatory expertise | Staff in various countries | Needs strongly dependent on sponsor size and strategy of globalization |
| Cost containment | Reduce clinical trial costs | Cost competitiveness to sponsor internal costs | Process expertise, system efficiency, volume efficiencies |
| Cycle time optimization | Reduce clinical trial cycle times (especially start-up, recruitment) | Commitment to timelines and efficiency | Process and local regulatory expertise, system efficiency, site networks |

This was to achieve risk minimization in terms of CRO failure (such as could due to late discovered lack of the CROs actual capabilities, or as another example due to financial instability, an aspect for which many sponsors had to pay high dues experiencing that their CRO and data vanished unexpectedly during critical development studies) and was usually addressed by more systematic CRO selection processes as well as according to cost control by better vendor oversight and control during both selection and study conduct.

### 17.2.2 Core versus Noncore Activities

The former mantra of keeping critical intellectual property (IP) and knowledge in-house, as well as to draw primarily or exclusively from in-house expertise to foster innovation, led to the concept of outsourcing "noncore" activities. Whether this was the actual conscious strategy behind such approach, or merely a fitting retrospective explanation of an established trend,[1] taken together with the outsourcing drivers shown in Table 17.1 led to primarily outsourcing the larger volume "noncore" task of clinical monitoring. Successively, clinical data management and clinical project management were added to the outsourcing portfolio, with monitoring and data management being indeed major cost drivers of clinical study conduct, and commonly high volume and repeat activities. The "noncore activity" classification was also applied to other areas, such as certain therapeutic areas

1) http://www.intetics.com/strategic-core-and-non-core-activities-what-to-outsource-for-best-results/.

(TAs) of development that were not the focus TAs of the organization's strategy, certain geographies usually driven by marketing considerations, or certain regulatory classifications of studies such as noncore studies being nonpivotal, supportive studies during development, regional regulatory studies, or postmarketing studies. With the exception of some noncritical supplemental clinical pharmacology studies, there is also a tendency to frame later development phases more as noncore than the data-critical and decisive early studies that are so crucial for the further portfolio evaluations.

Put simply, this approach of outsourcing "noncore" activities by definition put the CROs into a spot that was remote from the innovation process. The competitiveness of CROs had to be driven by being process experts rather than development or product experts. Our chapter would have likely ended here, as it concerns the contribution of CROs to innovations, would it not have been for the further maturing of the CRO industry as well as the outsourcing management by Pharma, as well as changing pressures on the pharmaceutical industry. In the following we will therefore attempt to develop the further fate of industry trends and specifically address the following questions:

- How has outsourcing innovated itself and what are the future trends?
- How does and will outsourcing contribute to innovation?

## 17.3 Genesis of Outsourcing in the Twentieth Century: From Commodity to Contribution

### 17.3.1 Outsourcing Portfolio and the Move to Full-Service Provision

The move to strategic outsourcing caused the creation of more professional outsourcing processes, together with establishing and increasing staff in the outsourcing departments of Pharma. Managing the outsourcing portfolio in terms of number of providers, as well as their particular strategic and contractual setting differs widely between companies. Accordingly, there are more than trivial or simple cause and effect relationships that will lead to strategy implementation and actual provider management. Nonetheless, the goal was mostly to make the outsourcing process more efficient in terms of price, service level, and management. The following are probably the most common drivers:

- A mandate to lower R&D spend (cost of outsourcing) while ensuring that quality and timelines are met
- A need for reduction of the number of providers for various reasons:
  - The potential to realize volume effects, optimize interfaces, and harmonize processes
  - To lower the effort for selection and oversight of the vendors on the sponsor side
  - The need for more specialized services, and more global geographies, leading to the selection of larger, global providers that can offer extended service

capabilities as well as a larger geographic footprint while still having to serve some needs with local or specialty providers
- And not the least the need to concentrate the outsourcing process to sell the leadership on the process by the outsourcing department within the own organization.

The genesis of these trends is by no means homogenous across the industry, as it is affected by the size and maturity of the sponsor organization and its processes, the particular outsourcing strategy, the needs of the development pipeline, growth, overall R&D spend and outsourcing degree, among others. Yet, starting probably around the mid-1990s and in fact still continuing to some extent, full-service outsourcing became more common as a result. This was due to classification of noncore studies and TAs that were handed in full to the CRO, as well as economic considerations that allow the CRO to offer efficiencies and hence potentially more value for money in full outsourcing due to its internally harmonized processes, as well as the need to reduce the interfaces between sponsor and various providers that quickly proved to be personnel-intensive and a risk factor on all sides. Having moved to full-service outsourcing in our historic review, in essence we have now progressed from the purchase of an "off-the-shelf" service to a usually reasonably sized contribution of the clinical trial process that is in the hand of the CRO. This is relevant as it puts the CRO more into the position of a potential contributor than a mere provider, be it that at this stage the contribution could be limited to "noncore" and hence somewhat deprioritized projects.

### 17.3.2
### Globalization and the Emerging Market Hype

The growing importance and focus on emerging markets in the early twentieth century prior to the 2008 crisis certainly has been a trend across virtually all industries. The decreasing margins and stalling growth rates in the established markets in the Pharma industry were probably the main drivers for the increasing focus on emerging markets from a marketing and sales perspective, which in turn had to be reflected in the development, and in fact to some degree also in the research process. In parallel, the costs to bring a drug to the market during this period approached the US$1 billion mark DiMasi and Grabowski (2007). If not already before, this made global development and later marketing of products a must, where global means here at least covering the main regions: United States, Europe, and Japan. This was further supplemented, depending on product and company strategy, by adding emerging markets such as the BRIC countries (Brazil, Russia, India and China) during the development stage. Their potential for improving timelines due to faster patient recruitment, coupled with considerable costs savings due to the lower cost base in such countries David *et al.* (2010), led to what can be called a *hype* to include emerging markets into clinical trials and the development process, in particular China and India. There have been reasonably large investments into these regions by Pharma, numerous alliances, and a very strong buildup of capacities and capabilities over the first decade of the twentieth century. It seems there were times where it was really difficult for a Division Head of Clinical Research in Pharma to explain to his board why any given study would not include at least either China or India. In fact, during this period, China and India, together with South Korea, had worldwide the highest growth rates in terms of the number of clinical trials initiated in the country per year. Yet this trend was rather suddenly reconsidered latest with the advent of the financial crisis in 2008 that led to stringent cost-cutting exercises by Pharma, a refocus on sales and marketing in the established markets as the backbone of Pharma income, and probably most important a stringent reconsideration and downsizing of the development portfolios. The overall more focused and lower level of clinical trial budgets and hence outsourcing spend during the crisis probably led to the need to leverage the well-proven marketing effect of conducting clinical trials where the money is – in the established markets. While this on a pure cost perspective per service unit of course may be more expensive, it probably promised more leverage from a marketing perspective, considering the need to distribute the limited moneys to spend available, and fastest possible market impact. An analysis of data in clintrials.gov (https://clinicaltrials.gov/) by the authors at the end of 2010 indeed shows that there was a trend in repatriation of clinical trials to the established markets (on a percent base of total) to the United States and Western Europe on account of emerging markets. Interesting is also to note that despite the considerable, double-digit growth rates in countries such as India and China in clinical trial activity, they had started at such low level that even in 2010 they were not to be found among the top 10 countries worldwide.[2] For Europe in particular, Central and Eastern Europe (CEE) has seen, and we believe will continue to see, relatively high growth rates and less of a crisis dip as a reflection of the European market expansion from West to East, while still maintaining considerable cost advantages (see also: Hajos and Schiewe (2010)).

Together with the attention to emerging markets, both Pharma but also particularly the CRO industry increasingly started to utilize offshoring of central services, in particular data management. The topic of offshoring, its advantages and disadvantages, has been reviewed in more details by the authors elsewhere (Hajos, 2010). In short, to create value an offshoring initiative needs a long-term perspective, whether it is realized in-house or via partners or CROs. We believe offshoring can create substantial cost savings and value creation but is by its nature of serving from a central offshoring location limited to certain services and expertise areas (such as central data services) and tends to suffer at least in the earlier stages of its implementation from potential communication and cultural issues. It would be for the limited service contribution and the communication and integration aspects that we do not consider offshoring a quick win in bringing innovation to the process. We do want to mention however that at least in theory, the potential

2) The authors' data indicate from the 2010 analysis that in fact India ranked just between Austria and Czech Republic, both being really small countries that would not be considered at the forefront of trial locations, and China being only slightly ahead of that.

of these very cultural differences as a source of diversity holds of course an element of sparking innovation once the model would be firmly established.

### 17.3.3 Procurement Takes over the Outsourcing Function

In parallel, and somewhat consistent with the developments of the crisis years with ample reengineering and cost-cutting exercises at Pharma and to a certain extent subsequently CROs, the past years have seen the (general) procurement departments of sponsors take over much or all of the management of provider management that was previously handled by "outsourcing departments" located in the R&D function. This change in many cases was probably "ordered" by the CEO or CFO in sponsor companies to ensure better cost control of R&D and make it more connected to the overall strategies of the organization, and brought to them by business consultants with checklists and excel sheets how to implement the cost-saving exercise that may not always been highly customized and aligned with the situation and strategy of that very company. It was nonetheless a logical extension of the professionalization of the outsourcing function that was needed and is advantageous as such. There has certainly been an interesting learning curve by both the sponsor's procurement departments and the CROs on how to handle those new negotiation partners, in order to arrive at a state that serves the intrinsic needs of Pharma in selecting and managing clinical providers well. Much of this has meanwhile been further improved and refined, not the least by contributions and discussions of all stakeholders, and via organizations such as the Pharmaceutical Contract Management Group (PCMG; http://pcmg.org.uk/), with PCMG as an example also more recently actively liaising with a procurement association (CIPS, http://www.cips.org/) with very positive results.

These developments bring us to the current or recent state of the outsourcing industry, summarized as follows:

- Preference of >50% of all sponsors to implement some form of preferred providers or alliances with their CROs.
- Large sponsors frequently have dedicated, often single, main alliances with global CROs that may include risk sharing models as well as allegedly financial benefits due to scale.
- The main decisive factors for selecting CROs are reported to be experience, data quality and compliance, and stability before cost savings and other factors.
- Outsourcing is commonly managed by the procurement function that often has recognized the need and implemented an alliance management function.
- Sponsors typically hold a portfolio of CROs as a mix of global CROs and regional or specialist providers. The number of CROs in the vendor portfolio of sponsors seems to have decreased but is stabilizing.
- Presence in the key markets for CROs is mandatory, while emerging markets and other locations are supplemental to ensure adequate patient recruitment or cover special market needs.

We would like to suggest that the diverse balance of the above-mentioned key drivers for selection of CROs, the diversity reflecting the very different priorities of sponsors from biotech to large Pharma, as well as their particular outsourcing and innovation strategies, ultimately lead to a positioning of whether a CRO is merely a provider (like a telecom provider), a development (and innovation) partner, or an alliance of some sort that may have any meaning somewhere between provider and partner. Naturally, there may then also be a noticeable difference between what the strategy was meant to be, and whether and how it is in fact implemented and lived on both sides, Pharma and the CRO.

## 17.4 Current and Future Trends in Outsourcing: From Contribution to Innovation

### 17.4.1 How Has Outsourcing Itself Innovated and What Are the Future Trends?

There is ongoing innovation in the trial processes, and it is probably fair to say that both sponsors and the CRO industry are actively contributing to these, ideally together with the regulatory agencies and other stakeholders, as is indeed sometimes the case. Examples would be the risk-based monitoring (RBM) initiatives started by the US-FDA (2013), then captured by TransCellerate (2013) and others, and in ongoing development as to its applicability by all stakeholders (Alsumidaie *et al.*, 2015). One could argue that RBM is however also another example of Pharma being a rather conservative industry, as it seems that the uptake and actual application of RBM by sponsors is fairly slow, similar to what had been seen as a process of almost two decades until EDC has become firmly established. In turn, this seems only natural, taken that nobody would want to take the risk of regulatory nonacceptance of processes, or impairing data quality – whether true or only due to lack of proof that this was not the case – and even less so if this could affect pharmacovigilance aspects.[3] Innovation in technology and in the use of technology in the clinical development process will nonetheless further innovate the process aspects of R&D and how sponsors and CROs work. Additional areas of innovating processes and approaches in the development process include the use of laboratory biomarkers and other diagnostic markers (such as diagnostic imaging) as surrogate endpoints, as well as further harmonization of regulatory procedures in various regions of the world such as, for example, the upcoming EU clinical trial regulation for a more centralized and harmonized EU-wide process of clinical trial applications. Not the least, we also feel that there has been a substantial improvement of more active communication and collaboration between the stakeholders involved in the development process. This includes more approachable and transparent regulators as such, as well as various

3) It has been said that the Pharma industry is one of the most regulated industries overall, only after the aviation industry, and more regulated than the banking sector, which is in itself remarkable.

initiatives of joint working parties or industry events (see, e.g., EUCROF,[4] DIA,[5] PCMG[6]) that bring together regulators, Pharma, and more recently also CROs.

There are many variations to the previously described standards of strategic outsourcing, vendor management, and global alliances. Some larger Pharma prefer the model of insourcing resources to integrate them on a project or functional level into their own organization, in order to have better control, and less interfaces when using their own processes and Standard Operating Procedures (SOPs). It may also provide with a more granular selection of the needed resource since it is frequently a more or less individual/single person selection. Such insourcing approaches tend to be focused on certain service areas, an approach that is referred to as a functional service provider (FSP) model. Roughly speaking, insourcing here is the in-house model, whereas offshoring can be viewed as the equivalent outsourced model. The downside of such insourcing models may however be the considerably higher complexity and the resulting higher need of Pharma's internal resources to manage staff as well as coordinate and oversee service delivery when compared to utilizing aligned full-service providers or in-house functions.

Other companies are putting an emphasis on the investigator site liaison by running a "site management" or site liaison organization internally in addition to the CRO conducting a study in order to circumvent the slight disconnect of the sponsor from the site. In the end the site is the customer, and in the case of high academic standing, key opinion leaders (KOLs) may be part of the innovation network of the organization.

There are various shades and degrees to which the outsourcing process and the collaboration with CROs and other stakeholders may vary. In the end, the innovation here is probably that the industry has learned to allow for certain flexibility for such varying models and is testing new ways of what works and what doesn't very much on a company-by-company level. One important aspect here is that such newer models allow for creation of and contribution in a network of parties that may include sponsor, development partners, or licensors such as a biotech, academic sites/KOL, other specialist consultants, and CROs and that make up such a network of diverse contributors which is an essential ingredient to allow for true innovation, as also discussed in more detail by Schuhmacher *et al.* in Chapter 15 in this book Garnier (2008), Hunter and Stephend (2010), Douglas *et al.* (2010), and Zhong and Mosley (2010).

These changes and innovations are on the process side and would contribute to a more efficient R&D process. They do so however largely in an indirect way, as they are not directly affecting the innovation of the drug or product itself, since the development process is here too far away from both discovery and the customer.

The procurement functions have certainly been creative to some extent, as well as innovative. While we have not yet seen CRO tenders on eBay, auctions on

4) http://www.eucrof.eu/.
5) http://www.diahome.org/en-GB/Meetings-and-Training.aspx.
6) http://pcmg.org.uk/.

electronic platforms for study tenders are not entirely uncommon. While preselection of the vendors admitted to the auction may have some benefit to secure the key objectives, we believe this defeats the purpose of selecting the right partner CRO as it ultimately reduces the selection to price. Or in other words, we have not yet seen marriage agencies use auctions on the referral fee to do the final matchmaking, but we may be wrong. If one selects a partner, it is unlikely to be done best on price alone. The reason to raise this here is not to prevent some CRO to close potentially a low margin contract but that procurement methods that would be strictly price driven may ignore the risk that the partner (CRO) needs to remain viable and financially healthy. Just as in nature, there needs to be some consideration of the principle of "live and let live," and in fact modern and proper procurement methods fully take account of such risks.

### 17.4.2 How Does and Will Outsourcing Contribute to Innovation?

Despite all due financial and business considerations, and the recognized professionalization of the outsourcing processes, the key to the relation between sponsor and CRO, just as it would be between a Pharma licensee and biotech licensor, or any other such collaboration, is trust. Such trust is indeed established on the individual and personal level, and it is also build on previous experience that "things worked beyond the honeymoon" to stay in our marriage example with some or many people in the organizations (sponsor and CRO). Sponsors will come back to those they found have answers to their questions or are creating value to find mutually beneficial solutions. We believe it is not the least for this reason that there is consensus that a balance between global alliances by large global CROs and their sponsors and small to midsize CROs that have specialist experience and/or particular regional expertise is healthy and keeping competition alive. This is not meant to say that specialist or regional expertise is not or less present in large and global CROs. It certainly is, and needs to be verified on a case-by-case basis, as the differentiator "specialty" is probably just about as flexibly used as the term *global*.

The recognition of the importance of smaller yet crucial functions such as medical management, pharmacovigilance, health economics and, while well established, regulatory and biostatistics are some examples of potential key contributors to innovation in the development and outsourcing process. Others that may be worth mentioning and with more an operational than a scientific and medical focus are probably quality assurance, clinical trial supplies, and lab and trial logistics that can all be a make or break for trial success. The participation of multiple and critical functions in more of a full-service model by the CRO will support contribution at critical decision points during the clinical trial and development process, as well as add value via oversight from different functional and experience perspectives. Prerequisite for such contribution is of course again a partner approach based on trust and recognition of the strength and experience of all team members on all sides. Such a team approach between sponsor experts, CRO personnel, investigational sites, KOLs, as well as other vendors and consultants is a basic ingredient and may make a good recipe for fostering an innovative climate among the network of contributors.

Having pointed out some of these more specialized functions, and having reviewed the history of putting clinical monitoring and clinical data management somewhat into a commodity or off-the-shelf service, we do need to recognize these functions here. Given there is a reasonably well-designed protocol that should meet the clinical and regulatory objectives, data quality is still primarily determined by having experienced Clinical Research Associates (CRAs) and Clinical Data Managers (CDMs) that are at the source of data and still manage the majority of the entire workload. As M. Martorelli puts it in his recent Pharmaceutical Outsourcing Monitor (Martorelli, 2015), "( ... ) journalists, analysts, and other outsiders who constantly blow the 'disruptive innovation' horn might be doing a disservice to thousands of drug development worker-bees. They (*the CRAs and CDMs*) are the ones who can make some real advances in the art and the science of drug development by just solving the little problems they confront every day."

## 17.5
## Discussion and Conclusion

The industry is torn between disparate forces of stringent financial control driven by procurement and the need to truly share risks and benefits between the partners to allow for commitment, value creation, and finally innovation needed to make development–outsourcing alliances work.

Drug development, and herein more specifically clinical development, has evolved over several decades to provide a wide spectrum of business models of how a pipeline is best managed to result in innovative new drugs and devices. Among the various stakeholders in the process are pharmaceutical companies, biotechs, venture capital funds, regulatory authorities, consultants, healthcare providers, know-how providers such as academic KOLs and investigational sites that are also the product customers, and finally patients. In the midst of it is outsourcing, and with it CROs, that have become a recognized participant in the industry. The genesis of drug development and outsourcing today allows for a wide variety of models of collaboration and positioning of what role each stakeholder may take in any given drug development project and sponsor company.

To allow for innovative processes to take place, various requirements have been recognized that are summarized in Table 17.2 and discussed in more detail in the following text. Innovation requires leadership (De Meyer and Garg, 2005) that defines the roles and provides the information and communication channels to enable true value contribution. CROs may be pure vendors delivering a defined unit of service where service charge is measured in minutes and according to dollars like a manufacturing step. This puts the staff and expertise at the vendor outside the innovation circle. As is true for many other stakeholders in the

**Table 17.2** Basic requirements and enablers for innovation in the pharmaceutical outsourcing process.

| Requirement | Enabler(s) |
|---|---|
| There is no innovation without leadership | Alliance management, project management |
| Innovation requires risk management | Outsourcing strategy, long-term commitment |
| Innovation is triggered by creativity and diversity | Globalization, outsourcing strategy |
| Innovation requires organizational integration | Outsourcing strategy, long-term commitment |
| Success in innovation requires excellence in project management | Alliance management, project management |
| Information is the crucial resource for effective innovation | Alliance management, project management |
| The results of creative efforts need to be protected | Procurement, alliance management, clinical operations |
| Successful innovation is rooted in a good understanding of the market and customer needs | Alliance management, project management |

Adapted by A. Hajos from De Meyer and Garg, 2005

network, contribution requires the recognition as a partner and potential contributor and furthermore means that the "service charge" allow creative individuals to at least take a few minutes to think, to elaborate, to interact, and to communicate. This margin to breathe and contribute would likely be defined by the alliance and resourcing model that can span from unit-based pricing to booked dedicated resources, with the resulting investment by the sponsor of "risk money" into contribution and/or innovation that may or may not pay off. This second element of innovation, risk management (De Meyer and Garg, 2005), is at the heart of pipeline management and the funding of drug development but is far away from at least the traditional models of outsourcing as a fee-for-service. While there have been various models and initiatives by sponsors and CROs alike of sharing risks, there seems to be still a dichotomy of business models and resulting collaboration to be overcome. We feel this is at least one of the important reasons for the observed discrepancies between theory and practice, that is, how alliances and sponsor–CRO relations are strategically designed versus how they are sometimes negotiated and lived. The diversity of sponsor needs taken together with the various strategic options of outsourcing models, and the fragmentation, specialization, and further maturing of the CRO market hold however the promise to allow for further organizational integration, flexibility, and creativity in various models of network interactions that again must be considered essential ingredients for innovation. Optimizing the flow and protection of information, as well as protecting the result of the creative efforts, is related to good project and alliance management. Being able to implement and live such good management, in the end this probably still primarily related to trust and a commitment of a long-term relation between the parties. Such cannot be just defined in a service contract but needs to be lived day by day by the partners.

Innovation of the development process requires the interaction of all stakeholders including the development team consisting of sponsor, CRO(s), and other vendors; the regulators; and the investigational sites. It would be futile to speculate where the best expertise and experience is located, or which ones are most important, as in this case the whole is clearly greater than the sum of its parts. While in a simplified viewpoint sponsors have at least previously been the product (drug) and TA experts, CROs have been process experts. With the delegation of even "noncore" activities and the diversity of needs of sponsors (from biotech to large Pharma), and not the least the increasing percentage of outsourcing overall, CROs and probably all other stakeholders in the network have accumulated experience and expertise in various further aspects. As mentioned before, the greater use of specialty and crucial CRO development functions such as those related to regulatory, medical, pharmacovigilance, biostatistics among others have also contributed. CROs and other vendors not being part of the sponsor organization may also have the advantage of less of a not-invented-here syndrome while facing various diverse clients and hence developments. While protecting the IP and knowledge of each one of these clients is mandatory and critical, these externals may apply learnings even from entirely different development projects. In other words, the exposure to a diverse "development market" by the external parties may sometimes also be an advantage versus the naturally focused viewpoint and experience of the pipeline holder.

Ultimately, the potential innovation of a product (be it the drug, its administration, formulation, indication areas, and alike), rather than innovation of the development process alone, is at the core of the innovation efforts. The market and product knowledge that is required for innovation usually resides within the sponsor organization, while CROs and several other network partners as such are rather remote from it as determined by their roles and business models. They are however still close to the customer: be it the principal investigator in an early trial in a phase 1 unit that has probably more a medical and scientific viewpoint on the potential of the drug and its factual application to subjects or the investigational sites with their entire study teams and including KOLs, patient advocacy groups, and alike in a global later phase trial, the CRAs and, somewhat more indirectly in a data-driven fashion, other CRO personnel are at the source of customer interaction. The well-taken attention of most sponsors on site relation (usually at least with KOL sites) and the site liaison initiatives of some Pharma companies are testimony to the importance of this interface. In turn, we have seen attempts by sponsors to keep CRO personnel away from interactive site management aside of pure process steps for monitoring and pushing forward, for example, local affiliates of the Pharma companies, to foster the interaction with these "customers" that seem contra productive to an open and creative information flow.

There are today ample options how to set up an innovative network of contributors for drug development. Globalization brings diversity in educational and cultural backgrounds of the individuals that may be part of such network. The newer models of pipeline and innovation management by Pharma, while adding complexities of interfaces, will foster a greatly enhanced coverage of different stakeholder perspectives. Also great is that technology will probably further allow to optimize ways of how we work together, communicate, and analyze our data, and it is very likely that CROs and other vendors will play an important role in such optimizations.

Despite all commercial considerations, innovation, and other boardroom strategies, we are all working in a highly ethical field in order to develop medications and healthcare solutions. The complexities of drug development require the collaboration of large teams in a mutually beneficial partner approach. Our motivation and commitment of being dedicated to the outcome are likely related to how we treat each other and is probably most strongly driven by enjoying joint team success during the little daily challenges as well as when taking larger key milestones in a project together. Further developing and improving how we all work together and how this can best be translated into outsourcing models as well as innovation network management is the path to further improvements in allowing for and managing innovation. To do this will still require trust and upfront investment not only of money but relationship building.

## References


Alsumidaie, M., Widler, B., Schenk, J., Schiemann, P., Andrianov, A., and Proupín-Pérez, M. (2015) RbM Guidance Document: Ten Burning Questions about Risk-Based Study Management. Applied Clinical Trials, http://www.appliedclinicaltrialsonline.com/rbm-guidance-document-ten-burning-questions-about-risk-based-study-management?topic=201%2C225%2C224 (accessed 12 August 2015).

David, E. *et al.* (2010) Pharmaceutical R&D: the road to positive returns. *Nat. Rev. Drug Discovery*, **8**, 609–610.

De Meyer, A. and Garg, S. (2005) *Inspire to Innovate: Management and Innovation in Asia*, Palgrave Macmillan, p. 15ff. ISBN: 13: 978 1-4039-9684-8.

DiMasi, J.A. and Grabowski, H.G. (2007) Managerial and decision. *Economics*, **28**, 469–479.

Douglas, F.L. *et al.* (2010) The case for entrepreneurship in R&D in the pharmaceutical industry. *Nat. Rev. Drug Discovery*, **9**, 683–689.

Garnier, J.P. (2008) Rebuilding the R&D engine in big Pharma. *Hav. Bus. Rev.*, **86**, 68–79.

Hajos, A.K. (2010) Offshore, Reshore or Rightshore: Perspectives on Offshoring to India, European Pharmaceutical Contractor, pp. 30–34, Samedanltd.com (accessed 12 August 2015).

Hajos, A.K. and Schiewe, A. (2010) Studien in Schwellenlaendern. *Transkript*, **7** (16), 36–37.

Hu, M. (2007) The Innovation Gap in Pharmaceutical Drug Discovery & New Models for R&D Success, Kellogg School of Management, www.kellogg.northwestern.edu (accessed 12 August 2015).

Hunter, J. and Stephend, S. (2010) Is open innovation the way forward for big Pharma? *Nat. Rev. Drug Discovery*, **9**, 87–88.

Martorelli, M.A. (2015) Fairmount Partners: Pharmaceutical Outsourcing Monitor, January 30, 2015.

Schuhmacher, A. and Betz, U.A.K. (2015) The pharmaceutical industry is opening its R&D boundaries, in *Value Creation in the Pharmaceutical Industry*, Chapter 15, Wiley-VCH Verlag GmbH & Co. KGaA, Weinheim.

TransCellerate Biopharma Inc. (2013) Model Approach for Risk Based Monitoring, http://www.transceleratebiopharmainc.com/our-initiatives/risk-based-monitoring (accessed 12 August 2015).

US-FDA (2013) Guidance for Industry: Oversight of Clinical Investigations – A Risk-Based Approach to Clinical Monitoring, http://www.fda.gov/downloads/Drugs/Guidances/UCM269919.pdf (accessed 12 August 2015).

Zhong, X. and Mosley, G.B. (2010) Mission possible: managing innovation in drug discovery. *Nat. Biotechnol.*, **25**, 945–946.

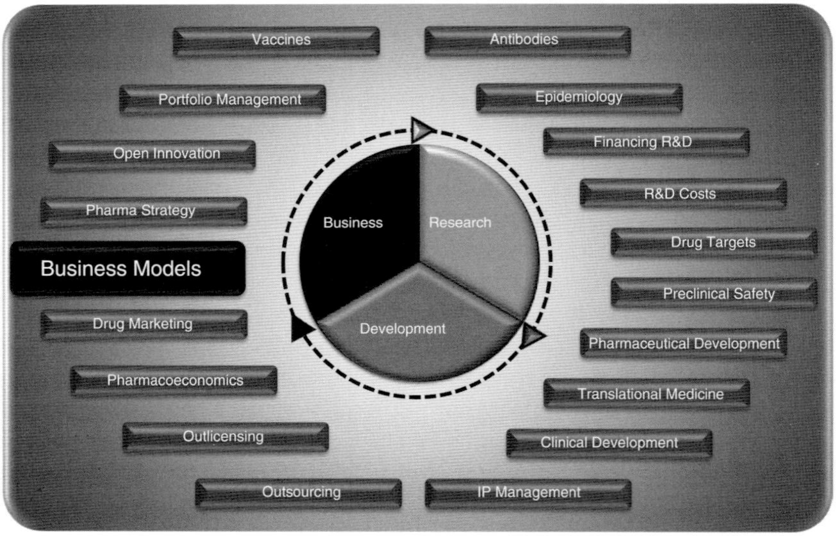
Vaccines
Antibodies
Portfolio Management
Epidemiology
Financing R&D
Open Innovation
R&D Costs
Pharma Strategy
Business
Research
Drug Targets
Business Models
Preclinical Safety
Drug Marketing
Development
Pharmaceutical Development
Pharmacoeconomics
Translational Medicine
Outlicensing
Clinical Development
Outsourcing
IP Management

# 18
# New Innovation Models in Pharmaceutical R&D

*Alexander Schuhmacher, Oliver Gassmann, and Markus Hinder*

## 18.1
## Introduction

The purpose of this chapter is to illustrate the consequences and measures taken in the past years in the pharmaceutical industry as a result of the historical low success rate in Research and Development (R&D). As already outlined in detail in our book chapter "a review of the pharmaceutical R&D efficiency: costs, timelines, and probabilities," the R&D efficiency of the pharmaceutical industry decreased continuously in the past years. Today, the costs for discovery and developing a new drug are enormously high, the probabilities to be successful are extraordinary low, and the timelines are very long.

A central observation of the past 25 years is the increase in R&D cycle times summing up to 14 years on average today. Paul *et al.* (2010) reported timelines per R&D phase of 4.5 years for discovery research, 1 year for preclinical testing, and 1.5, 2.5, and 2.5 years, respectively, for the three clinical development phases. Finally, the phase from submission of a market authorization to launch of a new drug lasts 18 months on average (Paul *et al.*, 2010).

The Pharmaceutical R&D Factbook 2013 of the Center for Medicine Research International (CMR) reported an average success rate of 4.1% from first toxicity dose to market approval. Based on the information of the participating 11 multinational pharmaceutical companies, the success rates for the preclinical phase, phase I and phase II of clinical development are 67, 46, and 19%, respectively (CMR, 2013). The reasons for the high attrition rates can be summarized as follows:

- Adverse pharmacokinetics and bioavailability
- Lack of efficacy and safety
- Lack of predictive animal models
- Difficulties in target validation due to the concept of target-based drug discovery
- Increased complexity of clinical trials in combination with the higher demands of regulatory authorities

*Value Creation in the Pharmaceutical Industry: The Critical Path to Innovation,* First Edition.
Edited by Alexander Schuhmacher, Markus Hinder, and Oliver Gassmann.

As a consequence, the long R&D timelines and the reported high failure rates result in enormously high costs for providing new molecular entities (NMEs). In 2010, the average total out-of-the-pocket costs for new drugs were calculated as USD 873 million, and the resulting total capitalized costs are USD 1.778 billion (Paul *et al.*, 2010). In consequence, fewer NMEs were launched to the market per USD billion invested and the return on investment (ROI) of pharmaceutical R&D is under pressure.

At least since the last decade, many efforts were made to reduce R&D costs, attrition rates, and cycle times to influence the R&D efficiency positively. This book chapter will summarize some attempts that have been taken, both on process and business model levels. In this context, this book chapter will answer the following key questions:

- How did the overall pipeline size in the pharmaceutical industry develop in the past?
- How did the investment in R&D in the pharmaceutical industry develop in the last years and which measures did pharmaceutical companies take to reduce their R&D costs?
- Which new R&D models were developed to increase the R&D efficiency?

## 18.2
## Some Attempts That Were Recommended in the Past

Numerous measures to increase the R&D efficiency, which is defined as the input/output ratio of the financial investment in R&D in relation to the number of NMEs launched to the market (Fernandez *et al.*, 2012), have been recommended. Some attempts will be reviewed here:

- For example, DiMasi (2002) reported, a reduction in the phase length of pharmaceutical R&D would reduce cost per approved drug significantly (DiMasi, 2002).
- Rawlins (2004) discussed a reduction of the number of clinical trials per drug candidate as an option to reduce costs of R&D. He suggested that one randomized, blinded dose–response phase IIb trial in combination with a single randomized controlled phase III trial would be sufficient to provide evidence of efficacy (Rawlins, 2004).
- Many other attempts were made to improve timing and output of R&D, such as the application of management techniques like project or portfolio management.
- Likewise, it was intended to improve the efficiency of preclinical research by the implementation of high-throughput and omics technologies, such as high-throughput screening (HTS), combinatorial chemistry, genomics, or proteomics.
- Today, licensing of drug candidates is a key cornerstone of pharmaceutical R&D. In this context, it has been reported that in-licensed drug candidates have a

higher PoS for phase I to submission than self-originated drugs (DiMasi *et al.*, 2010).

- Another direction has been proposed by Kneller (2010). He suggested to increase the financial funding of industry-wide R&D activities. In particular smaller companies often do not have the financial power of multinational pharmaceutical companies despite being critically important for discovering innovative drugs (Fernandez *et al.*, 2012; Kneller, 2010). The proposed mega-fund could finance drug discovery projects, such as target identification and validation. In combination with a broader externalization of pharmaceutical R&D to smaller and specified companies, this would help to mitigate the research-related technical risks associated with early drug research, while using the competences of pharmaceutical companies in preclinical testing and clinical development (Mullard, 2012). There is hope in respect to the large number of novel targets that, if investigated and clinically validated, they could be a basis for new, more efficacious, and safer drugs and, thus, improve the R&D efficiency (Bergren *et al.*, 2012; Scannell *et al.*, 2012).

Unfortunately, the efforts taken so far did not improve the overall R&D efficiency of the entire pharmaceutical industry substantially. Therefore, some pharmaceutical companies needed to take stronger measures. More specifically, some research-based pharmaceutical companies changed the elements with the greatest impact on productivity, namely, the R&D pipeline size, the overall R&D costs and their R&D model, and some companies even changed their overall business model. The following sections specify these trends in pharmaceutical R&D that arose in the last few years.

## 18.3
## The Increasing Pipeline Size

First, the global R&D pipeline, defined as the number of pipeline projects in the phases of preclinical testing to market launch, has increased enormously in recent years. Since 2001, the total number of projects listed in the pipelines of pharmaceutical companies worldwide has risen from 5995 to 11 307 (Citeline, 2015). In 2014, 5484 projects were listed in the preclinical testing phase, 1541 in phase I, 2011 in phase II, 744 in phase III, 170 in a preregistration phase, and 1074 in market launch (Citeline, 2015). The increase in the number of pipeline projects can also be observed when analyzing the individual pipeline size of companies. Within the top 25 research-based pharmaceutical companies with the most projects in their R&D pipeline, 17 companies increased their pipeline size between 2013 and 2014 (Citeline, 2015). Today, the corporate R&D pipelines of the top companies comprise more than 150 drug projects in development phases, with GlaxoSmithKline (GSK) (261), Roche (248), Novartis (223), and Pfizer (205) having 200 and more drug projects in their development portfolios.

## 18.4 The Reduction of R&D Investments

The total R&D expenditures of Pharmaceutical Research and Manufacturers of America (PhRMA) members increased enormously from 1995 (USD 15.2 billion) to 2007 (USD 47.9 billion) and have stagnated since then with a total R&D investment in the industry of 48.5 billion in 2013 (see Figure 18.1) (PhRMA, 2013).

We reviewed the R&D investments of pharmaceutical companies of the past years (see Table 18.1), which is based on the information that has been published in the annual company reports of 2009–2013. We found that a number of the research-based pharmaceutical companies, such as Amgen (+43%, USD 2.864 billion to USD 4.083 billion), Boehringer Ingelheim (+24%, USD 2.215 billion to USD 2.743 billion), Eli Lilly (+28%, USD 4.327 to USD 5.531), Merck & Co. (+28%, USD 5.845 billion to USD 7.503 billion), Novartis (+32%, USD 7.300 billion to USD 9.600 billion), and Sanofi (+17%, USD 4.091 billion to USD 4.770 billion) have significantly increased their R&D spending (in absolute figures) since 2009. Other companies however have stagnating or reduced R&D expenditures. In particular Pfizer has been active in reducing its R&D costs after running through two mergers since 2003 (http://www.fiercepharma.com/story/pfizers-post-megamerger-cost-cutting-record-51500-jobs-7-years/2014-04-29). After the acquisition of Wyeth in 2009 and in view of the loss of exclusivity (LOE) of their best selling drug Lipitor™ (Atorvastatin) in 2012, Pfizer cut its annual R&D expenditures from an all-time high of USD 9.4 billion in 2010 to USD 6.7 billion in 2013, bringing its R&D rate down to an all-time low of 12.9% (Cressey, 2011). Likewise, Takeda cut its R&D cots considerably by 28% from YEN 453.056 (2009) to YEN 324.292 (2013).

The R&D expenditures as a proportion of the total sales of a company (herein referred to as *RD rate*) can be used to estimate the strategic importance of R&D

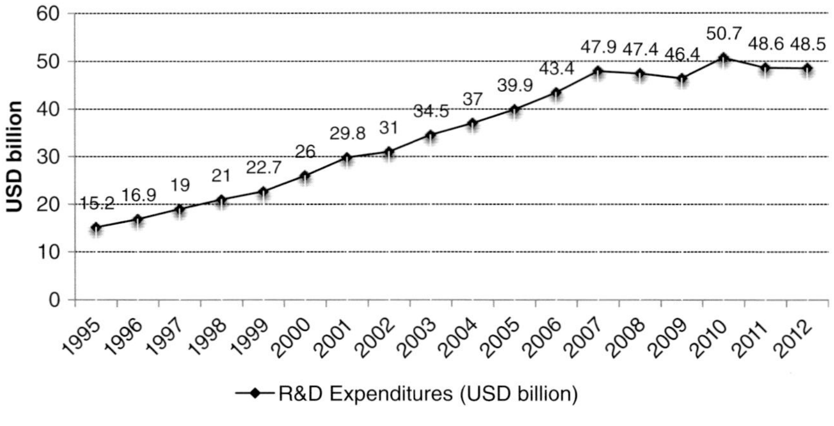


**Figure 18.1** Total R&D expenditures of PhRMA members in the years of 1995–2012. Data derived from PhRMA (PhRMA, 2013) Pharmaceutical Industry 2013 Profile, http://www.phrma.org/sites/default/files/pdf/PhRMA%20Profile%202013.pdf.

Table 18.1

| Year | Total sales (USD millions) | R&D expenditures (USD millions) | RD rate (%) | Total sales (USD millions) | R&D expenditures (USD millions) | RD rate (%) | Total sales (USD millions) | R&D expenditures (USD millions) | RD rate (%) |
|---|---|---|---|---|---|---|---|---|---|
| | Amgen | | | AstraZeneca | | | Boehringer | | |
| 2009 | 14 642 | 2864 | 19.6 | 32 804 | 4409 | 13.4 | 12 721 | 2215 | 17.4 |
| 2010 | 15 053 | 2984 | 19.8 | 33 269 | 5318 | 16.0 | 12 586 | 2453 | 19.5 |
| 2011 | 15 582 | 3167 | 20.3 | 33 591 | 5523 | 16.4 | 13 171 | 2516 | 19.1 |
| 2012 | 17 265 | 3380 | 19.6 | 27 973 | 5243 | 18.7 | 14 691 | 2795 | 19.0 |
| 2013 | 18 676 | 4083 | 21.9 | 25 711 | 4821 | 18.8 | 14 065 | 2743 | 19.5 |
| 2013/2009 (%) | 128 | 143 | 112 | 78 | 109 | 140 | 111 | 124 | 112 |
| | BMS | | | Eli Lilly | | | GSK | | |
| 2009 | 18 808 | 3647 | 19.4 | 21 836 | 4327 | 19.8 | 28 368 | 4106 | 14.5 |
| 2010 | 19 484 | 3566 | 18.3 | 23 076 | 4884 | 21.2 | 28 392 | 4457 | 15.7 |
| 2011 | 21 244 | 3839 | 18.1 | 24 267 | 5021 | 20.7 | 27 387 | 3687 | 13.5 |
| 2012 | 17 621 | 3904 | 22.2 | 22 603 | 5278 | 23.4 | 26 431 | 3979 | 15.1 |
| 2013 | 16 385 | 3731 | 22.8 | 23 113 | 5531 | 23.9 | 26 505 | 3923 | 14.8 |
| 2013/2009 (%) | 87 | 102 | 117 | 106 | 128 | 121 | 93 | 96 | 102 |

| | Merck & Co. | | | Novartis | | | Pfizer | | |
|---|---|---|---|---|---|---|---|---|---|
| 2009 | 27 428 | 5845 | 21.3 | 44 300 | 7300 | 16.5 | 50 009 | 7845 | 15.7 |
| 2010 | 45 987 | 11 111 | 24.2 | 50 600 | 8100 | 16.0 | 67 809 | 9413 | 13.9 |
| 2011 | 48 047 | 8467 | 17.6 | 58 600 | 9200 | 15.7 | 61 035 | 8681 | 14.2 |
| 2012 | 47 267 | 8168 | 17.3 | 56 700 | 9100 | 16.0 | 54 657 | 7482 | 13.7 |
| 2013 | 44 033 | 7503 | 17.0 | 57 900 | 9600 | 16.6 | 51 584 | 6678 | 12.9 |
| 2013/2009 (%) | 161 | 128 | 80 | 131 | 132 | 101 | 103 | 85 | 83 |
| | Roche | | | Sanofi/Aventis | | | Takeda | | |
| Year | Total sales (CHF millions) | R&D expenditures (CHF millions) | RD rate (%) | Total sales (EUR millions) | R&D expenditures (EUR millions) | RD rate (%) | Total sales (YEN millions) | R&D expenditures (YEN millions) | RD rate (%) |
| 2009 | 49 051 | 9874 | 20.1 | 25 823 | 4091 | 15.8 | 1 38 336 | 453 056 | 29.5 |
| 2010 | 47 473 | 9050 | 19.1 | 26 576 | 3884 | 14.6 | 1 65 965 | 296 392 | 20.2 |
| 2011 | 42 531 | 8073 | 19.0 | 27 890 | 4101 | 14.7 | 1 19 385 | 288 874 | 20.4 |
| 2012 | 45 499 | 8475 | 18.6 | 34 947 | 4905 | 14.0 | 1 08 932 | 281 885 | 18.7 |
| 2013 | 46 780 | 8700 | 18.6 | 32 951 | 4770 | 14.5 | 1 57 267 | 324 292 | 20.8 |
| 2013/2009 (%) | 95 | 88 | 92 | 128 | 117 | 91 | 101 | 72 | 71 |

as a growth driver in view of the overall business of a company. Amgen (+12%), AstraZeneca (+40%), Boehringer Ingelheim (+12%), Bristol-Myers Squibb (BMS, +17%), and Eli Lilly (+21%) all increased their relative R&D investments (or were not able to reduce their R&D costs in time in view of decreasing total sales). In the same period of 2009–2013, Merck & Co. (−20%), Pfizer (−17%), Roche (−8%), Sanofi (−9%), and Takeda (−29%) consolidated their financial R&D efforts in comparison to their total businesses.

Despite the consolidation in R&D that happened in the past years, several research-based pharmaceutical companies announced to further cut their R&D costs in the recent past. For example, Merck & Co. announced a 17% reduction in R&D personnel resulting in an USD 1.25 billion saving in R&D costs (http://www.fiercepharma.com/story/skinny-earnings-cost-cuts-boost-merck-bristol-myers-forest-fx-hits-sanofi/2014-04-29). The changes in R&D also include a refocus of therapeutic indications and the start of four innovation centers of external innovation.

In the same way, GSK published in 2012 that it wanted to "shake up" the R&D and manufacturing organizations to generate annual savings of GBP 1 billion by 2016 (http://www.pharmatimes.com/article/13-02-07/GSK_puts_faith_in_pipeline_and_cuts_costs_after_tough_2012.aspx).

And, Takeda announced to save YEN 30 billion in R&D by 2017 (http://www.fiercepharma.com/story/takedas-new-outsider-cfo-charged-1b-cost-cutting-plan/2013-11-18, last access: 29 March 2015).

Generally, a reduction in R&D costs is (in most cases) related to a release of R&D personnel, thereby focusing on lowering costs by reducing overhead functions in R&D. This is typically more prevalent in bigger organizations. It is also related to outsourcing to low-cost countries to reduce operational and infrastructure costs (David *et al.*, 2010). Another approach to reducing costs is a greater focus in project and portfolio management on project timing and costs and let drug candidates fail earlier and cheaper (David *et al.*, 2010). For example, DiMasi (2001) reported a decrease in the average time from the start of a research project to its abandonment in clinical trials by 30% from 4.7 to 3.3 years, indicating a trend toward earlier and more stringent decision making (DiMasi, 2001).

## 18.5 The Opening of the R&D Processes

Reducing the overall R&D costs of a pharmaceutical company is one measure to increase the R&D efficiency. This has often been combined with a second option, namely, the change in the R&D model. Two types are discussed herein that have been most prominent in the recent past:

First, the change from a centralized big in-house R&D organization to smaller, more focused, and more easily manageable R&D units. In this context, the biotechnology industry has been the model for some companies, such as GSK, to restructure R&D and to provide a more biotech like and entrepreneurial spirit in their R&D organizations (Douglas *et al.*, 2010; Zhong and Mosley, 2010).

Second, the change from an introverted R&D organization to an organization with more openness toward external innovation (Garnier, 2008; Hunter and Stephend, 2010). The opening of the R&D processes and organizations has been used to increase the output/input ratio of pharmaceutical R&D. Examples are the increasing number of drug projects acquired or licensed from external sources, the increasing importance of collaborations with academic institutions, open innovation initiatives, crowdsourcing platforms, open source drug discovery, and virtual R&D models (see also the chapter on "The Pharmaceutical Industry is Opening its R&D Boundaries").

With respect to open innovation, some of the research-based pharmaceutical companies have started to implement this new R&D paradigm by aligning their R&D organizations to access external innovation more efficiently. For example, GSK launched in 2007 its Center for Excellence for External Drug Discovery (CEEDD), an externally focused R&D center aimed at facilitating drug discovery alliances with external partners (http://www.out-sourcing-pharma.com/Preclinical-Research/GSK-opens-Centre-of-Excellence).

In 2010, Pfizer established the Global Centers for Therapeutic Innovation (CTIs), an open innovation model that aims at founding global partnerships between Pfizer and academic medical centers (http://www.pfizer.com/research/rd_works/centers_for_therapeutic_in-novation.jsp). Thereby, Pfizer is providing financial funding, human resources, phage display, and peptide libraries as well as associated technologies, while academics bring in their hypotheses of new drug mechanisms. Decision making is done in joint steering committees. The inventions are filed in the name of both partners with Pfizer having a right of first refusal (Ratner, 2011).

Additionally, as early as 2002, Eli Lilly began to establish the most radical concept among the global pharmaceutical companies (https://openinnovation.lilly.com/dd/), namely, the Fully Integrated Pharma Network (FIPNet), the Open Innovation Drug Discovery, the Phenotypic Drug Discovery Initiative, the Target Drug Discovery Initiative, and Chorus (http://www.choruspharma.com/about-us.html), all elements of a more open business model for pharmaceutical R&D (Alvim-Gaston *et al.*, 2014; Lessl *et al.*, 2011). In particular, Chorus has been a successful model for virtual R&D, where small groups of scientists discover and develop a new drug candidate purely relying on external resources. Eli Lilly is out-licensing drug candidates from its internal R&D to Chorus which is outsourcing 100% and is accessing the external resources flexibly on demand. Since 2002, the model helped to reduce costs and time of drug R&D significantly. Eli Lilly published that the productivity of Chorus is 3- to 10-fold higher than the one of its traditional pharmaceutical development arm (http://www.choruspharma.com/about-us.html).

A further example of an open innovation initiative in the pharmaceutical industry is the crowdsourcing platform Grants for Targets from Bayer, which was started in 2009 to access target proposals from external scientists (Strauss, 2010). Until today, 1110 applications for reviewing a new target proposal were filed, 13% of which were accepted (http://www.grants4targets.com/scripts/pages/en/status_report.php).

In addition, open source models, which might be defined as product development types wherein external participants have a universal access to contribute unrestricted, were also started in the pharmaceutical industry, such as the Open Source Drug Discovery (http://www.osdd.net/home) that aims at providing affordable healthcare for neglected diseases or the African Network for Drugs and Diagnostics Innovation (ANDI) that was launched in 2008 (http://www.andi-africa.org) (Munos, 2010; Roijakkers and Hagedorn, 2006). GSK together with Alnylam Pharmaceuticals and the Massachusetts Institute of Technologies (MIT) have formed the Pool for Open Innovation against neglected tropical diseases providing open access to 2300 patents in respect to the treatment of tropical diseases (http://investors.alnylam.com/releasedetail.cfm?ReleaseID=466757). Another example of open source models in the pharmaceutical industry includes the collaboration of GSK, Bayer, and Novartis with the Global TB Alliance (http://partnerships.ifpma.org/partnership/global-alliance-for-tb-drug-development-tb-alliance).

While some of these newer initiatives still need to show their proof of principle, more traditional models of open innovation, such as interfirm R&D partnering and collaborations with academic institutions were intensified in the past years (Melese *et al.*, 2009; Kneller, 2010). It has been published that academic institutes are a major source of drug targets and drug projects (Frye *et al.*, 2011). Academic collaboration partners can provide know-how in therapeutic areas and new technologies. For example, the Division of Signal Transduction Therapy (DSTT) is a collaboration between the University of Dundee, the Medical Research Council (MRC), and six pharmaceutical companies, namely, AstraZeneca, Boehringer Ingelheim, GSK, Janssen Pharmaceutica NV, Merck Serono, and Pfizer, to perform research on the development of new drug treatments for major global diseases (http://app.dundee.ac.uk/pressreleases/2012/may12/drugdiscovery.htm). The TB Drug Accelerator (TBDA) is a consortium of Abbott, AstraZeneca, Bayer, Eli Lilly, GSK, Merck & Co., and Sanofi together with Texas A&M University, Weill Cornell Medical College, and the Bill & Melinda Gates Foundation that was created to discover new treatments against tuberculosis (http://www.abbott.com/news-media/press-releases/seven-pharmaceutical-companies-join-academic-researchers-to-speed-tb-drug-discovery.htm).

Likewise, most of the research-based pharmaceutical companies broadened their portfolio breadth by accessing research projects and drug candidates from external sources (see Table 18.2) to supplement their in-house pipeline and to meet at least part of their growth objective by launching new drugs continuously. It has been reported that 50% of today's total R&D pipelines of multinational pharmaceutical companies stem from external sources (Schuhmacher *et al.*, 2013).

In our publication in Drug Discovery Today from 2013, we have analyzed the R&D models of 13 research-based pharmaceutical companies and categorized

**Table 18.2** Externally acquired R&D pipeline of research-based pharmaceutical companies.

| Company | Total externally sourced R&D pipeline (%) | Total externally sourced R&D pipeline from licensing (%) |
|---|---|---|
| Amgen | 52 | 13 |
| AstraZeneca | 57 | 27 |
| Boehringer Ingelheim | 18 | 18 |
| BMS | 59 | 18 |
| Eli Lilly | 29 | 15 |
| GSK | 44 | 34 |
| Merck & Co. | 56 | 24 |
| Novartis | 41 | 25 |
| Pfizer | 48 | 10 |
| Roche | 48 | 23 |
| Sanofi | 72 | 47 |
| Takeda | 38 | 49 |

Data derived from EvaluatePharma®, annual company reports, Datamonitor®

these companies in four new types of open innovators. First, the innovation model *knowledge creator* has been defined as a more traditional R&D model with an inbound preference for innovation management combined with a lower level of externally acquired R&D projects when compared with the industry standard. Boehringer Ingelheim and Novartis have been described as two examples of "knowledge creators," which built their R&D activities primarily on internal R&D projects, skills, and know-how and which is supplemented by focused licensing, university collaborations, and academic partnerships.

The second model, called *knowledge integrator*, has been defined as a preference of using externally generated innovation in a model that predominantly uses internal R&D resources. An example could be Sanofi, which creates value from their in-house expertise in R&D management, while they intensively license or acquire R&D projects from external sources. The third model, *knowledge translator*, has been defined as a preference to use resources and knowledge that come from external sources to support the progress of internally generated R&D projects. The "knowledge translator" model, exemplified by GSK, is characterized by R&D projects that have been initiated primarily by internal research, while they use outsourcing, collaborations, and other forms of partnerships to manage their R&D projects efficiently. The CEEDD that aims at facilitating drug discovery alliances with external partners is a main pillar of GSK's knowledge translator strategy.

We have presented in this previous report a fourth innovation model, namely, the "knowledge leverager," that combines open innovation aspects with the virtual R&D concept into one coherent strategy that helps to increase R&D efficiency (Schuhmacher *et al.*, 2013). Pharmaceutical companies, such as Shire, that follow this new strategy have a trim and extroverted R&D organization that is an almost virtual network with low overhead costs. The knowledge leverager focuses on externally generated innovation in combination with a predominantly extroverted way of innovation management. It acquires ideas, know-how, and technologies from the outside the company to discover and develop new drugs that were primarily sourced externally. This model provides the ability to reduce attrition by selecting the right portfolio of low-risk projects. It also provides the opportunity to manage the project pipeline effectively by accessing projects and resources from the outside flexibly and efficiently. Additionally, it provides the option to access resources cost efficiently, as resources can be accessed globally with low overhead costs. To become a "knowledge leverager," pharmaceutical companies need to increase their absorptive capacities by implementing extroverted R&D processes, hiring the right people that are intercultural, open-minded, and prepared with the mindset that innovation can be accessed globally. Secondly, the companies should improve their dynamic capabilities and people skills, which are gaining the same level of importance as scientific and technical know-how. Thirdly, the future knowledge leveragers need to form more strategic alliances and get more actively involved in innovation networks. Finally, fostering the managerial capabilities required to create value in the new world of open innovation is of highest importance. Despite all progress in medicine, the remaining medical need continues to be very high and still most diseases cannot be treated causally. With the right mindset to collaborate and work together in a global manner, living open innovation has the potential to start a new golden age of drug discovery.

In summary, the traditional method of pharmaceutical R&D that is dominated by in-house R&D has been complemented by licensing, acquisitions, and research collaborations at first. In a next step, some of the research-based pharmaceutical companies have started to implement tools and processes to increase the reach of their internal R&D organizations. Taking this into account, the authors rank virtual R&D, research alliances, and crowdsourcing as valuable options to improve the R&D efficiency. In particular, the model of the knowledge leverage Shire seems to be a new and interesting model for pharmaceutical R&D.

## 18.6
## The Challenge with the Return on Investment

As the overall efficiency of pharmaceutical R&D has been steadily declining in the past years, less NMEs were approved per USD invested in R&D. This decline in the R&D output/input ratio of pharmaceutical R&D could be compensated by an increase in the monetary value per NME launched to the market. In consequence and although the R&D efficiency is continuously low, the R&D productivity, defined as the financial output/input ratio, would be at least stable.

In this context, it has been published that the value of one NME, measured at the 5 years postlaunch sales, increased between 2010 and 2012 from USD 10 billion to USD 16 billion (Schulze, 2014). In contrast and published by another group, it is expected that the average peak sales per NME will decline from USD 900 million (2012) to USD 600 million (2015), because of the increased difficulty of offering benefits over existing treatments in light of the increasing price pressure and/or the development of new drugs for smaller and more targeted populations with higher response rates (Bergren *et al.*, 2012). In the mature markets of Europe and the United States, products face stronger competition and, thus, need to be developed for better profiled patients populations and to be launched to smaller market segments. In parallel, they face high cost pressure from the public and payors.

In the same context and in a third publication, it has been illustrated that the revenues of all NMEs launched between 2012 and 2016 is predicted to be USD 58 billion, whereas losses by patent expirations between 2013 and 2016 are forecasted to be USD 123 billion, showing that the new revenues based on NMEs approved for the pharmaceutical industry will not compensate the revenue losses resulting from patent expirations (Bergren *et al.*, 2012; Evaluate Pharma (2013)).

As a result, it turns out that the ROI of pharmaceutical R&D cannot be kept high by simply increasing the market potential of NMEs. Thus, pharmaceutical companies not only need to rethink their R&D strategies but need to find new ways of how to improve the R&D efficiency and, in parallel, how to further increase the financial value per drug launched.

## 18.7
## Changing the R&D Processes Is Not Enough

The reduced R&D efficiency and the challenges of increasing the value per drug launched make it necessary for some pharmaceutical companies which do not belong to the group of the top innovators and which do not follow a knowledge leverager strategy to keep an eye out for other growth options.

Two analyses independently demonstrated that growth in the industry will come primarily from the generics business (+USD 47 billion) and the emerging markets (+USD 150 billion). Both analyses forecasted an almost equal market potential of the emerging countries of USD 499 billion and USD 487 billion by 2020 (Cohen, 2005; KPMG, 2011). However, the challenge for the pharmaceutical industry is the low pharmaceutical sales per capita in emerging countries, which is 5–20 times lower than the pharmaceutical sales per person in developed countries (International Federation of Pharmaceutical Manufacturers and Associations, 2013). Presently, both healthcare system providers and private patients in the emerging countries find it difficult to pay for new medicines, in particular for the high-priced new biologics that are launched by the top innovators of the industry. Thus, the market share of generic drugs in the emerging countries is the dominant source for medication. In 2012, generic drugs had a market share in China of 76%, while off-patent drugs and innovative drugs with patent protection had a stake of 20 and 4%, respectively (IMS Institute, 2013). IMS Health has forecasted that generics will reach a higher market share than innovative medicines in developed and emerging countries by 2017

(IMS Institute, 2013). Consequently, some of the multinational pharmaceutical companies have changed their business model and are already generating today a major part of their total revenues outside the traditional main markets of Europe, the United States, Canada, and Japan by selling both innovative medicine and generic drugs (Cohen, 2005). By 2020, the emerging countries together may contribute as much to global pharmaceutical profits as the United States alone (KPMG, 2011). In consequence, the trend toward more generic drugs will be a growth option for some companies but will result in lower profit margins, in potentially lower investments in R&D in the future, and in an increased pressure on R&D organizations of these companies to improve their R&D efficiencies.

## 18.8
## What Is the Best R&D Model?

From what we have learnt in the past years, it is evident that the timelines of drug R&D are only partially under the direct control of pharmaceutical companies, as pharmaceutical R&D occurs in a highly regulated field and the actions taken to reduce cycle times were of limited overall success. On the other hand, companies can actively influence their strategy on what R&D activities they perform and how many resources they invest in R&D. Thus, pharmaceutical companies need to rethink both their R&D models and their overall business models.

First, the companies that aim at being at the top of the innovators in the pharmaceutical industry need to accept the enormously high costs for drug R&D. Several USD billion are needed per year to generate a continuous flow of new drugs, to identify the right growth strategy by launching innovation, and to build up the right core competences for drug R&D.

Next, other pharmaceutical companies that cannot afford the high R&D investments need to look for another R&D model that is more efficient for their setting. Undoubtedly, open innovation has proven to be a concept of significant interest in the pharmaceutical industry. In view of the challenges the industry is facing, there is a definite need for some research-based pharmaceutical companies to adopt a more open model of pharmaceutical R&D. The strategy of the knowledge leverager Shire seem to be an attractive option and R&D concept for some research-based pharmaceutical companies that do not have the critical mass to afford the model of a "knowledge creator," such as Novartis. To become a knowledge leverager, pharmaceutical companies need to increase their absorptive capacities by implementing extroverted R&D processes and hiring the right people that are intercultural, open-minded, and prepared with the mindset that innovation can be accessed globally. Secondly, the companies should improve their dynamic capabilities and interpersonal skills, which are gaining the same level of importance as scientific and technical know-how. Thirdly, the future knowledge leveragers need to form more strategic alliances and active involvements in innovation networks. Finally, the managerial ability to develop, manage, and exploit external partnerships for creating and capturing own value out of innovation is of highest importance.

Finally, companies that fail to establish the right core competences in R&D might even change their overall business model from a research-based pharmaceutical company to a more diversified pharmaceutical company that provides new drugs in therapeutic areas where they may have a competitive edge, complement prescription drugs by over-the-counter (OTC) medicine, and generate growth with a focus on emerging markets.

Another trend that has not been discussed here but which may have an enormous impact on the healthcare system and in particular on the pharmaceutical industry in the future is the emerging new competitors from the consumer electronics and information technology (IT) industries. They offer products and services that cater to societal trends and the needs of patients, namely, health awareness, mobile health, personal data collection, and analysis. These new players come from nonregulated industries and will continuously increase their presence in the healthcare sector, for example, in diagnostics and drug R&D. We briefly discussed the use of crowdsourcing platforms in drug research that help to increase the reach of pharmaceutical companies by identifying new drug targets but are limited by the number of participants. If the big players of the IT and consumer electronics industries that are already today collecting numerous data from their customers enter the field of the pharmaceutical industry, it is evident that this change will revolutionize elements of drug R&D process such as target identification, biomarker development, or patient recruitment. Until today, most of the pharmaceutical companies are not prepared for this scenario, and the question of whether the new players will increase the challenges for the research-based pharmaceutical companies or help to improve the low R&D efficiencies remains unanswered. The authors consider this development as an opportunity for the research-based pharmaceutical companies to follow the concept of the knowledge leverager as the model comprises the right skills to access the information and know-how that the new entrants might provide.

## References


Alvim-Gaston, M. *et al.* (2014) *Curr. Top. Med. Chem.*, **14**, 294–303.

Bergren, R. *et al.* (2012) *Nat. Rev. Drug Discovery*, **11**, 435–436.

Citeline (2015) Pharma R&D Annual Review, http://www.citeline.com/wp-content/uploads/2015/02/CITIF_RD_AnnualReview_031715.pdf (lastaccess: 05 November 2015).

CMR (2013) 2013 CMR International Pharmaceutical R&D Factbook, http://cmr.thomsonreuters.com/pdf/fb-exec-2013.pdf (accessed 07 November 2014).

Cohen, F.J. (2005) *Nat. Rev. Drug Discovery*, **4**, 78.

Cressey, D. (2011) *Nature*, **470**, 154.

David, E. *et al.* (2010) *Nat. Rev. Drug Discovery*, **8**, 609–610.

DiMasi, J.A. (2001) *Clin. Pharmacol. Ther.*, **69**, 297–307.

DiMasi, J.A. (2002) *Pharmacoeconomics*, **20** (Suppl. 3), 1–10.

DiMasi, J.A. *et al.* (2010) *Clin. Pharmacol. Ther.*, **87** (3), 272–277.

Douglas, F.L. *et al.* (2010) *Nat. Rev. Drug Discovery*, **9**, 683–689.

Evaluate, Pharma (2013) Biotech and Pharma 2012. Year in Review.

Fernandez, J.-M. *et al.* (2012) *Nat. Biotechnol.*, **30**, 964–975.

Frye, S. *et al.* (2011) *Nat. Rev. Drug Discovery*, **10**, 409–410.

Garnier, J.P. (2008) *Harv. Bus. Rev.*, **86**, 68–79.

Hunter, J. and Stephend, S. (2010) *Nat. Rev. Drug Discovery*, **9**, 87–88.

IMS Institute (2013) The Global Use of Medicine: Outlook Through 2017, http://www.imshealth.com/deployedfiles/imshealth/Global/Content/Corporate/IMS%20Health%20Institute/Reports/Global_Use_of_Meds_Outlook_2017/IIHI_Global_Use_of_Meds_Report_2013.pdf (accessed 01 November 2014).

International Federation of Pharmaceutical Manufacturers & Associations (2013) The Pharmaceutical Industry and Global Health – Facts and Figures 2012, http://www.ifpma.org/fileadmin/content/Publication/2013/IFPMA_-_Facts_And_Figures_2012_LowResSinglePage.pdf (accessed 01 November 2014).

Kneller, R. (2010) *Nat. Rev. Drug Discovery*, **9**, 867–882.

KPMG (2011) KPMG – Future Pharma, Five Strategies to Accelerate the Transformation of the Pharmaceutical Industry By 2020, http://www.kpmg.com/Global/en/IssuesAndInsights/ArticlesPublications/Documents/future-pharma.pdf (accessed 28 October 2014).

Lessl, M. *et al.* (2011) *Drug Discovery Today*, **16**, 288–292.

Melese, T. *et al.* (2009) *Nat. Med.*, **5**, 502–507.

Mullard, A. (2012) *Nat. Rev. Drug Discovery*, **11**, 735–737.

Munos, B. (2010) *Clin. Pharmacol. Ther.*, **87**, 534–536.

Paul, S.M. *et al.* (2010) *Nat. Rev. Drug Discovery*, **9**, 203–214.

PhRMA (2013) PhRMA – Pharmaceutical Industry 2013 Profile, http://www.phrma.org/sites/default/files/pdf/PhRMA%20Profile%202013.pdf (accessed 01 November 2014).

Ratner, M. (2011) *Nat. Biotechnol.*, **29**, 3–4.

Rawlins, M.D. (2004) *Nat. Rev. Drug Discovery*, **3**, 360–364.

Roijakkers, N. and Hagedorn, J. (2006) *Res. Policy*, **35**, 431–446.

Scannell, J.W. *et al.* (2012) *Nat. Rev. Drug Discovery*, **11**, 191–200.

Schuhmacher, A. *et al.* (2013) *Drug Discovery Today*, **18**, 1133–1137.

Schulze, U. (2014) Nature Reviews Drug Discovery, **13**, 331–332.

Strauss, S. (2010) *Nat. Biotechnol.*, **7**, 631–634.

Urquhart *et al.* (2012) *EvaluatePharma – Biotech and Pharma 2012 Year in Review.*

Zhong, X. and Mosley, G.B. (2010) *Nat. Biotechnol.*, **25**, 945–946.

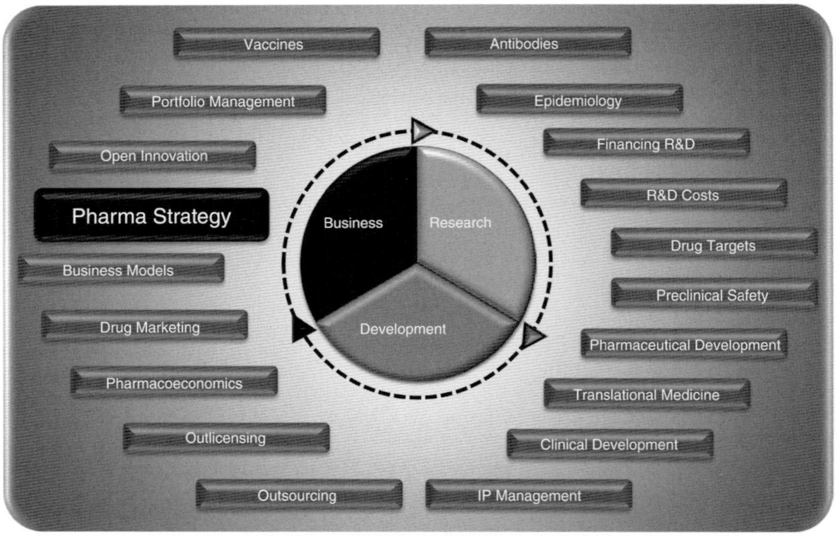
Vaccines
Antibodies
Portfolio Management
Epidemiology
Financing R&D
Open Innovation
R&D Costs
Pharma Strategy
Business
Research
Drug Targets
Business Models
Preclinical Safety
Drug Marketing
Development
Pharmaceutical Development
Pharmacoeconomics
Translational Medicine
Outlicensing
Clinical Development
Outsourcing
IP Management

# 19
# The Influence of Leadership Paradigms and Styles on Pharmaceutical Innovation

*Aubyn Howard*

## 19.1
## Introduction

The purpose of this chapter is to set the topic of Innovation firmly within the context of leadership and show how both collective leadership paradigms and individual leadership styles influence innovation within the pharmaceutical industry. We will show how the challenges facing this industry are contextualized within a wider process of transformation and evolution within organizations and society today. Finally we want to help the reader become more aware of how their own leadership paradigm and style impacts their capacity for enabling innovation.

It is hardly news that human factors have a considerable impact on innovation performance (e.g., Thamhain, 2003). The human dimension of innovation is arguably the key differentiating factor in any complex and competitive business environment today, and the pharmaceutical industry is no exception. Eventually scientific, technological, or business factors even out, the argument goes, and the human dimension proves decisive, which ultimately comes down to the quality, style, and effectiveness of leadership. Discussions about the importance of leadership, particularly in driving R&D innovation, have been widespread and frequent (e.g., Shaywitz, 2013) since industry innovation productivity started to level off and then dip more than 10 years ago. Moreover, most of the major industry players have backed this up with serious investment in leadership development, training, and coaching, often at multiple levels throughout their organizations. It is with good reason the pharmaceutical industry is the clear favorite for management consultants, leadership trainers, and executive coaches to work within. What is surprising therefore is how far away most organizations still seem to be in terms of getting on top of how leadership can make a real difference to innovation. This is a reflection of the apparent disparity between the enormous progress mankind has made in objective scientific knowledge over the last couple of centuries and our limited progress in understanding the more subjective mysteries of the human psyche and behavior. We appear to have mastered much of our immediate external physical universe, and yet comparatively, when it comes to understanding ourselves, we still seem to be fumbling about in the dark. The management consulting

*Value Creation in the Pharmaceutical Industry: The Critical Path to Innovation*, First Edition.
Edited by Alexander Schuhmacher, Markus Hinder, and Oliver Gassmann.

and leadership training business partly does so well because there are so many different answers and approaches to the leadership challenge. Alongside that, there is the more prosaic approach of the investment world (city, investors, shareholders, boards, etc.) for whom leadership is equated with "leader," and the answer to any leadership problem is to change the person. Hunches are taken based primarily on charisma and personality style, within the prevailing achievement-oriented paradigm which will be described later in the following text.

In this chapter, I will attempt to show the reader that there is some light at the end of the tunnel and that there is actually a growing body of psychological knowledge that could be applied to the leadership of innovation issues in the industry. However, there are (at least) two not inconsiderable barriers to this psychological understanding of the leadership domain being brought to bear in practice. First, the modern tendency toward oversimplification of complex issues by business leaders, politicians, commentators, media, and other influencers. The biases of the prevailing achievement-oriented leadership paradigm in particular mitigate against a more complex and multidimensional perspective of leadership becoming widespread. Secondly, the circular problem that senior leaders may themselves be the problem and unless they are willing to seriously look at themselves and take action to upgrade their own "operating model" (or even replace themselves if they are not fit for purpose), they may continue to be a block upon progress.

Leadership and the way that it impacts and influences the human dimension of innovation (or any business challenge for that matter) is a complex issue and cannot usefully be reduced to a few simple formulas or solutions. We need to at least provide the reader with an overview of the complexities, dimensionalities, and dynamics of this topic before applying it to the specific challenges of innovation within the pharma industry today.

Some of the questions we explore include:

- What is your concept or model of good leadership?
- What are leadership paradigms and styles and how do they relate to each other?
- How are leadership paradigms expressed through different organizational or operating models?
- How does leadership style influence innovation within the context of organizational cultures?
- How suitable are different leadership styles to different challenges within pharma R&D?
- Where is the future for pharma in terms of leadership paradigms and styles? What are the possible answers to the current innovation deficit?
- What should leaders be thinking about both in terms of their own development and more broadly, their organization's?

We will start by exploring what is meant by leadership, and then specifically by leadership paradigms and styles.

**Executive Summary**

Leadership style drives innovation by nurturing organizational cultures within which creativity can take place. Leadership styles are external expressions of internally held leadership paradigms, which are themselves both reflections of broader shifts of consciousness within society, as well as the process of individual human development. Complexity theory shows us that innovation is optimized at the "edge of chaos," where the optimum balance between structure and freedom is found. Early stage research requires a high level of freedom to experiment and explore opportunities through collaborative engagement with both internal and external networks. The challenge for many large pharmaceutical companies is to reproduce the kind of creative culture that exists in small biotech; however, many leaders do not understand how to do this. Others switch to in-licensing or buy-in strategies that also require an entrepreneurial outlook to be successful and that tends to be lacking in risk-averse pharma. Altogether, there is growing evidence of the limits of the prevailing leadership paradigm (achievement–pluralistic) in addressing the innovation deficit. The answers point toward nurturing the evolutionary leadership paradigm and building an "evolutionary bridge" of critical capabilities which make it possible to more fully embrace open innovation business strategies and models.

## 19.2 What Is Your Concept or Model of Good Leadership?

There are many different ways to approach the topic of leadership and describe the different styles or ways in which leader behave and operate. Current thinking has moved on from the "great man" approach, in which the characteristics and qualities of examples of successful or well-known leaders are generalized into some kind of idealized model of leadership for others to try and emulate. However, it can be useful to have a concept or general model of what we mean by good leadership as a starting point, and examples of these are provided in the works of Collins (2001), Kouzes and Posner (2002), Jaworski (1996) and Greenleaf (1977). Situational leadership (Hersey, 1985) offers a more dynamic concept of leadership, in that it shows leaders how to adapt their style to get the best results in different situations according to people's needs.

Other writers have focused on the distinction between leadership and management, for example, John Kotter (e.g., in Kotter, 2001) who emphasizes the leaders role in leading change. There is the simple descriptive model of leadership (e.g., Heron, 1999) as a continuum of options between the hierarchical/directive style and the democratic/facilitative style, which sometimes includes laisser-faire/empowering as a third locus on the scale. The classic Tannenbaum and Schmidt continuum (Tannenbaum and Schmidt, 1973) maps seven different combinations of authority or control and freedom or delegation as leadership behavior or style

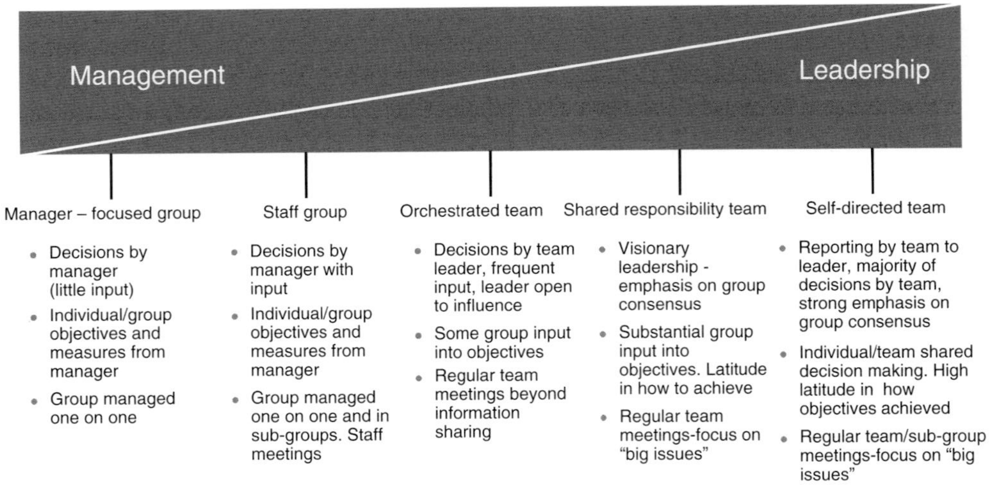


**Figure 19.1** Spectrum of leadership team options.

options. We can take this a step further to map a continuum of leadership style options giving rise to different types of team, as in the example in Figure 19.1 above (Creative Leadership Consultants, 2015).

The concept of distributed leadership (e.g., Spillane, 2006) is also useful in that it broadens the scope and relevance of leadership from the one to the many and from the responsibility of the person at the top to something that everyone might be involved in. All of these concepts or notions of leadership are useful in orienting us to what we mean by leadership and to the basic dichotomous style options.

## 19.3 Approaches to Leadership Modeling and Profiling

More sophisticated approaches today describe leaders using psychological or behavioral models and systems which set out the full spectrum of possible types, preferences, styles, or modes of operating. Psychometric tools are used to assess or profile individual leaders against the model, and these are widely used within all types of organization to provide a diagnostic starting point for leadership development, support team building, or profile potential candidates for a vacant leadership position. There are many such models and tools and different HR departments or consultants will have their preferred tools. There is surprisingly little good literature that compares and contrasts the relative strengths and merits or different approaches, so I will offer a brief overview here.

We can broadly group approaches into personality type, behavioral preference, and developmental stage models (and there are probably proponents of approaches that might argue that their particular tool fits either none or all of these three categories).

### 19.3.1 Personality Types

Personality type models include those based on the 16 personality types, the big five personality traits, and Jungian personality theory, such as Myers–Briggs (MBTI), DISC, and Insights. Other approaches in this category include HBDI (Hermann Brain Dominance Indicator), based upon brain hemisphere research, and stress drivers, which comes from Transactional Analysis. Many of these tools are used for a variety of purposes in society and organizations, including in a personal development context.

### 19.3.2 Behavioral Preferences

Behavioral preference models are mostly derived from observation or analysis of the way that people work, behave, and relate to each other. These include Belbin Team Roles, Thomas–Kilmann Conflict Mode Instrument, Firo-B, OCI, and Learning Styles. Some have emerged from the growing field of positive psychology and strengths-based literature, such as StrengthsFinder (Buckingham and Clifton, 2001). These behavioral approaches are usually descriptive of observable and measurable external behaviors and relational preferences, whereas the personality type models are attempting to describe the inner subjective workings of our minds and personalities that drive the behavior, although there is some overlap in many models.

### 19.3.3 Developmental Stages

Developmental stage models and associated tools have arisen from the field of developmental psychology, which is broadly concerned with how human beings grow and develop over time and along different dimensions. Early pioneers included Piaget and Erikson, and significant theories and models have been developed by the likes of Graves (1970), Loevinger (1998), Kohlberg (1984), Kegan and Lahey (2009), Wade (1996), and others. Howard Gardener's work on multiple intelligences (1983) can be included in this field, which links to specific works concerning emotional intelligence (Goleman, 1996), spiritual intelligence (e.g., Zohar and Marshall, 2000), and social intelligence (Goleman, 2007 and others). Developmental approaches are seeking to reveal the deeper psychological structures that underlie our personal styles and behavioral preferences and show how these evolve over time or adapt to different situations.

### 19.3.4 Competency Frameworks

At this stage it is also important to mention competency-based approaches and models (e.g., see Leslie and Palmisano, 2014). Alongside a leader's personality and preferences, their worldview and work style, as with anyone working within organizations, we can also consider and assess their skills, abilities, and competencies, as well as experience and knowledge, which for simplicity I will group together under competencies. Although important to include the competency dimension when working with leadership development (what I refer to as the horizontal aspect or dimension, e.g., see Torbert, 2014), this is not within the scope of exploration for this chapter. Most pharmaceutical companies already have very sophisticated and well-developed approaches to competency modeling and development for all their people. However, it is important to note that when working on individual leader development, both personal and professional aspects, horizontal and vertical development dimensions, leadership styles and competencies, come together in forming the individual leader development agenda. At the same time it is important to be aware of these distinctions, for example, when a leader is facing issues of working on their style, but mistakes this for a need to develop skills. We will return to this later.

## 19.4
## The Developmental Approach to Leadership Paradigms and Styles

Having provided this overview of the broad approaches, I am going to focus on the developmental stage model approaches and more generally draw from the field of developmental psychology in order to tackle the relationship between leadership and innovation in more depth. Apart from personal preference, the reason is that these approaches are more dynamic in several ways than the personality and behavioral approaches. Specifically they:

1) Show the connection between inner subjective perspectives (described as paradigms, worldviews, or value systems) and outer behavioral and relational preferences and recognizable leadership styles
2) Provide a dynamic view of how individuals develop over time and identity common patterns, recognizable stages, and transitions
3) Offer a framework for moving between the dimension of individual leadership styles and collective organizational cultures
4) Address the situational context of leadership style and show how it relates to different types of organization, stages of organizational development, and different situational challenges

We will explore these dynamics in more detail before bringing the whole discussion more toward the specific challenges of innovation in pharma and R&D in particular. To keep things simple I will follow the model of leadership paradigms described by Frederic Laloux in his recently published work "Reinventing Organizations" (2014, see Chapter 1.1 for a fuller treatment of this whole topic), which is based primarily upon Ken Wilber's (e.g., see Wilber, 1999) and Jenny Wade's (e.g., see Wade, 1996) meta-analyses of the various models that focus on different aspects of human development (e.g., ego identity, cognitive, moral, value systems, etc.) at individual and collective levels and are founded on validated scientific research. The research work by Clare Graves on the emergence of value systems (probing people's conception of adult full maturity) and that of Jane Loevinger on stages of development of ego identity have provided the foundations for later adaptation to the organizational domain – Spiral Dynamics (Beck and Cowan, 1996) builds upon Graves' work to describe value systems in organizational terms, and The Leadership Development Framework uses a language specific to the stages of leader development and associated crises of transition. Although their delineations of actual stages are slightly different, it is relatively easy to see how both these and other systems, derived from unrelated original research, are all describing the same essential unfolding pattern of human development and evolution but from slightly different perspectives.

Laloux describes seven organizational paradigms that broadly follow the emergence of human consciousness and societal worldviews over thousands of years of human history, but also mirror the developmental stages that individuals follow as they grow up and mature in adulthood (at least in potentiality). These are **Reactive**, **Magic**, **Impulsive**, **Conformist**, **Achievement**, **Pluralistic**, and **Evolutionary**. It may help to think of these as ways of thinking and operating in the world, which are more or less activated within an individual, group, organizational, or society depending upon history, circumstance, and situational factors.

Over the course of human history, we can trace the initial emergence of each new manifestation of consciousness and how the prevailing paradigm has then shifted from one to the other, but even in today's global society, there are widely different mixes and expressions of these paradigms in different societies, organizations, and people. Although an individual (organization or society) will generally have their center of gravity within one or other value system, they are always expressing a unique mix of more than one, as influenced by their story and personality. In addition, everyone has the capacity for all the value systems to emerge, although the way in which this happens over time will always broadly follow the primary sequence (e.g., you can't fully activate at Pluralistic until you have in some way activated at Achievement, and so on). Graves made a particular point that the emergence of value systems are responses to situational challenges to which the previous way of thinking or operating are no longer providing valid or useful solutions (akin to Einstein's quote about problems not being solved by the same level of thinking that gave rise to them in the first place). The transition from one paradigm to the next is often characterized by an inner or outer crisis of some kind, for example, a crisis of meaning for an individual for whom the Achievement paradigm no longer works or the crisis of survival for an organization which needs to outgrow a rigid hierarchical model if it is to compete successfully with new market entrants (as in the dynamic between big pharma and biotech in the case of the pharmaceutical industry). Turbulence, upheaval, or changing circumstances has driven human progress in the past (see the work of Diamond, 2007), and in the same way it is the challenges (e.g., from market forces and the R&D innovation deficit) facing today's pharmaceutical organizations that will give rise to the new evolutionary paradigm emerging through new styles of leadership, cultural orientations, and organizational models. At the same time, there is never a guarantee that this will happen, at least within today's organizations, and often progress follows the birth of new organizations that are more agile or adapted to present-day challenges, alongside the death of the old. Ken Wilber paraphrases Max Planck when he says "the knowledge quest proceeds funeral by funeral." The history of evolution in nature is littered with long forgotten extinct species and human evolution with disappeared civilizations (and organizations) that failed to adapt and evolve in response to a crisis.

## 19.5
## Inner and Outer Leadership

Let us now explore the dynamic between inner and outer aspects of leadership. This approach holds that leaders' inner orientation, which is made up of constellations of beliefs, assumptions, motivations, mindsets, and ways of thinking which we refer to as paradigms or worldviews, will broadly determine or give rise to their outer expression of leadership style, although the way in which this happens on an individual basis will be influenced by the leaders unique personality, preferences, skills, experience, and so on. If we were to ask a broad mix of people working in organizations what "leadership style" means to them, they will probably refer to styles in terms that describe their outer impact and way of operating, for example, controlling, egotistic, top-down, consensual, democratic, charismatic, motivating, encouraging, empowering, paternalistic, hierarchical, and so on.

The developmental system we will outline provides a way to see the pattern behind these different outer styles and realize that there is a progression from less sophisticated styles to more sophisticated and from styles suited to simple organizational systems and challenges to those needed to deal with the complexities of today. According to the Gravesian model, alongside this progression to greater complexity, there is also a cycling between individualistic value systems and collectivistic value systems, in other words, from those that emphasize individual values to those that place greater value on teams, the whole organization, and even the wider knowledge community. At each turn of the cycle, the collective orientation is always enlarging, that is, from tribe or group, to organization or nation, to society or human kind.

Table 19.1 provides a quick summary of the characteristics of each paradigm and how they translate into corresponding leadership styles. Again, we are skimming the surface here, and for a fuller treatment, I recommend reading the first part of Frederic Laloux's book or dipping into Ken Wilber's comparative analysis (e.g., Wilber, 2000).

Within each stage, there can be healthy and less than healthy expressions of leadership style, both depending upon the individual's level of psychological health and personal development, as well as the suitability of the leadership style to the

**Table 19.1** Leadership paradigms and styles summary (Howard, 2015).

| Leadership paradigm (inner orientation) + Graves thinking mode | Leadership style (outer impact) and other expressions | Individual or collective orientation and locus of attention + motivation | Primary motivations responding to each leadership style |
|---|---|---|---|
| Magic + *animistic* | Benevolent *Paternalistic* *Maternalistic* | Collective – tribe or family group | + Belonging |
| Impulsive + *egocentric* | Autocratic *Egotistic* *Dominating* | Individual – own needs | + Rewards |
| Conformist + *absolutist* | Hierarchical *Controlling* *Processing* | Collective – roles and rules within structures | + Responsibility |
| Achievement + *multiplistic* | Enterprising *Rational* *Driving* | Individual – individual performance in the team | + Material success |
| Pluralistic + *relativistic* | Social *Democratic* *Relational* | Collective – team culture | + Participation |
| Evolutionary + *systemic* | Evolutionary *Integrative* *Holistic* | Individual/collective – professional communities | + Learning/ transforming |

organizational situation and challenges. It has been successfully argued (e.g., Kets de Vries, 2006, p. 24) that in today's modern organizational environments (centered in the Achievement paradigm but with elements of Impulsive, Conformist, and Pluralistic, according to historical, market, and situational influences) that the people who rise to the top of hierarchical and competitive organizations are more likely to have some kind of personality pathology, certainly to be more self-oriented, egotistic, or narcissistic, than the average person. Business folklore is littered with stories of charismatic, powerful business people who, at least on the surface, have been highly successful, but around whom there has been much collateral damage in human terms. Only as the Pluralistic paradigm starts to become stronger is it more likely that genuinely empathetic, other-oriented, well-balanced, personality types are likely to become successful managers and leaders (Table 19.2).

## 19.6 Dynamics of How Leadership Paradigms Evolve

In our previous description, we have left out the first worldview or paradigm, which Laloux labels *Reactive*, because it is rarely found overtly in today's

**Table 19.2** Leadership style descriptions (Howard, 2015).

| |
|---|
| *Benevolent* – they lead as the guardian of a community. They tell stories, maintain traditions, and seek to preserve the wisdom of the past |
| *Autocratic* – they lead decisively and from the front, they are in charge. They control power and reward loyalty. Their way is the right way |
| *Hierarchical* – they lead by passing judgment according to a system of well-defined roles and responsibilities. They follow the established right way to do things |
| *Enterprising* – they lead by example and by creating opportunities for the team to succeed. They are driven to achieve goals. They constantly look for better ways to do things |
| *Social* – they lead by building consensus and providing opportunities for people to grow and develop. They know there is no universally right ways to do things |
| *Evolutionary–integrative* – they lead by responding to situations in whatever style is needed, seeking to create synergy within the wider system. The right way is the way that works |
| *Evolutionary–holistic* – they lead by guiding others to be leaders and by creating the context for growth within a healthy system. They are constantly evolving a new way or path |

organizations, although echoes of humankind's evolution as small bands surviving as foragers or hunter-gatherers between 100 000 and 20 000 years ago are still present in our deeper psyche and collective unconscious – for example, within our instincts for fight or flight in response to perceived danger. In the same way all human being are deeply impacted and imprinted by the corresponding early experiences of dependent infancy, even though these first few months of life are beyond our conscious memory.

### 19.6.1 Magic–Animistic

The Magic–Animistic leadership paradigm which is expressed through a Benevolent leadership style is more obviously present in some form within organizations. The paradigm emerged in human history as part of the move toward tribal society some 20 000 years ago, and is often expressed and embedded through tribal or family like metaphors or rituals within organizations or teams, and may become reactivated at times of threat to survival of the group. The organizational model and culture associated with this paradigm does not provide fertile ground for innovation or in fact progress of any kind at all, as the underlying though form is one of the maintaining cycle, as in the cycle of the seasons and the performing of traditions passed on from generation to generation. The Magic–Animistic paradigm is most usefully found today in traditional family businesses which have remained untouched by the influence of technological progress. However, this doesn't mean that distorted expressions of associated leadership styles (e.g., patriarchal, paternalistic, materialistic) are not present in all types of organization, usually as an consequence of psychological dysfunction of individuals who unconsciously meet unresolved psychological needs by taking up permanent parental roles beyond what is healthy in nurturing, mentoring, or protecting those in their charge. I can think of at least one leader in an R&D organization who became seen as benevolent father figure in this way. This may have helped stabilize the organization at a time of turbulence but did not engender the conditions for increased innovation once this was over.

### 19.6.2 Impulsive–Egocentric

In historical terms the Impulsive–Egocentric paradigm started to emerge with chiefdoms and eventually empires between 10 000 and 5000 years ago, and in psychological terms it represents the fully formed ego differentiating itself from parental symbiosis, and in this sense is the first truly individualistic (although very egocentric) worldview. The first Impulsive organizations appeared as small conquering armies, and this still represents a powerful organizational archetype today. Although street gangs and criminal organizations today can still be quite close to this model, variations can also be found in small businesses and start-ups that are driven by the energy and ego of the founder and where "their glue is continuous exercising of power in interpersonal relationship" (Laloux, p. 18). More commonly we find both healthy and dysfunctional expressions of power-oriented autocratic styles of leadership present in all types of relationship, often masquerading as more sophisticated styles (e.g., Achievement orientation – for good examples watch the Apprentice reality TV program in the United Kingdom). From professional observation, a generous dose of autocratic leadership style (founded on a degree of ego-narcissism or at least an inflated sense of self-worth) is very common in a typical start-up organization and can be seen to be part of the mix in many successful biotech organizations. The reason might partly be that entrepreneurs need a degree of inflated self-belief to break through initial barriers and keep their business or project going against the odds. This can also be associated with very creative times for a business, but usually in support of the original business idea, innovation, or purpose. Typically, such organizations arrive at a creative or innovative impasse at some point in their growth, unless the power-oriented leader is able to adapt his or her style or has the wisdom to allow a succession of leadership to take place. This is the first crisis of organizational growth (see The Five Phases of Growth, Larry Greiner, HBR May–June 1998 for a different perspective on this). Often this takes place by default at the point when the biotech business founders sell to large pharma or seek some kind of institutional investment, and whether willingly or not, the style of leadership changes and the organizational structure formalizes and develops to distribute responsibility away from the power leaders and toward smaller units. Sometimes this transition process to a more formal organizational model either inadvertently kills off the original creative culture (with people leaving) or leads to conflict between the start-up founders and the new parent leadership. From big pharma's point of view, this may or may not matter depending upon whether they think they are buying the golden goose (the innovative biotech culture and creative spark) or just the eggs it has already laid (discovery or research projects at various stages within the pipeline that can be adopted and transitioned within their existing organization). This later path is becoming a more explicit strategy for some pharma, but it is still not without its risks.

### 19.6.3 Conformist–Absolutist

Here we are describing the organizational transition from Impulsive–Egocentric to Conformist–Absolutist. The historical emergence of Conformist started about 4000 years ago, with the transition from chiefdoms to nation states and stable civilizations and subsequent founding of the world's great religious traditions. This has been the prevailing organizational model ever since until the last century or two and is still the backbone for much of what we would consider to be the establishment today – church, armed forces, government, and universities, for example. The Conformist–Absolutist organizational model is a significant progression from what comes before in that "organizations can now plan for the medium and long term and they can create structures that are stable and can scale" (Laloux, p. 20). In order to grow, many organizations draw upon the Conformist–Absolutist paradigm by establishing clear roles, responsibilities, and processes than enable this scaling. Many large pharma organizations will have significant elements of this leadership paradigm within their cultural mix, and it has been part of their growth DNA. Pharma also has a strong connection with the academic world where the conformist paradigm is often still prevalent. However, the buildup of bureaucracy, inertia, and conformity that characterizes this model as organizations become larger and larger (sometimes not through organic growth but compounded by acquisition) has certainly been part of the innovation deficit problem. Again, this paradigm contains within it the seeds of its own destruction or at least the death or organizations that fail to evolve beyond it in response to environmental change.

The accompanying change in leadership style is startling in its reversals away from autocratic self-orientation and opportunism – toward duty, responsibility, and professional dedication. The Leadership Development Framework developed by Fisher, Rooke, and Torbert (2000) and Rooke and Torbert (2005) delineates two distinct leadership styles associated with this paradigm – the dutiful Diplomat, who conforms to and enforces the expected norm and the professional Expert, who is dedicated to excellence within their discipline and leads as a role model for those starting out in their profession. This style of leadership is common within the scientific professions that are core to the pharma industry (Biology, Chemistry, Medicine, Data science, etc.), and it is all too common to find senior leaders that are still centered in the Expert style promoted to a senior leadership position for their scientific excellence rather their leadership suitability. Best practice these days is to provide clear parallel career paths within pharma (e.g., scientific, leadership, and business) so that good scientists don't feel they need to aspire to leadership positions for reward or progression. However, there is an underlying cultural drag in the industry that biases toward the best qualified scientist for senior leadership positions and also the Achievement–Materialistic paradigm that give rise to the scientific mindset.

### 19.6.4 Achievement–Multiplistic

We have already indicated that the prevailing organizational and leadership paradigm in Western society and certainly in business organizations is Achievement–Multiplistic. In historical terms, this emerged in the shape of the western enlightenment something over 200 years ago and has driven not just the explosion of scientific and technological discovery, the industrial revolution, and economic growth, but also the growth and dominance of prosperous modern democratic societies. Three big breakthroughs accompany the emergence of this paradigm which inform the Enterprising style of leadership – innovation, accountability, and meritocracy (Laloux, p. 26). In terms that we understand today, innovation really gets going with this paradigm and leadership style. Leaders operating from this paradigm can "live in the world of possibilities, of what is not yet but could be one day." There is no longer just one right way to do things but a multiplicity of possibilities. They challenge the status quo, always looking for better ways to do something, and they are open to change, uncertainly, and opportunity. Thus they challenged the morbidity and hierarchical inflexibility of Conformist–Absolutist organizations and invented departments that did not previously exist, including R&D, marketing, and product development, as well as the project-driven way of working that has superseded the purely process-driven operating model of Conformist organizations. This paradigm and the leaders who have embodied it are probably responsible for most of the achievements of the pharmaceutical industry over the last 50 years – driving projects that take early discoveries through the pipeline to an end result of effective and marketable drugs. Most of what is consider to be good leadership today is an expression of this paradigm and way of thinking – the Achiever leader encourages and enables teamwork, rewards and recognizes performance, and leads by example.

The growth of the pharmaceutical industry has been a magnificent demonstration of innovation in practice, and looking back has been one of the Achievement–Multiplistic's paradigms great successes! However, here's the rub – it has reached the limits of its effectiveness as the driver of innovation and as we all know diminishing returns have set in (hence the reason for this book), particularly where innovation has been kept broadly within the boundaries of the formal organization. The competitive mindset within the Achievement paradigm mitigates against opening up innovation to outside parties who are instinctively seen as a threat. When big pharma has tries to develop open innovation business models (see Schuhmacher and Betz, 2016), there are aspects of the Achievement–Enterprising leadership style that tend to hinder success (e.g., treating partners as suppliers, overelaborating contractual negotiations, not-invented-here attitudes). Even so, the achievement leadership paradigm, driving a vertically integrated closed innovation model, worked very well for a period of time, systematically working through the low-hanging fruit in terms of identifying chemical compounds or biologic processes that meet medical needs and driving through projects to establish their safety and efficacy and take them to market.

Analysis elsewhere in this book points to the specific reasons, but within Pharma's own terms industry decline set in and there is now a growing innovation deficit, which just doing more of the same within the confines of the old model (more targets! more projects! more performance!) will not solve. But there's another rub – the Achievement paradigm is itself now acting as a barrier to the emergence of the next leadership paradigm which we will discuss in the following. As I said in the introduction, the prevailing Achievement-oriented leadership paradigm mitigates against a more multidimensional perspective of leadership becoming widespread (it doesn't see the point, because it is confident in its own superiority as a way of looking at the world). So the Achievement worldview and style of leadership have become part of the problem: by overobsessing on the need for success in the short term (which feeds and is driven by the investment world); by focusing on success and winning rather than purpose; by overemphasizing the rational or cognitive in relationship to the emotional, social, spiritual, and ethical dimensions of human beings; and by maintaining the underlying limitations of the hierarchical system (as the hangover from the previous Conformist paradigm) and overfocusing on management – adding more and more layers of management as the answer to most problems, rather than stripping them away to release the creativity of those who are managed. There are interesting echoes here of the fault lines in our wider modern market capitalist society, which have become increasingly apparent since the start of the global economic crisis in 2008. The innovation deficit in pharma is mirrored by crises throughout business and society as a whole, from how to deliver better health services to how to respond to climate change, and increasingly people are beginning to realize that the answers to these crises may not come from the Achievement paradigm type of thinking.

If you are following this narrative, you might now be asking if there is a leadership paradigm representing a new shift in societal consciousness that addresses these issues of the prevailing Achievement paradigm. Yes there is, but before this arrives, there has been something of a diversion (albeit an essential and valuable one) – with the rapid emergence of the Pluralistic paradigm.

### 19.6.5 Pluralistic–Relativistic

The emergence of new paradigms is seemingly speeding up as part of the evolution of human consciousness, society, and culture. In the last 50 years or so, the Pluralistic worldview has developed at an astonishing pace and now pervades many spheres of society (e.g., the arts, academia, not-for-profit organizations, left-wing politics, etc.). This is the emergence of postmodernism, partly in reaction to materialistic modernism, of championing the people principle as a counterbalance to the profit principle and of the human perspective as antidote to the mechanistic clunk-and-grind economic progress of the Achievement paradigm. The Pluralistic – Relativistic paradigm brought three significant breakthroughs within organizations (Laloux, p. 32): empowerment and collaboration, values-driven culture, and the stakeholder perspective. These developments are now embedded in every modern large pharma organization, alongside Achievement – Multiplistic's principles of innovation, accountability, and meritocracy. The Pluralistic paradigm brought an openness to collaborative partnerships and a relationship orientation which made it possible for the open innovation paradigm to pick up and flourish within pharma, at least in the more traditional forms such as outsourcing, licensing, joint ventures, and research collaborations (see Schuhmacher and Betz, 2016).

The paradigm is expressed through a more collaborative, democratic, social, relational, and humanistic style of leadership. People are increasingly nurtured, developed, and consulted by leaders. This has certainly made organizations more human places to work and improved the experience of work for many people. There is one problem – although there are examples of value-driven businesses delivering improved shareholder value, there are also examples where it hasn't and even some cases where the Pluralistic paradigm and associated leadership style has become too dominant and performance has got worse, threatening the survival of the organization (e.g., Prudential in the early 2000s). This has contributed to distrust between Achievement – Multiplistic and Pluralistic – Relativistic leaders and to a clash of value systems between Profit and People orientations, waging away beneath the surface. The deeper problem is that despite the human tone that Pluralistic leadership style brings (listening, empowerment, engagement, 360 feedback, etc.), people still do not trust their organizations and will not bring their whole selves to their work (and therefore their full creativity and innovative edge). The fact that they need to be empowered by leaders, engaged by the organization, and enrolled in its purpose means by definition that at some point they have became disempowered (through the concentration of power at the top of organizations), disengaged, and alienated from a purpose that they were not involved in coming up with it in the first place.

Part of the problem is yet again, that the adherents of the Pluralistic paradigm and style of leadership do not see the whole picture, and set themselves against the excesses of the previous materialistic worldview. On one level they maintain that no one's viewpoint is more valid than anyone else's, and on another they also secretly believe that theirs is the right one. Importantly, however, despite its inherent contradictions, by bringing the human being back into the picture, the relativistic perspective lays the ground for the emergence of a truly transformative worldview, and evidence that this is finding its way into organizational and leadership expression has started to emerge.

### 19.6.6 Evolutionary–Systemic

Management writers, gurus, and consultants have been proclaiming the next great leadership paradigm for some time (e.g., Barrett, 2011) and largely been missing the target, typically by oversimplifying and conflating the old paradigms and by overidealizing or overidentifying with the new one. What makes Laloux's work (2014) groundbreaking is that he has carried out detailed research of a dozen large organizations where the new worldview has taken shape and describes the Evolutionary paradigm and associated leadership styles based upon evidence of what he found in common. Importantly, he properly draws out the sequence of previous paradigms, and shows how each new way of thinking and operating is both built on the gains of the previous one and is also an emergent response to its inherent limitations. He shows how evolutionary organizations can work radically differently (at once they are more effective, innovative, ecological, and human to work in) to the great majority that we know and experience today. Drawing from evidence, he identities three common characteristics or principles of evolutionary organizations – self-management, wholeness, and evolutionary purpose. He plays with the metaphor or organizations as living systems or organisms (also see Senge, *et al.*, 2005) with a purpose of their own, in contrast to the clunky and alienating machine metaphor's that identify the Achievement paradigm or the social and family metaphors that signal Pluralistic.

Laloux explores how the structures, practices, leadership styles, and cultures within evolutionary organizations reflect the principles of self-management, wholeness, and evolutionary purpose. He identifies the two necessary conditions for the emergence of Evolutionary organizations – a sufficient level of psychological development of the top leadership (e.g., CEO or founders) – and "enlightened" owners who are willing to embrace and trust the Evolutionary worldview of thee leaders.

The shift to an Evolutionary worldview is of a magnitude greater than any of the previous paradigm shifts we have described. It is variously called *second tier*, *higher order*, or *meta* something for this reason. From this perspective, the leader can work with the whole system of all the previous paradigms or worldviews and see the part they play in the evolutionary process. It is not just another worldview or paradigm, but one that can work with the health of the whole system of paradigms. More importantly for this topic, research has shown (e.g., by Rooke and, 2005) that Evolutionary leaders are by far the most successful at implementing large-scale corporate transformation programs. Clare Graves (1970) describes similar findings concerning creativity in finding creative solutions; when comparing groups of people operating from different paradigms and given complex tasks to perform, he found the Evolutionary group would find "unbelievably more solutions than all the other groups put together" and of "an amazingly better quality." Laloux's research comes to very similar conclusions – organizations with Evolutionary leaders are far more effective and innovative than similar organizations working under Achievement or Pluralistic leaders.

In the pharmaceutical industry the Evolutionary paradigm has started to emerge in the form of some of the newer concepts of open innovation, such as discovery alliances, innovation incubators, virtual R&D, crowdsourcing, and Open Source Innovation (see Schuhmacher and Betz, 2016). However, the Evolutionary paradigm is far from taking hold in the pharma industry, and these new ways of working may be hampered by senior leadership decision making that is still grounded in Achievement–Enterprising. None of the pharma companies I have worked with as a consultant nor any that I have knowledge about second hand could be said to have Evolutionary leadership at the most senior level nor in the leadership culture as a whole; they are mostly examples of more or less healthy expressions of a mix of Achievement and Pluralistic leadership, sometimes with unhealthy remnants of Conformist leadership. I have come across Evolutionary leaders working in isolated pockets, mostly a department, project, or lab team, but this is not the same thing, as we will explore in the following. It is more likely that there are examples of successful small biotech companies led by evolutionary leaders, but their challenge is to maintain such cultures as they grow or if they sell to big pharma.

## 19.7 Leadership at Different Levels within Pharma

So far we have been discussing leadership and innovation within quite abstract terms but with references to the innovation challenges within pharma. We can start to become more situational by first making the distinction between three levels of leadership:

- *Top leadership*: the CEO and possibly the C-suite team leading the organization
- *R&D leadership*: the Head of the Research and Development organization
- *Team or project leadership*: the next three or four levels reporting to the Head of R&D, as well as matrix organization leadership of projects and subprojects

According to Rooke and Torbert's research (2005), only about 5% of Leaders are centered in the Evolutionary leadership paradigm (equivalent to Strategist at 4% and Alchemist at 1% in the LDF system). My own findings from profiling leaders back this up. This means that effectively only 1 in 20 of leaders and managers are even ready to work from this leadership style and many may hold back from doing so because the conditions would not be favorable.

Only at the top level can any influence be directed toward transforming the entire organization through an expression of evolutionary leadership. However, this would appear to be a bit of a fantasy for large pharma, where senior appointments are mostly made within a very Achievement-oriented paradigm (with the possible exception being some private ownership) but might be quite possible within small- and medium-sized biotech, either at the point of start-up or at a key stage in their growth path.

R&D leadership is a more interesting opportunity – there is often a high degree of freedom for a leader in how he or she might run their organization, with relationships across the organization, (e.g., with commercial and manufacturing operations, etc.) working more in the style of collaborative partnerships between business partners than close-knit organizations. The degree of semiautonomy afforded to some R&D organizations can be high (at least in the good times – when money is tight this can rapidly disappear as I have witnesses in at least one Wall Street-oriented large pharma).

I worked with the Head of Discovery over a number of years within in a medium-large pharma, and within the wider context of aspects of an evolutionary leadership style provided by the Head of R&D, I witnessed a gradual transformation from a less than fully healthy and increasingly unproductive Achievement culture (with limiting influences from a deferential hierarchical legacy culture that was resting upon previous laurels) to a more vital, dynamic, and productive Achievement culture and style of operation. The Head of Discovery's leadership style was centered in Achievement (performance oriented, team-based accountability, open questioning, challenging, debating in team meetings, etc.), but he was sufficiently activated at Pluralistic (supportive of collaborative working, empathy and concern for his people, encouragement of professional development, etc.) and Evolutionary (passionate about meeting medical needs, excited about external collaborations and networking, etc.) to eventually bring the whole group toward greater alignment and innovation within a drug hunting culture. For example, target identification became something everyone in Discovery was involved in, project reviews were more frequent and invited wider input, and everyone became more commercial in their perspective. The Discovery business moved some way in its culture toward the entrepreneurial biotech business. However, the bumpy journey toward this also illustrated how important differences in national or regional cultures and expectations of leadership style can be (see more on this in the following text).

The Head of R&D during this period also introduced a flexible and dynamic project-oriented way of working across the whole of R&D, within which the key leadership role became the project leader rather than the line manager, and nurtured a culture which might be characterized as a mix of Achievement–Pluralistic–Evolutionary in style. The implementation of the new project-led model was hindered by resistance from traditional subcultures (Conformist) and old power bases (Impulsive) protecting their special interests, as well as by the usual communication and relationship issues that tend to accompany Achievement cultures in complex large pharma organizations. Bushe (2010) describes how "interpersonal mush" arises as the natural consequence of not checking out our fantasies and projections with each other as we build evidence for our different realities about what is going on, and shows how leaders need to engage their self-awareness and emotional intelligence, as well as encourage one-to-one practices (mainly Pluralistic) such as "learning conversations." Lencioni (2002) describes how good leaders take this a step further by supporting their teams to build trust, engage in open dialog, make commitments, hold each other accountable, and focus on results. Many of the project leaders and their teams (and subproject teams) during this period within this nameless organization went some way toward enacting these principles and consequently innovating the way that drug development projects could work. Speeding up decision making, running parallel development tracks, killing off cul-de-sac projects early, optimally mixing internal, outsourced, and other partner resources, and continuously involving stakeholders from both upstream and downstream all became part of the new way of working (demonstrating an optimized Achievement–Pluralistic operating model). At the same time, trust with top leadership was always on shaky ground (e.g., much needed investment was never forthcoming), and in the end this proved to be with good reason as the business was sold for a considerable profit to a larger pharma company. And despite the track record of successful progression of drug development projects, the project leadership teams were dispersed and absorbed within a new parent company whose culture and leadership style was a dysfunctional version of Conformist–Achievement (many layers of management, high levels of control, low transparency, rampant politics). As in most cases when a less evolved organization takes over a more evolved one, many of the best people left. The primary drivers behind the changes that eventually undid much of what had been created were of course short-term financial, rather than medium-term value creation (as often espoused from the Achievement perspective) or even long-term medical purpose. The point here is that this conflict of motivations is now endemic in an industry which is increasingly eating itself in frantic efforts to reverse the diminishing returns that have set in from an underexamined operating model and set of assumptions about what works.

Finally at the micro or local leadership level, a degree of autonomy or freedom can sometimes exist. A colleague and I interviewed a cross section of 50 R&D leaders and managers for a large US and European-based pharma in 2013 and found in many cases that although senior leadership had become increasingly autocratic, controlling, and reactive to short-term financial pressures over a period of about 10 years, there had initially been a period of 2–3 years when R&D management enabled enough freedom for some project leaders to establish a creative culture within their project team, which had given rise to exceptional innovation performance and rapid project progress. The overriding theme across nearly all the interviews, however, was how the style of senior leaders had increasingly become a barrier to the natural passion and creativity of the majority of a large body of dedicated scientists and project managers, most of whom had come from a legacy organization.

There is a complex interdependency of relationships between (i) collective leadership paradigms, (ii) the style of individual leaders, (iii) the prevailing organizational culture within their sphere of influence, as well as with (iv) the business strategies and models that leaders develop and implement (which are the focus of Chapter 15). Leadership drives innovation in a number of ways, most significantly by introducing dynamic organizational and business models and nurturing healthy cultures within which creativity can take place. Table 19.3 summarizes how different leadership styles give rise to different organizational models and

**Table 19.3** Leadership style and organizational culture, innovation, and situationality.

| Leadership paradigm (inner orientation) + *style (outer expression)* | Cultural orientation + *organizational model* | Structure/discipline versus freedom/ experimentation | Innovation edge | Situational suitability |
|---|---|---|---|---|
| Magic + *benevolent* | Family + *circle* | Low structure<br>Low freedom | Resistance to change, protective of what is best from the past | Stable family like groups or not-for-profit communities |
| Impulsive + *autocratic* | Power + *autocracy* | Medium structure<br>Low freedom | Supports owner- or founder-centered innovation | Start-ups, owner-managed businesses and autonomous units |
| Conformist + *hierarchical* | Role + *hierarchy* | High structure<br>Low freedom | Slow systematized or periodic innovation – for example, quality and process improvement | Stable bureaucracies and large process operations |
| Achievement + *enterprising* | Achievement + *adapted hierarchy* | Medium structure<br>Medium freedom | Incremental or step innovation in competitive environments – mostly in-house innovation | Performance oriented and business-like organizations |
| Pluralistic + *social* | Relationship + *flatter hierarchy* | Medium structure<br>Medium freedom | Step innovation through collaborating – for example, conventional forms of open innovation | Human-oriented organizations, communities of practice |
| Evolutionary + *systemic/holistic* | Evolutionary + *self-management* | Low structure<br>High freedom | Step or transformative innovation through collaborating – for example, new forms of open innovation | Self-managing teams, virtual projects, communities of practice |

cultural orientations, for example, autocratic (power orientation), hierarchical (role), enterprising (achievement), social (pluralistic), integrative (knowledge), and holistic (evolutionary), and what innovation looks like within these orientations.

The leadership style best suited to engendering R&D innovation will depend upon many situational factors. Organizations exhibit a mix of developmental stages (and this can vary between different parts of the organization, as we have seen previously), and leaders need to be sensitive to bringing about the optimum culture for their set of circumstances. An important characteristic of evolutionary leaders working in sub-evolutionary organizations is their ability to be flexible, adapt their style, and work within the prevailing leadership and cultural paradigm when necessary. This chameleon-like nature is a frequently observed facet of evolutionary leaders. They consider not just their own natural preferences or the prevailing organization-wide culture but the specific cultural mix and health of their part of the organization. This implies the need for a culture change or development strategy to bring about the optimum conditions for innovation to take place. The evolutionary framework we have introduced in this chapter can be used as the primary diagnostic and design tool for developing this, but with global teams and operations, it can sometimes be useful to think more deeply about the ingrained and unconscious aspects of national cultures.

A different approach to understanding organizational and national culture was developed by Geert Hofstede in the 1980s, (Hofstede, 1991), then at IBM and now Professor of Organizational Anthropology at the University of Limburg, Holland. He has identified five cultural dimensions in his research into cultural difference of 50 countries: Power Distance, Individualism/Collectivism, Masculine/Feminine, Uncertainty Avoidance, Long-term orientation. This framework has been used to profile national cultures and can be used to profile your organizational culture, as well as individual leaders (so that they can compare themselves to national and organizational cultures). In the example of the Discovery organization I cited previously, we were working within an interesting mix of US, European (primarily German), and Indian cultures. Challenges emerged from the contrasts between US individualism (speaking up comes naturally) and German collectivism (works councils and water coolers), between the Indian need for high Power Distance (you expect to be told what to do) and the German expectation of being consulted first, and between high Uncertainty Avoidance in Germany (a plan, process, and structure for everything) and more tolerance of flexible ad-hocracy in India. Of course the important thing is to generate awareness and dialog about these preferences so that a truly international, transcultural team can form, which it did over time.

## 19.8
## Optimizing Innovation in Different Organizational Models and Cultures

The critical challenge in relation to culture and innovation is in finding the right balance between structure (or discipline) and freedom (experimentation).

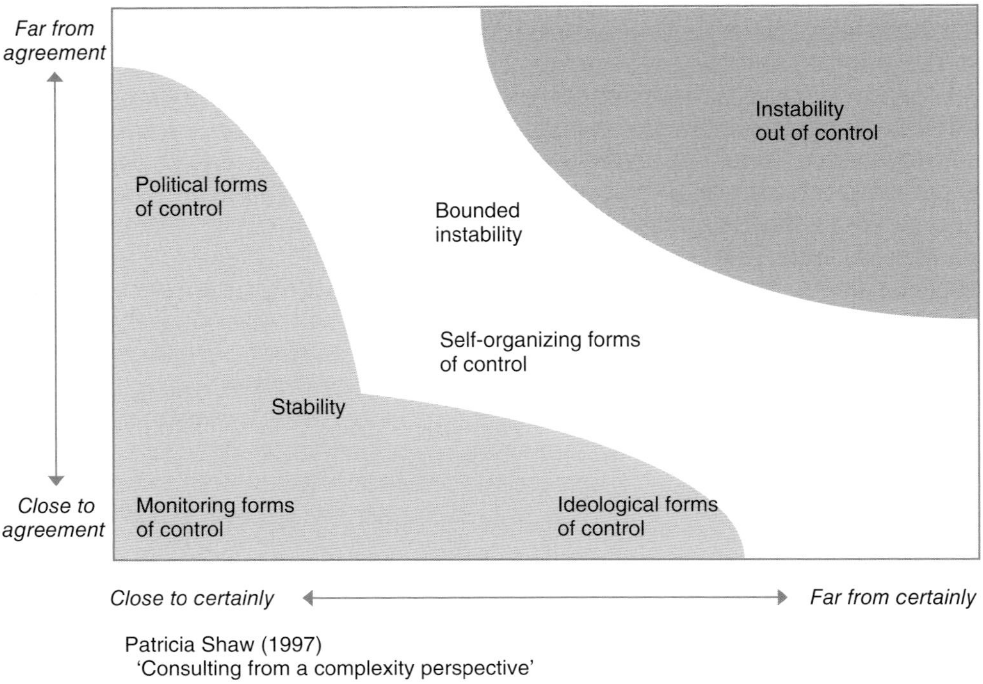


**Figure 19.2** Innovation at the edge of chaos.

Complexity theory shows us that innovation is optimized at the "edge of chaos," where there both is enough structure and enough freedom to allow emergence to take place (Shaw, 1997).

One of the most significant theoretical influences on organizational development thinking over recent years has been the embryonic field of chaos and complexity theory. This mirrors the emergence in practice of the Evolutionary leadership paradigm and organizational model and provides a theoretical body of knowledge that can be drawn upon to understand how self-organizing or complex adaptive systems work in the human organizational world.

Complexity theory shows us that innovation is optimized at the "edge of chaos," or what is called *bounded instability*. Another metaphor would be the "Goldilocks principle," where the leader needs to provide enough structure to get things going in the right direction but not too much so that it gets in the way of the creativity of passionate individuals and self-organizing teams (Figure 19.2).

Early stage research requires an enterprising and empowering leadership style that encourages freedom to experiment and the exploration of opportunities through collaborative engagement with both internal and external networks. The challenge for some large pharmaceutical companies is to reproduce the kind of creative culture that exists in small biotech, and many Achievement-style leaders do not understand how to do this. Others switch to an in-licensing or buy-in strategy that also requires an experimental and collaborative outlook that tends to be lacking in risk-averse pharma.

Later stage development requires more discipline and less freedom as well as a project-oriented culture that can drive performance in an adaptive way. The challenge here is to reduce the levels of organizational bureaucracy and overcomplexity that often accumulate in large pharmaceutical companies, particularly as a legacy of multiple mergers and acquisitions.

## 19.9 How Do We Support the Development of Evolutionary Leaders?

Laloux's work in "Reinventing Organizations" is more about the organizational model arising out of the Evolutionary paradigm than the style of leadership that supports it. He describes the leaders involved in his researched organizations as people and sketches out some commonalities in thinking, behaviors, and styles, but doesn't take this very far in terms of generalizing a leadership model. He illustrates how challenging it is for leadership working from this paradigm, showing how it is mostly about connecting with purpose, establishing (and continuously reestablishing) context, setting up the minimum key principles and processes for an Evolutionary paradigm to work, and then letting people get on with it, with buckets of trust, transparency, openness, and authenticity. Trust is critical – in many Pluralistic Pharma organizations, I have seen employees allowed to work on a pet project of their interest 1 day a month. In some Evolutionary organizations this is increased to 1 day a week. And in others this has been successfully extent to full time.

The challenge is for evolutionary leaders to let go of power, to no longer be the decision-making bottleneck, by letting decision be made at the level at which they are relevant (but always following the principle of decision making involving those impacted by the decision and those with relevant expertise). The top-level leaders in these researched organizations tend to spend their time doing whatever they are good at, getting involved in strategic research or developmental projects, being a role model for working in a different way, and offering a reference point when things go wrong. Their authority emanates from their moral leadership rather than their formal role, and people listen to them because they are respected for who they are and what they bring (and if they sense they are no longer relevant, they move onto something else). One instructive example is how the leader of a healthcare organization of some 7000 neighborhood nurses leads primarily by writing a weekly blog, which on average is voluntarily read by some 5000 of the nurses within a day of posting.

Laloux relates the life stories of these evolutionary leaders but doesn't discuss how to actively develop leaders toward an Evolutionary perspective or style. He doesn't believe in the possibility of influencing the vertical development of leaders (i.e., supporting a shift from one leadership paradigm to another), describing this process of what we call vertical development as a complex and mysterious process that happens from within and cannot be imposed on someone from the outside (p. 238). This is where I depart from Laloux in that I have witnesses many examples of leaders being supported by others to make some kind of transition of leadership paradigm. Not that I am suggesting that Managers, HR/OD practitioners, or coaches should drive the vertical development agenda for someone, but they can sense and support their emergent self and desire to grow, and they may be drawn toward people for whom the will for inner growth is strong. Admittedly this perspective will be a waste of time with many leaders for whom the best developmental path is still a horizontal one (e.g., becoming a better, more congruent, and healthy Achievement-style leader; developing core skills such as interpersonal awareness; change management; and team building).

So how might we develop leaders that show some readiness and capacity to be able to engage with and respond to this emerging evolutionary paradigm?

Much leadership development activity (coaching, education, training, development programs, etc.) is already about helping leaders increase their options for how they work with people, recognize and develop or strengthen alternative styles, and adapt their style to the needs of the situation (as touched upon previously in this chapter). Some leaders are more able to adapt their style to the needs of their situation, others are unlikely to change their style much over time and can only realistically be helped to modify or improve their style. Working in this dimension is the starting point for recognizing a leaders' capacity for vertical development, and their willingness and capacity for being situationally flexible. Of course it will also help to profile a leader to assess their mix of leadership paradigms and styles (alongside more conventional personality and behavioral assessments). However, it is more likely through a coaching or mentoring dialog that one will help draw out the leader's valency for this kind of development path.

## 19.10
## What Does It Mean to Operate from the Evolutionary Paradigm?

Evolutionary leaders are concerned with the development of their organization in relationship to uncertain and changing external environments and in pursuit of long-term relevance, viability, and sustainability. Evolutionary leadership involves developing an extra dimension, a kind of higher awareness or transcendent capacity for being concerned with the long view and the bigger picture (but not at the neglect of the immediate and the shorter term).

An evolutionary leader is grounded in the evolutionary worldview or paradigm. What does this mean? An evolutionary worldview is forward looking, dynamic, challenging, and emergent. The central tenet of this worldview (see Phipps, 2012) is that *we are moving* (or progressing) and that this movement is motivated by evolutionary purpose. Despite the occasional illusion of continuity or stability, everything in our modern world is moving and changing at an accelerating rate. Such a worldview not only embraces this but also seeks to consciously participate in and influence the direction of the evolutionary process. An evolutionary leader embraces this worldview in relationship to whatever they are leading – a group, an organization, community, society, network, and so on. This means that they are not just leading in their organization in the present but in dynamic relationship to an emergent future and their purpose for it. They are looking beyond their organization toward its place within a changing and evolving complex environment.

Evolutionary leaders are concerned with seeing the bigger picture and taking the long view. They are able to detect underlying patterns (meta-trends), they nurture a generalist capacity for weaving together different disciplines or approaches (integral thinking) and they are positive, proactive, and optimistic about engaging with the future (emergent awareness). These meta-skills allow them to respond appropriately and dynamically to an ever unfolding future that at the same time they are attempting to help create or influence.

## 19.11 Leadership and Personal Mastery

At the core of any approach to developing evolutionary leadership is personal mastery – and the individual's journey of self-development. Without a commitment to self-development, attempting to lead others can be a hit-and-miss affair. Further, personal mastery isn't something you get done and move on from, but part of the healthy ongoing development of effective leaders throughout their career. This is about the personal alongside the professional, and by personal we include the psychological, emotional, moral, cultural, social, and spiritual parts of ourselves. Developing personal mastery involves several different aspects, which include awareness, understanding, capacities, and skills. Building awareness, capacities, and skills comes from a mix of experience and personal development that is tailored to the individual and will most likely involve coaching, mentoring, group work, training, and development programs. Developing knowledge or understanding can also involve more conventional and self-directed learning.

Our description of evolutionary leadership may sound like a tall order for someone to embody to aspire toward. Can these capacities be learned and developed? Only if an individual has the desire and inclination, the passion for inner as well as outer learning and growth, and is willing to ask for help. The self-developing leader who in serious about progressing their development might start by assessing themselves (perhaps with the help of a coach or facilitator) against the following dimensions of leadership (see Evans, 2015 for a fuller exposition of the model I have drawn from) and related meta-skills:

1) *Self-**reflection** and self-awareness*:
   - Ability to **self-reflect** and build self-awareness
   - Building healthy sense of **self and identity**
2) *Awareness of difference in people and your **impact** on others*:
   - Awareness of your **impact on others and** being **authentic** in relationship with others

   - **Psychological mindedness** and **developmental awareness**
3) ***Systemic*** *and meta-awareness*:
   - Recognizing and holding **multiple perspectives**
   - Able to **move between different levels**, for example, personal and practical, business, and psychological
4) *Activating your* ***will*** *to make things happen*:
   - **Standing up** in the face of resistance, inertia, or the existing agreement
   - Using yourself as an **instrument of change**
5) *Openness to* ***help or support*** *from others*:
   - **Asking** for help and support
   - **Receptiveness** and responsiveness to feedback and input
6) *Connection with* ***purpose, meaning, and values***:
   - Distinguishing between ego needs and your **purpose** or vision
   - Creating context for **shared purpose**

This framework will look a little different to the usual leadership competency or style model used within modern Achievement-oriented pharma. Behavioral competency and skills approaches are still useful in developing leaders to a certain level (e.g., communication, interpersonal skills, team leading, motivating others, customer focus, change management, creative process management, etc.) but will not necessarily take you beyond the prevailing leadership paradigm. Dynamic approaches that include more of the whole human being (e.g., Dotlich *et al.*, 2006) provide an inspiring context but tend to reduce personal development to conventional skills improvement at the practical level. Working on skills is necessary and so is working on meta-skills, which involves exploring both deeper and higher aspects of the unconscious self. As I have suggested earlier, not everyone is ready for this work and you will know when you are. The usual signal that you are starting to make a transition from (being centered in) one leadership paradigm to another is the occurrence of some kind of personal or professional crisis. Personal crisis can take many shapes and forms, such as an inner crisis of meaning and values (what used to get you up in the morning doesn't any more) or can be triggered by an external breakdown of circumstances (e.g., losing your job, key relationships, or health). The important first step when you realize you are in a personal crisis is to find someone to talk to about it. The second is to realize that there may be opportunities for growth and learning inherent in your changing situation. The third is to make some commitment to your personal development as described previously.

## 19.12
## Building an Evolutionary Bridge to Release Innovation

In this last part, I also want to offer the reader a different and perhaps more pragmatic perspective on the evolutionary paradigm and the challenge of sourcing, enabling, and activating new strategies for more open innovation. How might leaders of large and complex pharma organizations establish the conditions to release a new and sustainable wave of open innovation, without completely deconstructing their existing organizational model? Is there an evolutionary bridge or platform that can be established to create the conditions for innovative pockets of enhanced innovativeness to emerge and experiments with new models to be successful? In other words, if your organization doesn't yet have the conditions (e.g., top leaders psychological development centered in Evolutionary thinking, enlightened owners supporting long-term purpose over profit, etc.) to give rise to the Evolutionary leadership paradigm in its full manifestation (self-organization, wholeness, evolutionary purpose, transparency, trust across the whole organization, etc.), what should be done in the meantime to release the latent innovative potential of the organization? What is the next best option? Another way to look at this is to ask, what can leadership do in terms of developing the organization to support both the emergence and chances of success of new business strategies, such as open innovation and business models based around open principles?

This approach explores the underlying conditions for achieving greater innovation within large organizations in terms of critical capabilities that emerge within an recognizable pattern under Achievement–Multiplistic and Pluralistic–Relativistic leadership. In short, innovation can be seen as one of six critical interdependent organizational capabilities. These are learning, change, innovation, collaboration, agility, and engagement. Top leadership may need to focus on developing one or more of the other capabilities in order to establish the ***evolutionary bridge*** or foundations for an innovative culture to more fully emerge, and in the pharma industry, this means supporting experimentation with new forms of open innovation.

Organizations evolve in dynamic relationship with their external environment (e.g., markets, customers, technologies, competitors, partners, and other stakeholders). Establishing an *evolutionary bridge* means having the capability to adapt or change with the times, to be sustainable, resilient, and enduring. It also gives expression to the proactive desire to influence the direction of change and to cocreate the emergent future. The test of sustainability is whether the organization can navigate successive periods of disruptive or turbulent environmental change, as well as succeed and prosper in the good times. Such an organization needs elements of evolutionary leadership, even if it continues to be anchored in the Achievement paradigm, and a culture that continues to attract or give rise to evolutionary purpose.

What are the critical and essential capabilities that leaders need to develop and nurture within their organizations to achieve evolutionary sustainability and to establish such an *evolutionary bridge*? Four years ago myself and three colleagues embarked upon exactly this inquiry and the outcome can be summarized as the "big six" critical organizational capabilities: **Learning**, **Change**, **Innovation**, **Collaboration**, **Agility**, and **Engagement**. The content of this list should not in itself be surprising – these are the most prominent themes that have come to the fore in organizational thinking and practice over the last 25 years, in roughly the order in which we have listed them. We would argue that other themes or topics have come and gone or can be fitted within one of these six, but that these six remain as the most important and essential capabilities an organization needs today. What is more surprising is how the emergence of these themes per se tells a story of organizational evolution as a response to increasing complexity and uncertainty over the years.

**Learning** is the lifeblood, the way that an organization benefits and builds from experience. **Change** is both about having resilience and responsiveness to events and an ability to move forward and adapt to a changing environment. **Innovation** is about moving onto the front foot and translating learning and change into growing value for the organization. **Collaboration** enables greater innovation (by easing the path from closed to open models of innovation) and builds sustainability in an increasingly complex world where you cannot survive alone. **Agility** is the ability to pull all these together in adaptive and responsive open or virtual business models that can stay ahead of the external pace of change. **Engagement** is the human piece, the need to be authentic, emotional, and relational.

These capabilities are separate but interconnected: for example, the ability to learn gives rise to the need to manage change, change is insufficient for sustainability without innovation, successful open innovation is dependent upon collaboration, collaboration opens the path to greater agility and experimenting with more open business models, but agility and virtuality without engagement lack the human connection needed to stay in relationship with today's employees, partners, and customers.

Many large pharma organizations today are currently focused on the latter two of these – Agility and Engagement. Agility (e.g., see Kotter, 2012) is mostly about creating virtual and flexible operational models that seamlessly optimize the involvement of external and internal resources according to changing needs, driven by strategies and projects, for example, crowdsourcing within communities of practice for complex problem solving and outsourcing to CROs and CMOs for different parts of the pharma value chain. Engagement (or reengagement of the people and their purpose) is trickier for reasons we have already discussed, and especially given the disruption that seemingly continuous mergers, reorganizations, and restructurings have upon a pharmaceutical company's workforce. The current trend seems to be to restructure toward smaller organizational units that people can identify with more easily.

If any of the previous capabilities have not been adequately established (or have waned or been neglected), then these need to be addressed as well. In fact it is likely that a process of continuous renewal is needed in relationship to all six. The key point here is to see their interdependence – and that innovation should not be addressed in isolation from an evolutionary leadership perspective. Building an *evolutionary bridge* can start with an appraisal of your relative strengths and developmental challenges for each of these "big six," followed by an exploration of how they need to come together for you organization to achieve innovative growth.

Our working definition of each of the six critical organizational capabilities is shown in Table 19.4.

**Table 19.4** Definitions of the six critical organizational capabilities.

| | |
|---|---|
| *Learning* | Learning and development as a continuous process at individual, team, and organizational levels |
| *Change* | Leading and managing change in response to external or unplanned events as well as proactively to transform or develop the organization |
| *Innovation* | Ability to continuously reinvent the organization and channel innovation in pursuit of sustainable value creation |
| *Collaboration* | Working collaboratively and partnering effectively both inside and outside the organizational boundary |
| *Agility* | Capacity to efficiently adapt and evolve in response to external opportunities and threats without needing to initiate disruptive change |
| *Engagement* | An organization's ability to engage authentically, emotionally, and ethically with its people, customers, and other stakeholders |

## 19.13 Conclusions

In this chapter we have taken a multiperspectival view of leadership and its impact on innovation in the pharmaceutical industry. This is a complex topic for which answers are clearly difficult to find, or it wouldn't attract such interest and frustration in equal measure. Rather than restricting our answers to the latest examples of best practices from the field (a very Achievement-style approach), we have sought to equip the reader with an appreciation of leadership as a whole and the developmental perspective in particular. We have described the basic developmental spiral following Frederic Laloux's recent work on organizational models and shown how leadership styles are external expressions of internally held leadership paradigms, which are themselves both reflections of broader shifts of consciousness within society, as well as the process of individual human development. At the same time, leadership style is always individualized and unique, so it helps to be aware of how personality and other factors influence the way that leaders operate.

We have applied all this within the situational context of the pharmaceutical industry's innovation deficit. Leadership style must adapt to situational need and cultural context. There is evidence of the limited ability of the Achievement – Enterprising leadership paradigm style to reinvigorate innovation, particularly through the vertically integrated closed innovation model. The rapid rise of the Pluralistic – Social paradigm over the last 20 years has brought more collaborative styles of leadership which enabled the emergence of conventional forms of open innovation, but this too may have peaked in terms of its impact on innovation productivity. The answers to the innovation deficit appear to point toward the Evolutionary leadership paradigm, either in terms of leaders evolving completely new ways for pharma organizations to work or by building an evolutionary bridge that supports new open innovation strategies and business models (that can work within existing big pharma organizations).

I have challenged the reader to reflect upon their own leadership development and how they might activate the evolutionary perspective for themselves. Coaches and HR professionals might think about how they support leaders to develop in this way. For organizations that are likely to remain subevolutionary in their organizational model, I have outlined how to build an evolutionary bridge based upon six interdependent capabilities to support the emergence of new forms of open innovation. For senior leaders who recognize the evolutionary paradigm within themselves, I wish you good luck with growing the next generation of biotech and pharma organizations.

## References


Barrett, R. (2011) *The New Leadership Paradigm*, Lulu.

Beck, D. and Cowan, C. (1996) *Spiral Dynamics*, Blackwell Publishing.

Buckingham, M. and Clifton, D. (2005) *Now Discover Your Strengths*, (StrengthsFinder), Free Press.

Bushe, G. (2010) *Clear Leadership: Sustaining Real Collaboration and Partnership at Work* (revised edn), Davis-Black, Boston, MA.

Collins, J. (2001) *Good to Great Leadership*, Random House.

Creative Leadership Consultants Limited (2015) Building Tomorrow's Leadership Mindset, http://www.clc.co.uk (accessed 11 August 2015).

Diamond, J. (1997) *Guns, Germs and Steel: The Fates of Human Societies*, W.W. Norton & Company.

Dotlich, D., Cairo, P., and Rhinesmith, S. (2006) *Head, Heart and Guts*, Mercer Management Journal.

Evans, R. (2015) *5DL: Five Dimensions of Leadership*, Creative Leadership Consultants.

Fisher, D., Rooke, D., and Torbert, W.R. (2001) *Personal and Organisational Transformations through Action Inquiry*, Edge/Work Press, Boston, MA.

Gardner, H. (1983) *Frames of Mind: The Theory of Multiple Intelligences*, Basic Books.

Goleman, D. (1996) *Emotional Intelligence*, Bloomsbury.

Goleman, D. (2007) *Social Intelligence*, Arrow Books.

Graves, C. (1970) Levels of existence, an open system theory of values, *J. Humanist. Psychol.* 1970. **10** 2, 131-155

Greenleaf, R.K. (1977) *Servant Leadership*, Paulist Press.

Heron, J. (1999) *The Complete Facilitators Handbook*, Kogan Page Limited.

Hersey, P. (1985) *The Situational Leader*, Center for Leadership Studies.

Hofstede, G. (1991) *Cultures and Organisations; Software of the Mind*, McGraw-Hill.

Howard, A. (2015) Background to the Evolutionary Leadership Profile, http://www.psychosynthesiscoaching.co.uk/resources-coaches/.

Jaworski, J. (1996) *Synchronicity: The Inner Path of Leadership*, Berrett-Koehler.

Kegan, R. and Lahey, L. (2009) *Immunity to Change*, Harvard Business School Press.

Kets de Vries, M. (2006) *The Leader on the Couch*, Jossey Bass, Chichester.

Kohlberg, L. (1984) *The Psychology of Moral Development: The Nature and Validity of Moral Stages*, Harper & Row.

Kotter, J.P. (2001) What leaders really do. *Harv. Bus. Rev.*, **79**, 85–96.

Kotter, J.P. (2012) Accelerate!. *Harv. Bus. Rev.*, **90** (11), 45–58.

Kouzes, J. and Posner, B. (2002) *The Leadership Challenge*, 3rd edn, Jossey-Bass.

Laloux, F. (2014) *Reinventing Organisations*, Nelson Parker.

Lencioni, P. (2002) *The Five Dysfunctions of a Team: A Leadership Fable*, Jossey-Bass.

Leslie, J.B. and Palmisano, K. (2014) *The Leadership Challenge in the Pharmaceutical Sector*, Center for Creative Leadership.

Loevinger, J. (1998) *Technical Foundations for Measuring Ego Development, Lawrence Erlbaum Associates.*

Phipps, C. (2012) *Evolutionaries: Unlocking the Spiritual and Cultural Potential of Science's Greatest Idea*, Harper.

Rooke, D. and Torbert, W.R. (2005) Seven transformations of leadership. *Harv. Bus. Rev.*, **83** (4), 66–76, 133.

Schuhmacher, A. and Betz, U.A.K. (2016) The pharmaceutical industry is opening its R&D boundaries, in *Value Creation in the Pharmaceutical Industry*, Wiley-VCH Verlag GmbH & Co. KGaA, Weinheim.

Senge, P. *et al.* (2005) *Presence: Exploring Profound Change in People, Organizations and Society*, Nicholas Brealey Publishing.

Shaw, P. (1997) Intervening in the shadow systems of organizations: consulting with a complexity perspective. *J. Organ. Change Manage.*, **10** (3), 235–250.

Shaywitz, D. (2013) Does Pharma's Innovation Difficulties Reflect Bad Leaders or Tough Circumstances? Forbes (Mar. 7).

Spillane, J. (2006) *Distributed Leadership*, Jossey-Bass, San Francisco, CA.

Tannenbaum, R. and Schmidt, W. (1973) How to chose a leadership pattern, *Harv. Bus. Rev.*, **51**, 162–180.

Thamhain, H. (2003) Managing innovative R&D teams. *R&D Manage.*, **33** (3), 297–312.

Torbert, W.R. (2014) "Vertical" vs. "Horizontal" Leadership Development, http://www.williamrtorbert.com/vertical-vs-horizontal-leadership-development/ (accessed 11 August 2015).

Wade, J. (1996) *Changes of Mind*, SUNY Press.

Wilber, K. (1999) *Integral Psychology*, Shambhala Publications.

Wilber, K. (2000) *A Theory of Everything*, Shambhala Publications, Inc.

Zohar, D. and Marshall, I. (2000) *SQ – Spiritual Intelligence, the Ultimate Intelligence*, Bloomsbury.

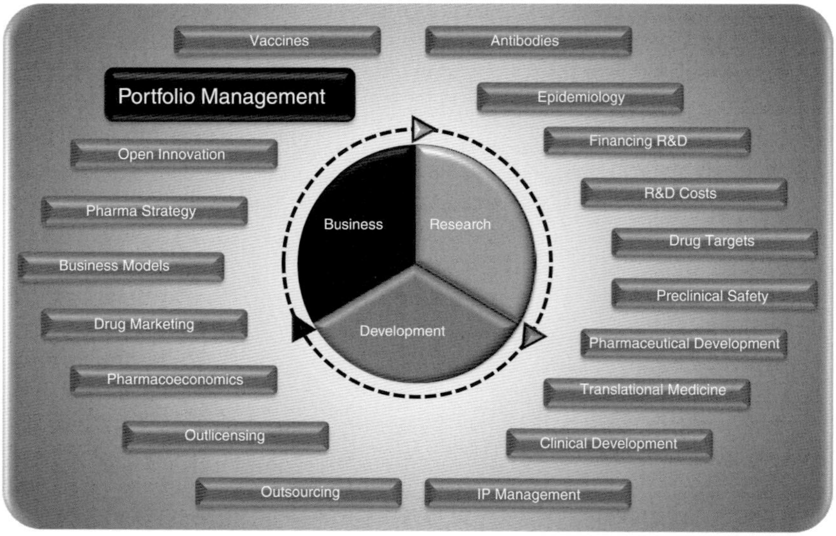
Vaccines
Antibodies
Portfolio Management
Epidemiology
Financing R&D
Open Innovation
R&D Costs
Pharma Strategy
Business
Research
Drug Targets
Business Models
Preclinical Safety
Drug Marketing
Development
Pharmaceutical Development
Pharmacoeconomics
Translational Medicine
Outlicensing
Clinical Development
Outsourcing
IP Management

# 20
# The Role of Modern Portfolio Management in Pharma Innovation

*Joachim M. Greuel and Axel Wiest*

## 20.1
## Introduction

The purpose of this book chapter is to introduce the reader to modern, value-driven portfolio management and to describe its role in pharmaceutical innovation. While the role of portfolio management is well established for managing financial assets, its relevance for pharmaceutical innovation is less obvious. However, as R&D productivity is declining relative to the financial investment required to launch a new product, pharmaceutical companies need to identify procedures that will allow them to market more high-value drugs for each dollar invested.[1] Implementing value-driven portfolio management is one possibility to enhance R&D productivity as only those projects will be supported that will likely add the most value to a firm's R&D portfolio.

The following issues are explored in this book: first, the origins of pharmaceutical portfolio management are outlined in the context of today's R&D challenges, followed by a description of goals and metrics of portfolio management. Next, we will provide arguments why value-driven portfolio management, among other factors, enables innovation; we will conclude with an outlook to the organization of future portfolio management that integrates elements of the distinct functions R&D, business development, and M&A (Mergers & Acquisitions).

As R&D productivity declines, companies seek measures how to increase R&D output. One possibility is the implementation of value-driven portfolio management. In value-driven portfolio management, projects are ranked according to value creation relative to the required investment or, in other words, according to projects' productivities. If budget limits are reached, projects with a low productivity may be licensed or abandoned. However, projects with negative net present values (NPVs) may not "automatically" get discontinued as classical

1) Productivity is generally defined as the ratio of output to input, typically applied to the production process; in the context of R&D, it is defined as the financial value gained (e.g., by launching new drugs) per dollar invested.

*Value Creation in the Pharmaceutical Industry: The Critical Path to Innovation*, First Edition.
Edited by Alexander Schuhmacher, Markus Hinder, and Oliver Gassmann.

finance theory would suggest[2]; management may analyze individual projects' future value paths and take portfolio decisions also considering potential future values. From a risk/return perspective, large portfolios are preferable not only because risk gets diversified away but also because portfolio value typically increases more than portfolio risk so that returns are better to predict and appear less erratic.

Furthermore, portfolio management enables innovation as it may show that groundbreaking projects that are often risky by nature and, thus, carry negative NPVs may well play a role within a portfolio of less innovative projects. Thus, they may not be abandoned on the basis of a negative NPV figure alone. Finally, an argument is made to combine traditional portfolio management (that analyzes and manages internal projects only) with business development and M&A. The integrated function would enable companies to view the entire world as "their laboratory", broadening their views beyond the obvious portfolio candidates.

## 20.2
## Challenges in R&D and the Origin of Pharmaceutical Portfolio Management

Pharmaceutical companies need to provide a financial return to investors. Although this may sound trivial, it is a quite challenging task given the high attrition rates observed in R&D. Once a target has been identified, the chance of obtaining a marketable drug that modulates a target's biological function is less than 5%, with associated capitalized costs above $1.7 billion (Paul *et al.*, 2010). In fact, unless yearly drugs sales exceed $500 million, many early-stage projects have negative NPVs.

Many factors contribute to high attrition rates, the most relevant being:

- Developing drugs for targets that have not been validated
- Poor predictability of preclinical disease models
- High proportion of patient nonresponders
- Patients suffering from serious adverse events and
- Suboptimal clinical study design (e.g., due to the lack of funds available)

In fact, Prinz et al. (2011) report that only 24% of key scientific findings that serve as a basis for target selection can be reproduced, a low figure that was later confirmed by Begley and Ellis (2012). Therefore, seemingly promising targets may later become devalidated, and companies investing in drugs modulating these targets may finally see them fail in clinical development. Other reasons for drug failure are the sometimes poor predictabilities of *in vitro* and *in vivo* models generally applied in preclinical research (Bracken, 2009) and poor drug–patient fit due to genetic variations that lead to either poor response rates or severe adverse

2) The NPV (net present value) is a measure of a drug's financial value; if positive, an investment will likely generate financial value for investors; if negative, investors will probably not receive the return expected. Thus, according to classical finance theory, positive NPV projects shall be pursued, negative NPV projects shall be abandoned.

drug reactions (Monte, 2012). Finally, suboptimal clinical development, for example a poorly designed phase III clinical trial due to inadequate phase II studies, may cause a clinical program to fail although the drug itself may be effective (Retzios, 2010).

For each project an R&D manager needs to ensure, to the extent possible, that a pharmaceutical target has been validated, patients selected for clinical trials are likely responders, and the design of clinical studies is not adversely affected by suboptimal funding. However, for each individual project, there will always be residual risk that cannot be eliminated irrespective of how well discovery and development have been designed and executed. Therefore, while individual projects may fail, a portfolio of drug development projects needs to be successful to provide the returns demanded by investors and patients alike.[3)] Without the high risk of failure observed in pharmaceutical development, portfolio management would be less relevant.

While modern portfolio management has its roots in finance,[4)] it was soon adopted by academics and by the pharmaceutical industry trying to apply the principles of financial portfolio management to master the challenges of pharmaceutical R&D (Blau and Pekny, 2004).

Major challenges are:

- A portfolio must be large enough to compensate for project failures, ensuring the highest possible financial return to investors.
- A portfolio must be small enough in order not to exceed the available budget.
- A portfolio must be well balanced such that a company's R&D resources are used in an optimal way, not producing major bottlenecks.

While there is no dispute on the advantages of financial portfolio management, the benefits of pharmaceutical portfolio management are less obvious. First, it is virtually impossible to correlate the financial success of a pharma company with portfolio management. This is because too many variables that cannot be controlled contribute to a company's profit and value generation. Second, even if one would find that companies who have implemented formal portfolio management do better than those who have not, portfolio management processes and practices vary widely so that firm conclusions cannot be drawn. Nevertheless, there is good reason to support the hypothesis that good pharmaceutical portfolio management will contribute significantly to a company's success.

## 20.3 Goals and Metrics of Portfolio Management

Apart from aligning strategy and operations, portfolio management seeks to:

- Minimize development and commercial risk to the extent possible

3) Patient "returns" may be measured as health benefits, for example, an increase in quality-adjusted life years (QALYs).
4) H. M. Markowitz was awarded the Nobel Memorial Prize in Economic Sciences in 1990 for his pioneering work in modern portfolio theory in the financial sector.

- Maximize the financial return within a given budget
- Maximize the financial return relative to the risk assumed
- Ensure optimum resource utilization

Toward the first goal, models have been suggested that are based on success probabilities and payoffs of each R&D project in a portfolio (Ringuest *et al.*, 1999). While there are multiple sources that provide benchmarks for success probabilities by individual R&D phase and also by indication, major challenges are to adjust those benchmarks to account for specific drug attributes (e.g., degree of innovation, small molecule vs. antibody) and to integrate project correlations into a model. For example, positive correlations exist between projects when they address the same mechanisms of action; if one project fails, the likelihood that the other one fails as well increases. Commercial risk exists as there are usually competing projects developed by other firms for the same indications, often targeting the same receptor. If typical attrition rates are applied to competing projects, an expected number of competing drugs can be estimated and patient share adjustments can be made accordingly. Those adjustments may be based on the number of competing drugs; they may also take drug attributes (in terms of efficacy and safety) into account. Once all relevant risks have been identified, a portfolio model can be developed that takes those risks as well as project correlations into account. Perfect modeling tools are add-on programs to Excel™ that allow Monte Carlo simulations of pharmaceutical portfolios. Further descriptions of project evaluation methods are provided elsewhere (Greuel, 2002; Bode-Greuel and Nickisch, 2008).

K. Girotra et al. state that "understanding the value of a product development project is central to a firm's choice of project portfolio" (Girotra *et al.*, 2007). As portfolio management strives to maximize returns to investors, and NPV is a measure whether or not a project provides a requested return, a portfolio's value is generally expressed as NPV. Typically, individual projects are plotted based on their "productivity" scores that are calculated by dividing each project's risk-adjusted NPV by its risk-adjusted future development and commercialization costs. "Risk-adjusted" means that a drug may not get launched and late costs may not be incurred in the event a drug fails early; the probabilistic nature of development needs to be reflected in the financial model. Once risk-adjusted project NPVs and cost figures have been established, their productivity scores may be plotted in ascending order (Figure 20.1). Projects with the "steepest" slope are typically the ones that are given highest priority. However, scale may also play a role as projects that do not provide a minimum value at a particular development stage may be licensed or abandoned.

*The optimum portfolio from a financial perspective is the one that provides the highest NPV for the assigned budget limit.* On project productivity charts, projects with a negative NPV have negative slopes and are usually projects that are not recommended for further development. However, very innovative projects typically carry high risks as, for example, their mechanisms of action may not yet be well validated. Once a negative NPV project overcomes the early development hurdles,

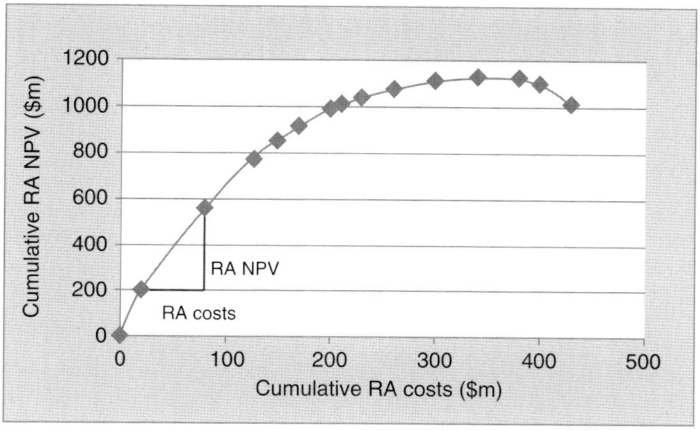


**Figure 20.1** Projects of a portfolio can be plotted in ascending order by their "productivities," that is, the ratio of risk-adjusted (RA) NPV and RA investment. Productivity is represented as the slope of the curve, and the steeper the slope, the higher the value contribution of a project relative to the investment required. If, for example, the budget limit equals $250 million, all projects to the right of the budget limit may be licensed out or, in the event the slope is negative (negative NPV), abandoned.

its value may rise sharply. In later development the values of innovative projects tend to be higher than the values of less innovative projects as, assuming anything else being equal, they usually have a larger commercial potential. Assume a portfolio of two projects, "A" and "B," with "A" having a positive NPV and "B" having a negative NPV at an early discovery stage. Further assuming that only one project could be developed, traditional financial wisdom would suggest to discontinue project "B" (Figure 20.2). However, if the value development by milestone is plotted, at a certain point (here: at phase I), project "B's" value may evolve higher than project "A's". If only one project could be developed further, the choice depends on the characteristics of the other projects in the portfolio as well as the company's strategic goals. *For any portfolio decision, it is important to know the potential future value paths of individual projects.*

When projects' risk-adjusted NPVs are calculated, when projects are ranked according to productivity, and, finally, when also future portfolio values are assessed, management will be in a good position to decide which portfolio is value-maximizing.

For portfolio management, it has been suggested to take advantage of Monte Carlo simulations, a method that calculates the values of thousands of scenarios (Blau and Pekny, 2004; Bode-Greuel and Greuel, 2005). Apart from showing managers the portfolio NPV and the full picture of financial risk and reward, Monte Carlo simulations also provide standard deviations ("SDs") of their output distributions, for example, a portfolio's value (Figure 20.3). SDs are measures of risk and can be assessed for any portfolio. Like in financial portfolio management, value and risk the portfolio's (SD) for different portfolios can be plotted and analyzed. One important parameter is the "value-to-risk" ratio that shows how much value is

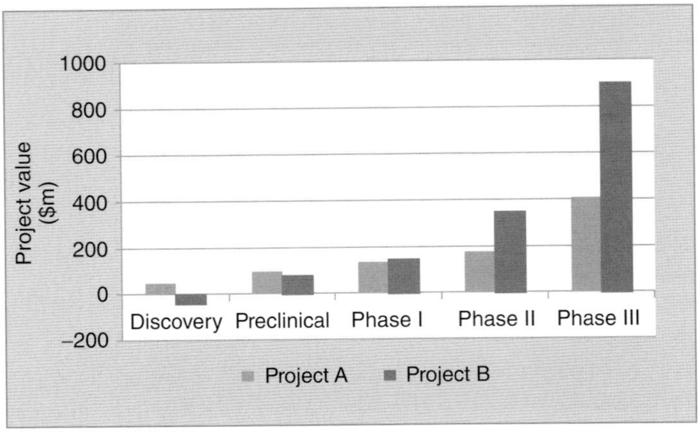


**Figure 20.2** Future project values can be calculated and plotted as a function of development phase. Innovative projects may add significant more value at later stages than less innovative projects. The information of how much value will likely be added once a particular milestone will be reached should contribute to decision making. For example, management may want to support and fund project "B" because of its value gain in later stages, although its NPV at the time of decision is negative.

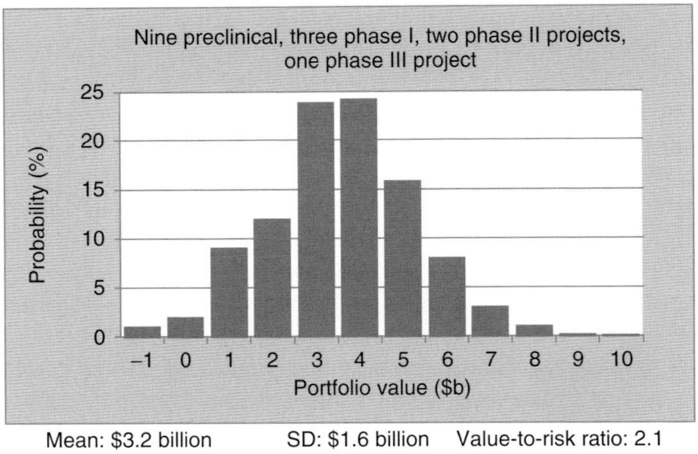


**Figure 20.3** Plot of a value distribution of a portfolio consisting of nine preclinical, three phase I, two phase II, and one phase III projects. The portfolio value mean is used as a single descriptor of the portfolio's value, the standard deviation (SD) being a measure of portfolio risk. The portfolio can be said to be reasonably diversified as the probability of portfolio failure (negative NPV) is in the order of 1%. Of interest is the "value-to-risk" ratio that can be obtained by dividing portfolio value by its SD. The more projects are added to a portfolio, the larger the ratio becomes, that is, more value is created per "unit of risk." Therefore, large portfolios have a much higher probability to provide a return on investment than small ones, and the return becomes bigger relative to the risk assumed.

**Table 20.1** Values, standard deviations (SDs), and value-to-risk ratios of value distributions representing portfolios that become increasingly rich (more projects) and mature (late-stage projects).

| Portfolio | Mean value ($m) | SD ($m) | Value-to-risk ratio |
|---|---|---|---|
| Two preclinical projects | 161 | 499 | 0.32 |
| Nine preclinical projects | 1061 | 1051 | 1.01 |
| Nine preclinical and three phase I projects | 1691 | 1315 | 1.29 |
| Nine preclinical projects, three phase I projects, two phase II projects, one phase III project | 3191 | 1550 | 2.06 |

Typically, larger portfolios have higher values and higher SDs compared with smaller ones. However, the increase in value is higher than the increase in risk. Therefore, from a risk/return perspective, companies should aim at building large portfolios.

generated per unit of risk. The ratio is obtained by dividing the mean of the value distribution by its SD. As a general rule, the larger the portfolio and the more late-stage projects, the larger the value-to-risk ratio (Table 20.1). That means a company gets the highest return relative to the assumed risk when it adds as many projects to its portfolio as possible.[5)]

Thus, from a risk/return perspective, companies should create the biggest portfolios possible, just limited by budget constraints. However, not all projects are statistically independent. Those addressing the same target have positive correlations, and those addressing different targets may either be neutral or even correlate in a negative way. Also, focus on one or two disease areas introduces a positive correlation between individual projects.

The portfolio that includes the largest number of projects may not necessarily be the one that offers the highest return relative to a portfolio's risk. Taking cost and risk measures as independent variables and value as the dependent variable, optimum portfolios lie on an "efficient surface," offering the highest values for a fixed budget and the desired exposure to risk.

Finally, the choice of the optimal portfolio also depends on the availability of resources to develop the projects further. If projects compete for the same resources, they may experience delays, and the projects' values may become negatively affected. Therefore, projects' consumption of resources has to be assessed and modeled. In the event bottlenecks are detected, they can either be managed (e.g., increasing the capacities) or the loss of value due to bottleneck-induced delays may be shown in the financial analysis. Typically, the choice of the optimal portfolio is an iterative process. Portfolio management will analyze which combination of projects may provide the highest value for the budget

5) The assumption applied here is that projects are independent from each other, that is, they are not correlated. If projects address the same mechanism of action or the same indication, this assumption does not fully hold. However, if only a few projects correlate and if, in addition, the correlation is weak, the assumption of general independence may be made without distorting the overall result.

available, followed by a thorough risk assessment. The risk assessment may lead to modifications of the portfolio composition, which is then analyzed for potential capacity constraints. For example, projects may compete for the same clinical development resources and may not be developed further in parallel, even when a simultaneous development would fit the budget. In those cases, management may modify the project portfolio composition or engage external firms such as CROs (Contract Research Organizations) to eliminate the bottlenecks.

## 20.4
## Portfolio Management as Enabler of Innovation

Portfolio management as described is not just a process alongside R&D or any other function. First, as shown before, innovative projects may have low or even negative NPVs. The reason is that innovative projects tend to be more risky than their less innovative counterparts. According to finance, projects with negative NPVs should be terminated as they do not earn the returns investors expect. Yet, in the context of a portfolio analysis, those projects may be important to pursue if value increases significantly after reaching future development milestones. Furthermore, the particular risk of innovative projects may be well balanced by other less risky projects.

Second, Monte Carlo analyses show that portfolios that are chosen based on the criteria described in the preceding text create more value than portfolios with projects assembled randomly. Although in real life portfolios are almost never chosen by a random process, they may also not emerge from a rigorous portfolio analysis that is based on processes and methods as described. For example, applying scoring and other semiquantitative systems appears to lead to poorer portfolio decisions when compared to quantitative, financially driven methods (Slater and Zwirlein, 1992). If companies that apply value-driven portfolio management achieve better commercial results and gain more value, more resources can be spent on R&D. Therefore, portfolio management indirectly supports innovation.

Third, effective portfolio management provides a tool for formulating and executing corporate strategy, matching internal projects with external opportunities, and creating transparency throughout the organization. Without transparency portfolio decisions may be hard to comprehend, and this affects an organization's capability to innovate in a negative way.

Fourth, value-driven portfolio analysis and management support senior executives in communicating value to external investors. Obtaining funds is key to innovation, and without a portfolio management process that is driven by R&D, marketing, and finance, portfolio value creation may not get assessed and, thus, may be difficult to communicate.

Taken together, modern, value-based portfolio management may be regarded as one enabler of innovation, among others, ensuring that scarce resources are used in the optimal way and that value creation is maximized. Companies without an operating portfolio management structure run the risk of suboptimal decisions and, thus, in the long run poorer performance than their competitors.

## 20.5 Modern Portfolio Management Integrates In-House R&D, Business Development, and M&A

It may be tempting to restrict portfolio management only to internal projects and already marketed products. However, if a gap in the portfolio has been identified, business development may search for opportunities outside the firm. Guidance what to seek for should come from strategy as well as from portfolio management; otherwise an organization runs the risk to look for opportunities that are not entirely in line with the company's goals and requirements. Likewise, in the event a company's project shall not be developed further, for example, due to poor strategic fit, it may be licensed out or sold. Therefore, portfolio management and business development will interact frequently in order to establish an optimum portfolio, considering drugs that shall either be in-licensed or out-licensed. It seems natural to take a holistic view, seamlessly integrating portfolio assessment and management of internal projects with potentially licensed external projects. Portfolio management can provide essential input to business development concerning the projects to look for, and in-licensing candidates can be integrated in a portfolio model well before agreeing on a deal. As a result, licensing decisions may not be based on a project's properties alone but also on its fit to the overall portfolio.

For already marketed drugs, life cycle management supports products during all commercialization stages. Patent expiries of major drugs are always a concern to senior executives, and one of portfolio management's roles is supporting the organization by (i) outlining the expected impact of patent expiries and (ii) suggesting strategies how to avoid the negative consequences of the anticipated sales loss. For example, a company may have prepared in time a successor with improved properties, be in licensing negotiations to fill the gap, or even consider acquiring a firm. As modern portfolio management concerns both, projects in development and already marketed products, its role should not be confined to solely manage R&D projects, nor should it be restricted to launched products; portfolio management has to represent projects before launch as well as products after launch.

Finally, even M&A, traditionally the domain of finance, may be an integral part of portfolio management. While the operational steps in an M&A transaction will be best managed by a company's finance department, the strategic fit of an acquisition target may, in agreement with senior management, largely be determined by portfolio management.

Modern portfolio management does no longer consider solely "internal" projects or products; in fact, portfolio management today analyzes a company's own projects as well as licensing and M&A opportunities worldwide to create an optimal portfolio, one that maximizes return, minimizes risk, and allocates

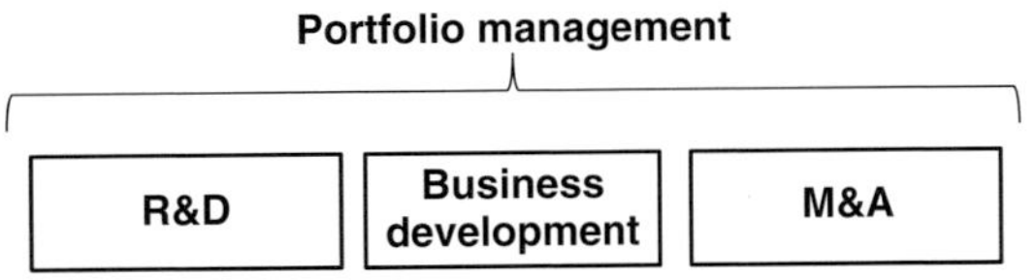


**Figure 20.4** Modern portfolio management spans internal R&D, business development, and M&A functions. This structure allows a company to take a bird's-eye view onto potential portfolios, whose individual project members are recruited for the entire world.

resources and capacities in the most efficient way. Due to portfolio management's new role that spans across several functions (Figure 20.4), it should report neither to R&D nor to finance or any other related functions. Instead, portfolio management may report directly to a company's head of strategy or to the CEO's office, depending on company size, as portfolio management's role clearly extends beyond individual functions.

## References

Begley, C.G. and Ellis, L.M. (2012) *Nature*, **483**, 531–533.
Blau, G.E. and Pekny, J.F. (2004) *J. Prod. Innov. Manage.*, **21**, 227–245.
Bode-Greuel, K. and Greuel, J. (2005) *J. Commer. Biotechnol.*, **11**, 155–170.
Bode-Greuel, K.M. and Nickisch, K. (2008) *J. Commer. Biotechnol.*, **14**, 307–325.
Bracken, M.B. (2009) *J. R. Soc. Med.*, **102**, 120–122.
Girotra, K. *et al.* (2007) *Manage. Sci.*, **53**, 1452–1466.
Greuel, J. (2002) *Curr. Drug Discovery July*, 37–42.
Monte, A.A. (2012) *J. Med. Toxicol.*, **8**, 43–51.
Paul, S.M. *et al.* (2010) *Nat. Biotechnol.*, **9**, 203–214.
Prinz, F. *et al.* (2011) *Nat. Rev. Drug Discovery*, 10–712.
Retzios, A.D. (2010) Bay Clinical R&D Services, pp. 1-46
Ringuest, J.L. *et al.* (1999) *Res.-Technol. Manage.*, **42**, 40–43.
Slater, S.F. and Zwirlein, T.J. (1992) *J. Manage.*, **18**, 717–732.

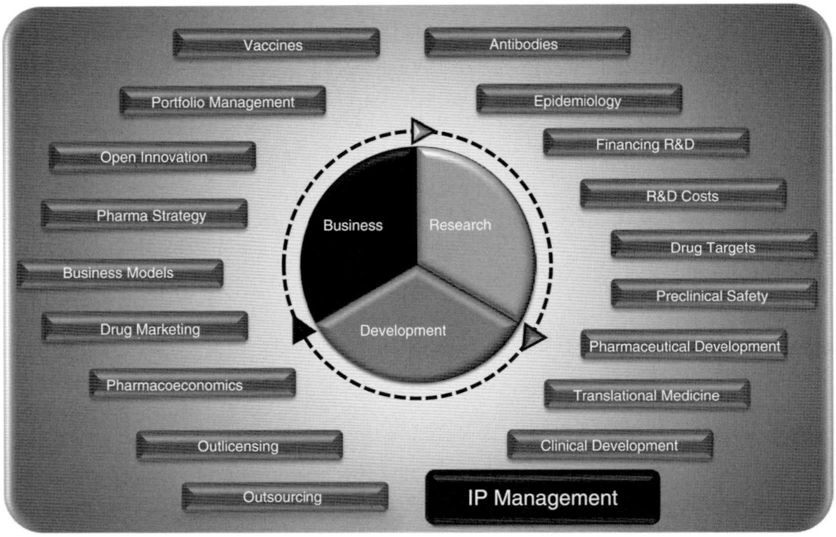
Vaccines
Antibodies
Portfolio Management
Epidemiology
Financing R&D
Open Innovation
R&D Costs
Pharma Strategy
Business
Research
Drug Targets
Business Models
Preclinical Safety
Drug Marketing
Development
Pharmaceutical Development
Pharmacoeconomics
Translational Medicine
Outlicensing
Clinical Development
Outsourcing
IP Management

# 21
# Patent Management Throughout the Innovation Life Cycle[1]

*Martin A. Bader and Oliver Gassmann*

## 21.1
## Introduction

The purpose of this book chapter is to outline which role patents play in the pharmaceutical industry. It provides best practice examples from industry leaders and deducts how patents can be managed in a research and development (R&D) environment that has long product cycles and reduced R&D efficiency.

The following questions are explored in this book chapter:

- What challenges do companies face when managing patents?
- How do best in class industry leaders manage patents?
- How can patents be managed throughout the innovation life cycle?

We begin with an overview of why patents have become a strategic factor in the pharmaceutical industry. Next we derive a patent life cycle management model and provide a best practice case study from Bayer. Lastly, we provide concluding remarks.

Effectively managing and optimizing the value of a patent portfolio is a major challenge for many firms, especially those in knowledge-intensive industries such as the pharmaceutical, biotechnological, and chemical industries. However, insights in effective patent portfolio strategies are rare. In this chapter, we investigate how firms successfully manage and optimize their patent portfolios to increase their overall competitiveness. We find that successful patent portfolio management is rooted in managing the patents throughout the innovation life cycle. Based on the findings of 10 case studies, we derive a holistic patent life cycle management model that reflects 5 distinctive phases of patent management: exploration, generation, protection, optimization, and decline. We conclude with commentary about how our findings can be used in practice.

1) This chapter is based on two previous publications: Bader et al. (2012) (Getting the most out of your IP – patent management along its life cycle. *Drug Discovery Today*, Vol. 17, No. 7/8, pp. 281–284) and Gassmann et al. (2008) (*Leading Pharmaceutical Innovation – Trends and Drivers for Growth in the Pharmaceutical Industry*. Springer).

*Value Creation in the Pharmaceutical Industry: The Critical Path to Innovation*, First Edition.
Edited by Alexander Schuhmacher, Markus Hinder, and Oliver Gassmann.

## 21.2 The Changing Role of Patents: From Legal to Strategic

The management of intellectual property (IP) has moved from a legal matter to a strategic issue (Smith and Hansen, 2002). Research and practice have recognized IP as a critical part of corporate strategy (Grindley and Teece, 1997; Sullivan, 2001) and a main source of competitive advantage (Hall, 1992). This has been underpinned by the growing number of patent applications over the last century. In the last two decades, the worldwide yearly patent filings more than doubled, and in 2012, 2.35 million patent applications were filed worldwide (WIPO, 2013). China filed the greatest number of patent applications (652 777 patent applications per year), followed by the United States (542 815 applications), Japan (342 796 applications), and the Republic of Korea (188 915 applications). At the European Patent Office (EPO), 148 560 patents were filed in 2012 (EPO, 2013). These numbers show that firms have increasingly accumulated patent rights and therefore face the challenge of effectively managing their growing patent portfolios.

Public law and regulation play a greater role in the pharmaceutical industry than in any other industry. The regulatory force impacts pharmaceutical innovation on several levels: R&D regulations and product registrations, price regulations and national healthcare systems, and IP rights (Gassmann, Reepmeyer, and von Zedtwitz, 2008). The overall purpose of patent law is to support research and ensure that all interests are satisfied. On the one hand, innovations should be made available for the public interest. On the other hand, innovators should be incentivized to innovate through assurances that their inventions are protected against unlawful imitation and replication of their knowledge. From a competitive perspective, patents are essential because it is relatively easy to ascertain the substances that comprise a drug, which means it is easy to copy or imitate pharmaceutical products. Studies have shown that patents are the most effective means of appropriation. On average, without patent protection, 65% of pharmaceutical inventions would not have been introduced to the market, as compared to a cross industry average of 8%.

Patent protection is still unclear for certain key areas of pharmaceutical R&D. For instance, at the time of writing, it is still unclear to what extent genes can be patented and thus "owned." Sometimes international patent law is only accepted if national interests are maintained. For example, over the last decade, Brazil has threatened several times to suspend domestic compliance with international patent rights for malaria drugs unless certain license fees were dropped.

The pharmaceutical industry appears to suffer from a lack of genuine innovation, and in many cases, pharmaceutical companies rely more on patent protection for existing drugs than on the invention of new drugs. Overreliance on patents, which are intended to encourage investments in innovation, has had the unintended consequence of reducing R&D productivity by diverting attention away from R&D and toward protecting existing product revenues from generic competition. Given the relatively low costs of post-patent expiration competition vis-à-vis the large investments required to develop a new molecular

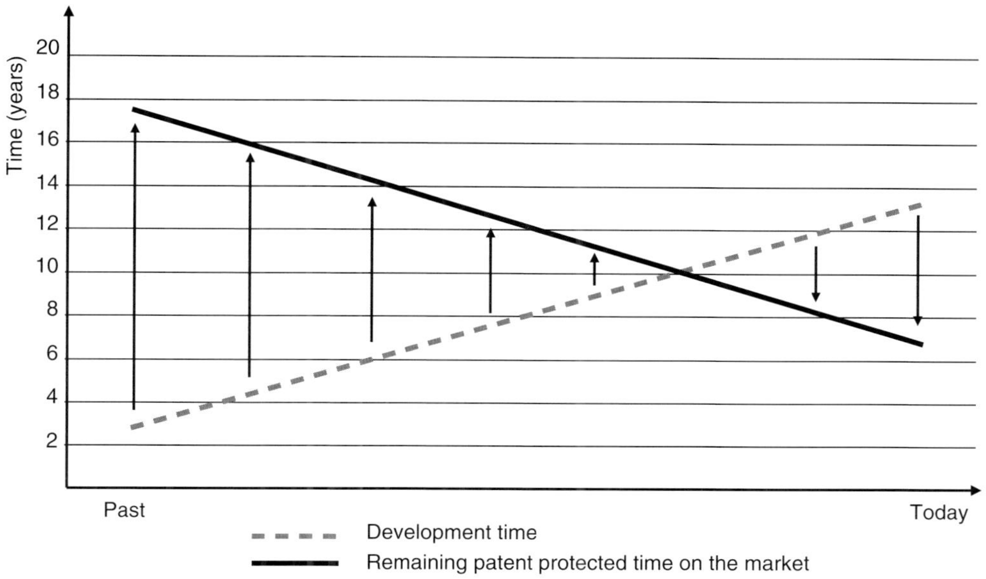


**Figure 21.1** The "innovation scissors" in the pharmaceutical sector.

entity, this is a successful short-term strategy. However, patent defense generates incremental revenue compared to the potentially huge gains generated by new innovative products, and it creates an overreliance on in-licensing to fill long-term revenue gaps.

It takes roughly 13 years for a pharmaceutical product to go from initial idea to market.[2] Thus, four out of five researchers retire or move on before they see any commercial impact of their work. As a result, incentive systems and employee motivation are very complex management issues in pharmaceutical R&D. Given the obstacles for future R&D performance and productivity, the entire drug discovery process faces structural challenges. Due to the unusually long development period, the effective patent protection duration of a drug on the market is very short. Some countries have therefore extended patent protection for pharmaceutical products by several years. Nevertheless, the trend toward longer development times means companies have an even smaller window in which to recoup their investment costs (i.e., currently 7–8 years, as compared with more than 17 years in the early 1960s). This phenomenon is referred to as the *innovation scissors* (Figure 21.1).

In addition to making inventions available to the public and IP rights, patents and trademarks fulfill another important role. They assign ownership to an individual and thus establish the inventor's legal basis to reap the commercial

2) See also chapter *"A Review of the Pharmaceutical R&D Efficiency: Costs, Timelines, and Probabilities."*

benefits of their invention. The enforcement of IP rights protects innovators against imitation and replication of their innovations and knowledge. This protection is crucial in the pharmaceutical industry; without it, nobody would invest in expensive and long-term drug development. Why should a firm invest $800 million in a new drug if generics manufacturers can copy the drug just a few months later? Even though successful drugs generate impressive profits, only a few drug candidates successfully leave the R&D pipeline and enter the market. This long development timeline and its associated large budget are part of the nature of the drug development process, in part due to the need for drug safety.

In the pharmaceutical industry, drugs and medicines can easily be copied or imitated because it is easy to analyze a pharmaceutical product and determine its components. Due to significant R&D costs and the high risks associated with new drug development, patent protection and IP management are particularly important to the pharmaceutical industry. Unless IP is protected with the outmost care, pharmaceutical innovation as we know it would not take place, and the overall quality of life would be significantly reduced. The negative impact for the society would be dramatic.

The 40 leading pharmaceutical firms worldwide have been granted an average of 5.8 patents per thousand employees. The United States accounted for approximately 45% of all pharmaceutical patents issued. Japan and Germany both accounted for roughly 10% of patents, followed by the United Kingdom (7%) and France (5%). US companies continue to dominate patent approvals in the United States. The proportion of US pharmaceutical patents issued to US and Japanese companies has increased over the last 20 years, whereas the proportion issued to European companies has declined: between 1980 and 1984, US companies were issued roughly 50% of patents, Japanese companies 13%, and EU companies 29%. However, between 1990 and 1994, the proportion of patents issued to US and Japanese companies increased to 55 and 15%, respectively, whereas the proportion issued to European companies fell to 24% (Gassmann, Reepmeyer, and von Zedtwitz, 2008). Worldwide, the leading therapy areas for patent approval were infectious disease (15%), oncology (14%), cardiovascular disease (10%), neurology (10%), and immune disorders (8%). The distribution of patents across therapy areas largely reflects the balance of the R&D pipeline and closely matches the relative unmet need and market opportunity.

International patent legislation typically encompasses four statutory classes of patentable inventions relevant to the pharmaceutical industry:

- Process patents
- Product patents
- Composition patents and
- Use patents

Process patents refer to the method used to produce a pharmaceutical product rather than to the chemical itself. As it may be possible to develop the same chemical through several different methods, it is often difficult to protect pharmaceutical products solely through process patents or to prove infringement of process patents. Product patents cover tangible products. Generally, these are commercially viable entities that are ready to be launched or already on the market.

In the pharmaceutical industry, patents are usually applied to medical devices, such as drug delivery mechanisms. Few manufacturers would want to wait until they perform clinical trials on a compound to apply for patent protection. Depending on the territory of filing, new substances or new processes receive patent protection for a period of 20 years (e.g., by the US Food Drug Administration). However, because the average drug development can take up to 13 years, the timing of the patent presents a major problem. If a company files for a patent too early, it only has a short period in which it can exclusively market and sell the drug. This means that the pharmaceutical company will only have a relatively short time to market their product and generate a high return on their initial investment. However, if a company files too late for a patent, it risks losing the invention to competitors. In practice, this means that the effective period of patent protection in the pharmaceutical industry is rarely more than 8 years. Sophisticated methods and techniques to deal with IP are needed.

The pharmaceutical, biotechnological, and chemical industries all have long product cycles; therefore patents are an important instrument for maintaining competitive advantage through temporary monopolies and are a major way of generating an increased return on investment (Levin *et al.*, 1987; Thumm, 2001). During the first months following patent expiration, the loss of sales of originator drugs to generic drugs can be as high as 50% (Figure 21.2).

Proactive patent management is therefore critical to avoiding price erosion of successful drugs (Gassmann and Bader, 2015). In general, effective patent management builds on:

1) Extension of patent protection:
   a. *Supplementary protection certificate (SPC)* for a limited patent extension for up to 5 years, depending on development and approval times with respect to the remaining base patent protection period
   b. *Orphan drug* regulations for niche drugs that were developed specifically to treat rare medical conditions, with an additional maximum protection length of 7 years in the United States and 10 years in the European Union, respectively
   c. Six-month extensions for *pediatric drugs* in the United States
2) Continuous enhancement of product value by adding new active pharmaceutical ingredients and finding new indications
3) Aggressive defense against patent infringements
4) Stabilizing sales after patent expiration by changing the drug status to over-the-counter (OTC) drugs or by the introduction of an internally developed generic drug substitute and
5) Out-licensing to third parties or by mergers and acquisitions

Many successful firms such as Dow Chemical Co., Roche, Novartis, and Bayer have established well-structured IP management processes and organizational structures and consider IP to be a major corporate asset. Despite the increasing

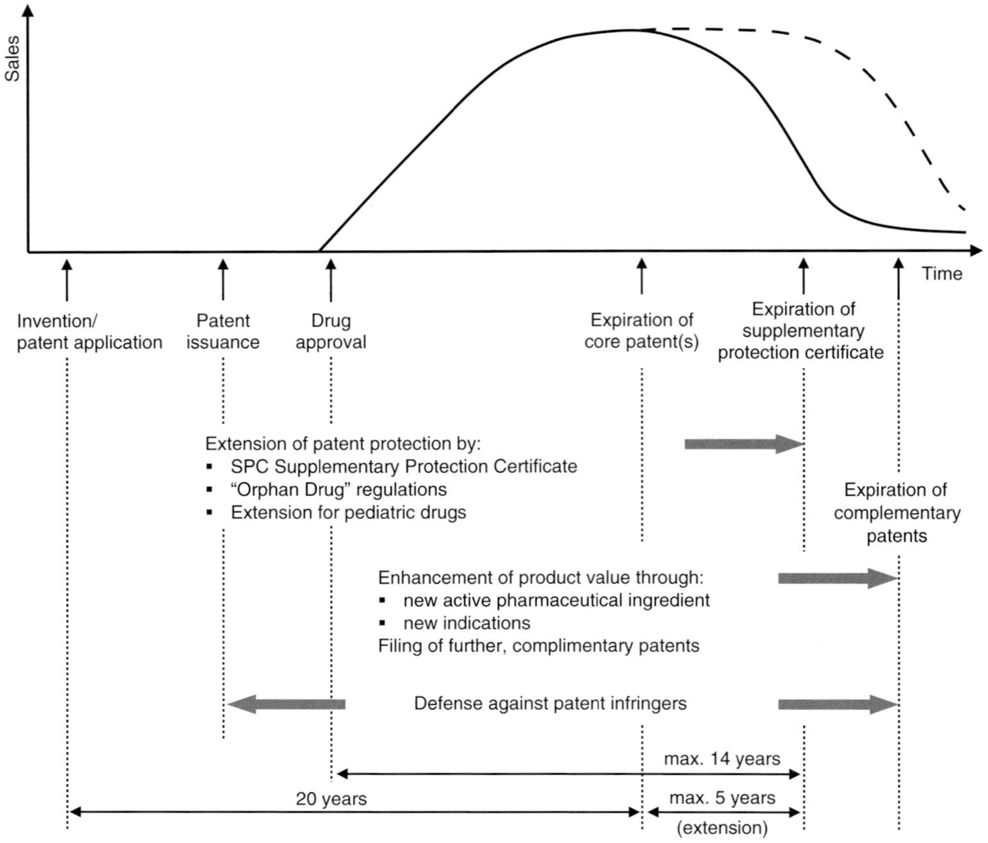


**Figure 21.2** Value effect of monopolizing patents in the pharmaceutical sector.

importance of IP management in the literature and in practice, insights into *how* firms manage their patents from a holistic, strategic perspective and how to optimize the portfolio value of patents remain scarce.

Smith and Hansen (2002) split the strategic management of IP into the activities *IP generation, protection,* and *valuation* and argue that firms must ensure that these activities are aligned with their overall business strategy. Carlsson et al. (2008) conducted a case study with 15 US-based technology-based firms to develop a general IP management process consisting of the phases *pushing for strategic IP, inventors' activities, screening techniques and checklists,* and *patent prosecution.* In a study on new technology-based firms, Lynskey (2009) based his research on a classic value chain model and applied this to the IP generation process. The IP value chain model that he developed consists of the following steps: *conception, primary document, review, formal document, legal document, patent prosecution,* and *IP portfolio* (Lynskey, 2009). All these IP management frameworks end with the application of the patent rights and do not include a comprehensive patent management model. To discover what a more holistic

**Table 21.1** Overview of the investigated firms.

| Firm | Industry | Turnover | Patents |
|---|---|---|---|
| Roche | Pharma | >39 billion EUR | 53 000 |
| Novartis | Pharma | >44 billion EUR | 25 000 |
| Bayer | Pharma | >40.2 billion EUR | 67 400 |
| Beiersdorf | Pharma | >6 billion EUR | 2 580 |
| Henkel | Chemicals | >16 billion EUR | 8 000 |
| Sika | Chemicals | >4 billion EUR | 1 600 |
| Prionics | Biotechnology | >25 million EUR | n/a |
| Cytos | Biotechnology | >1 million EUR | 450 |
| BIOTRONIK | Biotechnology/medical devices | n/a | n/a |

n/a = not applicable.

model could look like, we conducted 36 semistructured interviews with 10 firms from the pharmaceutical, biotechnological, and chemical industries (Table 21.1). The firms were chosen because they all possess well-structured IP management processes and organizational structures and thus provided the most detailed and insightful information.

Interview subjects were upper managers responsible for their firm's IP and R&D management. Based on the findings of our interviews, we argue that patent management should not end with the filing of the patents but rather that it should be better linked to the entire innovation process.

## 21.3 The Patent Life Cycle Management Model

Case study analysis revealed two major findings. First, we found that successful patent management follows technology management (i.e., the firms' patent management is strongly oriented to the life cycle of technologies, beginning with the discovery of ideas and continuing until a product is discarded from the firm's portfolio). Based on this, we identified five distinctive phases that reflect the patent life cycle management of firms that followed the technology life cycle: *exploration*, *generation*, *protection*, *optimization*, and *decline* (Gassmann and Bader, 2015). Second, our findings suggest that the way how patents are managed largely depends on both the patent's strategic value and the firm's internal resources. The patent's strategic value refers to the strategic value of the technology or patent relative to existing markets, competitors, and substitution technologies. The strength of internal resources refers to the firm's assets such as employees, know-how, and experience regarding a certain technology. When a company is in the first two phases (i.e., *exploration* and *generation*), it accumulates new competencies around a new technology, and these competencies remain at a high level in the phases *protection* and *optimization*. This is even true for the *decline* phase, although in the *decline* phase, the firm may decide to discard the patent. Thus, our framework displays patent life cycle management as a

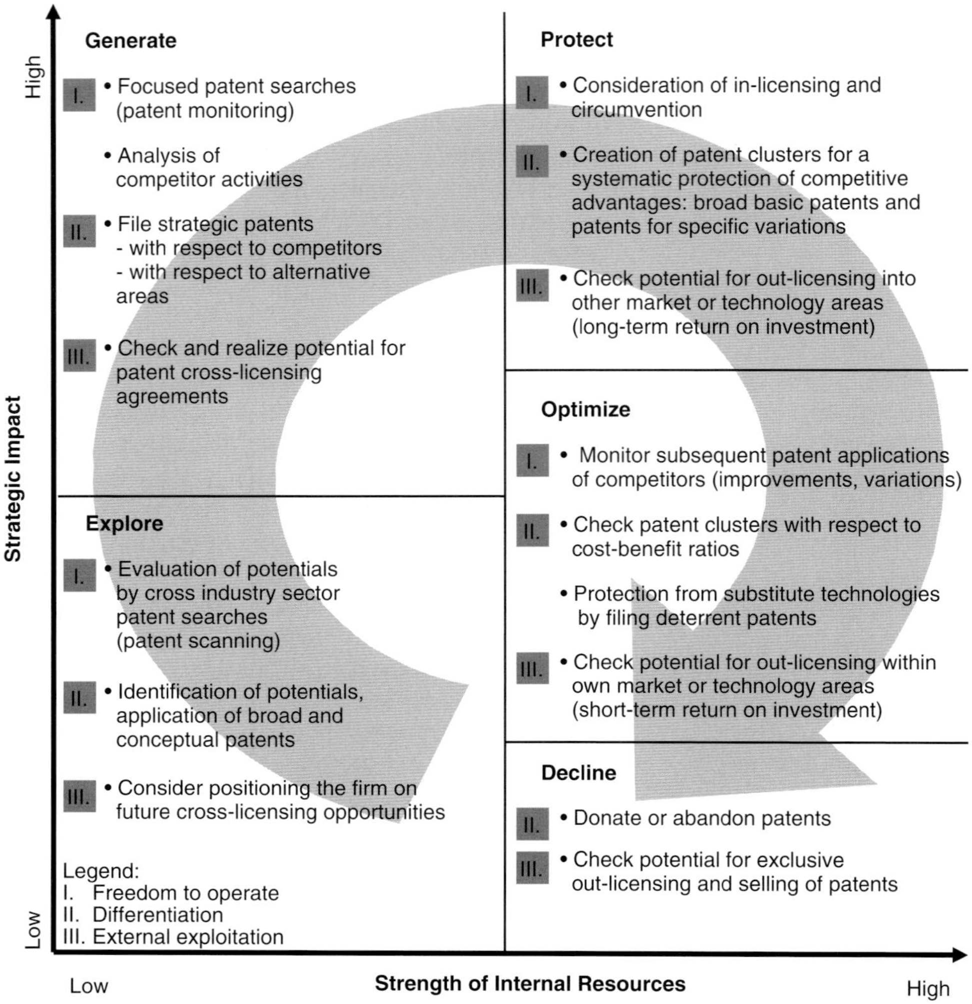


**Figure 21.3** The patent life cycle management model.

function of strategic impact and available internal resources in which each phase addresses three core dimensions of patenting: *freedom to operate, differentiation from competitors*, and *external patent exploitation* (Figure 21.3).

In the following sections, the phases of the patent life cycle management model are described and exemplified by the case studies (Gassmann and Bader, 2015).

### 21.3.1 Exploration

In the initial *exploration* phase, the firms collect ideas for new inventions. At this point, strategic impact is still low or unpredictable, and technological trends are explored through broad cross industry patent searches (e.g., patent scanning). The patentability of existing technologies and the freedom to operate are simultaneously checked during these patent scanning activities. The strategic positioning of the firm with regard to potential future cross licensing opportunities that enhance access to external knowledge is also considered. For example, Prionics uses the strategy where a new project always begins with a comprehensive patent search. For each new project, Prionics compiles an individual search profile, often with the help of external experts like the national patent and trademark office. Patent monitoring is conducted monthly by consulting the MEDLINE and Derwent databases. During this search, 400–500 potentially relevant literature citations and 75–120 potentially relevant patent citations are identified. An internal group of experts composed of R&D project leaders and product managers conducts a second screening to evaluate the search results and narrows the search to 30–50 relevant literature citations and 10–25 relevant patent citations. This search process is very successful at Prionics and is an integral part of their product development process.

### 21.3.2 Generation

Exploration activities are succeeded by the *generation* phase, where ideas are realized through new product development. For example, Henkel's strategy for its core competence areas is to strive for an exclusive protection of its products, technologies, packaging, and substances. Inventions in non-core competence areas often are not protected through patents but are published (e.g., in professional journals to prevent potential patenting by competitors). Cytos' strategy is to identify and patent new specific substances as early as possible and to partner with large pharmaceutical companies for further clinical development. Key considerations for Cytos are the maximization of the cost–benefit ratio and ensuring an appealing patent portfolio for partnering with large pharmaceutical companies.

Firms are increasingly opening up their research processes and acquiring external technologies to complement their internal technology portfolios (Chesbrough, 2003; Gassmann and Bader, 2006). In our case studies, Roche's R&D network, Bayer's cross licensing approach, and BIOTRONIK's technology in-licensing strategy were all examples of ways to complement internal know-how. Although Henkel is reluctant to open its innovation process, it does use cross licensing agreements with cooperation partners for certain parts of its portfolio. In these partnerships, Henkel tries to avoid financial compensations and instead agree upon a patent exchange that is based on the quality of the patents.

### 21.3.3 Protection

The *protection* phase is characterized by high strategic impact and high resource strength. Firms have accumulated comprehensive know-how in a field of competence with a high level of strategic importance. The potential for filing broad basic patents declines since public knowledge in these fields has already greatly increased. Instead, patent applications increasingly focus on more detailed and very specific embodiments, often with the motivation of building so-called patent fences around a core invention to forestall patenting of substitutes by rivals (Cohen *et al.*, 2000). Firms increasingly seek to create patent clusters in strategically important fields of technology. This involves generating patent portfolios that have a broad sweep but which are subsequently thinned out, once it is easier to estimate which ideas are technically and commercially viable. For example, Roche continuously builds up patent clusters by filing patents for backup candidates and follow-on patents in order to enhance the protection of their products. Roughly 1–10 basic patent applications and patents for backup candidates are typically filed per project. During the clinical development and early commercialization phases, roughly 3–30 follow-up patents are filed. Follow-up patent applications might include patents for polymorphs, salts, or alternative formulations, and so on.

### 21.3.4
### Optimization

In the *optimization* phase, the firm has a high level of competence in the respective technological field, but the strategic importance has declined (i.e., with regard to customers, markets, competition, or technology). Firms thoroughly monitor competitors' patenting activities and review their own patent clusters with respect to cost–benefit considerations. For example, firms regularly check for a potential reduction in the territorial coverage of the patents. Firms also look at potential risks posed by substitute technologies. If there is a risk that one of their competencies might be replaced by substitute technologies, then the firm may use their patents to block property rights and thus prevent a decline in value of the existing core technologies.

For research-based pharmaceutical firms, generic drug companies pose a particularly strong risk of IP infringement. Generic drug companies become competitors as the relevant patent's expiration day approaches. About 5 years before generic companies are able to legally enter the market and use the specific agent, Novartis will perform active competitive intelligence to identify potential infringement. Generic companies can begin regulatory readiness before the innovator exclusivity period has expired; thus Novartis keeps an active eye on all developments of "their" products. The first assessment is undertaken in the preclinical phase, and additional assessments take place at the entry of each development phase.

### 21.3.5
### Decline

When the strategic importance of a technology or competence has greatly declined, the corresponding patents are reviewed to determine whether they still add value to the firm and to define a divestment strategy. Before the patents are abandoned, firms check the patents for out-licensing, selling, or donation opportunities. BIOTRONIK conducts a yearly review on its patent portfolio and decides on how to proceed with obsolete patents. The IP review board, supported by R&D and management, is responsible for assessing obsolete patents with regard to an external exploitation via out-licensing. They check if the patent protects one of BIOTRONIK's products or if there is planned future internal exploitation. They also analyze whether competitors could potentially use BIOTRONIK's patents for their own technologies and whether the patents could be enforced in case of an infringement. Finally, the overall costs and efforts are estimated before BIOTRONIK decides on licensing or abandoning the patents.

## 21.4 Example: Managing IP Rights at Bayer

The Bayer case study illustrates how a pharmaceutical company can utilize and maximize the value generated by its IP by taking a proactive approach to commercializing its IP. In this context, Bayer views IP as a product. IP thus needs its own marketing plan. The IP products are typically spin-offs or "white-space" developments or technologies (i.e., devices or methods) that are no longer being used

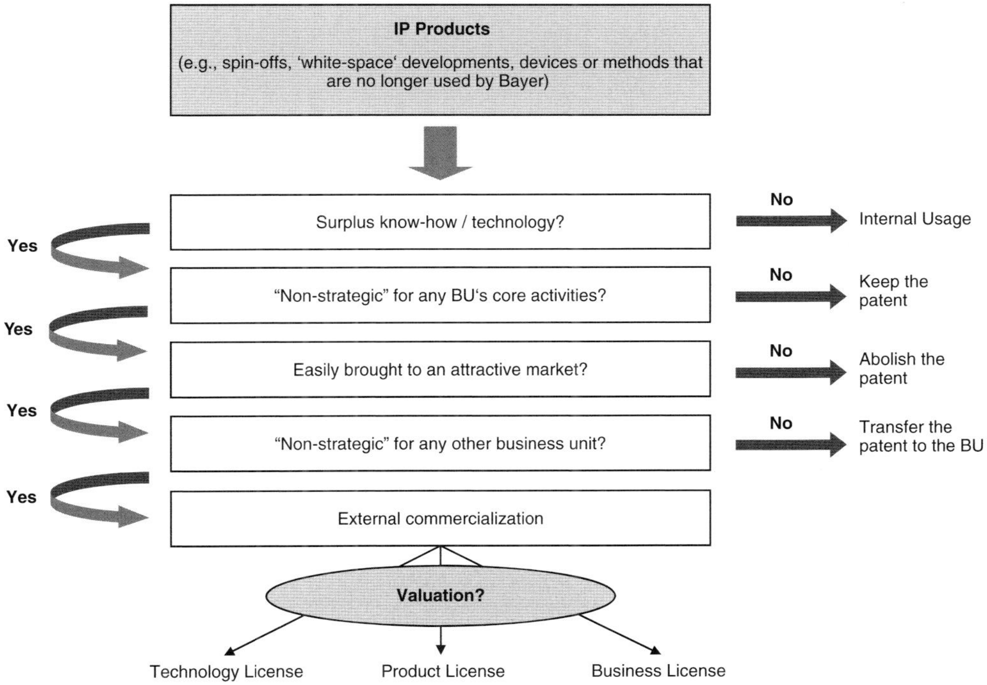


Figure 21.4 External exploitation of intellectual property at Bayer.

by Bayer's business units. When selling these products, Bayer adheres to the rule "don't try to sell any leftovers."

Bayer developed a four-stage process to decide if certain know-how or a certain technology can be utilized externally (Figure 21.4). First, Bayer asks if the respective know-how/technology is a surplus product. If yes, the second stage examines if the know-how is strategically valuable for any core activity of Bayer's business units. If it is not, the third stage analyzes if the respective technology could be easily brought to a potentially attractive market. If yes, the final stage analyzes whether or not the know-how is strategically valuable for another business unit within Bayer. If is it not, then the IP can be marketed outside of Bayer; otherwise it is retained in-house.

When assessing the valuation of IP, Bayer differentiates between business licenses, product licenses, and technology licenses. Business and product licenses deal with the entire businesses and/or products, and their value can easily be determined by using the scenario technique. The value of technology licenses is much more complicated to estimate and is done by looking at both technology maturity and commercial risk. This combination allows for a fairly good estimate. Finally, the marketing plan includes the IP utilization strategy, which could include cross licensing agreements, royalty payments, cash payments, or equity offerings.

## 21.5 Concluding Remarks

In general, patent protection has become particularly important in the area of biotechnology. Several ethical questions have arose that are not yet covered by existing patent laws and/or acts, including the ownership of genes and whether genes can even be patented. For example, the Swiss Ethics Committee on Non-Human Gene Technology recently concluded that intellectual achievements in the area of biotechnology are allowed to be protected. This is justified because the overall purpose of the patent act is to support research in the best interest of the public. While it is currently possible to receive patents for biotechnological inventions such as a gene, a genetically changed plant, a biotechnological process, or a microorganism, some inventions are excluded from patenting: processes regarding human cloning, processes regarding changes to the genetic identity of human beings, and the use of human embryos for industrial or commercial purposes.

This chapter describes a holistic life cycle management model for effective patent management and encapsulates best practices for pharmaceutical, biotechnology, and chemical industry firms to strategically manage their patents. The key message is that patent management should not be regarded as an isolated function, but rather as an integrated activity that considerably contributes to firm success because intellectual assets have become critical firm resources. Firms should therefore take a holistic view of patent management and consider the following:

1) Identifying new technological challenges is an important factor to fostering innovation. Thus, firms should establish active technology scouting and patent scanning processes. The earlier technological trends are identified, the better the firm can react and reach a first-mover advantage. It is especially important that these technology scouting and patent scanning activities are conducted and updated on a regular basis. Firms should also ensure that the appropriate employees, for example, R&D employees, patent managers, and business developers, are given access to the results.
2) During the development of new technologies and products, it is important to monitor competitor and market activities. Firms should therefore establish a patent monitoring system that regularly observes their environment. Special attention should be paid to identifying substitute technologies because these technologies might weaken the firm's temporary monopoly that had been gained through the patent protection.
3) External leveraging of patents through strategies such as out-licensing, cross licensing, sale, strategic alliances, and joint ventures may enable firms to generate additional returns on investment and to induce strategic benefits. Therefore, consideration of external exploitation opportunities at all stages of the patent life cycle should be a standard patent management activity.
4) Firms are advised to conduct regular (e.g., yearly) patent audits to assess the value of their patent portfolio. These audits should also be used to monitor how each patent is exploited (i.e., which patents protect which products or technologies, which patents have a blocking function, which patents are out-licensed and to whom, and which patents are currently not used for any competitive advantage or financial benefit). Based on this audit, decisions on when and where new patents should be filed and which patents could be out-licensed to generate additional income can be made, and further steps for implementing these activities can be defined.
5) Issued but unused patents cause unnecessary maintenance fees. Hence, firms should make proactive patent divestment decisions to avoid accumulating unused patents. First, potentially obsolete patents should be evaluated with respect to the benefits and costs for the firm. If the patent reveals potential attractiveness for other firms, then out-licensing, sale, or donation should be considered.
6) Finally, patent life cycle management should be considered as an interdisciplinary task. Thus, the aforementioned recommendations should be implemented by a small group of senior executives that includes the heads of IP, R&D, business development, product development, and marketing.

## References

Bader, M.A., Gassmann, O., Ziegler, N., and Rüther, F. (2012) Getting the most out of your IP – patent management along its life cycle. *Drug Discovery Today*, **17** (7/8), 281–284.

Carlsson, B. *et al.* (2008) Intellectual property (IP) management: organizational processes and structures, and the role of IP donations. *J. Technol. Transf.*, **33** (6), 549–559.

Chesbrough, H.W. (2003) The era of open innovation. *MIT Sloan Manage. Rev.*, **4** (3), 35–41.

Cohen, W.M. *et al.* (2000) Protecting their intellectual assets: appropriability conditions and why U.S. manufacturing firms patent (or not). NBER Working Paper No. 7552.

EPO (2013) Annual Report 2012, European Patent Office.

Gassmann, O. and Bader, M.A. (2006) Intellectual property management in inter-firm R&D collaborations. *Taiwan Acad. Manage. J.*, **6** (2), 123–140.

Gassmann, O. and Bader, M.A. (2015) *Patentmanagement – Innovationen erfolgreich nutzen und schützen*, 4th edn, Springer, Berlin, Heidelberg.

Gassmann, O., Reepmeyer, G., and von Zedtwitz, M. (2008) *Leading Pharmaceutical Innovation – Trends and Drivers for Growth in the Pharmaceutical Industry*, Springer, Berlin, Heidelberg.

Grindley, P.C. and Teece, D.J. (1997) Managing intellectual capital: licensing and cross-licensing in semiconductors and electronics. *Calif. Manage. Rev.*, **39** (2), 8–41.

Hall, R. (1992) The strategic analysis of intangible resources. *Strateg. Manage. J.*, **13** (2), 135–144.

Levin, R.C. *et al.* (1987) Appropriating the returns from industrial research and development. *Brookings Pap. Econ. Act.*, **3**, 783–831.

Lynskey, M.J. (2009) Aligning strategy and intellectual property to maximise business value: a proposal for new technology-based firms. *Int. J. Intellect. Prop. Manage.*, **3** (4), 301–325.

Smith, M. and Hansen, F. (2002) Managing intellectual property: a strategic point of view. *J. Intellect. Cap.*, **3** (4), 366–374.

Sullivan, P. (2001) *Profiting From Intellectual Capital*, John Wiley & Sons, Inc..

Thumm, N. (2001) Management of intellectual property rights in European biotechnology firms. *Technol. Forecasting Social Change*, **67** (2–3), 259–272.

WIPO (2013) WIPO IP Facts and Figures. WIPO Publication No. 943E/13, World Intellectual Property Organization.

# Index



*Value Creation in the Pharmaceutical Industry: The Critical Path to Innovation*, First Edition.
Edited by Alexander Schuhmacher, Markus Hinder, and Oliver Gassmann.

***p***